THE HORSE

J. WARREN EVANS
Texas A & M University

ANTHONY BORTON
University of Massachusetts

HAROLD F. HINTZ
Cornell University

L. DALE VAN VLECK
University of Nebraska, Lincoln

THE HORSE

SECOND EDITION

WITH CHAPTERS BY

Jay R. Georgi, *Cornell University*
R. P. Hackett, *Cornell University*
John E. Lowe, *Cornell University*
Gary D. Potter, *Texas A & M University*
B. F. Yeates, *Texas A & M University*

W. H. FREEMAN AND COMPANY
NEW YORK

Library of Congress Cataloging-in-Publication Data

The Horse / J. Warren Evans . . . [et al.]. —2d ed.
    p. cm.
  ISBN 0-7167-1811-1
  1. Horses.   I. Evans, J. Warren (James Warren), 1938-

SF285.H748   1990   89-17027
636.1—dc20   CIP

Printed in the United States of America

Tenth printing, 2000

CONTENTS

PREFACE

There have been many changes since the publication of the first edition of this book in 1976. The authors have certainly gotten older and perhaps wiser. The amount of information available on horses has greatly increased because of the increased amount and sophistication of equine research. For example, at the 1975 Equine Nutrition and Physiology Society meetings, 21 papers were presented compared to 110 papers at the 1987 meeting. Equine exercise physiology has particularly advanced in the past decade. We have incorporated many recent findings into this edition. For example, the 1989 National Research Council's subcommittee report on nutrient requirements was used extensively in the preparation of this edition.

Of course, many things have not changed. Many of the principles of horse management are still the same. The importance of common sense and the "eye of the master" in horse management cannot be overemphasized. We hope this edition's blend of old and new information will be useful to the experienced horseperson as well as to the novice student of horses.

September 1989

J. WARREN EVANS
ANTHONY BORTON
HAROLD F. HINTZ
L. DALE VAN VLECK

PREFACE TO THE FIRST EDITION

The horse ceased to be an important source of agricultural power in the United States approximately 30 – 40 years ago, but popular interest in horses appears to be ever increasing. More people own horses, enjoy watching them, enjoy reading about them, and wish to learn more about them today than at any other time in American history.

Although horses have not received as much attention from scientists as have the other species of livestock, many significant findings about horses have been made in recent years. These findings, combined with the results of early equine studies, information extrapolated from experiments with other species, and observations of experienced horsemen, form the basis of most recommendations for modern horse management. This reference and text-book on the horse represents an attempt to combine these sources of information.

July 1976

J. WARREN EVANS
ANTHONY BORTON
HAROLD F. HINTZ
L. DALE VAN VLECK

THE HORSE

1

THE HORSE—
A UNIQUE ANIMAL

CHAPTER 1

INTRODUCTION

The horse is an animal unlike any other—an animal of uncommon beauty, grace, sensitivity, and athletic ability; a fascinating animal to study, for it is unusual in its development, history, anatomy, physiology, and movement; and most of all, an animal with a singular relationship to mankind. For the development of most modern civilizations has been inextricably interwoven with the domestication of the horse.

From ancient to modern times, the horse has served as a beast of burden, a draft animal, and a means of transportation; has helped wage wars; and has provided recreation, companionship, and even food.

> Look over the struggle for freedom
> Trace your present day strength to its source
> You'll find that a man's pathway to glory
> Is strewn with the bones of a horse.
> —*Anonymous*

According to Charles Russell (1927), the cowboy artist and humorist,

> It was this animal that took 'em from a cave. For thousands of years the hoss and his long-eared cousins furnished all transportation on land for man an' broke all the ground for their farmin'. He has helped build every railroad in the world. Even now he builds the roads for the automobile that has made him nearly useless, an' I'm here to tell these machine-lovers that it will take a million years for the gas wagon to catch up with this hoss in what he's done for man.

Fortunately, the automobile has not rendered the horse useless, and today this sensitive, intelligent, athletic animal is enjoying unprecedented popularity in this country, where its ancestors originated some 60 million years ago.

1.1   History and Development

The horse family is one of the classic examples of evolution for two reasons. First, enormous changes took place in the size and structure of the animal in response to environmental changes, such as changes in climate and food sources. Second, the fossil remains of the horse have been well preserved in the river valley clays, sands, and sandstones of the rich paleontologic beds of the American West. The association of fossil remains of prehistoric horses with the various geological times furnishes one of the best-documented examples of the evolutionary changes of an animal species.

The horse as we know it today is descended from a small, primitive, four-toed animal that inhabited the river banks of the early Eocene epoch. Approximately fifteen million generations later, *Equus* evolved as the first "true horse." In the Ice Age *Equus* migrated from North America throughout the world over existing land bridges. Fossil remains of *Equus* have been discovered in Asia, Europe, and Africa as well as throughout North and South America.

Thus, *Equus* originated in North America a million years ago and migrated throughout the rest of the world, which was fortunate, because there were no horses in the western hemisphere when it was discovered by Europeans. The extinction of the horse in North America, after it had flourished for four geological epochs and roughly 60 million years, remains one of the unsolved mysteries of history. The extinction cannot be attributed directly to the effect of glacial cold, because parts of the continent were safe from glaciation, and the horse actually survived the ice ages, only to disappear as the ice was retreating. It should also be noted that companion grazing animals, such as the bison, survived and thrived in the New World. On the other hand, there seems to be no adequate explanation of why other specialized mammals of North America also disappeared during the Pleistocene epoch and more recently (for the rhinoceros, camel, saber-toothed tiger, elephant, and mastodon, like the horse, all succumbed).

Insects, disease, the acts of early man, and the disappearance or depletion of food sources have all been considered as the sole cause of the extinction of horses throughout North and South America, and all have been discounted—a state of affairs that caused George Gaylord Simpson (1951) to reflect that "this seems at present one of the situations in which we must be humble and honest and admit that we simply do not know the answer. It must be remembered too that extinction of the horses in the New World is only part of a larger problem. Many other animals became extinct here at about the same time."

In the 1 million years since the first appearance of *Equus* and its mysterious disappearance from the American continent, man has become civilized, and numerous animal species, including the horse, have been domesticated, but modern *Equus* is still an athletic animal, adapted for running and grazing

over vast areas of plains, subsisting solely on grasses, and escaping its enemies with bursts of speed—a superb, intelligent, graceful, athletic animal
little changed by man but one that has changed the life of all mankind.

The tangled web of the recent history of the horse is more obscure and
controversial than the horse's evolutionary development. We know that
Equus spread throughout the world from North America. Various distinct
forms of *Equus* then developed in different areas of the world at different
times. The horse's development was probably significantly affected by wide
variations in altitude, climate, soil, and forages. However, the early wild
horses exhibited great adaptability—they flourished in steppes, forest, deserts, and tundra. Some studies of prehistoric horses have classified them by
racial groups according to the environment in which they developed.

Domestication

Little is known about either the early development of specific strains of
horses or the first domestication of the horse by man. However, modern man
and modern horse developed together. Undoubtedly, the earliest association
between man and horse was a one-sided one in which man hunted and
subsisted on the flesh of horses. The bones of 40,000 horses that existed
25,000 years ago, found outside a rock shelter at Solutré, France, provide
evidence of the cave man's dependence on the horse. Perhaps the Paleolithic
inhabitants of North America were partly responsible for the extinction of
the horse on that continent, because the appearance of man there closely
coincided with the disappearance of the horse.

Early man must have discovered other virtues of the horse even as it was
providing him a source of food. Although the precise date of domestication
remains unknown, the walls of numerous caves throughout the Old World
contain drawings that indicate that man's dependence on the horse dates
back to the most ancient times (Figure 1-1).

Charles Russell (1927) has written an entertaining account of man's first
domestication of the horse. He describes a cave man who is looking down
"from his ledge to the valley below where all these animals is busy eatin' one
another, an' notices one species that don't take no part in this feast, but can
out-run an' out-dodge all others. This cave man is progressive, an' has
learned to think. He sees this animal is small compared to the rest, an' ain't
got no horns, tusks or claws, eatin' nothin' but grass." Later, when man has
succeeded in capturing a horse, "he finds out that though this beast ain't got
horns or claws, he's mighty handy with all four feet, and when Paw sneaks
home that evenin' he's got hoof marks all over him an' he ain't had a ride yet.
Sore as he is, he goes back the next day an' tries again."

The date and place of the initial domestication of the horse are under-
standably the subject of considerable disagreement. Simpson (1951) logically suggests that there were probably a variety of races of wild horses and

Figure 1-1
Prehistoric cave paintings of horses from the cave gallery at Lascaux, France.
(Photograph courtesy of Caisse Nationale des Monuments Historiques et des Sites,
Paris.)

that initial domestication undoubtedly occurred in several parts of the world
at approximately the same time. Certainly two of these areas must have been
China and Mesopotamia between 4500 and 2500 B.C. In any event, it was in
the Near East, the cradle of European civilization, that the horse was rapidly
integrated into a way of life that relied upon its use as a draft animal. By
1000 B.C., the domestication of the horse had spread to almost every part of
Europe, Asia, and North Africa.

Primitive man now enjoyed a freedom he never had before. The horse
provided a means of travel across lands and distances heretofore impossible.
The age of exploration began. Soon, however, came periods of conflict. In
the final analysis, the horse's greatest contribution to human history may be
its use as a tool of warfare. It was the horse that changed pastoral nomads
from the steppes of central Asia into mounted warriors that with relative
ease conquered the great civilizations of the Mediterranean. From ancient
times until just recently the horse has played a significant role in all warfare.
Even today, in some parts of the world, the horse remains the most effective
means of transporting personnel and supplies into rebel camps.

Remarkable as the horse is as a riding animal, it was the harnessing of the
horse that had the greatest significance in building civilization, for the horse

has no equal as a draft animal. Oxen are better for working small plots in some of the more rugged terrains, but the strength, tractability, and speed of the horse is so remarkable that even today we evaluate the performance of tractors, automobiles and trucks in terms of horsepower. Draft horses were used to farm the land and move heavy loads to the population centers. The overland transportation system relied on the horse. As cities developed and roads were built the carriage horse became a mode of transportation between them, and the trolley horse provided transportation within them.

Return of Horses to America

The first horses to reach the North American continent were the well-bred mounts of the Spanish conquistadores. It is a myth that the vast Mustang herds that roamed the West by the 1880s were strays from the expeditions of Cortez, in Mexico in 1519; Coronado, from Arizona to Kansas in 1540; and DeSoto, in Florida and the Southeast in 1541. The exploits of these early Spanish explorers are fascinating and the horse's importance to them was undeniable, but the evidence does not support the contention that these expeditions brought the horse to the Indians and the continent. Rather, the Spanish missions that followed the Spanish explorers into the Rio Grande Valley in the early 1600s brought with them large numbers of livestock: goats, sheep, cattle, and, of course, horses. More specifically, Juan de Onate established a large settlement in what is now Santa Fe, New Mexico, in 1594. A series of other missions were established (24 in New Mexico), where the Indian children undoubtedly learned farming and were exposed to the breaking, training, and use of the horse. These Indians, thus instructed in horsemanship, probably passed on their knowledge and skills, and even some of the Spanish horses, to other Indians. The century from 1650 to 1750 was a period during which the Spanish horses were dispersed over the plains, and among the Indians a great "horse culture" developed (Figure 1-2). The profound effect that the horse had on the life-style and culture of the Plains Indian in a relatively short period resulted in a population and cultural explosion that enabled the Indians to effectively slow the white man's takeover of the West.

It took only 200 years for the great "wild" horse herds of the Plains to become adapted to a region where, 60 million years earlier, the ancestors of the wild horse had begun their development at the dawn of history. The history of the development of the vast horse, bison, and long-horned cattle (also Spanish in origin) herds in the American West is one of the most colorful aspects of the history of the horse in the United States. However, it was the early colonists who settled the East Coast who were responsible for developing many of the American breeds of horses.

Figure 1-2
Plains Indians hunting bison from horseback. This painting by George Catlin, an American ethnologist who studied the Indians of western North America during the 1830s, clearly depicts the impressive riding skill of the Plains Indians. (Photograph courtesy of The Mansell Collection, London.)

The East Coast Indians were introduced to horses by the Spanish explorers at about the same time as their western cousins. However, they were farmers and trappers, and their needs for horses were quite different. Their society did not develop around the horse to the same extent that western society did, for the horse was used by these Indians primarily as a pack animal to haul hides to the coast. However, their horses were no less fine and were also of Spanish origin, although they were referred to as "Chickasaw" horses. These horses came from a series of Franciscan missions that were established in the Southeast (Georgia) at the same time that Juan de Onate was establishing his mission in the Southwest. Many of the early colonists bought Chickasaw horses from the Indians to use on their farms. These horses then provided a "Spanish" base in the "native" herds that were later used to breed the Quarter Horse, the American Saddle Horse and the Tennessee Walking Horse.

The colonists along the Atlantic seaboard, the English in Virginia, the Dutch in New York, and the French in Quebec brought horses with them. The colonists subsequently imported more horses, but most of them came not from the Old World but from the horse-breeding farms established by the Spanish in the West Indies. The Spanish basis of the light-horse stocks of the New World was well established in the Mustangs of the West, the Chickasaw and other Indian horses of the Southeast, and the mounts and farm horses of the colonists.

Horses were little used in colonial New England. The small, hilly, rocky fields were better suited to oxen than to draft horses, and the Puritan ethic militated against the expense and frivolity of keeping riding horses. Horse racing was socially unacceptable because it was too closely associated with the landed gentry in England, whom the colonists had sought to escape by coming to America. Later on, however, horse breeding became a popular enterprise, as the market developed for riding horses, for coach horses, and for work horses in the cities and in the West Indies, where they were used on the sugar plantations. The development of harness racing created a further strain on the colonists' puritanical values, as a demand was created for some of New England's fast strains of harness horses.

As the colonies developed, the farmers in New York and Pennsylvania had need of heavy horses that would not only till the soil but also haul their products to the markets in Philadelphia and New York City. The native horses were a little small and too light for this rugged work, so the colonists naturally turned to the horses of their homelands and imported the Belgian, Percheron, Shire, and Clydesdale breeds. These draft stallions were mated to the native mares, and although many fine, heavy, coach and wagon horses were produced for use in America's developing cities, no American breed of draft or coach horse was ever developed. A typical tall, rangy, active Conestoga horse was bred in southern Pennsylvania and was highly sought after by the freight haulers, but no breed registry was established, and the Conestoga horse eventually disappeared as imports from the Old World increased. In the following chapter the breeds of horses are discussed in more detail.

Number of Horses in the United States

The agricultural development of the United States relied on the work horse. Draft horses thrived on America's farms—by 1918 the United States held 21 million head of horses, most of them draft horse type. The advent of the combustion engine meant the demise of the draft horse, as it was replaced by the tractor. In the 1950s the number of horses had dropped to near 2 million head and few people expected the horse to be a significant factor in the economy again. None of the nation's agricultural colleges were conducting equine studies.

Today, the increase in personal income and free time has resulted in an unprecedented popularity of the riding and driving horse as a sport and recreational animal. Because most horses are no longer used for farming, there is no reliable census of the current horse population in the United States; estimates range from 5.2 million to over 10 million. Whatever their exact number, the purebred breeds are enjoying their greatest growth in history and the annual economic impact of the horse industry has been estimated by the American Horse Council at a staggering $15 billion (see Table 1-1). The widespread popularity of horses and their importance to the agricultural economy has resulted in increased academic research and attention. Many veterinary colleges are emphasizing equine medicine, agricultural colleges have established equine studies programs and research in nutrition, physiology, genetics, and reproduction is being conducted as never before.

1.2   The Horse and Mankind

Use

The remarkable versatility of the horse is exemplified by the wide variety of ways in which the horse is used by mankind. The horse excels in such diverse events as racing, rodeo, endurance riding, competitive carriage driving, Combined Training, horse show classes, fox hunting, dressage, polo, gymkhana games and, of course, pleasure riding and driving. It can be fitted with a variety of bridles, bits, saddles, harnesses and handled in many different ways while still performing amazing athletic feats.

Training

It is a happy coincidence of nature that the horse, an animal of great strength and athletic ability, should possess the disposition, sensitivity, and intelligence that enables it to be trained for a variety of uses. Training horses adds an additional dimension to their management — for a horse must not only be carefully and patiently trained, it must also be handled and managed in a manner that will realize its potential usefulness. The industry is labor-intensive, as the horse must be handled and trained individually.

Table 1-1.
U.S. horse industry Gross
National Product 1985

Region	Total GNP
California	1903
Colorado	343
Florida	714
Illinois	608
Kentucky	487
Louisiana	300
Michigan	398
Minnesota°	181
Maryland	393
New Jersey	589
New York	1257
Ohio	557
Oklahoma°	233
Pennsylvania°	614
Tennessee	228
Texas	1013
Virginia	331
Washington	441
Pacific Mountain	1354
West Central	905
Eastern	1223
South Central	1008

°The amounts above indicate each state's and region's contributions to the $15 billion horse industry. The asterisks indicate those states where racing receipts have been excluded due to disclosure laws. The U.S. total, however, includes the earings of racing in those states. The rodeo industry's GNP is also accounted for.

SOURCE: American Horse Council. 1987.

Intelligence

When compared with other animals the horse does not rank high on tests of intelligence. But while a horse does not have an ability to reason, it does learn readily to respond to even very subtle "cues" or "aids," and also possesses an outstanding memory. More complete information on the training and intelligence of the horse is found in Chapter 18.

Soundness

A horse must not only be trained; it must also remain healthy and capable of performing. Used as it is as an athlete, a horse consequently must be conditioned and remain sound (free from injuries). An unsoundness is a defect in form or function that interferes with the usefulness of a horse. Equine Sport Medicine is a rapidly growing field that emphasizes conditioning, measuring fitness, and improving injury rehabilitation techniques. Soundness is one of the topics of Chapter 4.

Special Senses

The horse is readily trainable because of its ability to perceive and respond within its environment. It has survived in the wild because it feels, hears, smells, and sees danger. While experts do not agree on which of the horse's senses are most highly developed, the integration of the senses creates an animal that is alert and sensitive. This same integrated sensory system makes the horse readily trainable. The horse's senses are discussed in more detail in Chapter 3.

Relationship Between Horse and Mankind

The final consideration in the relationship between mankind and the horse is an emotional one. The dog enjoys perhaps the closest relationship of any animal with the human but the horse's role is also unique. The bond between the horse and mankind is built on a working relationship that has served man well in many different ways. People using horses routinely trust their lives to the horse's judgment, responsiveness, and athletic ability, so it is only natural that strong emotional bonds have developed between man and the horse. "There is something about the outside of the horse that is good for the inside of mankind." (Anonymous.)

This emotional attachment and the remarkable strength, agility, and athletic ability of the horse have been so appreciated that the horse is one of the most cherished subjects of art. For centuries mankind has revered the horse in sculptures, carvings, drawings, and paintings.

1.3   The Science of Horses

The horse is a unique animal that must be studied and understood if it is to be used and managed effectively. In addition to the subjects already mentioned, this book contains the most recent information on equine anatomy, physiology, nutrition, health, genetics, reproduction, and management.

The structure of the horse is unique and responsible for its athletic ability. The zoological classification of the horse is of the order Perissodactyla—odd-toed, nonruminating, hoofed animals (including the horse, tapir, and rhinoceros). The horse (and its immediate relatives) has evolved as the only single-toed animal in the world. The evolutionary process has resulted in an animal designed for speed, with long slender legs, that is permanently on its toe. The bones of the limbs have elongated, increasing leverage so that the muscles located in the upper legs provide a maximum of motion with a minimum of contraction. Lateral flexion of the joints has been lost in favor of one fluid motion forward and backward, without wasted deviation. The head and neck have elongated so that the animal can graze while on its feet; the eyes are placed high on the head so that vision was possible over a long distance while grazing. The large capacity of the circulatory and respiratory systems provides physiological support for the massive muscles of the equine. The unusual structure and physiology of the horse is further described in Chapters 3 and 4.

The horse is a nonruminant herbivore that has a rather small, simple stomach and a large functional cecum located at the junction of the small and large intestines. This system, specialized for a forage eater, creates some interesting nutritional considerations that are discussed in Chapters 5, 6, 7 and 8.

The horse wears its teeth in clipping and grinding tough siliceous forages but its teeth are high crowned and grow throughout its life so it enjoys a long

Table 1-2
Zoological classification of the horse

Kingdom:	Animalia	Includes all animals
Phylum:	Chordata	Animals with backbones
Class:	Mammalia	Warm blooded animals that give milk, have hair
Order:	Perissodactyla	Odd-toed, nonruminating, hoofed
Family:	Equidae	The horse family in the broadest sense, including evolutionary ancestors
Genus:	Equus	Living members of horse family and close ancestors—horses, zebras, asses
Species:	Equus caballus	Domestic horse and close wild relatives

productive life as a grazer. The wearing and continued growth of the horse's teeth over time cause changes that enable the teeth to be used to determine the animal's age. See Chapter 3.

The unique digestive tract of the horse is not without consequences. Colic, a general term for abdominal pain, is a major cause of death in horses. The simple-stomached horse is particularly susceptible to molds, toxins, bacteria, and poisons in the feed supply that can cause illness. Feeding and management programs must constantly consider these problems. Internal parasites are one of the major causes of colic and undernourished horses. The horse is particularly susceptible to a wide range of parasites and any management plan must control this problem. Internal parasites of the horse are addressed in Chapter 17, while colic and other health and first aid problems are covered in Chapters 8, and 16.

The horse breeder has the responsibility of selecting and planning matings with the goal of making genetic improvement. A knowledge of the principles of genetics, the heritability of traits, the importance of relationship and the methods of selection are essential. Knowledge of the mode of inheritance of coat color is also necessary in some breeding programs. Chapters 12, 13, 14, and 15 discuss the current knowledge of equine genetics.

The horse has flourished and reproduced in a variety of environments despite the fact it has one of the lowest reproduction rates among domestic animals. The horse is seasonally polyestrus, responding to changes in day length. The most unique feature of equine reproduction is the tremendous variability in all aspects of the reproductive cycle. Also, because of the large investment in breeding horses, most mating is strictly controlled to prevent injury. Thus there are techniques of estrus detection (teasing), mating, or artificial breeding that have been specifically developed for the horse. Chapters 9, 10, and 11 discuss equine reproduction in detail.

Finally, horses, being large, powerful, athletic animals, present some unique problems in the stables and fences it takes to contain them. Chapter 20 considers a wide variety of facilities for maintaining horses and the general management of a horse farm is presented in Chapter 21.

REFERENCES

American Horse Council. 1987. *The Economic Impact of the U.S. Horse Industry.* Washington, D.C.: A.H.C.

Crowell, Pers. 1951. *Cavalcade of American Horses.* New York: McGraw-Hill.

Denhardt, Robert Moorman. 1948. *The Horse of the Americas.* Norman: University of Oklahoma Press.

Dobie, J. Frank. 1952 *The Mustangs.* Boston: Little, Brown.

Epstein, H. 1971. *The Origin of the Domestic Animals of Africa*. New York: Africana.

Gianoli, Luigi. 1969. *Horses and Horsemanship through the Ages*. New York: Crown.

Haines, Francis. 1971. *Horses in America*. New York: Crowell.

Morris, Pamela MacGregor, and Nereo Lugli. 1973. *Horses of the World*. New York: Crown.

Russell, Charles U. 1927. *Trails Plowed Under*. Garden City, New York: Doubleday.

Ryden, Hope. 1970. *America's Last Wild Horses*. New York: Dutton.

Simpson, George Gaylor. 1951. *Horses*. New York: Oxford University Press.

Summerhays, R. S. 1961. *Horses and Ponies*. London: Frederick Warne.

Willoughby, David P. 1974. *The Empire of Equus*. Cranbury, New Jersey: A. S. Barnes.

CHAPTER 2

BREEDS IN
THE UNITED STATES

Each country claims many breeds of horses as its own, and the United States is no exception. Although all foundation stock for American breeds was imported, usually from Europe, American breeders have developed many distinctly American breeds from these stocks. Only a few breeds have been transferred directly to this country.

The usual definition of a breed is that it is a group of animals that have certain distinguishable characteristics, such as function, conformation, and color. Breed registries depend on correct identification to ensure accuracy of ancestry and to establish standards for the fair exchange of horses between buyer and seller. The first appendix to this chapter gives a brief outline of identification procedures based on color and color pattern.

A rather arbitrary distinction exists between horses and ponies. Ponies, measured at the withers, are less than 14 or 14.2 hands in height (14.2 means 14 hands and 2 inches), where a hand is 4 inches. Many horse breeds, however, include small members that fall below the dividing line.

Most of the major breeds of horses, ponies, and asses that maintain registries in the United States are briefly described in this chapter. More detailed accounts are available in the many books that have been written on the history and development of single breeds or that deal specifically with several breeds. The following discussion of breeds follows no particular order except for some consideration of the historical development and recent popularity of each breed.

2.1 The Arabian

No breed of horse has influenced the development of breeds of light horses in America more than has the Thoroughbred, but the Thoroughbred was developed largely from the Arabian. Although the history and origin of the Arabian horse are not always agreed upon even by experts, there is no question that the Arabs have been breeding and selecting for improved Arabian horses for 2000 years or more. The stock for this selection may have existed for as long as 3000 years in the Mideast or northern Africa before Arabian horses were first bred on the deserts of the Arabian peninsula.

Some historians believe that the Barb horses of northern Africa (the Barbary States) were ancestors of the Arabian, whereas others believe the Arabians were used in development of the Barb. In any event, the Arabians, Barbs, and Turkmene horses all developed in the same general region of the world. Most of the so-called hot-blooded horses of the world can be traced to these three ancestors.

In spite of the long history of the Arabian horse, the first registry for recording the breed in America was not organized until 1908. The name of the registry, the Arabian Horse Club Registry of America, was changed in 1949 and was shortened in 1969 to the Arabian Horse Registry of America (AHRA). The first Arabians apparently were imported to America shortly before the Revolutionary War, but the major expansion of the breed occurred about 1906 when Homer Davenport imported 27 horses from the deserts of Arabia. Records show that 39 other horses had been imported between 1760 and that year. This stimulus resulted in the organization of the first breed registry. Before 1908, American Arabians were registered by the Jockey Club, which discontinued registration of Arabians and Anglo-Arabs (crosses between registered Thoroughbreds and Arabians) in 1943 and currently registers only Thoroughbreds.

In the 1960s, Polish Arabians became popular, and many have been imported (Figure 2-1). Since the 1970s, breeders have imported breeding stock from several countries including Poland, England, Egypt, Spain, Russia, Sweden, and Australia.

In 1950, the International Arabian Horse Association was formed to promote the Arabian in America. In 1951, that association took over the registration of Half-Arabians and Anglo-Arabs from the American Remount Association. Half-Arabians have a registered Arabian either as sire or dam. Thus, the ancestry of the Half-Arabian will be 50 percent or more Arabian.

The Arabian is a general-purpose, light horse with an unsurpassed reputation for endurance. Arabians generally stand 14.1 to 15.1 hands at the withers and weigh between 800 and 1000 pounds—somewhat smaller than most general-purpose riding horses.

Figure 2-1
A Polish Arabian stallion, °Bask, imported in 1963 by Lasma Arabians — the asterisk preceding horse's name indicates importation. (Photograph courtesy of International Arabian Horse Association.)

The head of the ideal Arabian is distinctive: relatively small, dished, and triangular, with a small muzzle, wide-set eyes, and a chiseled appearance. The neck is also distinctive — long and highly arched, and set high on the shoulder. The tail is often arched above croup level while the horse is walking or trotting.

The colors for Arabians preferred by most breeders are generally solid: bay, brown, chestnut, gray, and black. Bays (Figure 2-2) and grays (Figure 2-3) have been particularly popular. White Arabians are grays that have turned white with age.

Half-Arabians resulting from crosses with Appaloosas, Paints, or Palominos, as shown in Figure 2-4, may be variable in color while maintaining some desired Arabian characteristics. Naturally, Half-Arabians vary greatly also in size and type as well as color, depending on the cross.

Figure 2-2
A National Champion Arabian
stallion, Khemosabi, owned by Haifa
Arabian Horses. (Photograph by
Polly Knoll, courtesy of Dr. and Mrs.
B. P. Husband.)

Figure 2-3
°Dornaba, a Champion Arabian mare. Owned by Dr. Howard F. Kale. (Photograph
courtesy of International Arabian Horse Association.)

(a)

(b)

Figure 2-4
Half-Arabians: (*a*) Candyhorse Kachina,
Appaloosa-Arabian of Candyhorse Farm
(Alexander photo); (*b*) Shoshone's Fancy,
Pinto-Arabian, Richard and Sherrie
Koehler, owners; (*c*) My Mystic Mirage,
Palomino-Arabian gelding owned by Mr.
and Mrs. George Albin. (Photographs
courtesy of International Arabian Horse
Association.)

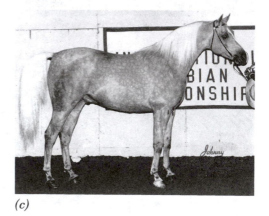

(c)

2.2 The Thoroughbred

The Thoroughbred has been developed for speed at intermediate distances.
No other breed can match the Thoroughbred at racing distances of 6 fur-
longs (¾ mile) to 1½ miles. In addition, Thoroughbreds have been popular
as polo ponies, hunters, and jumpers, as well as for pleasure riding. For many
years, Thoroughbreds and Half-Thoroughbreds were popular with the
United States Cavalry. In fact, General George A. Custer was mounted on a
Thoroughbred, Vic, by Austerlitz, at the Little Big Horn River on June 25,
1876.

The Thoroughbred provided foundation stock for many of the light
horse breeds of the United States, including the Standardbred, the American
Saddle Horse, the Morgan, and the Quarter Horse.

The history of the Thoroughbred as a breed began in England. Native horses had been crossed with light horse mares imported from Spain, Turkey, and Italy. Then, from the late 1600s until 1750, Arabians, Turks, and Barbs (Oriental sires as they were called) were imported for the purpose of increasing the speed of horses used for the popular sport of racing. Three of these 174 stallions became most famous and eventually became known as the basis of the three stallion lines to which nearly all Thoroughbreds can be traced.

A horse known as the Byerly Turk (foaled in 1679) was taken to England by Captain Byerly in 1689; Herod, a great-great-grandson (1758), was a founder of one of the three stallion lines. The Godolphin Arabian (some claim that he was a Barb), foaled about 1724 on the Barbary Coast, found his way to France and later to England, where he was the property of Lord Godolphin. His grandson was Matchem (1748), another foundation sire. Nearly 90 percent of all Thoroughbreds trace to Eclipse (1764), a stallion that was unbeaten in 26 starts. This foundation sire was the great-great-grandson of the Darley Arabian that was foaled in Syria in 1700 and imported to England in 1704. Bulle Rock is traditionally regarded to be the first Thoroughbred imported to America (in 1730 at the age of 21 years).

A fourth foundation sire, the Curwen Bay Barb, should be added to the three stallions usually listed as major contributors of genes to current Thoroughbreds. Mahon and Cunningham (1980) reported that twelve foundation ancestors accounted for 55 percent of the genes in the British Thoroughbred — led by the Godolphin Arabian with 14.6 percent, the Darley Arabian with 7.5 percent, the Curwen Bay Barb with 5.6 percent, and the Byerley Turk with 4.8 percent.

The first recording of Thoroughbreds in England was in 1791 by James Weatherby, Jr., in his *Introduction to a General Stud Book*. Volume 1 of the *General Stud Book* appeared in 1793, and revisions were published in 1803, 1808, 1827, 1858, and 1891. All English Thoroughbreds must trace to animals included in the *General Stud Book*. This requirement naturally excluded many Thoroughbreds in America. Volumes 1 and 2 of *The American Stud Book* were published in 1873. The rights to this registry were purchased in 1894 by The Jockey Club, which continues to publish *The American Stud Book* and to register American Thoroughbreds. Only horses whose sire and dam are registered in *The American Stud Book* or similar stud books of other countries are eligible to be registered.

The ideal Thoroughbred is difficult to describe. The oldest axiom in racing is "They run in all shapes and sizes." The most complete measure of the racing Thoroughbred is the stopwatch. Performance under racing conditions is the essence of racing. A superior racer will have acceptable conformation, but superior conformation does not necessarily lead to even adequate speed. Most Thoroughbreds, however, tend to have a long forearm and gaskin and display considerable length from the hip to the hock. They

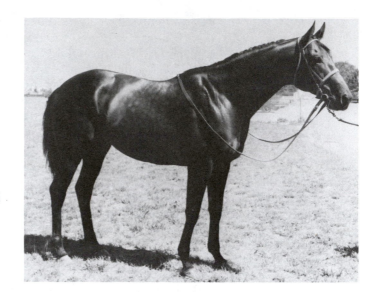

Figure 2-5
Boldwood, a Thoroughbred
son of Bold Ruler and
grandson of Bull Lea, at
stud at Matron Farm.
(Photograph courtesy of
James K. Thomas,
Lexington, Kentucky.)

are noted for long, smooth muscling. The rear, or propelling, quarters are especially powerful. Thoroughbreds excel at the run or extended gallop. The usual range in size at racing condition is 15.1 to 16.2 hands and 900 to 1150 pounds. The modern Thoroughbred (Figure 2-5) is nearly 2 hands taller than the foundation Thoroughbreds of about 1750.

The world record of 1:32⅕ seconds for the mile around one turn was set by Dr. Fager (Figure 2-6) in 1968 at Chicago's Arlington Park while carrying a heavy weight of 134 pounds. In 1973, Secretariat (Figure 2-7) became the first horse in 25 years to win all the Triple Crown races for 3-year-olds. His amazing performance was climaxed in the final race at Belmont, a 31-length victory, where he broke the track record of 2:26⅗ by 2⅗ seconds. His feats, however, only slightly dull the performance of the original Big Red, Man-O-War, that was a winner in 21 of 22 starts but was not entered in the Kentucky Derby.

Thoroughbreds are performance horses, so their color is not important. Colors and markings are recorded at registration, however, for purposes of identification. The Jockey Club recognizes black, dark bay or brown, bay, chestnut, gray, and roan. The description of roan is really a nonblack gray, as true roan is rarely seen in Thoroughbreds. White Thoroughbreds are generally gray early in life, and with age may turn white. Native Dancer, shown in Figure 2-8 at his retirement, was dark gray when he excited television fans but later turned snow white. Dun is a rare color, but white at birth is even rarer. The first white United States Thoroughbred, a filly named White Beauty, was registered in 1963. Another was born in France in the same

Figure 2-6
Dr. Fager, Champion
American Thoroughbred
Racehorse of 1968 and
holder of the record for one
mile, 1:32⅕ seconds
carrying 134 pounds at
Arlington Park, Illinois, on
August 24, 1968.
(Photograph by Jim Raftery
Turfotos, courtesy of the
owners, Tartan Farms.)

year, and yet another had been foaled in Germany more than 100 years earlier.

In colonial America, raising and racing racehorses was centered in Virginia. The center of Thoroughbred racing then moved to the bluegrass country of Kentucky and Tennessee. More recently many breeding farms have been established in warm-weather states such as California and Florida.

2.3   The American Quarter Horse

Admirers of the Quarter Horse claim it was the first breed developed in the United States, even before the Thoroughbred was developed and long before Paul Revere made his historic ride — on, it is said, a Quarter Horse. The quarter running horse (quarter of a mile) is said to have run in colonial America on the short, flat stretches of towns and villages. These horses probably would have developed from crosses with horses that had been brought to Florida earlier by the Spaniards. Whether this is accepted history or whether the development of the cow horse in the southwest range country during the middle and late 1800s should be considered the beginning of the breed does not really matter to the many proud owners of the modern Quarter Horse.

Even if this version is not accepted by all historians, there is general agreement that a Thoroughbred imported in 1752 had a lasting influence on

Figure 2-7
The 1973 Triple Crown winner, Secretariat, (NYRA photograph by Bob
Coglianese, courtesy of Mrs. Penny Renquist.)

the development of the Quarter Horse type. Most of the foundation sire lines
trace to him. Janus, a grandson of Godolphin Arabian, was noted for speed at
distances of 4 miles. Yet he sired many quarter running horses that had
exceptional short-distance speeds. He also sired many famous Thoroughbred
stallions and mares. In fact, more than one winner of the Kentucky Derby

Figure 2-8
Native Dancer (*left*) retiring at Belmont with E. Guerin up. (Photograph courtesy of Keeneland-Morgan, Keeneland Library, Lexington, Kentucky.)

traces to Janus. Thus, he truly represented the meaning of his name—to look in opposite directions.

Many stallions were bred during the movement west and were mated to horses of Spanish ancestry in the Southwest during the development of the working Quarter Horse. Their names are colorful and evoke memories of times past. Steeldust, foaled in Illinois in 1843 and moved to Texas in 1846, was perhaps the most famous horse of the early Quarter Horse type. Until about 1938, horses of the quarter type were called Steeldusts. Copper Bottom and Old Shiloh were two of his contemporaries.

In 1895, Peter McCue was foaled in Illinois and became the most important of all sires in the development of the breed. More than 20 percent of the

Quarter Horses registered prior to 1948 traced on the male side to Peter McCue. The next most important horse, Traveler, had only one-third as many similar descendants.

Old Sorrel, foaled in 1915, a grandson of Peter McCue, deserves special mention. The King Ranch of Texas decided that he most nearly fit their ideal of the working cow horse. A linebreeding program that had hardly ever been used with any other kind of livestock was initiated to fix his type (Rhoad and Kleberg, 1946).

The American Quarter Horse Association was the first organization formed to register Quarter Horses. The bases for registration were type, pedigree, and performance. This and most rival organizations were merged in 1950 and issued a more nearly closed stud book.

Breeders of the Quarter Horse are somewhat divided as to the performance objectives of the breed. The increasing demand for and stakes in Quarter Horse racing have directed some to breed primarily for speed by increased introduction of Thoroughbred breeding. Go Man Go, a leading sire with regard to earnings and register of merit winners, shown in Figure 2-9, was sired by Top Deck, a Thoroughbred. For many years Three Bars, a Thoroughbred, was also a leading sire. Other breeders are more interested in maintaining the image of the shorter-coupled, more muscular front- and rear-ended cow horse, which has the dexterity and tenacity the rancher needs. This type of horse is illustrated by Wimpy P-1, shown in Figure 2-10 in retirement at the King Ranch. Wimpy was awarded the first permanent

Figure 2-9
A head study of Go Man Go, a son of the Thoroughbred Top Deck, whose Quarter Horse get have earned more than $4.5 million. (Photograph by Bob Taylor, courtesy of Jack McReynolds, Purcell, Oklahoma.)

Figure 2-10
Wimpy P-1. The holder of the first permanent registration number in the American Quarter Horse Association is shown in retirement at the King Ranch. (Photograph courtesy of King Ranch, Incorporated, Kingsville, Texas.)

registration number by virtue of being named the grand champion stallion at the 1941 Fort Worth Exposition. Wimpy is a grandson of Old Sorrel, a grandson of Peter McCue. Whether both groups should be maintained under one registry is debatable, but the breed may be numerous enough to allow the luxury of two distinct types of Quarter Horses. Some horses combine ability in racing and excellence in other areas of performance (Figure 2-11), although such general ability is not always found.

Figure 2-11
Deck Jack, a Champion Quarter Horse stallion, owned by Robert Kowalewski, East Aurora, New York. (Photograph by Darol Dickinson, courtesy of Robert Kowalewski.)

Color of the Quarter Horse is stated to be of no particular importance except for personal preference, but animals with spots or markings that indicate Paint, Pinto, Appaloosa, or American Albino breeding are not eligible for registration.

2.4   The Standardbred

The Standardbred, once called the American Trotting Horse, was developed from Thoroughbred, Norfolk Trotter, Barb, Morgan, and Canadian pacing ancestors. The name Standardbred comes from the practice that began in the 1800s of registering horses that trotted or paced the mile in less than a "standard" time. Over the years, since Yankee first trotted the mile under saddle in less than 3 minutes at the Harlem racetrack in 1806, the standard has been lowered. The standards first officially set in 1879 were 2:30 for trotters and 2:25 for pacers. The current standard is 2:20 for 2-year-olds and 2:15 for older horses, although horses are now only rarely registered on the basis of these standards.

Many breeders trace the Standardbred to Messenger, a gray Thoroughbred imported to Philadelphia from England in 1788. Messenger traces to all three foundation sires of the Thoroughbred breed. Although neither Messenger nor any of his sons trotted or paced, he traces to a stallion, Blaze, that sired the foundation horse of the Norfolk-Hackney Trotters, Old Shales. Messenger appears three time in the pedigree of his great-grandson, Hambletonian 10 (Figure 2-12), foaled in 1849. Hambletonian, number 10 in the

Figure 2-12
Nearly all Standardbred horses trace to this stallion, Hambletonian 10, which was owned by Bill Rysdyk. Rysdyk bought the bay as a foal for $125 and eventually collected $300,000 in breeding fees. (Photograph of a painting courtesy of United States Trotting Association, Columbus, Ohio.)

first numbered stud book, is perhaps the greatest name in the history of the breed; approximately 99 percent of all Standardbreds trace to him. Part of his predominance may be due to his prolificacy — 1331 living foals. His sons founded the four predominant sire lines of the present-day breed — The Direct and The Abbe lines of pacers and The Axworthy and Peter the Great lines of trotters — although Hambletonian himself never raced. The imported Bellfounder, which had Norfolk Trotting ancestry, was the maternal grandsire of Hambletonian. Tom Hal was the founder of a Canadian pacing line that, along with other Thoroughbreds, Morgans, and native horses, contributed to the development of the breed.

The official stud registry of the breed is the *Sires and Dams Book*, now administered by the United States Trotting Association. J. H. Wallace prepared the forerunner of this book in 1871. The stud book is now closed to all horses except those sired by registered sires and dams. Horses that have a registered sire and that meet the current standards of performance may also be registered.

The measure of performance is speed, usually for the 1-mile distance. Parimutuel racing has changed the original format of racing. The classic harness race was originally to win 3 heats; it later became the best 2 of 3 heats. Winners would occasionally need to race 11 or 12 heats so that endurance was emphasized as well as speed for the mile. Some major races still require the winner to win 2 heats. Now most races are decided by a single 1-mile "dash," which furnishes better entertainment for race fans.

Dan Patch was one of the most famous and the most enduring star of pacers. He held the record from 1903 until 1938, when Billy Direct lowered the record by only ¼ of a second — and that record lasted for another 22 years. Dan Patch was literally four or five generations ahead of his time. His best time was 1:55¼ in 1905. He is shown with the equipment of that era in Figure 2-13.

Greyhound (Figure 2-14) held the trotting crown for nearly as long, from 1937 until 1969, when Nevele Pride bettered Greyhound's time of 1:55¼ by less than half a second. Greyhound was also the holder of records for trotting under saddle.

Niatross (Figure 2-15), in a time trial of 1:49.1 in 1981, smashed the pacing record for the mile by over 2 seconds. Time trials are races against time, not against other horses. Records made in time trials are denoted as, for example, TT 1:49⅕.

Early breeders favored trotters and did not appreciate pacers. Now, however, except for some of the classic races, pacers predominate at most parimutuel betting tracks. Usually 8 of 10 races on the card are for pacers, which are, on the average, only marginally faster. Pacers do not race with trotters. Genetic factors and, more importantly, training and shoeing methods, are used to determine or to change the gait. A few horses have the ability to race well at either the trot or the pace — in separate races, of

Figure 2-13
Dan Patch, considered by many to be the greatest harness horse of all time, shown with driver H. C. Hersey. Dan Patch held the pacing record for 35 years. His fastest time of 1:55¼ was beaten by only ¼ second 33 years later by Billy Direct, whose record stood for 22 years thereafter. (Photograph courtesy of the United States Trotting Association, Columbus, Ohio.)

Figure 2-14
Greyhound, shown with trainer-driver Sep Palin, helped increase the popularity of harness racing. His trotting record of 1:55¼ in 1938 stood for 31 years. He still holds the world trotting record for geldings. (Photograph courtesy of the United States Trotting Association, Columbus, Ohio.)

Figure 2-15
Niatross, shown with Clint Galbraith in the bike, may be the finest Standardbred ever. The son of Albatross won all but two of 39 races and became the first harness horse to pace the mile in under 1:50, TT 1:49.1. (Photograph courtesy of United States Trotting Association, Columbus, Ohio.)

course. The trot is a two-beat diagonal gait; the opposite front and rear feet push off and land at the same time. The pace is a two-beat lateral gait; the front and hind feet on the same side start and land together.

Some breeders believe the tendency to trot is inherited as a single dominant trait, but noted breeding authority James Harrison (1968) asserts that the tendency to pace is dominant. He has observed that pacers mated to pacers produce pacers 99 percent of the time, whereas pacers mated to trotters usually produce pacers, and trotters mated to trotters frequently produce pacers. Actually, his observation also supports the theory that the gene for trotting is dominant over the gene for pacing. Probably neither view is completely correct since the traits are undoubtedly affected by more than one genetic factor.

The measure of performance of the Standardbred is speed. Conformation may contribute to freedom from injury and breakdown, but it is not primarily important. The range in height is often 14.2 to 16.2 hands; the range in weight is 850 to 1150 pounds when the horse is in racing condition. Early breeders tended to favor horses whose length exceeded their height. However, Greyhound may have changed the minds of many breeders because he was taller than he was long.

Bay is the predominant color, but chestnut, brown, black, and, of course gray, are also seen. There seems to be no color discrimination if the horse is fast.

2.5 The Appaloosa

The Palouse River country of the northwestern United States has given its name to a distinctive breed of horse. The name "Appaloosa" was derived from the slurring of "A Palouse" to form Apaloose, which later became Apaloosie and is now Appaloosa (ap-pah-*loose*-ah).

Horses with the colorful characteristics of the Appaloosa appear in Chinese art dating from 500 B.C. and in Persian and European art of the fourteenth century.

Spanish horses that were brought to Mexico about 1600 apparently have formed the basis for the present-day Appaloosa. The Spanish horses and their descendants spread northward and, by 1730, had been acquired by the Nez Percé tribe in the Palouse country. Because of their colorful markings and riding characteristics (endurance, surefootedness), the Nez Percé bred the Appaloosa for rugged mountain traveling for the next 100 years. As a type, the Appaloosa nearly disappeared after the surrender of Chief Joseph and the Nez Percé to the United States Army in the Bear Paw Mountains of Montana in 1877.

The Appaloosa Horse Club was formed in 1938 to preserve, improve, and standardize the spotted horse of the Nez Percé. Since the few foundation descendants of the Nez Percé horses were registered in that year, the registry of the Appaloosa has grown rapidly.

Three distinctive characteristics are required of all Appaloosas: (1) the eye is encircled with white like the human eye (Figure 2-16); (2) the skin is mottled irregularly with black and white (parti-colored), particularly around the nostrils (Figure 2-16) and genitalia; and (3) the hooves are narrowly striped vertically in black and white (Figure 2-17). The color patterns vary widely, as shown by the 8 different patterns in Figure 2-18. Any combination and many variations of these patterns may occur. Because the genetics of the color patterns is not well understood, standardizing the breed, and even predicting the results of particular matings, is difficult. The color patterns may not be apparent at birth and may change with age. The mane and tail of many Appaloosas are sparse. This is called the rat-tailed condition. For breeding purposes some solid-color horses are registered in a separate section, but are not eligible to show as Appaloosas.

A definite effort has been made to standardize the breed to a general-purpose riding horse that can be used for pleasure, parade, rodeo, western show, and racing. Horses with pony or draft horse breeding are not eligible for registration. In an attempt to protect the distinctive color pattern, horses with Albino, Pinto, or Paint breeding or markings are also excluded, as are horses with excessive white or misplaced spots.

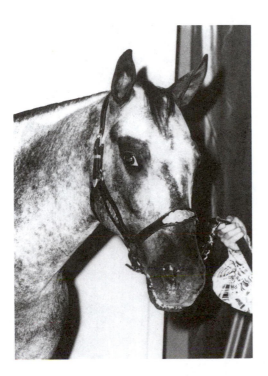

Figure 2-16
Photograph of Joker's Starlight.
Registered Appaloosas must have
white encircling the eye, and the
skin must be mottled, which usually
shows on the muzzle or genitalia.
(Photograph by Johnny Johnston,
courtesy of Howard C. Hanson, Jr.,
Blair, Nebraska.)

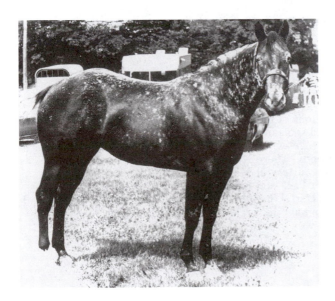

Figure 2-17
The hooves of registered
Appaloosas must show vertical
stripes of black and white, as
shown in this photograph of
Hands Up Cowgirl. (Photograph
courtesy of Ken Friday, Jefferson,
Iowa.)

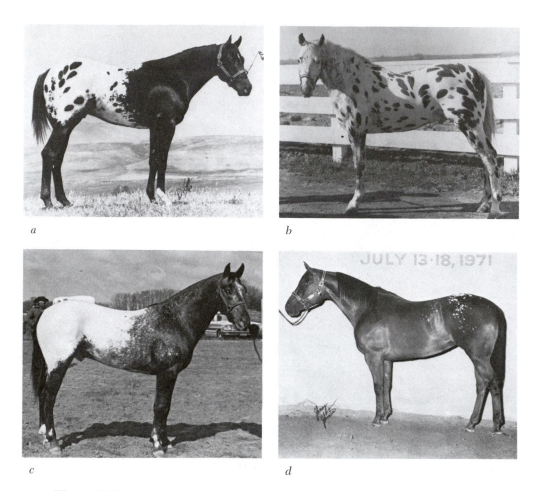

Figure 2-18
Illustrations of eight different patterns of Appaloosa markings. There are many variations and other combinations of these patterns. (*a*) Most typical is the colored pattern with white blanket and spots over loin and hips (Stud Spider; photograph courtesy of Jane Brolin, Los Angeles, California). (*b*) The leopard pattern — white with colored spots over the body. There is much possible variation in size of spots (Joker's John E.; photograph courtesy of Mr. and Mrs. E. D. Pollard, Caldwell, Idaho). (*c*) White blanket over back and hips, no spots in blanket (Domino's Le Don; photograph courtesy of Darrel Steenblock, Fremont, Nebraska.) (*d*) Colored body with white spots over loin and hips (Dash's Charm; photograph by Johnny

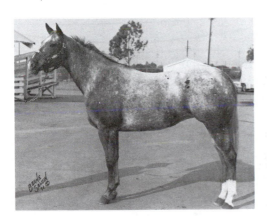

e

f

g

h

Johnston, courtesy of Howard C. Hanson, Jr., Blair, Nebraska). (*e*) Colored body, speckled white over entire body (Spots Souviner; photograph courtesy of Kent E. Ellsworth, Ontario, California). (*f*) Colored body, slight roaning with dark spots over loin and hips, no blanket (Good John; photograph courtesy of L. J. and Zora Estes, Milton-Freewater, Oregon). (*g*) Roan body, light over loin and hips (Bar My Hart; photograph by Berne Salvin, courtesy of Don F. Craib, Jr., Lake Forest, Illinois). (*h*) Roan body, white blanket with spots over loin and hips (War Reed's Echo; photograph by Ric Robinson, courtesy of Ray Jensen, Manager, Windswept Acres, Woodstock, Illinois).

2.6   The Morgan Horse

Morgan is the only breed named after a horse. No other horse has such a distinction as Justin Morgan, the foundation sire of the Morgan Horse. Actually, the stallion was called Figure as a foal, and took his adult name from his owner (a Massachusetts schoolteacher), as was the custom in the late 1700s. He followed Justin Morgan from Massachusetts (where the stallion was foaled) through and around Vermont.

Justin Morgan, the horse, became famous because of his outstanding progeny and because of his ability, according to stories that still persist, to outrun, outpull, outwalk, and outtrot all competition. Horses with Morgan blood became popular before 1850 for their all-purpose ability—on the farm, at the trot, and under the saddle. After 1850, the developing Standardbred breed, to which Morgan Horses contributed substantially, replaced the Morgan on the race tracks. Later, motor vehicles replaced the Morgan on the farms and roads. Since then the Morgan has been used primarily for pleasure riding and more recently as a show horse.

The ancestry of Justin Morgan is not clear, although Colonel Joseph Battell, who became a prime benefactor of the Morgan breed, came to believe that the horse was sired by a Thoroughbred called True Briton (also called Beautiful Bay) and out of a mare of Arabian breeding. Since Thoroughbreds of that time were closely related to Arabians, Justin Morgan had and passed on many characteristics of the Arabian—especially the refined head and raised tail when on the move.

Justin Morgan's progeny and grandprogeny were used in establishing the Standardbred and American Saddle Horse breeds. Many Quarter Horses also trace to Morgan breeding.

Colonel Battell founded the Morgan Horse Register in 1894. The Register was taken over by the Morgan Horse Club in 1930, some 21 years after the Club was organized. Colonel Battell also gave a large farm near Middlebury, Vermont, to the United States Department of Agriculture to be used to preserve and improve the breed. The farm was turned over to the University of Vermont in 1951.

The Morgan today is popular for riding (Figure 2-19) and for shows. The size has increased, from Justin Morgan's 14 hands and less than 1,000 pounds to a usual range of 14.1 to 15.1 hands and 1000 to 1200 pounds.

The American Morgan Horse Association, as it is now named, registers only horses that have registered sires and dams. All colors are acceptable except white. Spotted horses and those with Appaloosa patterns are ineligible. Palominos, duns, and buckskins are acceptable. The dark liver or black chestnut color is found more in the Morgan than in any other breed.

Figure 2-19
Tara's Delight, a Morgan mare shown in an English pleasure class. (Photograph by
Paul Quinn, courtesy of Dr. and Mrs. V. Watson Pugh, Raleigh, North Carolina.)

2.7   The American Saddlebred

The American Saddlebred originated in the United States, and the type
evolved as the needs of the country changed. The residents of the bluegrass
region of Kentucky and areas of Tennessee, Virginia, West Virginia, and

(later) Missouri desired an easy-riding, general-purpose type of horse for the plantations and hilly grazing areas. The first name given to horses of this type was the Kentucky Saddler. This easy-gaited horse developed from Thoroughbred, Canadian pacer, American trotter, Morgan, Arabian, and other ancestors. In 1901 the American Saddle Horse Association listed ten foundation sires, but in 1908 the list was reduced to the Thoroughbred stallion, Denmark—undoubtedly the most important. The others were given registration numbers and the status of Noted Deceased Sire. Another famous family is the Chief Family, which traces to Mambrino Chief, a trotter that in turn traces to the imported Thoroughbred, Messenger. Early development of the breed preceded Denmark, although the era of its popularity was after Denmark and the Civil War.

The American Saddlebred Horse Association was formed in 1891 with the current name adopted in 1980.

As needs changed, the American Saddlebred developed into what has been called the peacock of the horse world. Breeding has been largely for horse-show purposes. The Saddlebred is in demand for three- and five-gaited classes, for fine harness, and for combination saddle and harness classes, although many of these horses are also used for pleasure riding (Figure 2-20).

The show class Saddlebred has emphasized the flashy and exaggerated but controlled gaits, high carriage of the head, and distinctive set of the tail

Figure 2-20
An American Saddlebred, Genius Ebony Lady, being used for Western Trail Riding. (Photograph by Joan S. Bryne, courtesy of American Saddlebred Horse Association.)

Figure 2-21
Wing Commander, six-time winner of the $10,000, five-gaited championship at the Kentucky State Fair. (Photograph by John K. Hasst, courtesy of the American Saddle Horse Breeders Association.)

(Figure 2-21). The angle of the tail is partly determined by how the trainer "sets the tail."

Dark colors often are preferred but the coloring can be bay, brown, black, chestnut, gray, or roan. Large, white markings are avoided by many breeders and trainers. The Pinto Horse Association registers spotted horses of the Saddlebred type.

The American Saddlebred can be three- or five-gaited. The Saddlebred is trained to perform each gait distinctly with considerable action, to go without hesitation from one gait to the other, and to change lead at the canter from left to right on command. The basic three gaits are the walk, trot, and canter. The three-gaited horse must go from the trot, a high-action, two-beat diagonal gait, to the slow, springy four-beat walk, and then to the slow, rhythmic, smooth canter, a three-beat gait. Additional training and ability are required for the trained gaits of the five-gaited horse. The slow gait is a high-stepping four-beat gait. The rack is a fast, flashy, four-beat gait (sometimes called the single foot) that is free from any pacing motion. This gait is easy on the rider but tiring for the horse and receives much emphasis in the judging of five-gaited horses.

2.8   The Tennessee Walking Horse

As did the Saddlebred and the Morgan, the Tennessee Walking Horse developed as a general-purpose breed for riding, driving, and farm work. The history of the breed, which originated in the Middle Basin of Tennessee, traces to contributions of Thoroughbreds, Standardbreds, American Saddle Horses, Morgans, and Narragansett and Canadian pacers, as well as less well-recorded stock.

The Tennessee Walking Horse naturally overstrides; when performing the running walk, a good show horse of the breed will place the back hoof ahead of the print of his fore hoof as much as fifty inches or more. Although special shoeing will accent the running-walk gait, inhumane practices such as soring are not necessary to produce the gait.

When the Tennessee Walking Horse Breeders' Association was formed in 1935, Allan F-1 (also known as Black Allan) was designated as the official foundation sire. Allan F-1, foaled in 1886 of mixed Standardbred, Morgan, and Narragansett pacers, sired several sons that were bred to other saddle breeds. One of the more famous sons was Roan Allen F-38, who was reportedly able to trot in harness and win, then do five gaits and win, then return and win the walking classes. The influence of Allan F-1 was nearly as great in establishing the Walking Horse type as the influence of Justin Morgan was in establishing the Morgan breed. Interestingly, the Tennessee Walking Horse (at one time more popularly known as the Plantation Walking Horse) developed naturally, as did the Morgan, to meet the work needs of its region and not the desires of fanciers, although more recent demands of the show-ring have not been as natural.

The first Saturday night in September climaxes the most exciting week of the year for Tennessee Walking Horse breeders, for it is then that approximately 25,000 people gather to see the Grand Champion Walking Horse of the world crowned at The Tennessee Walking Horse National Celebration at Shelbyville (Figure 2-22). The best of the Tennessee Walkers compete at

Figure 2-22
The smooth-riding Tennessee
Walking Horse is equally popular as
a sure-footed trail mount or a
high-stepping show-ring performer.
(Photograph by Harold Twitty,
courtesy of the *Voice of the
Tennessee Walking Horse Magazine*.)

this horse show, which began in 1939 shortly after the breed registry was
chartered in Tennessee.

Walking Horses come in all solid colors. White markings are common.
Gray and roan are not undesirable. In the 1940s and 1950s, roan was very
popular, but since the 1960s and 1970s blacks and dark colors have been
most popular.

2.9   The Fox Trotting Horse

The Fox Trotting Horse developed in the nineteenth century in the Ozark
Mountains of southern Missouri and northern Arkansas to meet the need of
that area for a riding horse that could travel long distances with a comfortable gait at a speed of 5 to 8 miles per hour. With the resurgence of the
popularity of riding in the 1960s, these horses were well suited for pleasure
and cross-country trail riding.

The characteristic gait that developed, the fox trot, is a major requirement for registration in the Missouri Fox Trotting Horse Breed Association,
which was incorporated in 1948.

Melvin Bradley, a Missouri extension horse specialist, has described the fox trotting gait as one that

> starts out as a simple trot; that is, diagonal feet leave the ground at the same time. The back diagonal foot, however, comes down later than the front foot. This makes a four-beat gait instead of the hard two-beat square trot. The back foot does not come down in a hard step, but actually appears to slide a little bit or contact the ground softly. The body is rising in front and lowering behind in unison. This keeps the rider hinged in the middle with a very soft ride.

The Fox Trotting Horse traces first to Arabians, Morgans, and plantation horses. There was a subsequent infusion of the American Saddlebreds, Tennessee Walkers, and Standardbreds. The names of early Fox Trotting families (Copper Bottoms, Diamonds, Chiefs, Steel Dusts, Cold Decks, and others) suggest strong ties to the early Quarter Horses.

Nearly all colors are common, but palominos, blacks, sorrels, and blue and red roans are the most popular. A full mane and long, flowing tail are desirable. The preferred body type is somewhat intermediate between the Quarterhorse and the American Saddlebred or Tennessee Walking Horse. A World Champion Fox Trotter Show and Celebration is shown in Figure 2-23.

2.10   The Pasos: Paso Fino and Peruvian Paso

Most of the light-horse breeds in the United States have developed in the United States from varied ancestral breeds. In the middle 1960s, however, two breed registries were incorporated that register somewhat similar

Figure 2-23
Danney Joe W., a World
Champion Fox Trotter.
(Photograph courtesy of the
owner, Dale Wood, Nebo,
Missouri.)

horses: Pasos imported from Peru (the Peruvian Paso) and Pasos imported from Puerto Rico and Colombia (the Paso Fino). Both groups trace to similar ancestors and are noted for the smoothness of their natural gait, called the *paso*. The same rhythm is maintained for all speeds of the gait.

The paso gait is essentially a broken pace, that is, a four-beat lateral rather than a diagonal gait. The sequence of movement of the hooves is: right rear, right fore, left rear, left fore. The hind foot touches the ground a fraction of a second ahead of the front foot, and this helps to eliminate the jarring effect of the true pace so that the rider has little up and down movement.

The Pasos are descendants of the Spanish horses of Andalusion, Barb, Spanish Jennet, and Friesian breeding brought to Central and South America. The Spanish Jennet was a light, agile horse developed by the Spanish with Arab and Barb breeding. The refined Spanish Jennet was used in the development of the English saddle horses. Above all, the Spanish Jennet should not be confused with Spanish donkeys which are asses rather than horses. These horses, and similar ones that were imported later, became the foundation stock for the remount station of the Conquistadores. Salazar took some of them to Puerto Rico in 1509, and when Velasquez invaded Cuba in 1511, he took others. Pizzaro used Dominican horses when he seized Peru in 1533.

The Paso Fino

The Paso Fino Owners and Breeders Association registers all strains and crosses of Paso Fino Horses. Most imported Paso Fino horses registered by PFOBA have come from Puerto Rico (Figure 2-24) or Colombia. The American Paso Fino Pleasure Horse Association also promotes and registers Paso Fino horses. Paso Finos show all solid colors and roans, as well as spotted, creme, buckskin, and palomino patterns.

The Peruvian Paso

The Peruvians are proud that their Paso developed solely from horses brought to Peru by the Spanish Conquistadores. Horses of pure Peruvian blood are eligible for registration by the Peruvian Paso Horse Registry of North America (Figure 2-25). Pintos and whites or cremellos are not eligible for registration. The Peruvian Paso is known for the *termino*, a flowing movement in which the forelegs roll to the outside as the horse strides forward. The American Association of Owners and Breeders of Peruvian Paso Horses also registers Peruvian Pasos.

Figure 2-24
Picasso Lace, a Paso Fino gelding.
(Photograph courtesy of Paso
Fino Horse Association, Inc.)

Figure 2-25
Rizado, a National
Champion Peruvian Paso
gelding. (Photograph by
Foucher Equine
Photography, courtesy of
Hacienda de la Solana,
Guerneville, California.)

2.11   The Galiceño

The Galiceño (gal-i-*sehn*-yo) is a small, sturdy horse of 12 to 13.2 hands—
the size of a pony but in all other ways a horse. These horses have been
imported from Mexico since 1959. The breed probably originated in six-
teenth-century Spain in the ancient province of Galicia, of Spanish Jennet
and Barb breeding.

The Galiceño Horse Breeders Association was formed in 1959.

The Galiceño was a contemporary of the Paso Fino stock of Central and
South America but has not been crossed with larger breeds, so the Galiceño
is now much smaller—600 to 700 pounds (Figure 2-26). The Galiceño
comes in all solid colors; albinos, pintos, and crosses are not eligible for
registration. The Galiceño has a running walk similar to that of the Tennessee Walking Horse and the Pasos. The other standard gaits—walk, trot, and
canter—are also natural. Most owners have thought the Galiceño to be ideal
for children or for young adults who want something larger than a pony.

2.12   The Morab

Crosses between Morgans and Arabians (Morabs) have been made since the
1800s to complement the strengths of the two breeds. After World War I the
cross gained popularity as a family pleasure and ranch horse. The Morab
Horse Registry, however, was not established until 1973 (Figure 2-27). The

Figure 2-26
A registered Galiceño mare, dun
with black dorsal stripe.
(Photograph courtesy of the
owners, Mr. and Mrs. Robert I.
Kinsel, Jr., Hamilton, Texas.)

Figure 2-27
Tezya, the first horse registered
by the Morab Horse Registry.
(Photograph courtesy of North
American Morab Horse
Association.)

goal was to emphasize the characteristics of the 50 : 50 proportion of the two
breeds, as crosses with more than 75 : 25 or 25 : 75 proportions tend to
become more like the breed contributing the larger fraction of genetic
material.

2.13   Paint, Pinto, and Spotted Horses

Many breed registries will not register horses with body spots. Since color is
not related to function, several registries whose function is to register spotted horses have developed.

The terms *paint*, *spotted*, and *pinto* are synonymous since "pinto" is
derived from a Spanish word that means "paint or painted or spotted." All
can be used to describe horses with body markings of white and another
color.

The English have two other words to describe the spotted horse. *Piebald*
refers to a horse that is black with white spots; *skewbald* denotes a spotted
horse of white and any color other than black.

Two types of spotting are the Overo (o-*ver*-o) and Tobiano (toe-be-*an*-o)
patterns. These patterns describe the general location of white on the horse
and not the amount of white.

The *Overo* pattern (Figure 2-28) is basically colored with white spots. The usual guidelines are:

1. White does not cross the back.
2. One or more legs are dark.
3. The head is often bald, apron-, or bonnet-faced.
4. The white body markings are irregularly spotted or splashy.
5. The tail is usually one color.

Figure 2-28
Yellow Mount, the first American Paint Horse Association Champion, showing the Overo pattern on a red dun background. (Photograph courtesy of the owner, Stanley Williamson, Iowa Park, Texas.)

The *Tobiano* pattern (Figure 2-29) is basically white with colored spots. Other guidelines are:

1. White crosses the back.
2. The head is marked like that of a solid-color horse — solid or having blaze, strip, star, or snip white markings.
3. All legs are white, at least below hocks and knees.
4. Body spots are regular, oval-shaped, and distinct.
5. One or both flanks are usually dark.

Either pattern may be mainly colored or mainly white, but the ideal for both is approximately equal parts white and colored. Tobiano is thought to be a dominant genetic trait and Overo is thought to be primarily a recessive genetic trait. Many genes probably operate to develop both patterns. Crosses between horses with Overo and Tobiano patterns add to the variation in spotting. Neither of the two associations registers spotted horses of pony or draft characteristics. Glass eyes are acceptable because this is a common characteristic of horses with spotted breeding — especially when the white spotting extends over the area of the eye.

The American Paint Horse Association

The American Paint Horse Association, formed in 1965, was organized partly as the result of the failure of the American Quarter Horse Association to register spotted horses. The APHA registers primarily stock and Quarter

Figure 2-29
A saddle type Pinto Horse filly, She's a Desperado, with the tobiano pattern. Owned by Richard and Kay Patterson, Sisters, Oregon. (Photograph courtesy of the Pinto Horse Association of America, Inc.)

Horse type horses. Some of the more important requirements of the American Paint Horse Association for the regular registry are similar to those of the American Quarter Horse Association.

The Pinto Horse Association

Horses registered in the Pinto Horse Association of America, founded in 1956, generally belong to four conformation types: the stock horse of Quarter Horse breeding, the hunter type with Thoroughbred breeding, the pleasure type with Arabian and Morgan bloodlines, and the saddle type, which traces to the American Saddlebred, Hackney, or Tennessee Walking Horse.

2.14   The Palomino

The Palomino is truly the golden horse. Palominos are registered according to color and not as to type except that pony and draft breeds are excluded. The color (within three shades) is approximately that of an untarnished United States gold coin. The mane and tail are white or near white, with no more than 15 percent darker hairs.

Within the color and light-horse limits, the uses of the Palomino are widely varied: cutting horse, Quarter Horse racing, parade horse, pleasure horse, trotting and pacing, harness classes, three- and five-gaited classes, and Tennessee Walking Horse contests (Figure 2-30). In fact, many Palominos

Figure 2-30
Mack's Wonder Boy, Palomino stallion of Saddle Horse type. (Photograph by Jean Whitesell, courtesy of owner, Franklin L. Hersom, Curlew, Iowa.)

are double-registered with their appropriate breed associations (Figure 2-31).

The earliest myths and legends of both Eastern and Western cultures referred to the golden horses with silver manes and tails. In the Spain of Queen Isabella, such horses became known as Golden Isabellas. One version of the origin of the American name Palomino is that it is derived from the color of the golden grape of California, the Palomino grape. The ancestors of the American Palomino, which developed primarily in Mexico and California, were undoubtedly introduced from Spain by Cortez and the early Spanish explorers.

The first registry of Palominos was private and began in 1932. The Palomino Horse Association was formed in 1936. In 1941, the Palomino Horse Breeders of America became the primary registry for the Palomino. The requirements for registration of the PHBA and PHA are similar except that horses of either light or dark skin can be registered in the PHA but only those with dark skin can be registered in the PHBA. The general color rules are that the body coat color must approximate that of a United States gold coin and the mane and tail must be white with not more than 15 percent dark, sorrel, or chestnut hair in either.

Since eligibility of a horse to be registered as a Palomino depends on obtaining the correct golden color, some mention of the difficulties should be made even before the discussion of genetics in a later chapter. Research has shown that a simple genetic factor for dilution of chestnut color results in the palomino color. Animals-carrying two dilution genes will be diluted to a near-white or off-white called cremello. Thus, the expected results from various matings to produce palominos are:

Figure 2-31
Hoppy's Own, a Palomino Horse Breeders Association Champion and double-registered American Quarter Horse. (Photograph courtesy of owner, Don Scroggins, Pearl River, Louisiana.)

palomino by palomino: one-fourth chestnut + one-half palomino + one-fourth cremello

chestnut by palomino: one-half chestnut + one-half palomino

palomino by cremello: one-half palomino + one-half cremello

chestnut by cremello: all palomino

Thus, there is no way palominos can breed true or produce on the average more than half palominos from matings among themselves. The mating of chestnut with cremello should give all palominos, but if cremellos are classified as albinos, then that is not a legal mating. In fact, true albinos are not known in the horse since all white horses have colored rather than pink eyes.

2.15   Buckskins

Since 1963, two registries have developed for Saddle Horses with buckskin, dun, and grulla colors: the American Buckskin Registry Association, Inc., and the International Buckskin Horse Association. Both registries have nearly the same description of these colors. Following is an abridged classification of eligible colors of the International Buckskin Horse Association:

Buckskin   The body coat of the Buckskin is predominantly a shade of yellow, ranging from gold to nearly brown. Points (mane, tail, legs, and so on) are black or dark brown. On the true Buckskin the dorsal stripe, shoulder stripe, and barring on the legs is always present. However, the dorsal stripe is *not necessary* for registration of the Buckskin.

Dun   The Dun differs from the "Buckskin" only in the respect that the body color is of a lighter shade. [Genetics journals and dictionaries, however, refer to buckskin as a light, clear shade of dun.]

Grulla   Smoky blue or mouse colored, with black points. The Grulla (*grew*-yah) has no white hair mixed in with darker hair, as is seen in the roan or gray. The name "Grulla" comes from the Spanish, meaning "Blue Crane." Grulla hair is a solid mousy blue or slate color.

Red Dun   The Red Dun is just that—red. Body coat may vary from a yellow to a nearly flesh color. Points are dark red. Dorsal stripe must be present.

NOTE: The Grulla, Red Dun and some shades of Dun must have the dorsal stripe to be eligible for registration. Dorsal stripe is not a requirement for the Buckskin.

A Buckskin mare with the dorsal stripe and shoulder striping is shown in Figure 2-32. A golden Buckskin is shown in Figure 2-33.

Buckskin breeders trace the ancestry of their horses to a true-breeding Buckskin of Spain, the Sorraia, and to the Norwegian Dun. These were crossed with Barb and Arabian horses in Spain. During the following 700-year period, Spain developed many fine horses, many with buckskin and grulla patterns. Some of these were introduced by the Spanish explorers to Mexico and the United States early in the sixteenth century. Many of the modern western Buckskins probably trace to these animals. The buckskin pattern, however, appears in most breeds.

The dun or buckskin pattern, with lighter body and dark points and dorsal stripe, is similar to that of the Tarpan, a wild horse of central Europe, and of Przewalski's horse, which represent nearly true breeding for color. Discussion of the inheritance of buckskin, dun, and grulla patterns is found in a later chapter.

The American Buckskin Registry Association, Inc., was formed in 1963 and the International Buckskin Horse Association began registering horses late in 1971.

The requirements for registry are similar for both registries. Conformation can vary in each, although draft type animals are not eligible.

Figure 2-32
Nicky Dean, a Buckskin mare, registered with the American Buckskin Registry Association and the American Quarter Horse Association, showing the black dorsal stripe, black points, mane, tail, and shoulder stripe. Owned by Clayton E. Gillette, Wessington Springs, South Dakota. (Photograph courtesy of the American Buckskin Registry Association, Anderson, California.)

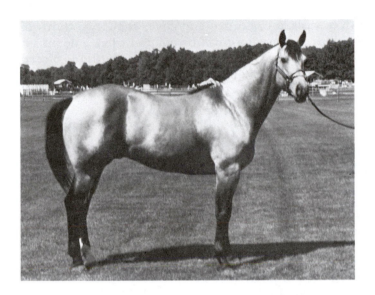

Figure 2-33
Hank's Billy Van, a golden Buckskin, registered with the International Buckskin Horse Association. (Photograph courtesy of IBHA, St. John, Indiana.)

2.16   Whites and Cremes: The American White Horse

The white horse is the horse of heroes, history, art, and fiction. Children who grew up during the 1940s remember the Lone Ranger's horse, Silver. Fewer people, however, remember Old Whitey, the mount of General Zachary Taylor, who pastured the horse on the White House lawn after he became President.

The Thompsons of Naper, Nebraska, contributed to the history of the white horse when in 1918 they purchased Old King, a white stallion of Arabian and Morgan breeding. He was the foundation sire of the American Albino registry set up by the Thompsons in 1937. Many of his white foals were from solid-colored Morgan mares.

Although albino is a synonym for white, the choice of name was probably unfortunate because in many species the true albino has serious problems because of lack of pigment in the eye. As previously mentioned, no true albino is known in the horse. The horses of the Thompson's White Horse Ranch had eyes with blue, brown, or hazel pigment. The skin is pink and the hair is clear white, but some small, colored spots occasionally occur.

The dominant white of the American White Horse is lethal when both gene units are the dominant white. Thus, no dominant white horse can breed true. Since one-fourth of foals are resorbed and never seen, two-thirds of the foals born will be white and one-third will be colored when a dominant white is mated to a dominant white. Matings of dominant whites to colored horses produce one-half white and one-half colored.

The name of the registry was later changed to The American White and Creme Horse Registry.

Only dominant white horses and ponies (ponies if less than 14.2 hands) were registered until 1949. Then the cremellos and perlinos (near-white horses), which result from a double dose of the dilution gene that produces palominos and buckskins, were also registered as albino types A and B and later as cremes.

There are no conformation standards because this is a color registry. All types are accepted. Many are used in troupes and in parades (Figure 2-34). The dominant white has appeared in most breeds of horses including, for the first time in 100 years, the Thoroughbred breed—one in Kentucky and one in France, both in 1963.

Figure 2-34
R. R. Snow King, an American White Horse in parade equipment, ridden by Rose Simmering. (S & E photography courtesy of American Albino Association and Ruth White.)

2.17   The "Native" Horses

Many of the horses of the early Spanish explorers escaped or were captured by Indians. Some of these were bred and selected by Indian tribes to form such types as the Appaloosa Horse in the Northwest and the Chickasaw Horse in the Southeast. Others ran free, especially in the Southwest and West, and became feral and semiferal in the harsh world of the survival of the fittest.

Chickasaw Horse

The first horses of the Chickasaw Indians of Tennessee and North Carolina were captured from the members of the 1539 expedition of DeSoto. These small, short-coupled, well-muscled horses were popular with early colonists for general-purpose use although not for distance running. The Chickasaw Horse was utilized in cross-breeding to develop the early colonial quarter-mile horse, and it is one of the ancestors of the modern Quarter Horse. The Chickasaw Horse Association registers horses of the early Chickasaw type with a height range of 13.1 to 14.3 hands.

The American Indian Horse

The American Indian Horse Registry was incorporated by a native American Indian in 1961 to collect, record and preserve pedigrees of the horses the Indians originally obtained from Spanish herds. Descendants of these horses figured prominently in American history, particularly that of the southwestern United States. The registry is dedicated to the preservation of these horses, as are those registering similar horses collectively called mustangs.

Spanish Mustangs

Robert Brislawn, Sr., and his brother, Ferdinand Brislawn, are given credit for forming the first and oldest mustang registry in 1957 in Sundance, Wyoming—the Spanish Mustang Registry. Since the 1920s the Brislawn brothers had been collecting a foundation stock of the purest wild and semiwild Spanish Mustangs. These last remnants of the naturally selected wild horses of Spanish-Barb and Andalusian ancestry were to be preserved and perpetuated for posterity as a living heritage of frontier America. To ensure the authenticity of future registered animals, the rules state that if at any time . . . any or all of the directors of the registry attempt to improve or in any way change the genotypes of the registered mustangs by hybridization with any other breed or breeds of horses or hybrids thereof, then the registry and its name should become null, void, and defunct.

Only horses that can reasonably be shown to be authentic are registered, and only after inspection. The typical mustang is approximately 13.2 hands, shortbacked, and wiry, and weighs 800 to 900 pounds. Mustangs have developed in the many colors and patterns of duns, solids, whites, palominos, appaloosas, and pintos. One of the most unusual is the Medicine Hat pattern, which was especially favored by the Cheyenne Indians who thought the Medicine Hat had supernatural powers of protection and invincibility.

The Spanish-Barb

The Spanish-Barb Breeders Association was formed in 1972—also to promote and perpetuate the mustang as a breed but with emphasis on breeding for, and breeding back for, the ideal of the original Spanish-Barb Horses. Only authentic horses may be registered, and crossbreeding is not practiced. A Spanish-Barb gelding is shown in Figure 2-35. A future goal is to re-establish the Spanish-Barb in a sanctuary in the western plains—to allow them again to run wild without interference from man.

Figure 2-35
Taw-ka Chi Who-ya, a Spanish-Barb gelding in action. (Photograph by Susan Banner, courtesy of the Spanish-Barb Breeders Association, Colorado Springs, Colorado.)

The American Mustang

Since 1957, the American Mustang Association, has been dedicated to pre-serving and continuing the best specimens of the American Mustang, which descended from the early Spanish horses. The association has begun a stud book and, unlike many mustang registries, holds a national and several local mustang shows (Figure 2-36). Inspection for suitable conformation by an authorized inspector or a licensed veterinarian is required before registration. To be registered, a mustang must be between 13.2 and 15 hands. Any color is acceptable.

American Bashkir Curly

Most horses have straight, smooth hair. A recessive gene, when homozygous (both genes of the pair alike), results in a curly coat that is very distinctive. The coat sheds out to nearly straight or slightly waved in the summer. The mane and sometimes the tail shed out in the summer to grow in again in the fall. The fall coat can be said to resemble a permanent "permanent." The breed in the United States is named for horses with curly coats that were raised in the Bashkir region of the Ural Mountains of Russia. The breed originated in the United States, however, from three curly-coated feral horses found in 1898 on a ranch in Nevada.

Figure 2-36
Taric, a National Merit American Mustang stallion and National Grand Champion Get of Sire Award winner. His dam was carrying him when captured from a wild herd in Utah. (Photograph by Robin Bock, courtesy of Diane and Robin Bock, Mustang Manor, Costa Mesa, California.)

The Rangerbred Horse

The Colorado Ranger Horse Association, chartered in 1938, registers horses that trace to three foundation stallions: Linden Tree (a blue-gray, pure Barb) and Leopard (a dapple-gray, pure Arabian), which were given to General U. S. Grant in 1878 by the Sultan of Turkey, and Max, white with black leopard spots, owned by Governor Oliver Shoupe of Colorado. These stallions or their descendants were crossed with native western mares, some of which may trace to the Nez Percé horses, to produce working ranch horses for the western plains. Although the registry stresses breeding and has no color requirement, most Rangerbreds are spotted and, therefore, are often confused with the Appaloosa. For 30 years the association was restricted to 50 members, which severely limited the expansion of the breed. Many horses otherwise eligible for registry were instead registered as Appaloosas. After membership was opened in 1968, the breed spread eastward and overseas.

2.18   The Shetland

When most people think of a pony, they have in mind the Shetland. The technical dividing line between horses and ponies is at 14 or 14.2 hands. However, Shetlands are much smaller. Their maximum height is 46 inches (11.2 hands), and most are approximately 40 inches tall. As is well known, these ponies developed in the Shetland Islands approximately 100 miles north of Scotland and 350 miles from the Arctic Circle. The name "Shetland" derives from an old Norse word meaning "highland." These small islands provided a harsh, rugged environment for the development of a hardy breed of ponies, which began before the Norsemen settled the islands about 850. The shaggy, furry coat worn by Shetlands in winter months and by some foals until they are 2 years of age must have developed to withstand the rigors of the North Sea winters and storms. The native pony of the Shetland Islands was a miniature draft horse, as contrasted with the refined American Shetland (Figure 2-37). They became popular in England and Scotland for work in the mines because of their strength and small size.

The Shetland came with the English settlers to the United States, as did the Thoroughbred. The Arabian and Barb breeding, by way of the Hackney, as evidenced in the modern American Shetland, apparently resulted from crosses that were made in the 1880s. Other than those crosses, the Shetland has bred relatively true—at first because of the isolation of its native islands, and later because of the desire to maintain a small children's pony.

The modern American type of Shetland is the result of selection from the draft or "Island" types that were imported to America. The types have

Figure 2-37
A young Shetland pony mare,
Fernwood Frisco Caroline II.
(Photograph by Carlin H.
Brearley, courtesy of the owner,
Harry K. Megson, Chazy River
Farms, Champlain, New York.)

diverged so much that today the native Shetland can no longer be registered
in the American Stud Book. Many nonregistered Shetlands, however, show
more of their ancestral type than do the registered show-class Shetlands.

The American Shetland Pony Club was organized in 1888.

Ponies come in all colors. Black, dark brown, bay, and chestnut predominate, especially for the show-ring, where white markings have not recently
been preferred although color fads change. Spotted ponies are popular for
children's mounts. A color unique to Shetlands is the silver dapple—a
dappled chestnut with silver or white mane and tail.

The show-ring accommodates the versatile Shetland in breeding and
harness classes (Figure 2-38). As with most pony breeds, the classes are
divided according to height. The dividing line of 43 inches appears to nearly
equalize the number in the "over" and "under" classes.

2.19   The Pony of the Americas

Black Hand No. 1, born in 1954, the foundation sire of the Pony of the
Americas, was the result of a mating of an Appaloosa mare and a Shetland
pony stallion (Figure 2-39). The characteristics of Black Hand served as an
inspiration for his owner, Leslie L. Boomhower of Mason City, Iowa, to
establish a breed association for an "in-between" size, western type of pony
that would be small enough for children, yet large enough for adults to break
and train. This registry, the Pony of the Americas Club (POAC), was orga-

Figure 2-38
A Shetland pony in action in a fine harness class. (Photograph by Morris, courtesy of the American Shetland Pony Club, Fowler, Indiana.)

Figure 2-39
Black Hand No. 1, the foundation sire of the Pony of the Americas, was the result of mating an Appaloosa mare and a Shetland stallion. (Photograph courtesy of the U.S. Department of Agriculture.)

nized in 1955. The foremost purpose of the organization was to establish and promote a children's working pony.

The size limits of 46 to 56 inches were met by crossing pony breeds with Appaloosas, crossing Quarter Horses and Arabians with ponies having Appaloosa markings, and importing from Mexico and Central and South America, small horses of the required size and Appaloosa colors.

The ideal for the Pony of the Americas (POA) has been described as a cross between a Quarter Horse and an Arabian in miniature that has the Appaloosa color patterns. The eligibility requirements for color are essentially the same as those for the Appaloosa. The white sclera is required but does not have to encircle the eye. Striped hooves are desirable but are not required.

2.20 The Welsh Pony

The ancestors of the Welsh Mountain Pony developed in the severe terrain of Wales even before the Roman legions invaded the British Isles. In the early 1800s, Thoroughbred, Arabian, and Hackney blood was incorporated. In the late 1800s, some Welsh Ponies were imported to the United States, where the Welsh Pony and Cob Society was incorporated in 1906. (The Welsh Cob is a larger version of the Welsh Pony.) The breed expanded for some time, then declined. There were virtually no registrations during the depression of the 1930s. A renaming and reorganization took place in 1946, and "Cob" was dropped from the society's name. Most Welsh Ponies trace to importation from England after 1947.

The Welsh Pony is intermediate between a Shetland and most riding horses and is useful for children who have outgrown ponies of Shetland size (Figure 2-40). Welsh Ponies are also used as hunters for children.

A frequent color of the Welsh is gray, especially in England (Figure 2-41). Any color is acceptable, but spotted patterns are not.

The Welsh has frequently been crossed with larger riding-horse breeds to produce ponies whose height ranges between 13 and 14 hands.

2.21 The Hackney Pony

The elegant, fiery Hackney has been called the Prince of Ponies. Used almost exclusively for showing in harness classes, this aristocrat of the pony world is an adult's pony—too small to be ridden by an adult but too spirited and frisky for most children.

Figure 2-40
Liseter Shooting Star, a Welsh
Pony champion and sire of
champions. (Photograph courtesy
of the owner, Mrs. J. A. duPont,
Newton Square, Pennsylvania.)

The Hackney Pony derives from the same ancestors as his bigger
brother, the Hackney Horse. The ancestors of the Hackneys were bred for
riding and driving in England, where the terms "Hackney" and "Roadster"
were applied to them. The Norfolk trotters were the most famous. Crosses of
these horses with the Thoroughbred led to the Hackney that pulled the
British Hackney coaches of the eighteenth century. They were selected to
show flashy action at the trot. Their hallmark was, and remains, extreme

Figure 2-41
A gray Welsh stallion and
Section B champion, Cusop
Sheriff, was imported from
England. (Photograph by
Tarrance Photos, courtesy of
the owner, Mrs. Karl D.
Butler, Ithaca, New York.)

flexion of knees and hocks at the trot. The Hackney Horse was not large—14.2 to 15.2 hands and 900 to 1100 pounds. A pony version of the Hackney was made up of the small Hackneys and resulted also from crosses with the Welsh Pony. Selection for a smaller size completed the establishment of the Hackney Pony. There are no major differences between the Hackney Pony and the Hackney Horse except size. The pony may more closely approach the Hackney ideal in exaggerated action and flash. The Hackney Pony is indeed a horse in a small package.

Hackney's were imported to America, particularly in the late nineteenth century, mainly for show and fancy driving. Today both are registered in the United States by the American Hackney Horse Society. Only in shows are they separated; the pony is classified as having a height of 14.2 hands and under.

At horse shows, Hackney Ponies are divided into Cob-Tail (Hackney) or Long-Tail (harness) classes (Figures 2-42 and 2-43). The name "Cob-Tail"

Figure 2-42
Whyworry Lobelia, shown in Hackney (Cob-Tail) pony class. (Photograph by Fallow, courtesy of Mrs. W. P. Roth, San Mateo, California.)

Figure 2-43
An international champion harness (Long-Tail) pony, Johnny Dollar, shown by
Mrs. David LaSalle. (Photograph by Tarrance Photos, courtesy of David LaSalle,
North Scituate, Rhode Island.)

derives from the English custom of docking the tail to a 6-inch length for
carriage style. Most Hackney Ponies are dark. They are predominantly bay,
but there are also some blacks, browns, and chestnuts. White socks and
stockings as well as white face markings are allowed. Spotted ponies seldom
exist.

Approximately 95 percent of the Hackneys registered each year are
ponies. The American Hackney Horse Society was founded in 1891. Only
ponies with dams and sires registered in the American, English, and Canadian Hackney Stud Books are eligible for registration.

The unique action of the Hackney Pony that so fascinates and excites
spectators and exhibitors is the reason for its dominance of the show-ring.
The extreme action of the front legs — the "rainbow" arch that they form
on the way to the ground — and the matching high hock action, together
with the well-flexed neck holding the head high and proud, have maintained
the status of the Hackney as the Prince of Ponies.

2.22   The Connemara

The Connemara, famed for its ability as a jumper (Figure 2-44), originated in the Connemara region in County Galway, Ireland. The ancestry of the breed includes Spanish-Barbs, Jennets, and Andalusians of the sixteenth century that were crossed with native ponies. Later, Arabians were crossed with the descendants of these horses. During the last half of the nineteenth century, there was much crossing of native Connemara Ponies with other breeds. To prevent the breed from being crossed out of existence, the Connemara Pony Breeders' Society was formed in 1923. It selected the finest of the available stock on which to base the breed.

Figure 2-44
A gray Connemara mare, R. H. F. O'Harazan, in action. (Photograph by Paul A. Quinn, courtesy of the owner, Gilnocky Farm, Windsor, Vermont.)

The first Connemaras were imported to America for breeding purposes 28 years later in 1951. The owners, although few, were spread across the country. The American Connemara Pony Society was founded late in 1956. All colors are permitted except that neither spotted nor blue-eyed cream ponies may be registered. Many Connemaras now exceed 15.0 hands, and there is a tendency to breed for the larger type of Connemara. The Irish Connemara remains pony size.

2.23   Other Ponies

A synthetic breed is started by crossing two or more breeds and then mating among the first generations of crossbred animals. The first cross may give the desired result with considerable uniformity as dictated by the laws of genetics. Matings among crossbreds, however, also follow the laws of genetics and result in considerable variability, ranging in extremes between the parent breeds. Continued *inter se* mating of future generations of crossbreds and selection is required to develop a desirable and reasonably uniform breed. Some other ponies are miniature versions of their larger brothers and are really small horses.

The Americana

The Americana is a synthetic breed that was begun in 1962 by crossing Shetland and Hackney Ponies to produce a miniature Saddlebred type of show pony. The disposition and conformation were to come from the Shetland; the action, animation, style, and slightly larger size were to come from the Hackney.

The American Walking Pony

The American Walking Pony originated in the 1950s at Browntree Stables in Macon, Georgia. The owner, Mrs. Joan Brown, chose to cross the Tennessee Walking Horse and the Welsh Pony to form a breed of walking pony to be used as a pleasure and show pony.

The Walking Pony

Just as the American Saddlebred comes in pony size, so does the Tennessee Walking Horse. The pony-size Walking Horse has the same colors, conformation, and gaits as do larger Walking Horses.

The American Quarter Pony

Quarter Horse types that could not be registered because of lack of height led to the establishment in 1964 of the American Quarter Pony Association and in 1975 of the National Quarter Pony Association, which functioned to register riding and show ponies of the Quarter Horse type.

The Trottingbred

The Trottingbred breed is a synthetic breed that originated in the 1960s from crosses between Standardbreds and Shetland, Hackney, and Welsh ponies for sulky racing. Some are trotters and others are pacers as are Standardbreds. The distance raced is ½ mile. The maximum height measured at the withers is 51 inches. The International Trotting and Pacing Association is the registry for the breed.

Miniature Horses

Miniature horses are rare. Only about one in several thousand horses qualifies as a miniature. The maximum height of a miniature horse is usually defined as 32 inches, although some assert it to be 36 inches. Miniatures are used as pets, in circuses, and even as curiosity pieces. The true miniature is simply a small pony or horse and is not a malformed dwarf. Miniatures are just as healthy as regular horses, although because of their size they are much more susceptible to accidents and to attacks by packs of dogs.

There are no miniature breeds, but the American Miniature Horse Registry registers horses under 34 inches. If successful, a breed may develop from this effort.

Another group that may be considered a private breed are the Falabella Horses of Argentina. The Falabella family began in 1868 to breed for miniature size. Falabellas are derived from Shetland as well as larger stock. Many have been imported into the United Sates since 1962. The Falabella Horses are noted for the excellence of their conformation, which is often faulted in many miniatures. The smallest Falabella is 15 inches tall and weighs only 27 pounds.

Several breeders in the United States have collected and developed herds of miniature horses (Figure 2-45). Some of these horses are as small as 20 inches and weigh only 40 pounds. Colors generally cover the same broad range as the colors of the Shetland, from which most miniature herds have developed.

Tom Thumb was a rather famous small horse that was 23 inches tall and weighed 45 pounds at 8 years of age. He was exhibited in sideshows with a

Figure 2-45
Joel R. Bridges, surrounded by part of a band of miniature horses—the smallest
26½ inches tall and the largest 30 inches tall as adults. (Photograph by Pat
Canova, courtesy of the owner, Kokoma Ranch, Newberry, Florida.)

small mare, Cactus, that measured 26 inches. Both, however, were dwarf
horses, not miniature horses. Both were from full-size parents and were
found to be sterile.

2.24   The Draft Breeds

Descendants of the "Great Horse," the horse of knights in armor of the
Middle Ages, make up the five major draft breeds in the United States. These
breeds all developed in northern Europe—Scotland, England, France, and
Belgium. All are named for their regions of origin: the Percheron from La
Perche, an ancient district south of Normandy, France; the Belgian, a de-
scendant of the great horse of Flanders; the Clydesdale from the Clyde River
area known as Clydesdale in Scotland; the Shire, named for the English areas
of Lincolnshire and Cambridgeshire; and the Suffolk from the agricultural
county of Suffolk, also in England.

These horses, so frequently described in history and legend, developed into heavy-carriage, draft, and farming horses. Although their early history is obscure, the Percheron is thought to be the only one of these so-called cold-blooded horses to have had an infusion of Arab and Barb (hot-blooded) ancestry, a legacy of the Moorish invasion of Spain and western France in the early eighth century.

Nearly all the draft breeds are large and heavily muscled; they stand approximately 16 to 17 hands and weigh 1600 to 2200 pounds depending on sex, age, and condition.

The Percheron

Percherons are known to have been imported to the United States in 1839. After 1851, imports were numerous as the Percheron became the most popular among the draft breeds. The forerunner of the Percheron Horse Association of America was first organized in 1876.

The Percheron is known as the breed of blacks and grays because approximately half are black and half are gray (Figure 2-46). Other colors are known, and white markings are common although not extensive.

Despite its immense size, the Percheron is active and light on its feet. It displays considerable knee action and a bold trot, both of which make it a popular horse in draft-horse shows. Its popularity as a draft horse was due to its speed at the walk and the lack of feathering about the fetlock.

The Belgian

From 1910 through the 1930s, the Belgian sorrels and roans surpassed the blacks and grays of the Percheron in popularity among American farmers. The predominant color soon became sorrel (many light-horse breeders call it chestnut) with white mane and tail (Figure 2-47). The lighter shades are known as blonde sorrels. The characteristics of the Belgian suited the needs of the American farmer, and thus Belgian stallions were much in demand for crossing with native draft type horses and particularly for crossing with Percheron mares. The uniformity of color of Belgians was popular and they were known for ease of management. The Belgian has been distinguished in pulling contests and is very quiet and docile, but is somewhat slow-motioned and does not show much high-leg action. The influence of the imported roan stallion, Farceur, provided more action and quality (that is, better body and leg conformation).

The origin of the Belgian Draft Horse Corporation of America traces back to 1887. The current name was adopted in 1937. In recent years, more Belgians have been registered each year than all other draft breeds combined.

(a)

Figure 2-46
The Percherons are known as the
breed of the blacks and grays. (*a*) A
black stallion, Don-A-Tation, with
braided mane and tail, in a halter
class. (Photograph by John M.
Briggs, Ithaca, New York.) (*b*) A
gray stallion, Shady Creek Carnot,
with braided tail. (Photograph by
James M. Barnhart, Butler, Missouri.)   *(b)*

The Clydesdale

The advertising (including television commercials) of the Budweiser brew-
ery (Figure 2-48) and other companies having six- and eight-horse hitches
make the Clydesdale one of the most well known of American draft horses.

Figure 2-47
Sunny Lane Tamara, a champion Belgian mare. (Photograph by Leonard C. Novak, courtesy of the owner, Leo J. Fox, David City, Nebraska.)

Figure 2-48
The Budweiser Clydesdale eight-horse hitch showing the noted action and dramatic feathering of the breed. (Photograph courtesy of the owner, Anheuser-Busch, Inc., St. Louis, Missouri.)

The extensive white face and leg markings, together with the "feather" (long, silky hair on the lower legs), which creates a bell-bottomed effect, give the Clydesdale a unique appearance to go with the noted action of the breed—long, springy strides with extreme flexion of the knees and hocks.

The Clyde is somewhat lighter on the average than the other breeds although it is just as tall. The bones of these horses are cleaner and flatter than those of other breeds. Clydes were considered more nervous than the other draft breeds and were difficult for most American farmers to handle.

The Clydesdale Breeders Association of the United States is the registry for the breed.

The Shire

The Shire, like its ancestral cousin, the Clydesdale, developed from the English Great Horse. Robert Bakewell (1726–1795), who developed many improved breeds of livestock, also initiated the improvement of the Shire as a draft horse. The Shire is known as one of the tallest of the draft horses. Early imports to Canada date from 1836, although most activity occurred between the 1880s and 1930s.

Black is a common color of Shires, although the colors of bay, dark brown, gray, and chestnut are also present. (Figure 2-49). White markings on the face and legs are common. Like the Clydesdale, the Shire has feathering on its legs, although the feather is finer and silkier now than when it retarded the acceptance of the Shire in America. The American Shire Horse Association is the registry.

Figure 2-49
Jim's Chieftain, a black Shire stallion, showing the feathering and extensive white markings of the breed. (Photograph courtesy of the owner, Arlin Wareing, Blackfoot, Idaho.)

The Suffolk

This chestnut breed of draft horse was developed primarily for farm use. Due to their rather rounded, punched-up appearance, they were originally called Suffolk-Punches. The Suffolk differs from the other two British breeds in that the leg is not feathered. The color has been standardized to a true-breeding chestnut or sorrel; the mane and tail are often lighter (Figure 2-50). Like the Percheron, the white markings of the Suffolks, if present at all, are unobtrusive on both face and legs.

Because they were too small to produce draft animals that would be large enough when crossed with the lighter breeds found in America, relatively few Suffolk horses were ever imported or bred in the United States. The American Suffolk Horse Association is the registry.

2.25 Recent Introductions of European Breeds to the United States

Several associations register breeds that originated in Europe. Many are well known internationally for their all-around performance for national equestrian teams.

Figure 2-50
A six-horse hitch of Suffolks at the trot. (Photograph by Edward T. Gray, courtesy of the owner and driver, D. F. Neal, Slippery Rock, Pennsylvania.)

Hanoverians

The American Hanoverian Society registers Hanoverians and Hanoverian crossbreds and is dedicated to promoting and improving the breed in North America. The breed, which originated in the section of Germany known as Hanover, is the most numerous in Europe, including about 30 percent of mare registrations in Germany. They excel as heavy hunters, as dressage horses and as show jumpers.

Holsteins

Most Americans think of black-and-white cattle when the name Holstein is mentioned. Nevertheless, the Schleswig-Holstein area of Germany is the native home of a tall, light horse known as a Holstein. Not as numerous as Hanoverians, they have had a major impact at international equestrian events, being renowned in dressage and as jumpers. The American Holstein Horse Association began selective registrations in the late 1970s. Contrary to common expectation, these horses are not black-and-white but solid-colored, usually bay or seal brown.

Trakehners

Trakehners—tall, light horses, also from Germany—are well known for their performance in dressage and show events. The North American Trakehner Association promotes and registers these horses on the west side of the Atlantic.

Haflingers

This small, general-purpose, mountain horse from the south Tyrol region near the town of Hafling in Austria is registered in the United States by the Haflinger Association of America. They are light-horse size (1000 to 1200 pounds) with a long, easy stride that somewhat resembles a draft horse. When first seen they resemble small Belgians, mostly because they are often a light shade of chestnut with flaxen mane and tail.

The Gotland

The first Gotlands were brought to the United States in 1957 from Sweden where their ancestors had existed since the Stone Age, as evidenced by excavations from a cave on the Swedish Island of Gotland. The Gotland, a small horse called Skogsruss by the Swedes, served as mounts for the Goths, Vikings, and other Swedish warriors and generally fall into the category of less than 14 hands. They are of good temperament and uniform type because of selective breeding by the Gotland Pony Club, which acted to preserve this historic breed from possible extinction by wholesale exportation to the mines of England, Germany, and Poland. Naturally, Gotlands share some of the characteristics of the wild Tarpan horses of northern Europe, because they have descended from them without much crossing except for a few matings to Oriental blood that were made about 1850 to provide a strong genetic base.

These small horses are used as children's mounts (Figure 2-51) and have been shown in halter, equitation, harness, and hunter classes. Gotlands also compete in trotting races as well as endurance and competitive trail rides.

The American Gotland Horse Association maintains a closed herdbook to prevent further dilution of the breed.

2.26 The Long Ears: Burros, Donkeys, and Mammoth Jacks

A close relative of the horse (*Equus caballus*) is the ass (*Equus asinus*). Just as horses come in all sizes and colors, so do asses. The male ass is known as a *jack* and the female is known as a *jennet.* The most noticeable differences

Figure 2-51
Kronas Kometen II, a sorrel Gotland gelding ridden by Marsha Price in a trail class. (Photograph by William Stinson, courtesy of the owner, Marsha Price, Bonner Springs, Kansas.)

between an ass and a horse are that an ass has longer, larger ears, a sparser mane and tail, a more cowlike tail, and smaller hooves. The muzzle and underbelly area usually are light in color, and there are no chestnuts on the inner sides of the legs. The characteristic bray contrasts with the whinny of the horse. The gestation period of the jennet is approximately 30 days longer than the 11 months of the mare.

During the early history of the United States, jacks were bred to use on horse mares to produce *mules* (Figure 2-52). Mules were very popular as work animals. The reverse cross between the jennet and the stallion is known as a *hinny* (Figure 2-53). The mule is said to be somewhat more like the ass and the hinny is a little more like the horse. Mules and hinnys of both sexes generally are sterile, but their sexual instincts are normal.

Donkeys are small asses. The name derives from a diminutive of the English word "dun," which describes the usual color. The Spanish translation of "donkey" is "burro," so donkey and burro are synonymous. *Burro* is primarily now used to denote feral asses. *Donkey* is generally used for domestic animals.

As the demand for draft animals declined in the United States, the popularity of asses also declined, until the 1960s when they became more popular as pets and curiosities. There are now several registries for different types of asses.

Figure 2-52
Mr. Tips of Oregon, an unusual leopard Appaloosa mule from the cross of a jack and an Appaloosa mare. (Photograph courtesy of Urban J. Woida, Veneta, Oregon.)

Figure 2-53
A saddle hinny from a cross of a
Morgan stallion and a burro jennet.
Owned by
E. C. Porter, Safford, Arizona.
(Photograph courtesy of American
Donkey and Mule Society,
Indianapolis, Indiana.)

The American, Mammoth, or Standard Jack and Jennet

The American Jack, also known as the Mammoth Jack or the Standard Jack,
was bred for crossing with mares to produce mules. The American Jack
is a blend of various stocks imported from southern Europe and the
Mediterranean.

George Washington was one of the first breeders of Jacks; he crossed
strains of asses received as gifts from the King of Spain and General Lafayette
of France. Henry Clay imported Maltese stock. The famous Jack, Imported
Mammoth, imported from Catalonia in 1819, was crossed with the Clay
imports. Since most Jack stock traces to Mammoth, it is natural that one of
the alternate names for American Jacks is the Mammoth Jack.

The Standard Jack and Jennet Registry and the newer (1969) American
Donkey and Mule Society register American Jacks and Jennets. The ADMS
registers all breeds and types of asses and was formed to promote interest in
the long-eared species.

Any color is permitted, although most Jacks are dark or red sorrel with
white points.

The Burro

The Burro is the small, so-called native, feral ass of North America and South America. Most are 40 to 50 inches in height. The burro or donkey of the Americas is a blend of many breeds of European and Middle Eastern countries. The ancestors of the burro were probably brought to the Americas with the Spanish expeditions. Burros come in all sizes and colors (Figure 2-54). They can be registered in the stud book of the American Donkey and Mule Society.

The American Spotted Ass

The American Council of Spotted Asses was formed in about 1967 to register spotted or pinto asses of primarily burro size. All animals must be inspected and approved before registration. The color pattern may be white with colored spots or colored with white spots (Figure 2-55). The qualifying spots must be above the knees and hocks and behind the throat latch. Stockings and face markings do not qualify as spots.

The Miniature Donkey

A Miniature Donkey Registry was formed in the Midwest in 1958 to provide an imported strain of miniature donkeys less than 38 inches high. The organizers included the owners of the St. Louis Cardinals and Cincinnati

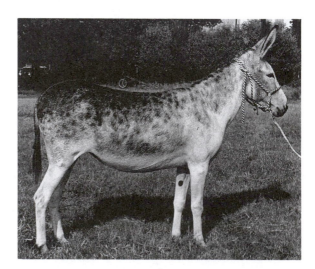

Figure 2-54
An unusual dappled gray American Burro jennet owned by Carl A. Wilson, Indianapolis, Indiana. (Photograph by D. Bennett, courtesy of Paul and Betsy Hutchins, Denton, Texas.)

Figure 2-55
Shenandoah, a young American
Spotted Ass jennet owned by E.
Diane Hunter. (Photograph courtesy
of American Donkey and Mule
Society, Indianapolis, Indiana.)

Reds baseball teams. The original animals were imported from the Mediterranean (chiefly Sicily and Sardinia), and were 28 to 38 inches in height.
Import restrictions have curtailed further importation.

The characteristics of the Miniature Donkey make it an attractive pet
(Figure 2-56). Colors range from a light gray to a dark, almost black, brown.
The cross on the back, a well-known trademark, is formed by a dorsal stripe,
running from the mane to the tail, intersected by the arms of the cross at the
withers. The cross is said by legend to be the reward for carrying Mary to
Bethlehem and Jesus to Jerusalem.

Figure 2-56
A registered Sicilian Miniature
Donkey owned by Danby Farm,
Omaha, Nebraska. (Photograph
courtesy of American Donkey and
Mule Society, Indianapolis, Indiana.)

APPENDIXES

Identification

The purposes of breed associations can be satisfied only by reliable identification of animals and their sires and dams. Nearly all breeds require a written description of colors and color patterns together with a sketch or picture of the animal. Even more positive identification is required by some registries: lip tattoos, photographs of a horse's "fingerprints" (unique patterns of the chestnuts (night eyes) that are found on the insides of the legs), and blood typing. Muscular dimples, cowlicks (hair swirls), scars, or brands should also be described and located on the sketch.

Color is an obvious characteristic of a horse although many horses have the same color. Color terms may also have different meanings for different people. Colors are discussed and illustrated in Chapter 13. There are five basic colors together with variations that result from graying, roaning, spotting, and dilution.

Basic Colors

All colors described are in addition to white markings on the head and legs.

1. *Black*   The entire coat is black, including the muzzle, flanks, and legs. Some black horses will fade or have a smoky appearance.

2. *Brown*   Many brown horses are dark enough to appear black except that close examination will reveal brown or tan hairs about the muzzle or flanks. The mane, tail, and legs are always black. A dark brown horse is sometimes called a seal brown, and a light brown horse is sometimes called a dark bay.

3. *Bay*   The body color may range from a light golden red to a dark mahogany color. The body color is similar to that of chestnuts except that the lower legs, mane, and tail of a bay horse are always black.

4. *Chestnut*   The range of chestnut runs from light golden red, sometimes called sorrel, to a very dark chocolate shade called liver or black chestnut. The legs never have black hairs, and often a lighter shade appears on the lower legs. The shade of the mane or tail may be the same as the body, lighter or darker than the body, but never black.

5. *White*   A true white horse is born white. Most white-appearing horses are grays that become progressively whiter with age. Some near-white horses are light ivory or cream; these horses are difficult to distinguish from the whites.

Modification of Basic Colors

The major variations simply modify the basic color in some way.

1. *Gray* The foal coat of a gray horse will be solid color. Each new coat adds more and more white hairs until the horse appears white. The gray pattern (intermixture of white hairs with colored hair) may occur with any background color or pattern. Gray on black, liver chestnut, or seal brown will be a blue or steel gray; gray on bay or chestnut will be a shade of rose gray. Dappling occurs within any color pattern but is more obvious on a gray background.

2. *Roan* Roan and gray are often confused. The roan horse, however, is born with the same proportion of white hairs as will be present in each successive coat. As with gray, the roan pattern may be present on any background of colored hair. Often patches of roan will not be uniform over the body. The head, neck, and lower legs in particular may be more solid-colored than the remainder of the body. Red roan comes on a bay background, strawberry roan comes on chestnut, and blue roan comes on a dark (black, brown, or liver chestnut) background.

3. *Dilutions* The dilution colors come from a lessening of the intensity of the basic color in each hair, not from mixing with white hairs. Chestnut dilutes to palomino of various shades — from bright copper to light yellow with lighter-to-white mane and tail. Double dilution causes the chestnut to go to a cream. Bay dilutes to buckskin or dun of shades similar to the palomino, but the mane, tail, and points are black whereas a Palomino has no black hair. Duns often have a black dorsal stripe and sometimes have black stripes on legs and withers. A double-dilute bay is called a perlino and is ivory white with slightly darker, rusty-appearing points.

A red or claybank dun has a dorsal stripe that is a darker red (not black) on a diluted chestnut background. A dilute black is known as grulla (*grew*-yah), which is also described as a mouse color. The coat has no white hairs and appears to have a blue or yellowish tone. The points are always black.

4. *Pinto-Paint* The body is spotted — either color on white or white on color. This spotting is in addition to the usual white markings on the head and lower legs. White spotting of any size should be indicated. True white areas are present at birth, grow out of pink skin, and do not change with age.

Head and Points

Head markings may consist of some combination of a star, strip, snip, or blaze, as shown in Figure 2-57.

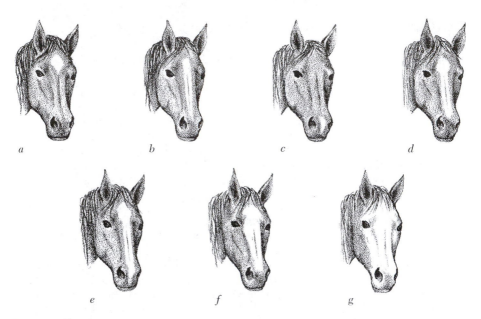

Figure 2-57
White head markings: (*a*) star; (*b*) stripe; (*c*) snip; (*d*) star and stripe; (*e*) connected star, stripe, and snip; (*f*) blaze; (*g*) bald face.

1. *Star*—a white mark on the forehead.

2. *Strip*—a narrow patch of white down the face from the forehead to the muzzle.

3. *Snip*—a narrow patch of white down over the muzzle.

4. *Blaze*—a wider patch of white down the face covering the full width of the nasal bones.

5. *Bald face*—a white marking covering the front of the face and extending over the sides of the face.

6. *Wall-eye*—(watch eye, glass eye)—a light blue or hazel iris resulting from a lack of brown pigment. (Horses' eyes are usually brown with no white around the edge.) In some areas the term "wall-eye" indicates a defective eye, and in other regions a horse that has white spotting covering the eye is called a wall-eye. A requirement for Appaloosas is the white ring of sclera around the iris, which is rare in some breeds.

Leg Markings

Hooves are usually white at birth but attain their adult color as growth occurs. The nomenclature of white on the feet and legs refers to the area covered by white as shown in Figure 2-58.

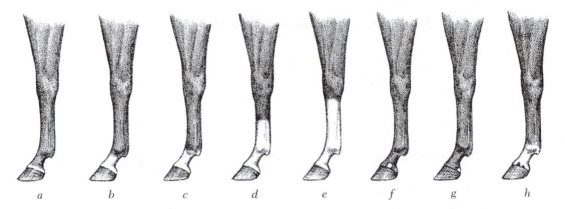

a   *b*   *c*   *d*   *e*   *f*   *g*   *h*

Figure 2-58
Description of white on the feet and legs: (*a*), coronet; (*b*) pastern; (*c*) ankle; (*d*) sock or half-stocking; (*e*) stocking; (*f*) white spots on coronet; (*g*) white spots on heel; (*h*) distal spots (dark on white coronet band).

1 *Coronet*—a white strip covering the coronet band.

2 *Pastern*—white from the coronet to the pastern.

3 *Ankle*—white from the corone to the fetlock.

4 *Sock* or *half-stocking*—white from the coronet to the middle of the cannon.

5 *Stocking*—white from the coronet to the knee.

6 *White marks* or *spots*—white about the front of the coronet or the heel.

7 *Distal spots*—dark spots on a white coronet band.

Variations and extensions of these should be indicated.

Mane and Tail

The mane and tail of many horses is the same color as the body, although sometimes a lighter or darker shade occurs. Bays and seal browns have black manes and tails. Flax or flaxen refers to a straw yellow or off-white color caused by a mixture of dark and white hairs in the mane and tail. A silver mane or tail is mostly white with few dark hairs. True white manes and tails include only white hairs. A heavy, coarse, full tail is referred to as a broom or bang tail. The tail of a rat-tailed horse is sparsely furnished with hair.

The careful reporting of these characteristics as part of positive identification is important in establishing sound breeding programs and in maintaining integrity in the buying and selling of horses.

Breed Associations and Registries

The American Horse Council located in Washington, D.C., maintains a current list with addresses of breed associations and registries.

REFERENCES AND FURTHER READINGS

American Horse Shows Association Rule Book. New York: The American Horse Shows Association, Inc. [Annual.]

American Racing Manual. Chicago: Triangle Publications. [Annual.]

American Stud Book. New York: Jockey Club. [Every 4 years.]

Bailey, L. H., ed. 1908. *Cyclopedia of American Agriculture.* New York: Macmillan. [Contains articles on many pre-1905 breeds.]

Briggs, H. M. 1969. *Modern Breeds of Livestock.* 3rd ed. New York: Macmillan.

Castle, W. E., and J. L. King. 1947. The Albino in Palomino breeding. *Western Horseman* 12:24 (December).

Davenport, Homer. 1909. [Republished 1947.] My quest of the Arabian horse. *Arabian Horse Registry of America.*

Denhardt, R. M. 1948. *The Horse of the Americas.* Norman: University of Oklahoma Press.

Deutschbein, Liz. 1971. Albino horses: A horse of history. *American Horseman* 39 (September).

Dinsmore, Wayne, and John Hervey. 1944. *Our Equine Friends.* Horse and Mule Association of America, Inc. Chicago: Drivers Journal Press.

Edwards, Gladys Brown. 1971. *Know the Arabian Horse.* Omaha, Nebraska: Farnam Horse Library.

Estes, J. A., and Joe H. Palmer, 1942. *An Introduction to the Thoroughbred Horse.* First revision, 1949, by Alex Bower; second revision, 1972, by Charles H. Stone. Lexington, Kentucky: *The Blood-Horse.*

Fletcher, J. Lane. 1945. A genetic analysis of the American Quarter Horse. *J. Heredity* 36:346.

Fletcher, J. Lane. 1946. A study of the first fifty years of Tennessee Walking Horse breeding. *J. Heredity* 37:369.

Gazder, P. J. 1954. The genetic history of the Arabian horse in the United States. *J. Heredity* 45:95.

Gilbey, Sir Walter. 1900. *Ponies, Past and Present.* London: Vinton.

Glyn, Richard. 1971. *The World's Finest Horses and Ponies.* Garden City, New York: Doubleday.

Goodall, D. M. 1965. *Horses of the World.* London: Country Lite Ltd.

Gremmels, Fred. 1939. Coat color in horses. *J. Heredity* 30:437.

Griffen, Jeff. 1966. *The Pony Book.* Garden City, New York: Doubleday.

Haines, Francis. 1963. *Appaloosa, The Spotted Horse in Art and History.* Fort Worth, Texas: Amon Carter Museum of Western Art.

Haines, Francis, Robert L. Peckinpah, and George B. Hatley. 1957. *The Appaloosa Horse*. Lewiston, Idaho: R. G. Bailey Printing Co.

Harrison, James C. 1968. *Care and Training of the Trotter and Pacer*. Columbus, Ohio: U.S. Trotting Association.

Hayes, M. H. 1904. *Points of the Horse*. London: Hurst and Blackett, Ltd.

Hervey, John. 1947. *The American Trotter*. New York: Coward-McCann.

Houser, Helen B., and Leslie L. Boomhower. The heritage of POA. *Pony of the Americas*.

Jones, W. E. 1965. *The phenotypic effects of the D gene in the American Quarter Horse*. M. Sc. thesis. Colorado State College, Greeley.

Knight, L. W. 1902. *The Breeding and Rearing of Jacks, Jennets, and Mules*. Nashville, Tennessee: Cumberland Press.

Mellin, J. 1961. *The Morgan Horse*. Brattleboro, Vermont: Stephen Greene Press.

Miller, Robert W. *Appaloosa Coat Color Inheritance*. Bozeman: Montana State University.

Nye, Nelson. 1964. *The Complete Book of the Quarter Horse*. New York: A. S. Barnes.

Osborne, W. D. 1967. *The Quarter Horse*. New York: Grosset & Dunlap.

Patten, John W. 1960. *The Light Horse Breeds*. New York: A. S. Barnes.

Reese, H. H. 1956. *Horses of Today, Their History, Breeds, and Qualifications*. Pasadena, California: Wood and Jones.

Rhoad, A. O. 1961. The American Quarter Horse. *Quarter Horse Journal* (March).

Rhoad, A. O., and R. J. Kleberg, Jr. 1946. The development of a superior family in the modern Quarter Horse. *J. Heredity* 37:227.

Robertson, W. H. P. 1965. *The Hisory of Thoroughbred Racing in America*. Englewood Cliffs, New Jersey: Prentice-Hall.

Salisbury, G. W., and J. W. Britton. 1941. The inheritance of equine coat color. II. The dilutes with special reference to the Palomino. *J. Heredity* 32:255.

Sanders, A. H., and Wayne Dinsmore. 1917. A history of the Percheron horse. Chicago: *The Breeders' Gazette*.

Savory, Theodore H. 1970. The Mule. *Scientific American* 223:102 (December).

Savitt, Sam. 1966. *American Horses*. Garden City, New York: Doubleday.

Sires and Dams Book. Columbus, Ohio: U.S. Trotting Association. [Annual.]

Speelman, S. R. 1941. Breeds of light horses. *U.S.D.A. Farmers' Bulletin* 952.

Steele, D. G. 1944. A genetic analysis of the recent Thoroughbreds, Standardbreds, and American Saddle Horses. *Kentucky Agricultural Experiment Station Bulletin* 462.

Stetcher, R. M. 1962. Anatomical variations of the spine of the horse. *J. Mamm.* 43:205.

Taylor, Louis. 1961. *The Horse America Made: The Story of the American Saddle Horse*. New York: Harper & Row.

Telleen, Maurice. 1972. The draft breeds. *Western Horseman* 37:80 (October).

Trotting and Pacing Guide. Columbus, Ohio: U.S. Trotting Association. [Annual.]

The Welsh Pony. West Chester, Pennsylvania: The Welsh Pony Society of America, Inc.

Wentworth, Lady. 1945. *The Authentic Arabian Horse and His Descendants*. London: George, Allan, and Unwin, Ltd.

Widmer, Jack. 1959. *The American Quarter Horse*. New York: Scribner's.

Zaher, A. 1948. *The genetic history of the Arabian horse in America*. Ph.D. thesis. Michigan State University, East Lansing.

2

BIOLOGY OF
THE HORSE

CHAPTER 3

ANATOMY AND PHYSIOLOGY

3.1 Introduction

The horse's body is a complex mechanism, and an understanding of its structure and function will result in more intelligent care and management of the animal. Anatomy and physiology are the sciences of the relationship of form to function, and knowledge of these subjects separates the skilled horseman from the amateur.

The body of the horse is made up of the following systems:

1. Skeletal (the bones and joints)
2. Muscular (the muscles)
3. Respiratory (the lungs and air passages)
4. Circulatory (the heart and vessels)
5. Digestive (the gastro-intestinal tract and urinary system)
6. Nervous (the brain, spinal cord, associated nerves, and special senses)
7. Endocrine (the ductless glands, responsible for the chemical control of the body)
8. Reproductive (the ovaries, testicles, and associated organs)
9. Integumentary (the skin and associated structures)

3.2   Skeletal System

The skeleton of the horse (Figure 3-1) consists of the trunk (skull, spinal column, ribs, and breastbone) and limbs. The skeletal system includes the bones and ligaments, which bind the bones together to form joints. It provides the framework that gives the body form, supports the soft parts, and protects the vital organs. The bones act as levers, store minerals, and are the site of blood cell formation.

The skeleton of the horse is made up of 205 bones as follows: vertebral column, 54; ribs, 36; sternum, 1; skull, 34; thoracic limbs, 40; pelvic limbs, 40.

Bones are classified as long, short, flat, and irregular. The *long bones* function chiefly as levers and aid in support of weight and locomotion. The *short bones* absorb concussion. They are found in the complex joints such as the carpus (knee), tarsus (hock), and fetlock (ankle). The *flat bones* enclose the cavities containing vital organs: skull (brain) and ribs (heart and lungs). The flat bones also provide large areas for the attachment of muscles. The *irregular bones* are the bones of the spinal column; they protect the central nervous system.

The periosteum is a tough membrane that covers the bones throughout the body except at their points of articulation. The periosteum protects the bone and is the site of healing should there be a fracture. Abnormal growth in the periosteum is termed *exostosis*. In the horse, the response of the periosteum to injury may result in undesirable bone growths, such as splints, spavins, and ringbone (see Chapter 4). The articulating surface of the bone is covered with a thick, smooth cartilage that diminishes concussion and friction.

Bones are held together by *ligaments*, whereas muscles are attached to the bones by *tendons*. The layer within the joint capsule is sealed by a delicate layer of synovial membrane, and the joint is lubricated by a secretion termed *synovial fluid*.

Skull

The bony framework of the head consists of 34 irregularly shaped flat bones jointed by immovable joints. The *cranial cavity* encloses and protects the brain and supports many sense organs. The facial portion of the skull consists of orbital, nasal, and oral passages. The *orbital cavity* is the bony socket that surrounds and protects the eye. The *nasal cavity* is the passageway to the respiratory system. It contains scroll-shaped turbinate bones that serve as baffles to deflect and warm inspired air as it passes over the vascular, mucous membrane that lines the entire cavity. This mucous membrane also contains

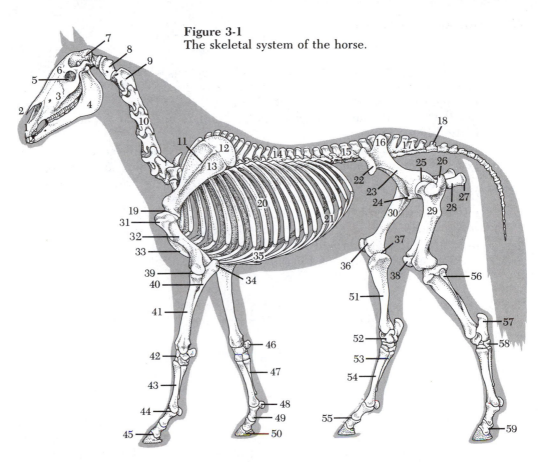

Figure 3-1
The skeletal system of the horse.

1 Incisive bone (premaxillary)	20 Ribs (forming wall of thorax: there are usually 18 ribs)	42 Carpus
2 Nasal bone		43 Metacarpus
3 Maxillary bone		44 Fetlock joint
4 Mandible	21 Costal arch (line of last rib and costal cartilages)	45 Coffin joint
5 Orbit		46 Accessory carpal bone (pisiform)
6 Frontal bone	22 Tuber coxae	
7 Temporal fossa	23 Ilium	47 Small metacarpal bone (splint bone)
8 Atlas (first cervical vertebra)	24 Pubis	
9 Axis (second cervical vertebra)	25 Hip joint	48 Proximal sesamoid bone
10 Cervical vertebra (there are 7 of these, including the atlas and axis)	26 Femur, greater trochanter	49 First phalanx
	27 Tuber ischii	50 Distal phalanx (third phalanx)
	28 Ischium	
11 Scapular spine	29 Femur, third trochanter	51 Tibia
12 Scapular cartilage	30 Femur	52 Talus (tibial tarsal bone) (astragalus)
13 Scapula	31 Humeral tuberosity, lateral	
14 Thoracic vertebrae (there are usually 18 of these)	32 Humerus	53 Small metatarsal bone (splint bone)
	33 Sternum	
15 Lumbar vertebrae (there are usually 6 of these)	34 Olecranon	54 Metatarsus
	35 Costal cartilages	55 Pastern joint
16 Tuber sacrale	36 Femoral trochlea	56 Fibula
17 Sacral vertebrae (sacrum) (there are usually 5 vertebrae fused together)	37 Stifle joint	57 Calcaneus (fibular tarsal bone)
	38 Patella	
	39 Elbow joint	58 Tarsus
18 Coccygeal vertebrae	40 Ulna	59 Middle phalanx (second phalanx)
19 Shoulder joint	41 Radius	

the sensory nerve endings of the olfactory nerve, which is responsible for conveying the sense of smell. The *oral passage*, buccal cavity, or mouth is the passageway to the digestive tract. The maxillae are the bones of the upper jaw that carry the upper cheek teeth (pre-molars and molars). The mandible is the hinged lower jaw.

The age of a horse can be determined with considerable accuracy up to 9 years of age by examining the incisors (Figure 3-2). There are 6 incisors in the upper jaw and 6 in the lower jaw. In the male horse, canine teeth (tushes) erupt in the interdental spaces, but these 4 extra teeth are usually missing in the mare. There are 6 premolars and 6 molars in each jaw of both sexes. Small pointed teeth that sometimes appear at the base of the first premolar tooth are termed *wolf teeth*.

The dental formula of a mature horse is:

$$\text{Male: } 2\left(I\,\frac{3}{3}\ C\,\frac{1}{1}\ P\,\frac{3\text{ or }4}{3}\ M\,\frac{3}{3}\right) = 40 \text{ or } 42$$

$$\text{Female: } 2\left(I\,\frac{3}{3}\ C\,\frac{0}{0}\ P\,\frac{3\text{ or }4}{3}\ M\,\frac{3}{3}\right) = 36 \text{ or } 38$$

In the formula, *I*, *C*, *P* and *M* refer to the incisors, canines, premolars and molars, respectively. The number of upper and lower teeth are indicated above and below the line, i.e., $\frac{3}{3}$. Because the formula within the parentheses is for one side of the jaw, it is multiplied by 2.

The foal often does not have any incisors visible at birth, but they erupt at regular intervals during the first 6 to 10 weeks. The young horse has 24 deciduous or temporary milk teeth. The 12 incisors are all replaced by the time an animal is 4½ years old. The eruption and wear of the permanent incisors are used as follows as a method of determining age of the horse:

Permanent incisor	Age at eruption	Years of wear
1st incisor	2½ years	3
2nd incisor	3½ years	4
3rd incisor	4½ years	5
Canine	4 to 5 years	

The permanent premolars and molars are also erupting during this period but they are not used to determine the age of a horse because of their inaccessibility. The age at eruption of premolars and molars is as follows:

Premolars and molars	Age at eruption
1st premolar	5–6 months
2nd premolar	2½ years
3rd premolar	3 years
4th premolar	4 years
1st molar	9–12 months
2nd molar	2 years
3rd molar	3½–4 years

With experience, one can determine age between 6 and 9 years by using the wear or smoothness of the indentations (cups) in the incisors as a guide. The cups disappear from the lower central, intermediate, and corner incisors at 6, 7, and 8 years, respectively, and from the upper central, intermediate, and corner incisors at 9, 10, and 11 years, respectively. After 9 years of age, other subtle changes occur in the incisors that aid in determining the age of older horses. At 10 years of age, Galvayne's groove (see Figure 3-2) appears at the gum line of the upper corner incisor. It is halfway down the incisor at 15 and all the way down at 20 years of age. By the age of 30, it has disappeared from the corner incisor. As the horse ages, the slant of the incisors changes as indicated in Figure 3-2. Also, as a horse ages, the chewing surfaces of the teeth change in appearance from oval to triangular. Since environmental conditions markedly affect the wear of the incisors, determining the age of a horse is only an estimate. For example, horses raised and kept on soft feed may appear younger than they actually are and horses kept on sandy soil pastures may appear older because sand tends to wear down the teeth at a faster rate.

Abnormalities in the jawbone can result in an overshot jaw (parrot mouth) or undershot jaw (monkey mouth). These conditions are congenital and may be inherited (see Section 4.3). Because they can interfere with successful grazing, they are undesirable traits.

The upper lip is a sensitive, strong, mobile organ of prehension. In grazing, the lip places food between the incisor teeth to be cut. In the manger, the lip and the tongue aid in collecting loose food.

Simple up and down movement of the premolars and molars is inadequate to grind tough, coarse feed, and therefore the horse is capable of lateral grinding movements of the jaw. The upper jaw is slightly wider than the lower jaw. The lateral grinding movement of the jaws develops chisel-sharp surfaces on the inner edge of the lower and outer edge of the upper molars. These edges can cause damage to the soft tissues of the mouth. The inner edges of the lower teeth can injure the tongue, and the outer edges of the upper teeth can lacerate the cheek. Consequently, it is sometimes necessary to file the sharp edges of molars and premolars. This process is termed

Figure 3-2
Guide to determining the age of a horse by its teeth.

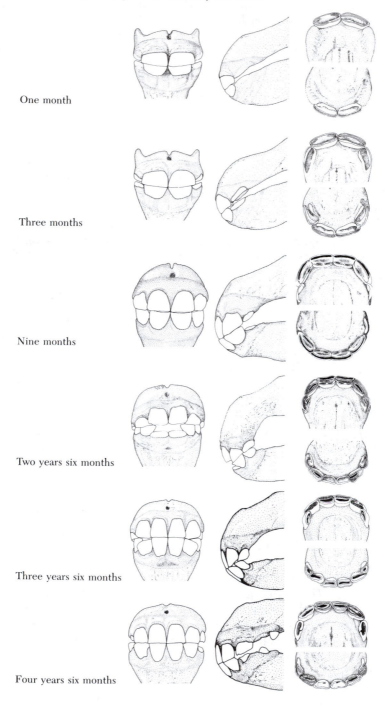

One month

Three months

Nine months

Two years six months

Three years six months

Four years six months

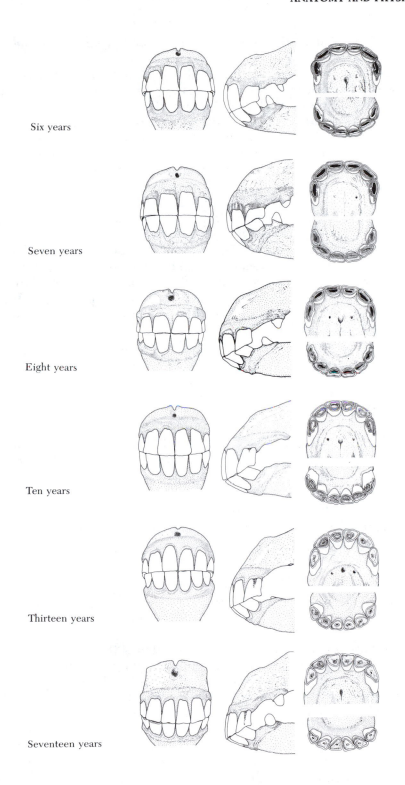

Six years

Seven years

Eight years

Ten years

Thirteen years

Seventeen years

floating the teeth. The tooth of the horse is composed of veins of enamel that keep the surfaces rough.

The development of sharp edges, or hooks, is not recognized by many horse owners. The presence of sharp edges can result in a loss of weight and a change in a horse's eating behavior. When eating, it will drop feed out of its mouth and hold its head to one side while chewing. The side of its face becomes sensitive to pressure because of the sore spots on the inside of its cheeks. Pressure from a bit may cause it to "throw its head." Good management requires annual inspection of the premolars and molars for these sharp edges and, if they are present, floating the teeth.

There are three pairs of salivary glands in the horse: the parotid gland, located below the ear, which is the largest; the submaxillary gland, located primarily between the jaws; and the sublingual gland, located in the mucous membranes beneath the tongue. These glands can become the site for an acute contagious bacterial infection termed *equine distemper* or *strangles* (see Chapter 16). This infection causes a fever of 104 to 106°F and an intense swelling of the submaxillary and parotid lymph nodes, which may abscess and rupture. If untreated, the infection may spread to the abdominal lymph system and the disease may become fatal.

The lateral portion of the skull articulates with the atlas joint of the cervical vertebrae.

Spinal Column

The spinal or vertebral column is comprised of a series of irregularly shaped bones that stretch from the base of the head to the tip of the tail. The regions of the spinal column are:

1. Cervical (neck): 7 vertebrae
2. Thoracic (back): 18 vertebrae usually, but sometimes 19 or 17
3. Lumbar (loin): 6 vertebrae usually, but sometimes fused; 5 are not uncommon
4. Sacral (croup): fusion of 5 vertebrae
5. Coccygeal (caudal or tail): varies considerably in number; average is approximately 18

The cervical region is the most flexible portion of the vertebral column. The initial vertebra of the cervical region is the *atlas;* it is attached to the skull to permit extension and flexion of the head and neck. The second vertebra is the *axis;* it permits side-to-side articulation with the atlas. The atlas-axis joint is commonly referred to as the "yes-no" joint. The remaining cervical vertebrae form an S shape when viewed from the side, and the lengthening or shortening of the neck can be accomplished by changing the amount of this curvature.

The 18 vertebrae of the thoracic region have articulating surfaces on their dorsal processes for attachment of the 18 pairs of ribs. The spinous process (upper surface of the spine) is of increased height in the third, fourth, and fifth vertebrae; these form the withers, the highest part of the horse. There is very little movement of the thoracic vertebrae. The thoracic cavity is formed by the vertebrae on the top, the ribs on the side, and the sternum on the floor. This cavity contains the vital organs of the respiratory, circulatory, and digestive systems. The first 8 pairs of the 18 pairs of ribs are the *true* ribs, which are attached to the sternum by means of cartilage. The last 10 pairs are *false* ribs, which are connected to each other by cartilage and then to the sternum.

The lumbar region forms the loin of the horse and sometimes is missing a vertebra. Horses of Arabian breeding often have only 5 lumbar vertebrae, but a number of purebred Arabians also have 6.

There is considerably more movement in the lumbar vertebrae than in the thoracic or sacral regions.

The sacrum contains 5 vertebrae that are fused together and underlie the croup of the horse. The sacrum is jointed securely to the hip bones (pelvis) on either side.

The tail region is made up of a varying number of coccygeal vertebrae. They become reduced in size as they proceed caudally, and the last one is pointed. The spinal canal is very narrow in this portion of the spinal column. The tail of driving and working horses was often cut off (docked) to prevent its interference with the driving lines.

Limbs

The front limb of the horse is not attached directly to the vertebral column. The *scapula* (shoulder blade) is attached by a muscular sling that supports the thorax and reduces concussion. The slope of the scapula, and therefore the angle formed by its junction with the humerus (arm), provides additional shock absorption and has much to do with the smoothness of gait of a riding horse. The *radius* is the main bone of the forearm, and the *ulna* is fused to the upper part. The *carpus* is the knee and consists of 8 carpal bones arranged in 2 rows. Therefore, the carpus joint or knee joint actually consists of 3 articulating surfaces. The *metacarpal* bones are 3 in number. The large middle metacarpal (cannon bone) extends from the carpus to the fetlock joint. The 2 smaller metacarpals on each side of the cannon are termed *splint bones* and are vestiges of additional toes.

The hind limb is attached to the vertebral column at the sacrum by the pelvic girdle that is formed by the union of the hip bones. The *femur* is the bone of the thigh, and the patella is a small bone of the stifle joint corresponding to the knee cap in man. The *tibia* is the main bone of the gaskin, and the *fibula*, a small rudimentary bone, is fused to it. The *tarsus* or hock contains 7 bones and corresponds to the ankle and heel in man. The 3 bones of the cannon on the hind limb are the *metatarsals*. As in front, there are a large middle metatarsal and 2 splint bones.

The structures of the front and hind foot of the horse are basically the same (Figure 19-1). At the base of the cannon bone there are 2 small bones called the *proximal sesamoids*, which form the back part of the fetlock joint. These bones are completely surrounded by connective tissue and provide a bearing surface for the flexor tendons.

Below the fetlock joint are the *first phalanx* (long pastern), *second phalanx* (short pastern), and *third phalanx* (coffin or pedal bone), which is surrounded by the hoof. In back of the coffin bone is the *distal sesamoid* or *navicular bone*, which provides an articulating surface for the pedal bone and an important bearing surface for the deep flexor tendon. An inflammation of the distal sesamoid results in *navicular disease* (see Chapter 4). The coffin, navicular, and short pastern bones are the 3 bones of the foot. The coffin bone is attached to the hoof wall by the sensitive laminae (see Chapter 19 for more information about the hoof).

Joints of the Leg

Movement of the horse is dependent upon the contraction of muscles and the corresponding articulation of the joints. In the front leg are 6 joints: the shoulder, elbow, carpus, fetlock, pastern, and coffin joints. The hind limb has 7 joints: the sacroiliac, hip, stifle, hock, fetlock, pastern, and coffin. These numerous articulations are the source of many of the unsoundnesses in the horse. The joint capsule is a fluid-filled sac (synovia). If an excess of this fluid is produced, a puffy, soft swelling such as a bog spavin occurs. Ligaments hold the joints together (attach bone to bone) and have a limited blood supply. If there is an injury to a ligament, such as a sprain, the ligament tends to heal slowly and often incorrectly. Sesamoiditis is an example of such an injury to the suspensory ligament, and a curb is an example in the plantar ligament, which holds the back of the hock together. Muscles are attached to bone by tendons. The horse has no muscle below the knee or hock, and consequently many leg muscles have long tendons that pass down the leg over joints where there are protective tendon sheaths or "tendon bursa." Chronic irritation of these bursa can result in excess fluid production and soft swellings. When they appear above the fetlock, these swellings are called wind puffs or wind galls. They rarely result in a distinct lameness. Distinct lameness can result when the tendons bear against bones that are roughened from fracture or exostosis, as in navicular disease. Abuse of the legs can result in the contraction, rupture, or bowing (straining) of tendons.

3.3   Muscular System

The muscles are the largest tissue mass in the horse's body (Figure 3-3). They perform their work by contracting to allow for locomotion and the performance of vital functions. The muscles are classified as *smooth muscle, cardiac muscle,* and *skeletal muscle.* The smooth and cardiac muscles are involuntary or automatic in their contraction and are active in the digestive tract, respiratory, circulatory, and urogenital systems. The skeletal muscle is voluntary and functions in the movement of the horse.

The activities of the multitude of muscles in the body are complex. A comparison of a horse grazing with a horse performing in a Grand National racing competition emphasizes the diversity of muscular activity that is possible. However, the basic principle of muscular activity is the same for all muscles: a period of contraction (shortening of muscle fibers) followed by a period of relaxation (lengthening of muscle fibers).

The skeletal muscles of a horse consist of a variety of fibers that have distinct functional and metabolic characteristics. Fibers within a group in-nervated by the same nerve have the same properties. In one system of muscle-fiber classification, two broad classifications of fiber types are used. Type I fibers are slow-twitch fibers, are resistant to fatigue, and have low glycolytic activity. These fibers are used for long-term, low-power work such as standing and other nonstrenuous types of movements. Also, they provide endurance. Type II fibers are fast-twitch fibers that fatigue much more quickly and have high glycolytic activity. Within the typing, type II fibers are further classified as IIA, IIB, and IIC according to their myosin structure. Type IIC fibers are transitional fibers, which can be recruited to type IIA or IIB based upon training and are generally found in young horses. Their numbers relative to types IIA and IIB are small in mature horses. The type IIA fibers are used for speed at longer distances while type IIB fibers are used for quick bursts of speed for short distances. Another system of classification of muscle fibers is: slow-twitch, high-oxidative (ST), which corresponds to type I; fast-twitch, low-oxidative (FT), which corresponds to type IIB and fast-twitch, high-oxidative (FTH), which corresponds to type IIA. Their characteristics are given in Table 3-1.

Muscle fibers are primarily composed of contractile proteins called myofilaments. Thin filaments (actin) are pulled towards the center of a sarcomere by the thick filaments (myosin) when a muscle contracts. Sarco-meres are the areas between the Z lines, which are the areas where the thin filaments are connected forming cross striations. Cross-sectional area of type II fibers is related to force output. Larger areas are associated with larger force outputs. Thus, cutting horses require larger muscles for explosive activity. Quarter Horses running short distances require larger muscles than Thoroughbred horses running distance races. Muscle fibers are bound to-gether by fibrous connective tissue to form bundles and the bundles are arranged in parallel bands to form muscles (Figure 3-4).

Figure 3-3
The muscular system of the horse.

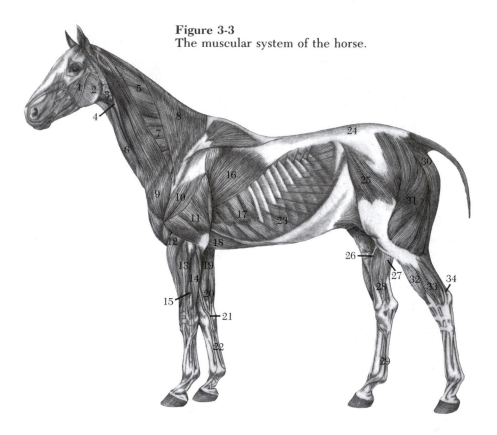

1 Facial nerve	20 Ulnar carpal flexor muscle
2 Masseter muscle	(flexor carpi ulnaris muscle)
3 Parotid salivary gland	21 Cephalic vein
4 Jugular vein	22 Digital flexor tendons
5 Splenius muscle	23 External abdominal oblique muscle
6 Sternocephalic muscle	24 Gluteal muscles
7 Serratus ventralis muscle	25 Tensor fasciae latae muscle
8 Trapezius muscle	26 Saphenous vein
9 Brachiocephalic muscle	27 Gastrocnemius muscle
10 Deltoid muscle	28 Long digital flexor muscle
11 Triceps muscle	29 Digital flexor tendons
12 Radial carpal extensor muscle	30 Semitendinosus muscle
(extensor carpi radialis muscle)	31 Biceps femoris muscle
13 Common digital extensor muscle	32 Long digital extensor muscle
14 Ulnaris lateralis muscle	33 Lateral digital extensor muscle
15 Lateral digital extensor muscle	34 Achilles' tendon
16 Latissimus dorsi muscle	(also called the hamstring: consists
17 Serratus ventralis muscle	of tendons of gastrocnemius, biceps
18 Pectoral muscle	femoris and superficial digital
19 Radial carpal flexor muscle	flexor muscles attaching to calcaneus)
(flexor carpi radialis muscle)	

Table 3-1
Characteristics of muscle fiber types in the horse.

	Type I	Type II	
	ST	FTH	FT
Speed of contraction	Slow	Fast	Fast
Max. tension developed	Low	High	High
Myosin ATP-ase activity pH 9.4	Low	High	High
Myosin ATP-ase activity pH 9.4 (after pre-incubation pH 4.35)	High	Low (IIA) Intermediate (IIB)	Intermediate (IIB)
Oxidative capacity	High	Intermediate to high	Low
Capillary density	High	Intermediate	Low
Glycolytic capacity	Intermediate	High	High
Lipid content	High	Intermediate	Negligible
Glycogen content	Intermediate	High	High
Muscle fibers per motor unit	Low	High	High
Fatiguability	Low	Intermediate	High

ST = Slow-twitch, high-oxidative fibers; FTH = Fast-twitch, high-oxidative fibers; and FT = Fast-twitch, low-oxidative fibers.
SOURCE: Snow, Persson and Rose, 1982.

The energy for muscle contraction is provided by high-energy phosphate bonds in ATP. Within the first 10 seconds of exercise, the main source of energy is from that stored in creatine phosphate. It serves to rephosphorylate ADP to ATP. Within a few seconds after the start of exercise, the ATP is generated from two other sources that serve as the main sources for muscle contractions. Glucose is metabolized in the absence of oxygen to yield ATP via the glycolytic metabolic pathway. This anaerobic form of energy production has a low yield of ATP, but it can be turned on very rapidly and will form ATP at a high rate. Lactic acid is the end product of glycolysis. The second source of energy is referred to as *aerobic energy production.* Sources of its fuel are glucose, fatty acids, and protein. Advantages of energy formed from aerobic oxidation are: a high ATP yield, the absence of toxic waste products (since carbon dioxide and water are the end products), and that the energy can be produced for long periods of time. Disadvantages to forming energy in this way are the need for oxygen and that it takes over a minute before appreciable amounts of ATP can be produced. The increased oxygen demand must be met by a series of cardiovascular and respiratory system changes. If these are inadequate to meet energy demands during strenuous exercise, anaerobic sources are called upon. Thus, muscle fibers are selectively recruited in a specific pattern that varies according to gait, speed, and duration of exercise. As speed increases, fibers are recruited in order of type (I through IIB).

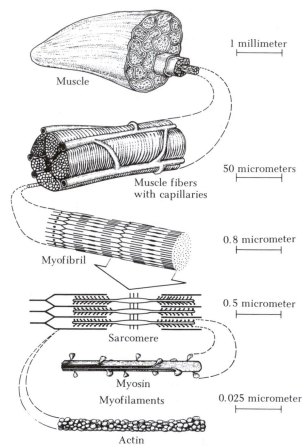

Muscle

1 millimeter

Muscle fibers
with capillaries

50 micrometers

Myofibril

0.8 micrometer

Sarcomere

0.5 micrometer

Myosin

Myofilaments

0.025 micrometer

Actin

Figure 3-4
Diagrammatic representation of the structural composition of skeletal muscle tissue. One micrometer corresponds to $\frac{1}{1000}$ of a millimeter or $\frac{1}{25,000}$ of an inch. (Adapted from Hoppeler and Mermod, 1987).

The distribution of fiber types between muscles is not uniform and is related to functions of the muscle. Also, the distribution is not uniform within the same muscle between breeds (Table 3-2). Within a given muscle, compartmentalization occurs so that the distribution is not uniform throughout the muscle.

Influence of Training   Muscle is an extremely adaptable tissue and the adaptations are related to the specific type of training stimulus.

The goal of training a horse for endurance events is for it to be able to exercise for an extended period of time. Thus, the energy must be derived from aerobic metabolism. An increase in the number of mitochondria in the muscle cell so that more oxygen can be utilized per unit of time and an increase in the oxygen delivery system to the muscle is desirable. These

Table 3-2
Fiber composition in the middle gluteal of different breeds of horse. Percent
fiber type (mean ±SEM).

	n	ST	FTH	FT
Quarter Horse	[1](28)$_x^a$	8.7 ± 0.8	51.0 ± 1.6	40.3 ± 1.6
Thoroughbred	[1](50)$_x^a$	11.0 ± 0.7	57.1 ± 1.3	32.0 ± 1.3
	[1](22)$_ø^a$	7.3 ± 0.9	61.2 ± 1.5	28.8 ± 1.5
Arab	[3](6)^a	14.4 ± 2.5	47.8 ± 3.2	37.8 ± 2.8
Standardbred	[2](8)	24.0 ± 3.6	49.0 ± 3.1	27.0 ± 3.3
	[1](9)$_ø^a$	18.1 ± 1.6	55.4 ± 2.2	26.6 ± 2.0
Shetland pony	[3](4)^a	21.0 ± 1.2	38.8 ± 1.9	40.2 ± 2.7
Pony	[3](8)^a	22.5 ± 2.6	40.4 ± 2.3	37.1 ± 2.8
Heavy Hunter	[3](7)^a	30.8 ± 3.1	37.1 ± 3.3	37.8 ± 2.8
Donkey	[3](5)^a	24.0 ± 3.0	38.2 ± 3.0	32.1 ± 3.4

ST = Slow-twitch, high-oxidative fibers; FTH = Fast-twitch, high-oxidative fibers; and FT = Fast-twitch,
low-oxidative fibers.
[a]Out of work, øElite stallions at stud, [x]Elite broodmares, [1]Snow and Guy (1981), [2]Lindholm and Piehl (1974),
[3]Snow and Guy (1980).

SOURCE: Equine Exercise Physiology, edited by Snow, Persson and Rose.

changes are induced by modifying training intensity and duration. Training
intensity should be such that it fully taxes the aerobic system and does not
induce the anaerobic system. Since the transition from the aerobic to the
anaerobic system is known to occur when a specific plasma lactate concentration is observed (4 millimoles per liter) and the plasma lactate concentration is closely correlated to heart rate, the horse can be trained by closely
monitoring the heart rate and keeping it within the predetermined optimal
rate. An accurate assessment of endurance capacity can be obtained by
determining the speed at which lactate begins to accumulate in the blood.
The greater the speed attained before the lactate starts to accumulate, the
fitter the horse.

Training for quick bursts of high-intensity exercise involves training for
strength. This involves increasing muscle mass by increasing the number of
sarcomeres in parallel to existing sarcomeres. High-intensity exercises for
short periods of time increase strength. Adaptive mechanisms for endurance
training will override those for strength when both are trained for at the
same time. Thus, no strength is gained.

Most performance-horse events last for time periods between 30 seconds and 4 minutes. During this time of high-intensity performance, anaerobic energy sources predominate. Thus, training to tolerate very high levels
of plasma lactate during this period is important. To train for these activities,
repeated bouts of high-intensity, short-duration exercise (sprints less than 40

to 45 seconds) are used. Results of this type of training program may include: no significant change in the proportion of type I fibers, an increase in the ratio of type II high-oxidative to type II low-oxidative fibers, and a significant increase in speed at which the plasma lactate concentration does not exceed 4 millimoles per liter. Thus, the horse responds to intensive sprint training by increasing the oxidative capacity of its skeletal muscles.

Muscles are attached to bone by *tendons*, composed of dense connective tissue. The tendons may be short, as at the shoulder blade, or long, as in the legs. The action of bone levers, joint hinges, and tendon cables, and the contraction of muscles make motion possible. Skeletal muscles have counterparts that produce opposite effects. Thus, in the leg of the horse are a group of muscles that cause flexion of a joint (*flexor muscles*) and an opposing group that extend or straighten a joint (*extensor muscles*).

The contractive process is a chemical reaction within the muscle that produces heat in addition to performing work. The heat of contraction and recovery is important in body temperature regulation. In cold weather, *shivering* is a spasmodic muscle contraction that produces heat to help maintain body temperature. In hot weather, the heat of exertion must be dissipated by sweating and radiation.

Overexertion of a muscle without adequate conditioning can lead to muscle fatigue because of the depletion of muscle stores of glycogen and the accumulation of metabolic waste products (lactic acid). Paralytic myoglobinuria (Azoturia, Monday Morning disease, Blackwater) and the tying-up syndrome are complex metabolic disorders called exertional myopathics that affect the muscles of the horse. A careful conditioning program combined with a sound nutritional regime will prevent these muscle disorders. There are also myopathics that result from muscle disuse. Muscle atrophy can have a variety of causes but often denervation of the muscle tissue is a main cause. For example, trauma to the suprascapular nerve can result in disuse and atrophy of several shoulder muscles, resulting in a *shoulder sweeney* (see Chapter 4).

3.4   Respiratory System

The respiratory system is vital to life; its primary function is to oxygenate the blood so that oxygen can be carried to the tissues. In exchange for oxygen, carbon monoxide is removed from the tissues and eliminated from the body in expired air. Secondary functions include temperature regulation and phonation.

The science of equine exercise physiology is making great strides in developing methods of training performance horses to achieve a high degree of physical fitness and preventing unsoundness problems. An understanding of the physiology of respiration will aid the horse trainer in properly conditioning a horse and evaluating its physical condition. Also, the anatomy of the upper air passages (Figure 3-5) must be understood to correctly pass a stomach tube into the esophagus and avoid many potential problems, such as passing fluid into the lungs and rupture of blood vessels in the nasal area.

Anatomy

The respiratory system is a series of air passages that connect the lungs with the external air. The air passages are referred to as the *upper airways,* the *tracheobronchial airways*, and the *lungs.* Initially, air enters the nasal cavities through the nostrils. As it passes through the nasal cavities and past the two turbinate bones in each nasal passage, it is warmed very close to body temperature and becomes fully saturated with water vapor.

Each nasal passage is divided by two turbinate bones into three channels: the dorsal, middle, and ventral nasal meatuses. The *dorsal meatus* is a narrow passage bound dorsally by the roof of the cavity and ventrally by the dorsal turbinate bone. The *middle meatus* is located between two turbinate bones. The *ventral meatus* is located below the middle meatus and is the direct passage between the nostril and the pharynx. About eight inches inside the ventral meatus is an elevation of the hard palate. Above this elevation is an opening into the middle nasal meatus.

The *pharynx* is a cavity located at the back of the nasal passages and the mouth. From the pharynx, the *trachea* leads to the lungs and the *esophagus* leads to the stomach. The *epiglottis* is a cartilage trap that serves to prevent food from entering the larynx when swallowing occurs. The *larynx* (Figure 3-6) serves to regulate the flow of air into the trachea and to prevent the inspiration of foreign objects. It also contains the vocal cords and thus is the voice organ. *Vocal cords* are two muscles that project across the lower part of the larynx. Paralysis of one or both of the muscles results in a roaring sound when the horse inspires air—an unsoundness referred to as "roaring." The trachea is about 2 inches in diameter and made up of rings of cartilage. It is lined with two kinds of cells: mucus-secreting cells, which lubricate it, and ciliated cells lined with tiny hairs that continually beat upward, pushing impurities up and out of the trachea. The trachea divides into the left and right *bronchi.* They subdivide into pulmonary bronchi which further subdivide into interpulmonary bronchi. Further branching occurs, finally resulting in the gaseous exchanging subunits of the lung called *alveoli* or *air sacs.* The paired lungs are lobed organs consisting of three lobes for the right lung and two lobes for the left lung.

When passing a stomach tube, care must be taken not to pass it into a lung. The tube is inserted into the nostril and below the middle nasal meatus

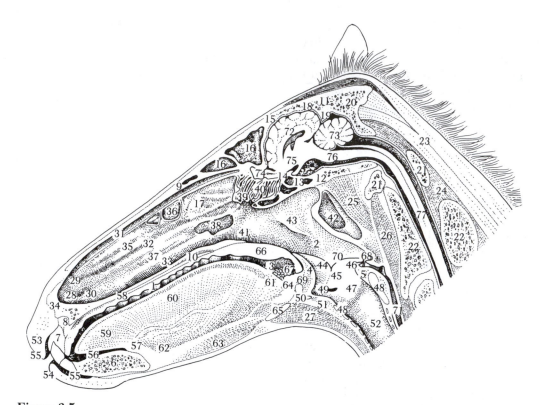

Figure 3-5

Sagittal section of the head. (1) oral cavity; (2) nasopharynx; (3) oropharynx; (4, 5) laryngopharynx; (6) incisive part of mandible with first incisor; (7) incisive bone with first incisor; (8) incisive duct; (9) nasal bone; (10) osseous palate; (11) interparietal bone; (12) sphenoid bone; (13) sphenoid sinus; (14) ethmoid bone; (15) frontal bone; (16, 17) conchofrontal sinus; composed of (16) frontal sinus and (17) sinus of dorsal nasal cocha; (18) parietal bone; (19) tentorium cerebelli osseum; (20) occipital bone; (21) atlas; (22) axis; (23) funicular part of ligamentum nuchae; (24) rectus capitis dorsalis; (25) longus capitis; (26) longus colli; (27) sternohyoideus and omohyoideus; (28) nasal vestibule; (29) alar fold; (30) basal fold; (31) dorsal meatus; (32) middle meatus; (33) ventral nasal meatus; (34) nasal septum; almost entirely removed; (35) dorsal nasal concha; (36) conchal cells; (37) ventral nasal concha; (38) sinus of ventral nasal concha; (39) middle nasal concha; opened; (40) ethmoid conchae; (41) choana; (42) pharyngeal opening of auditory tube; (43) right gutteral pouch, opened; (44) epiglottis; (45) aryepiglottic fold; (46) arytenoid cartilage; (47) comiculate process; (48) cricoid cartilage; (49) entrance to lateral laryngeal ventricle; (50) thyroid cartilage; (51) cricothyroid ligament; (52) trachea; (53) upper lip; (54) lower lip; (55) labial vestibule; (56) sublingual floor of oral cavity; (57) frenulum linguae; (58) hard palate with venous plexus and palatine ridges; (59) apex; (60) body of tongue; (61) root of tongue; (62) genioglossus; (63) geniohyoideus; (64) hyoepiglotticus in glossoepiglottic fold; (65) basihyoid with lingual process; (66) soft palate with glands and muscles; (67, 68) rostral and caudal boundaries of intrapharyngeal opening; (69) free border of soft palate; (70) caudal end of the palatopharyngeal arch; (71) esophagus; (72) cerebrum; (73) cerebellum; (74) olfactory bulb; (75) optic chiasma; (76) brainstem; (77) spinal cord. (Adapted from Nickel et al., 1973.)

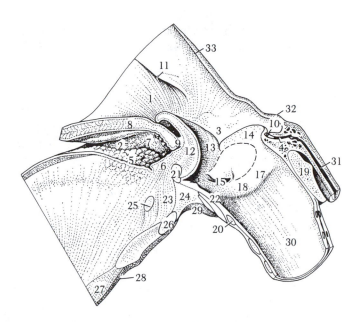

Figure 3-6
Sagittal section of the pharynx and larynx. (1) nasopharynx (2) oropharynx; (3, 4)
laryngopharynx; (5) root of tongue with tonsillar follicles (lingual tonsil); (6) glossoepiglottic
fold; (7) palatoglossal arch; (8) soft palate; (9, 10) boundaries of intrapharyngeal opening;
(9) free border of soft palate; (10) palatopharyngeal arch; (11) pharyngeal opening of
auditory tube; (12) epiglottis; (13) aryepiglottic fold; (14) corniculate process of arytenoid
cartilage; (15) entrance to lateral laryngeal ventricle, the broken line marks the extent of
the ventricle; (16) median laryngeal ventricle; (17) vocal process of arytenoid cartilage;
(18) vocal fold; (19, 20) cricoid cartilage; (21) thyroid cartilage; (22) cricothyroid ligament;
(23) hyoepiglotticus; (24) fat; (25) hyoideus transversus; (26) basihyoid and lingual
process; (27) geniohyoideus; (28) mylohyoideus; (29) sternohyoideus; (30) trachea; (31)
esophagus; (32) caudal pharyngeal constrictors; (33) hyopharyngeus. (Adapted from Nickel
et al., 1973.)

into the ventral meatus. The tube should have some rigidity and curvature to
it so that its tip can be directed downward as it passes over the elevated
portion of the hard palate. Otherwise, the tip may be deflected upward into
the middle meatus where it may strike the ethmoid bone and cause severe
bleeding. When the tip gets to the pharynx, the tube is rotated so that the
curvature is pointing upward. This makes it easier to keep it away from the
laryngeal opening and to insert it into the esophagus when the horse swallows. Swallowing can be encouraged by slightly bumping the tip against the
back of the pharynx. The following things are helpful in determining if the
tube is in the esophagus and not in the trachea: seeing the tube move, feeling
the end of the tube, blowing into the tube and getting back stomach gases
that are not coordinated with respiratory movements, and shaking the tra-

chea, which causes the tube to vibrate if it is in the trachea. Passing fluid into the lungs by error can kill the horse.

Respiration

Movement of air into and out of the lungs is referred to as *respiration*. *Inspiration* is achieved by the contraction of the *diaphragm* and *intercostal muscles*. Expiration is a passive process while the horse is at rest; it becomes an active process during exercise by contraction of ancillary skeletal muscles, which contract the rib cage, and by contraction of abdominal muscles, which force the diaphragm forward. Contraction of the abdominal muscles to aid in expiration is markedly evident for horses suffering from chronic obstructive pulmonary disease (COPD), commonly called *heaves*. The condition is caused by a breakdown in the elasticity of the air sacs and loss of elastic recoil of the lungs. The typical "heave line" is a result of hypertrophy of the abdominal muscles running parallel to the last rib, which contract to force the air out of the lungs.

At rest, the normal horse breathes at a rate of 8 to 16 times per minute. Several factors will influence respiration rate. Exercise, temperature, humidity, fever, distress, and anxiety will increase the respiratory rate.

Pulmonary volumes and capacities are illustrated in Figure 3-7. The volume of air contained in the lung is divided into the tidal, inspiratory

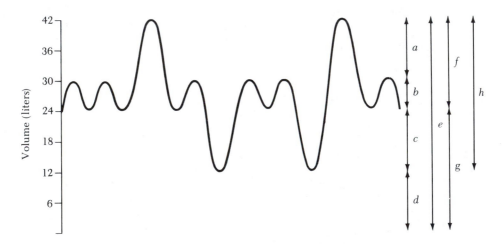

Figure 3-7
Pulmonary volumes and capacities of the horse: (*a*) inspiratory reserve volume; (*b*) tidal volume; (*c*) expiratory reserve volume; (*d*) residual volume; (*e*) total lung volume; (*f*) vital capacity; (*g*) functional residual capacity; (*h*) inspiratory capacity.

reserve, expiratory reserve, and residual volumes. *Tidal volume* is the volume of air inspired or expired during each breathing cycle and is approximately 6 liters. *Inspiratory reserve volume* is the amount of air that can be inhaled above the tidal volume and is approximately 12 liters. *Expiratory reserve volume* is the volume of air that can be forcibly exhaled at the end of the tidal volume. It is approximately 12 liters. After a maximal forced expiration, approximately 12 liters of air, the *residual volume,* always remain in the lungs. Inspiratory capacity is the amount of air that can be inhaled from the end of a normal expiration to full inspiratory capacity and is about 18 liters. It is the sum of the tidal and inspiratory reserve volumes. The amount of air remaining in the lungs after a normal expiration is the functional residual volume and is the sum of the expiratory reserve and residual volumes. It is approximately 24 liters. *Vital capacity* is the maximum amount of air that can be forcibly exhaled from the lungs after a maximal inspiration. Thus, it is approximately 30 liters and is the sum of the inspiratory reserve, tidal, and expiratory reserve volumes. *Total lung capacity* is the maximum volume of the lungs and is the sum of all four lung volumes. *Minute respiratory volume* is the product of tidal volume and respiratory frequency. At a respiratory frequency of 15 breaths per minute and a tidal volume of 5.8 liters, the minute respiratory volume is 87 liters per minute. The normal range is 65 to 90 liters per minute. *Maximum breathing capacity* is the product of tidal volume and respiration rate when oxygen consumption is maximal. Values of up to 1,500 liters per minute have been recorded during strenuous exercise. Since gaseous exchange does not occur in the airways but only in the alveoli, the *minute alveolar ventilation* is the volume of air that moves in and out of the alveoli each minute. The *dead space air* (the volume of air in the airways) is subtracted from the tidal volume and is approximately 25 to 30% of tidal volume. Oxygen uptake is about 1.8 liters per minute in the resting horse.

Effects of Exercise   The equine respiratory system is well adapted to function during exercise. Respiration rates are related to exercise intensity. As intensity increases, the rate increases. However, respiration rate is correlated with stride frequency when Standardbreds are racing at the trot and when other horses are cantering and galloping. When trotters are working at submaximal intensities, the ratio of respiration to stride frequency is 1:1, but as they work at maximal intensities, the ratio changes to approximately 1:2 so that they can breathe more deeply, with tidal volumes of 20 to 25 liters. During the canter and gallop, the ratio is 1:1. As the horse extends its body, it inspires and as it contracts its body, it expires. While at a fast gallop, horses are forced to respire up to 130 to 140 times per minute with a low tidal volume (10 to 12 liters). However, during the initial stages of a fast sprint, the horse may hold its breath for about 30 seconds. Some Quarter racing horses do not breathe during a race. During certain forms of maximal

efforts such as jumping, respiration is inhibited. Immediately after a strenuous exercise bout such as racing, the respiration rate exceeds the previous
stride frequency. Thus, oxygen consumption is not maximal nor is alveolar
ventilation sufficient to prevent oxygen debt during strenuous exercise.
Environmental temperature has a marked effect on respiration rate. Evaporation of water from the lungs is a means of body-temperature regulation.

Oxygen consumption and pulmonary ventilation are linearly related to
riding velocity and oxygen consumption reaches approximately 90 percent
of its steady-state value within one minute. Maximum oxygen uptake of
well-trained horses is in the range of 70 to 80 liters per minute but only 60
liters for untrained horses. During maximal exercise bouts, pulmonary ventilation may increase 23-fold, tidal volume 2.6-fold, respiratory frequency
10-fold, and oxygen uptake 33-fold.

Three general aspects of respiratory muscle function have been considered as possible limits to respiration: muscle strength and endurance, speed
of muscle contraction, and neural control of muscle activity. Although these
factors have not been studied in detail, it is assumed that respiratory muscles
do not limit pulmonary function. However, endurance-race horses that are
less physically fit and are exhausted from exercise do lose control over
contractions of the diaphragm. The debilitating condition is referred to as
thumps or *synchronous diaphragmatic flutter* and is a rhythmic contraction of
the diaphragm that occurs in rhythm with atrial depolarization of the heart.
Thus, prolonged exercise in hot and humid conditions such as those often
encountered during an endurance race, respiratory fatigue may develop and
play a role in the exhausted-horse syndrome.

Effect of Training   It is generally assumed that the respiratory system undergoes little or no adaptation to exercise training. However, respiration
rates are relatively lower for similar exercise intensities and return to normal
rates faster following exercise training. Trained horses reach maximum oxygen uptake at faster speeds than untrained horses. This is due to changes in
the cardiovascular system, which are discussed below.

Common Respiratory Problems

Several respiratory ailments afflict horses. Infections caused by various bacteria, viruses, and fungi are the most common. Parasites and noxious chemicals also cause serious problems. *Bronchitis* is a lower respiratory infection
located in the bronchi. When an infection is localized in the lungs, it is
pneumonia, which is characterized by the walls of the alveoli being irritated
and the alveoli filled with fluid. If located in both areas, it is *bronchopneumo-
nia* chronic obstructive pulmonary disease, commonly called heaves, usually
leads to permanent lung damage as a result of obstruction of the bronchi and

bronchioles by inflammation. This leads to a buildup of air in the lungs and swelling and bursting of the alveoli. *Pleurisy* is an inflammation of the membranes, or *pleura*, that surround the lungs. It commonly results from the spread of bacteria or viruses from inside the lungs or from a puncture wound that penetrates the chest wall. *Pulmonary hemorrhage* is bleeding from lung tissue. It is a common problem during strenuous exercise such as racing. Horses afflicted with the problem are referred to as "bleeders."

3.5   Cardiovascular System and Blood

The *cardiovascular system* of the horse is of interest because of the changes it undergoes during exercise, the monitoring of which can be used to determine physical fitness. The cardiovascular system of the horse consists of the heart, arteries, capillaries, and veins; it functions to perfuse the tissues of the body with blood.

Heart, Arteries, Veins, and Capillaries

The *heart* serves as the pump for the cardiovascular system; its anatomy is shown in Figure 3-8. The heart is composed of four chambers. Blood is pumped from the heart to the lungs by the right ventricle and to the periphery by the left ventricle. The muscle surrounding the left ventricle is much stronger than the muscle for the right ventricle. Oxygenated blood returning from the lungs (pulmonary circulation) enters the left atrium and, when the heart muscles surrounding the atrium contract, is forced into the left ventricle. Unoxygenated blood returning from the peripheral circulation enters the right atrium and then travels to the right ventricle before leaving the heart by way of the pulmonary artery.

Heart size is related to body size. The average heart-weight to body-weight ratio is 0.94 per cent. Size can be estimated by the sum of the QRS durations for leads 1, 2, and 3 from the electrocardiogram (ECG). The sum of the time intervals is referred to as the heart score, which has been shown to be associated with successful racing performance.

The heart beats an average of 35 beats per minute when the horse is at rest. A number of factors influence heart rate: exercise, physical condition, environmental temperature, excitement, pathological conditions, age (Table 3-3), and so on. The stroke volume, the quantity of blood ejected by the heart during each contraction, is approximately 1 liter. Since cardiac output is the quantity of blood pumped per minute (stroke volume × heart rate), its average is 35 liters.

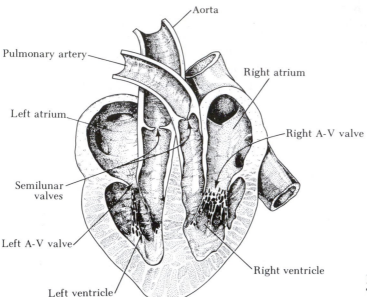

Figure 3-8
The equine heart.

Arteries are thick-walled, muscular vessels that conduct blood from the heart. As the blood gets further from the heart and the arteries branch, the arteries decrease in size, forming arterioles, and, finally, the capillary bed is formed. In the capillary beds, nutrients and oxygen escape to the tissues and products of metabolism and carbon dioxide enter the capillaries to be eliminated from the body. The vessels leading to the heart increase in size as the capillaries unite to form veinules and finally veins. Fluid that does not reenter the capillary network is picked up by the lymphatic system, which transports it to the venous system. *Lymph,* a clear colorless fluid, is moved

Table 3-3
Age and corresponding pulse rate of the horse

Age	Pulse rate (beats per minute)
8–10 weeks	60–79
6 months	60–71
10–12 months	50–68
2 years	44–65
3 years	39–62
4 years	36–59
5 years	36–57

toward the venous system by pressure formed when muscles contract and by a series of valves. As it moves through lymph nodes prior to entering the venous system, it is filtered to remove bacteria and foreign material. Horses kept in confinement and not exercised may develop *leg edema* as a result of insufficient muscular exercise to pump the excess fluid (lymph) back into the circulation. The edematous condition is called *stocked up* and is usually corrected by mild exercise. Other causes of interference with lymphatic drainage are injury, venous valve incompetence, wrapping legs too tight with bandages, and pregnancy.

Effects of Exercise   The changes in the heart rate before, during, and following exercise are shown in Figure 3-9. Prior to exercise, heart rate increases slightly as a result of anticipation. As exercise starts, there is a rapid increase in heart rate, which reaches a peak in approximately 30 to 45 seconds. Then a slight decrease occurs, followed by a period of steady state. After cessation of exercise, the heart rate drops rapidly for 1 or 2 minutes and then gradually returns to its pre-exercise rate within the next few minutes to over an hour. A number of factors are discussed below which influence the peak heart rate, steady-state rate, and rate of decline.

Heart rate is correlated with exercise intensity, since it increases linearly with increased work when the horse is running or swimming until the maximum heart rate is approached. Maximum heart rates are highly variable and normally range between 212 and 240 beats per minute during running and up to approximately 210 during swimming. During exercise, physical

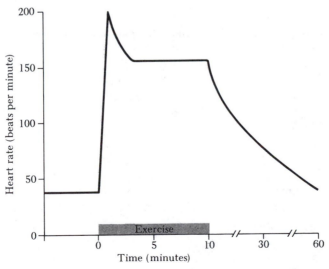

Figure 3-9
Heart-rate response to a 10-minute exercise bout.

factors such as exercise conditioning, duration and intensity of the exercise bout, ambient temperature, and humidity, as well as psychological factors, affect the maximum heart rate. Stroke volume and cardiac output are increased during exercise due to an increase in heart rate and in stroke volume. A cardiac output of 253 liters has been recorded from a horse with a heart rate of 199 during tethered swimming. As exercise intensity increases, heart rate and cardiac output increase linearly over the range of 110 to 210 beats per minute. At heart rates of 240 or greater, observed during very strenuous exercise bouts, the stroke volume may decrease due to insufficient filling time for the ventricles.

During exercise, blood pressure increases from the normal value of 140 mm mercury for the left ventricular systolic pressure. It increases to about 220 during treadmill exercise and in excess of 300 during strenuous swimming. Carotid blood pressure is the only blood pressure that is highly correlated to work effort. At rest, it averages 180 mm mercury.

The pattern of blood distribution changes during exercise to accommodate the increased metabolic demand of exercising muscles. Increases of 75-fold above resting values have been observed for the fore- and hindlegs during severe exercise. The diaphragm muscle has the largest increase in blood flow during exercise. Blood flow to the skin increases to improve the efficiency of heat loss. Distribution of blood to nonexercising organs such as the kidneys and the digestive tract are dramatically reduced. Contraction of some of the large-capacity veins shifts blood into the central circulation. Contraction of the spleen, a large storage depot for red blood cells, results in a dramatic rise in the hematocrit and the hemoglobin concentration.

Training Effect   Changes in heart rate and stroke volume in response to exercise training have not been completely resolved. Generally, exercise training results in no change or a slight decrease in resting heart rate. Aerobic training for as little as 5 weeks results in a significant decrease in the heart rate for a given exercise intensity when compared with the rate before conditioning. Stroke volume increases so that cardiac output can be maintained at the lower heart rate. Training has no effect on maximal heart rate.

Heart size is affected by training. A maximal increase in heart size (approximately 0.2 percent increases in the heart-weight to body-weight ratio) occurs after a limited training stimulus.

Training induces an increased capillarization of the muscles. The increased size of the capillary beds results in a longer transit time for the blood to flow through the capillaries, thus aiding in the exchange of gases, metabolic products and nutrients. It normally takes 3 to 4 months of training for increased capillarization to occur.

Assessment of Fitness and Fatigue   An assessment of the fitness of performance horses can be obtained by taking advantage of the relationship of work intensity and heart rate and the effect of degree of physical fitness on the relationship (Figure 3-10). To evaluate a horse, a standardized exercise test must be used and an "on board" heart-rate monitor must be connected to the horse. When a treadmill is used, either it is set at a slight incline and the speed progressively increased at timed intervals until the maximum heart rate is attained or it is set at a high rate of speed while the heart rate is monitored until fatigue. On the race track it is difficult to control the conditions affecting the horse, and the test is not as accurate or as easy to administer. Standardized performance tests for cutting horses have been developed and are being utilized to assess fitness. During the test, the cutting horse works a mechanical cow during repeated 90-second exercise bouts of simulated cutting work at 240 meters per minute, including 22 stops and turns. Each bout is followed by a 30-second recovery period. Fitness is assessed by the number of exercise bouts that the horse can perform before fatigue, which is defined as when the heart rate at the end of the exercise period and/or at the end of the 30-second recovery period is increased for two consecutive times. A typical response to the test is shown in Figure 3-11.

Evaluation   The *electrocardiogram* is a graph of the average electric potential (volts) generated in the heart muscle versus time during the different phases of the cardiac cycle (Figure 3-12). Combinations of electrodes attached to specific areas of the body, referred to as *leads,* measure the

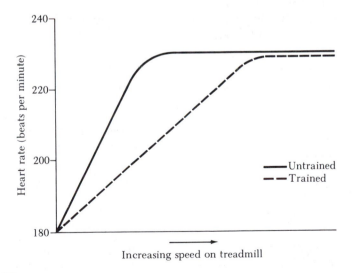

Figure 3-10
Influence of training on heart rate. Note that it takes faster speeds for the heart rate of a trained horse to reach maximum value and at a given speed, the heart rate of a trained horse is lower.

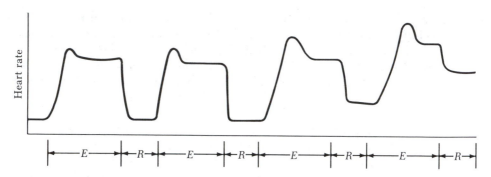

Figure 3-11
Heart-rate response to a cutting horse exercise test to determine physical fitness. Note that fatigue was evident when the heart rate did not recover to the resting value after the third and fourth exercise bouts and the peak heart rate was increasing during the third and fourth exercise bouts.

electrical activity from different angles in the body. The ECG from each lead is composed of continual changes in the voltage (i.e., the P-wave results from atrial depolarization, the QRS complex represents ventricular depolarization and the T-wave depicts ventricular repolarization). The ECG data are used as diagnostic tests to detect and define cardiac arrhythmias, electrical conduction disturbances, chamber sizes, and pathological states. Ultrasonography is also used for diagnostic purposes. The beating heart can be visualized so that cardiac lesions and diseases can be identified and quantified.

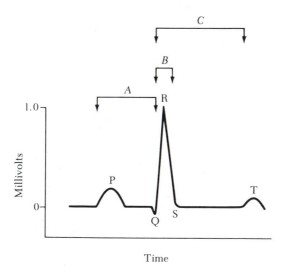

Figure 3-12
Lead II ECG tracing. Wave forms consist of P, QRS, and T. Negative electrode is on the right front limb and positive electrode on the left rear limb. The P-wave is indicative of atrial depolarization, the QRS complex is ventricular depolarization, and the T-wave is ventricular repolarization. A = P-R interval, B = QRS complex, and C = Q-T interval. Normal values in seconds are: P-wave duration, 0.14 + .02; P-R interval, 0.29 + .005; QRS duration, 0.113 + .01 and Q-T interval, 0.5 + 0.04.

Blood

Blood consists of two major components, plasma and cells. *Plasma* consists of water, ions, proteins (albumin, globulins, and others), and lipids. The *cells* are classified as *esinophils, erythrocytes, monocytes, neutrophils,* and *thrombocytes.*

When the *splanchnic reservoir* (blood stored in the spleen) is taken into account, blood volume varies from 12 to 14 percent of body weight (i.e., 12 to 14 liters per 100 kilograms body weight). Failure to mobilize the red blood cells from the spleen results in much lower values ranging from 6 to7 percent. Plasma volume ranges between 4.1 and 4.9 percent of the body weight in resting horses. It is slightly increased when the spleen is contracted. The red-blood-cell volume ranges between 1.8 and 2.8 of the body weight at rest and up to 8 percent upon splenic mobilization. The packed cell volume is slightly larger than the red-blood-cell volume since it includes the white blood cells.

Erythrocytes (red blood cells) contain hemoglobin and thus transport oxygen in the blood. They remain in the system approximately 120 to 150 days and are then removed by the spleen and liver. *Neutrophils* are produced in response to infections but other factors such as excitement, exercise, and stress, as well as changes in the rate of their production by bone marrow or their half-life in the blood also alter their concentration in the blood. Little is known about the remaining cells present in the blood of the horse.

Effects of Exercise   The intensity and duration of exercise determines the type of hematological response. High-intensity, short-duration bouts of exercise result in rapid mobilization of red blood cells from the spleen (within 1 to 2 minutes) to increase the oxygen-carrying capacity of blood. Thus, the packed cell volume and the hemoglobin concentration increase. Blood viscosity increases as a result of an increase in plasma viscosity and hematocrit.

Endurance horses working at slow speeds for long periods experience a different hematological response. Spleenic mobilization is not complete, so the hematocrit does not increase as much. Dehydration is a common occurrence during elevated temperatures, along with loss of electrolytes by sweating. Thus, plasma volume decreases and the hematocrit increases.

Effects of Training   The horse that is well conditioned for high-intensity exercise bouts is characterized by an increased red-blood-cell and hemoglobin concentration and an increased blood volume. Well-trained endurance horses have very little change in red-cell parameters.

Evaluation of Performance Potential   No correlation between racing performance and any hematological parameter has been found. However, racehorses with resting hematocrits above 0.47 and below 0.36 are less likely to perform up to expectations. This is associated with the overtraining syndrome observed in Standardbred horses and is usually corrected by rest periods of several months. Horses with a hematocrit that is consistently below 0.36 usually perform below their potential and are commonly referred to as being anemic. This results in a lowered oxygen-carrying capacity in the blood.

Fluid Volumes

There are several fluid compartments in the horse's body. Changes in their size and constituent concentrations affect cardiovascular function and may affect athletic performance. The amount of water in the horse's body (total body water) ranges between 600 to 710 milliliters per kilogram of body weight and accounts for approximately 66 percent of body weight. The daily water turnover is 26 to 32 liters per day, but this is affected by many factors such as body weight, diet, temperature, humidity, and exercise. The *extracellular fluid volume* is the space consisting of the plasma, interstitial fluid, lymph, and transcellular fluids, which include the fluid content of the gastrointestinal tract. Normally, it ranges from 20 to 30 percent of body weight. *Intracellular fluid volume* is the difference between total body water and extracellular fluid volume. Very little is known about intracellular fluid volume changes in the horse.

Effects of Exercise   A Thoroughbred may lose up to 15 liters of water during a one- to two-mile race. Endurance horses lose 10 to 15 liters per hour during a race and may have a water deficit of 20 to 40 liters by the end of 80- to 160-kilometer races. The initial fluid deficit resulting from sweating during an endurance race is in the plasma volume. Then fluids shift from other components of the extracellular fluids, and probably a small amount from the intracellular fluid space. The fluid content of the gastrointestinal tract serves as a major fluid reservoir to replace plasma fluid loss.

Effects of Training   Training does not result in changes in the amount of body water. However, today body water as a percentage of body weight does increase during training as a result of a decrease in body fat. No significant changes occur in the extracellular fluid volume with training.

3.6   Digestive System

The gastrointestinal tract is a musculo-membranous tube that extends from the mouth to the anus. In the horse, this mucous-membrane-lined tube is approximately 100 feet long and functions in ingestion, grinding, mixing, digestion, and absorption of food, and elimination of solid waste. The digestive organs of the horse are the mouth, pharynx, esophagus, stomach, small intestine, cecum, large intestine, and anus (Figure 5-1). The digestive system is discussed more fully in Section 5.1.

3.7 Urinary System

The urinary system consists of a pair of kidneys, the ureters, the bladder, and the urethra. The kidneys provide a blood-filtering system that is responsible for the excretion of many waste products from the body. The kidneys control water balance, pH, and the levels of many electrolytes. They cleanse the blood and are responsible for the stabilization of blood composition. The kidneys are located in the loin region of the horse, and many stiff and sore backs have been attributed to kidney disease. In fact, the horse suffers very few diseases of the urinary system, and renal (kidney) diseases are extremely rare. Azoturia is a metabolic disease, severe cases of which can result in renal failure, but it is not primarily a kidney disease (see Chapter 16).

The kidney filtrate is urine. It is conveyed to the bladder by two muscular tubes, the ureters. The bladder is a flexible, distensible, muscular storage organ for urine. At the time of micturition (urination), the muscular walls of the bladder contract and urine is carried to the exterior by the urethra. The external organ of the urinary system is the penis in the stallion and gelding and the vulva in the mare.

3.8 Nervous System

The nervous system is an extensive control mechanism that can perceive and immediately react to changes in the external and internal environment of an animal. The nervous system also stores and associates sensations in the memory for future use. The equine nervous system is highly sophisticated, as evidenced by the outstanding coordination of motor activities.

The horse is unusually sensitive, and its acute tactile perception, coupled with an unusual learning ability, have enabled it to be domesticated and utilized by mankind. Its superb athletic ability is fortunately combined with a willingness and desire to please, but it is the sensitivity of the horse to touch and pressure, its low tolerance for pain, and its outstanding memory that make the horse the most tractable of the large domestic animals.

The functional divisions of the nervous system are the central nervous system, the peripheral nervous system, and the specialized sensory organs.

The *central nervous system* consists of the brain, brain stem, and spinal cord. The brain is enclosed and protected by the cranial cavity, and is not unusually large for an animal of its size. As a result the reasoning intelligence of the horse is somewhat limited, but the horse is amazingly trainable and possesses a remarkable memory. The central nervous system coordinates the activities of the peripheral nervous system, as it is the brain that processes, integrates, and stores sensory information. The lower brain (brain stem) subconsciously coordinates and controls many of the life processes (respiration, blood pressure, and so on).

The spinal cord connects the brain with the peripheral nerves, and at this level many of the simpler motor responses occur as "spinal reflexes" without directly involving the higher "conscious" levels of the nervous system. For example, the basic patterns of locomotion appear to occur as a spinal reflex and do not require conscious coordination on the part of the horse. The panniculus muscle of the horse underlies the skin and is capable of shaking off a fly that lands on the skin. This reflex action exemplifies the tactile sensitivity of the horse and the action of the nervous system below the conscious level.

The *peripheral nervous system* provides a network of communication between the internal or external environment and the central nervous system. The peripheral nervous system consists of the spinal and cranial nerves and their sensory or motor endings. Pain is an example of an internal stimulus. A horse is extremely sensitive to pain and appears to have a low tolerance for internal pain. The horse's ability to detect vibrations on the ground long before the cause can be seen or heard as an example of response to an external stimulus. This stimulus from the external environment is transmitted by the peripheral nervous system to the central nervous system. In this instance, there is no unconscious spinal reflex. Rather, the impulse is transmitted to the brain. The animal then becomes aware of approaching danger and makes the appropriate response.

The autonomic nervous system is the involuntary portion of the peripheral system that is associated with control of the glands, blood vessels, heart, sweating, body temperature, gastrointestinal motility, urinary output, and smooth muscle activity. The autonomic system provides involuntary control of body actions below the level of consciousness, particularly in time of stress.

Special Senses   The horse has survived throughout history because of its ability to perceive its environment and respond accordingly. Because the horse had to run away from danger rather than fight its enemies, survival was often based on its perception of sight, sound, or smell. The eyes, ears, and nose are *specialized sensory organs* of the nervous system that are little studied and often misunderstood in the equine.

Hearing   The horse has a well-developed sense of hearing and is capable of hearing sounds at frequencies above those perceived by man. Knowing the range of sounds that horses can hear can be useful in identifying sounds that a horse might respond to or be disturbed by. For example, the horse might be disturbed by ultrasonic rodent or pest repellers, depending on how close the repeller is to the horse and how loud it is. While most sounds audible to horses are also audible to humans and vice versa, several differences in hearing sounds do exist. Beginning at the low-frequency end of sound, auditory sensitivity of the horse improves gradually from a lower limit of 55 hertz (Hz) to about 1 kilohertz (kHz) at an intensity of 60 decibels. Man is more sensitive in that the lower limit of auditory sensitivity is about 29 Hz. The best range of hearing for the horse is from 1 to 16 kHz while it is from 500 Hz to 8 kHz for man. Above 16 kHz, sensitivity decreases rapidly until an upper limit of about 33.5 kHz. Thus, the practical hearing range for the horse is 55 Hz to 33.5 kHz. Above 8 kHz, horses are better able to hear sounds than humans, because the hearing limit of humans is 19 kHz. As horse grow older, they lose sensitivity at higher frequencies. For unexplained reasons, stallions seem to hear better than geldings and mares.

Horses have a good ability to localize the source of sound. There are ten muscles controlling each of their ears so that they turn their ears in almost any direction. There is good coordination between movements of the eyes and ears as a horse directs its attention toward something. Usually, the horse can detect the direction of sound before it can visually detect the source and can hear sounds from great distances.

Olfaction   Very little is known about a horse's sense of smell but it is generally agreed that its sense of smell is well-developed. Horses use smell to aid in locating food and to identify each other and humans. Thus, it is part of the normal behavior of mares to smell their foals when the foals start to suckle. Horses also like to approach and smell a person for identification purposes. They are able to associate a medicinal smell with veterinarians and their procedures. If the associations have been unpleasant, the horse may become quite nervous and difficult to control when approached by a veterinarian.

The *vomeronasal organ* is part of an accessory olfactory system containing olfactory receptors. In the horse and donkey, the opening into the vomeronasal organ is located in the nasal cavity caudal to the alar cartilages. The organ's function seems to be related primarily to sexual and social activities and is most likely to be involved in behavioral urinalysis by the stallion. When a mare is teased or subjected to urinalysis to determine if she is in estrus, a curling of the upper lip occurs. This response is called the *Flehmen response.*

Vision   The eyes of the horse are unique. Their anatomy (Figure 3-13) and the physiology of equine visual mechanisms are not fully understood.

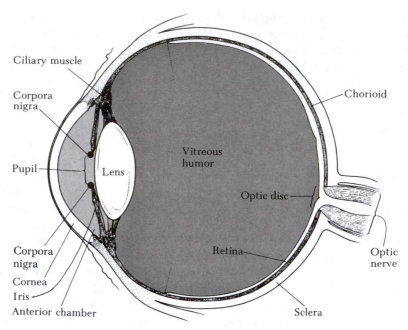

Figure 3-13
Horizontal section of the equine eye.

In horses as well as other mammals, light rays enter the lens of the eye and are focused on the retina. The shape of the lens is altered by the ciliary muscle surrounding it, so that the eye can focus on objects at variable distances. The horse has a fairly continuous focus from a distance of about 3 feet forward. At distances of less than 3 feet, the ability to focus may be limited. Inside 1.5 feet, fine focusing power may be limited and result in blurring. However, this does not prevent the horse from accurately responding to the demands of man. It doesn't need much capacity to see up-close details.

Visual acuity in the horse is the poorest of all the domestic species, since horses have the lowest concentration of nerve cells relaying image information from within a centrally located field. In the equine retina as in other mammalian retinas, the ganglion cells as well as their nervous components are of a higher density within a centrally located field, the visual streak. This central area is shaped like a narrow horizontal bar in the horse. It lies above the optic disc and extends in both nasal and temporal directions. A field of nerve cells of relatively higher density, which serves as an area of higher visual acuity, is found within the temporal arm. This area is closely related to the region used for binocular vision and provides a particularly acute image of part of the environment that is important to the horse. The location of the visual streak coincides with the lower part of the reflective layer of the tapetum. Functionally, this means better exploitation of the incoming light, but, generally, loss of resolution by scattered light.

Because the retina is transparent, only a fraction of the light reaching it is absorbed. The rest of the light passes on to be absorbed without contributing to retinal stimulation. The horse possesses an anatomic device, the *tapetum,* for reflecting this otherwise wasted light back onto the receptors. So, they receive almost twice the light stimulation that they would receive without a tapetum. What is not absorbed by the retina, after the light has passed back by the receptors for a second time, passes out through the transparent tissues on a forward path through the pupil and out of the eye again. This returned light is what is seen when light shines directly into a horse's eye in the dark and it appears to glow.

If one reduces the many refracting elements of the eye to a single refracting curve at the corneal surface, a reversed and inverted image will form on the retina. When the image is not formed at the plane of the retinal receptors, then a visual error or refractive error occurs. The error is named according to whether the image falls in front of the retina (*myopia*) or behind it (*hypermetropia*). As horse approach maturity, they become increasingly hypermetropic.

The entire spatial area from which the complete visual image of an eye is formed is known as the *field of vision.* In the horse, it is about 215 degrees for each eye. The wide-set eyes of the horse enable it to enjoy a panoramic field of vision (Figure 3-14), even to the extent of seeing everything around itself with slight head movements. Only what is immediately behind the horse's hindquarters is outside its field of view, until it moves its head and neck. Thus, it is almost impossible to approach a grazing horse with its eyes above the grass without the horse knowing it. By taking advantage of its monocular vision capabilities, the horse is capable of an independent view from each eye, enabling it to watch a show-ring steward with one eye and spectators in a stadium with the other eye. It usually takes experience in the show-ring before the horse will focus its attention forward and use binocular vision.

There is an area forward of the head where the visual fields overlap. The impression of the images must superimpose perfectly in the cerebral cortex where these fields overlap, or there will be double vision. This area of overlap forms a zone of simultaneous binocular vision that is intimately involved with depth perception and judgment of position. Because it has little overlap of the visual fields (60 to 70 degrees) for a three-dimensional view, the horse has poor depth perception. However, horses can detect movement at great distances.

The visual field is influenced by the shape of the head, size of the eye and jaw, and position of the head. The visual field is restricted in certain areas as the size of the nostrils and muzzle and the length of the head

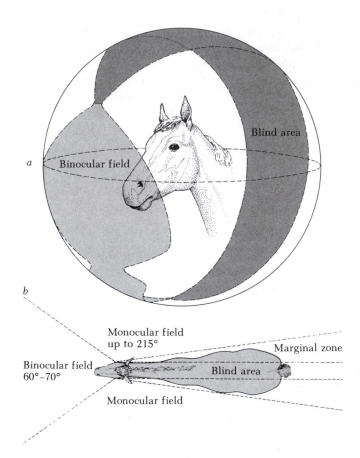

Figure 3-14
Visual field of the horse. (*a*) Ophthalmoscopically defined ocular field; (*b*) panoramic visual field. (Adapted from G. H. Waring, *Horse Behavior*, Noyes Publications,)

increase. It is increased as the eyes become set wider apart. Larger eyes increase the visual field. The so-called "pig-eyed" condition, in which the eyes are small and set back into the head, will restrict the field of vision. Wider and deeper jaws will restrict the visual field in the area located below the head. Because of all these factors, the horse will have a blind spot on the ground from the front legs to about four feet forward of the head.

Horse's eyes adjust to changing light much more slowly than those of humans. They have a rectangular-shaped, horizontally oriented pupil, which cannot constrict very much, even when a beam of light is directed at the eye. When a horse moves from an area of bright light into an area of low light, or vice versa, the horse takes longer to adjust its vision than the handler does. Thus, one should give horses time to adjust to light changes before asking them to move into unfamiliar places such as stalls or trailers. *Corpora nigra* or *granula iridica,* are black nodules that are found on the upper and lower margins of the pupil so they are on the iris. They may function to decrease the amount of light entering the eye.

The question of the horse's ability to see color has not been completely answered. One test has indicated that horses can differentiate between shades of yellow and green, but not red and blue. Other tests indicate they can see blue and can distinguish between a color and gray of equal brightness.

The eye is lubricated by lacrimal fluid from the *superior* and *inferior lacrimal glands* located above the eyes (Figure 3-15). This fluid is drained from the eye via two tiny holes. The ducts leading from these two holes come together to form the *lacrimal duct*. The lacrimal duct drains into the nasal

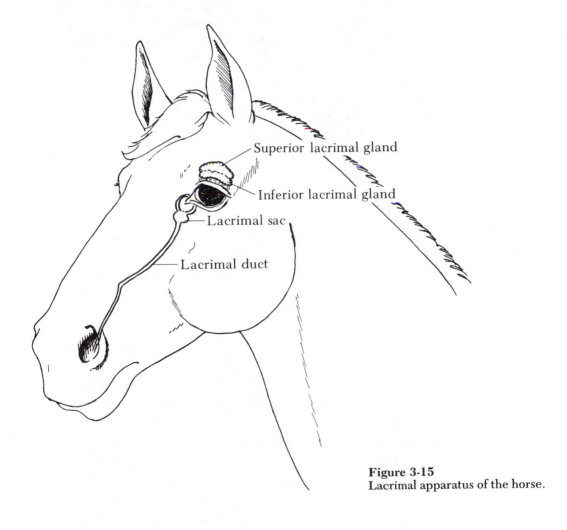

Superior lacrimal gland

Inferior lacrimal gland

Lacrimal sac

Lacrimal duct

Figure 3-15
Lacrimal apparatus of the horse.

cavity at the inside lower posterior margin of the nostril. If tears from the eye pour out on the horse's face, the duct may be plugged with dirt or debris and may need to be back flushed to open it up. The lacrimal fluid in conjunction with the nictitating membrane located at the inside corner of the eye keep the eye cleansed of foreign matter. Contraction of the *nictitating membrane* moves it across the eye, wiping foreign substances off the cornea.

3.9   Endocrine System

Whereas the nervous system provides immediate response to the environment, the endocrine system exercises long-range control over the body systems. The endocrine system is the name applied to a number of ductless glands of the body that produce chemical substances called *hormones,* which are transported by the circulatory system and can influence a variety of body functions. The endocrine system, in combination with the nervous system, is important in maintaining internal body stability or homeostasis.

Endocrinology is a sophisticated science whose detailed analysis is beyond the scope of this book. However, it should be recognized that the endocrine glands control such important functions as growth, reproduction, metabolism, and digestion. The endocrine tissues, the hormones secreted, and their major physiological actions are listed in Table 3-4.

Table 3-4
Endocrine tissues in the horse: The hormones and their action

Endocrine tissue	Hormone secreted	Tissues influenced	Action
Hypothalamus (brain)	Releasing factors	Anterior pituitary	Releases specific tropic hormones
		Posterior pituitary	Releases oxytocin, vasopression
Pituitary			
Anterior	Growth hormone (GH)	All	Promotes increase in protein and fat in body; promotes growth of muscle and bone
	Thyrotropin (TSH)	Thyroid gland	Maintenance of the secretion of the thyroid gland
	Adrenocorti-cotropin (ACTH)	Adrenal cortex	Maintenance of the secretion of the adrenal cortex
	Follicle stimulating hormone (FSH)	Ovary	Stimulates follicle development

Table 3-4
Endocrine tissues in the horse: The hormones and their action (*Continued*)

Endocrine tissue	Hormone secreted	Tissues influenced	Action
		Testis	Stimulates spermatogenesis and inhibin secretion
	Luteinizing hormone (LH)	Ovary	Ovulation and formation of corpus luteum
		Testis	Stimulates testosterone production and spermatogenesis
	Prolactin (PRL)	Mammary gland	Milk formation in alveoli
Posterior	Oxytocin	Smooth muscle	Stimulates smooth muscle contraction in uterus and mammary gland
	Vasopressin	Blood vessels	Vasoconstrictor; antidiuretic
Pancreas	Insulin	All	Controls glucose level in body
Thyroid	Thyroxine	All	Controls metabolic rate
Parathyroid	Parathyroid hormone	Kidney, bone	Maintains calcium and phosphorus levels
Adrenal Glands Cortex	Corticoids	All	Controls salt, glucose, and water levels of body
Medulla	Epinephrine	Skeletal muscle	Controls blood flow and strength of muscular contraction Mobilizes energy stores
Ovaries Follicle	Estrogens	Mammary gland	Duct development
		Genital tract	Estrus behavior/prepares for fertilized ova
Corpus luteum	Progesterone	Mammary gland	Alveolar development
		Genital tract	Diestrus/maintains pregnancy
Testes	Androgens	Sex organs	Sperm production, sex drive
		Body	Secondary male characteristics
Uterus	Pregnant mare's serum gonadotropin (PMSG)	Ovary, uterus	Complements other gonadotropin action
	Prostaglandin (PGF$_{2\alpha}$)	All, corpus luteum	Numerous functions/ inhibits corpus luteum action

3.10   Reproductive System

The male reproductive system consists of the testicles, the accessory glands and ducts, and the external genital organ (see Chapter 10). The female reproductive system consists of the ovaries, oviducts, uterus, vagina, and external genitalia (see Chapter 19).

The primary sex organs, the testicles and ovaries, produce the germinal or sex cells (ova and spermatozoa) as well as the sex hormones. The sex cells contain the genetic material that unites during fertilization to form a new individual. The mare's ovaries produce ova or eggs and the female sex hormones estrogen and progesterone. The testicles of the stallion produce spermatozoa as well as the male sex hormone testosterone. Because of the complexity and importance of the reproductive process, Chapters 9, 10, and 11 of this book are devoted specifically to that topic.

3.11   The Integument

The integument is the skin and hair that covers the horse's body and forms the boundary between the animal and its environment. The integument provides protection from mechanical, chemical, and physical agents. The skin is important in thermoregulation and in the sensation of heat, cold, pain, and touch. The mane, tail, and body hair provide effective protection against insects and play a prominent role in temperature regulation.

The skin of the horse is remarkably strong and highly sensitive. The skin consists of two layers, the *epidermis* or outer layer and the underlying *dermis*. The epidermis consists of stratified squamous eipthelium cells and is divided into a deep, growing layer and a hard, tough, cornified surface layer. The surface has no blood supply, and as a result these flat, dry cells are continuously being shed.

The hair and hooves of the horse grow directly from the epidermal layer, and the epidermis is modified to form hornlike growths on the inside of the legs (chestnuts) and at the back of the fetlock (ergots). Chestnuts are located on the inside surface of the front legs above the knees and on the rear legs on the bottom inside portion of the hocks. The chestnut is a modification of only the epidermal layer and there is no evidence for the theory that represent vestiges of missing digits from extinct species of horses. Ergots are located at the back of the fetlocks, and their size differs between breeds. The ergot is hidden in the fetlock hair (feathers) and often goes undetected.

The *dermis* is the deep connective tissue portion of the skin that attaches to the panniculus carnosus, a sheet of skeletal muscle that separates the rest of the body tissues from the skin. The dermis contains arteries, veins, capillaries, lymph veins, sensory nerve endings and fibers, hair follicles, and sweat and sebaceous glands. It is the active portion of the skin that is vital for a healthy, pliable, elastic skin and rich, bright hair coat. The skin of the horse is highly sensitive because it contains many sensory endings. There are also sensory nerve connections in the hair follicles and on the tactile hairs of the muzzle.

The sweat glands are located over the entire body of the horse except the legs. The evaporation of sweat and the piloerection of the hair are the horse's primary means of temperature regulation. A horse sweats more readily than any other farm animal. The sweat glands produce a salty, alkaline, watery fluid that is discharged directly on the skin surface and imparts a characteristic "horsey" odor. A horse sweats most readily at the base of the ears and on the neck, chest, and flanks. The evaporation of sweat from the body has a cooling effect on the surface of the body.

Hair covers most of the skin area except underneath the tail, around the genitals, and on the inside of the thighs. The fibrous and bulky nature of the hair provides excellent protection against cuts and abrasions and is an effective insulator against the elements. In the dermis are bundles of smooth muscle fibers, called *erectores pilorum,* that attach to the hair follicle and the surface of the skin in such a manner that their contraction causes an erection of the hair. The piloerection of the hair increases the insulating effect by reducing air convection currents across the skin as well as heat loss in cold weather.

There are two types of body hair, the dense undercoat and the less prevalent long "guard" hairs. The mane, tail, eyelashes, and tactile hair of the muzzle are permanent but the general body hair is shed twice a year, in the spring and fall. Hair color is due to pigmented *melanin* granules. The variation in the hair color of different horses is due to differences in the amounts and location of the melanin granules. These color differences are under genetic control (The inheritance of coat color is explained in Chapter 13). The skin is pigmented in all coat colors except under some white markings or spots.

The sebaceous glands are located in the same places as the hair follicles and open directly on the surface of the skin. They produce *sebum,* an oily, waxy secretion that coats the hair, protects it from overwetting, and increases its insulating ability. The sebum adds sleekness and luster to a horse's coat; vigorous brushing and rubbing can increase secretion and make the coat "bloom."

REFERENCES

Barnett, K. C. 1972. The ocular fundus of the horse. *Equine Vet. J.* 4(1):17–20.

Bechtel, P. J., and K. H. Kline. 1987. Muscle fiber type changes in the middle gluteal of Quarter and Standardbred horses from birth through one year of age. *Equine*

Exercise Physiology 2. J. R. Gillespie and N. E. Robinson, eds. Davis, California: ICEEP Publications.

Bone, J. F., ed. 1963. *Equine Medicine and Surgery.* 1st ed. Wheaton, Illinois: American Veterinary Publications.

Carlson, G. P. 1987. Hematology and body fluids in the equine athlete: a review. *Equine Exercise Physiology 2.* J. R. Gillespie and N. E. Robinson, eds. Davis, California: ICEEP Publications.

Catcott, E. J., and J. F. Smithcors, eds. 1972. *Equine Medicine and Surgery.* 2d ed. Wheaton, Illinois: American Veterinary Publications.

Cavalry School. 1935. Horsemanship and horsemastership. *Animal Management.* Vol. 2, Part 3. Fort Riley, Kansas.

Dukes, H. H. 1955. *The Physiology of Domestic Animals.* Ithaca, New York: Cornell University Press.

Edwards, Gladys Brown. 1973. *Anatomy and Conformation of the Horse.* Croton-on-Hudson, New York: Dreenan Press.

Frandson, R. D. 1965. *Anatomy and Physiology of Farm Animals.* Philadelphia: Lea and Febiger.

Fregin, G. F. 1988. Cardiovascular and respiratory responses to exercise and training in the horse. *Association of Equine Sports Medicine Quarterly* 3(1):7–9.

Gunn, H. M. 1987. Muscle, bone and fat proportions and the muscle distribution of Thoroughbreds and other horses. *Equine Exercise Physiology 2.* J. R. Gillespie and N. E. Robinson, eds. Davis, California: ICEEP Publications.

Heffner, H. E., and R. S. Heffner. 1983. The hearing ability of horses. *Equine Practice* 5(3):27–32.

Houpt, K. A., and L. Guida. 1984. Flehmen. *Equine Practice* 6(3):32–35.

Jones, W. E. 1988. The cardiovascular system. *Equine Sports Medicine.* W. E. Jones, ed. Philadelphia: Lea and Febiger.

Rose, R. J. 1987. Cardiopulmonary system. *Equine Exercise Physiology 2.* J. R. Gillespie and N. E. Robinson, eds. Davis, California: ICEEP Publications.

Kays, D. J. 1969. *The Horse.* Rev. ed. New York: A. S. Barnes.

Kline, K. H., L. M. Lawrence, J. Novakofski, and P. J. Bechtel. 1987. Changes in muscle fiber type variation within the middle gluteal of young and mature horses as a function of sampling depth. *Equine Exercise Physiology 2.* J. R. Gillespie and N. E. Robinson, eds. Davis, California: ICEEP Publications.

Knill, L. M., R. D. Eagleton, and E. Haver. 1977. Physical optics of the equine eye. *Amer. Vet. Med. J. Res.* 38:735–737.

Lindsay, F. E. F., and F. L. Burton. 1983. Observational study of "urine testing" in the horse and donkey stallion. *Equine Vet. J.* 15:330–36.

Montavon, S., and F. Barrelet. 1986. Towards a better knowledge of vision in the horse. *Pratique Veterinaire Equine* 18(4):177–182.

Oldberg, F. O. 1978. A study of the hearing ability of horses. *Equine Vet. J.* 10(2):82–84.

Prince, J. H., et al. 1960. *Anatomy and Histology of the Eye and Orbit in Domestic Animals.* Springfield, Illinois: Charles C Thomas.

Rose, R. J., and D. L. Evans. 1987. Cardiovascular and respiratory function in the athletic horse. *Equine Exercise Physiology 2.* J. R. Gillespie and N. E. Robinson, eds. Davis, California: ICEEP Publications.

Rossdale, Peter D. 1972. *The Horse.* Arcadia, California: California Thoroughbred Breeders Association.

Sivak, J. G., and D. B. Allen. 1975. An evaluation of the "ramp" retina of the horse eye. *Vision Research* 15:1353–56.

Snow, D. H., S. G. B. Persson, and R. J. Rose. 1982. *Equine Exercise Physiology.* Cambridge, England: Granata Editions.

Swenson, Melvin J. 1970. *Duke's Physiology of Domestic Animals.* 8th ed. Ithaca, New York: Cornell University Press.

Way, Robert F., and Donald G. Lee. 1965. *The Anatomy of the Horse.* Philadelphia: Lippincott.

Wilson, R. G., J. R. Thornton, S. Inglis, and J. Ainscow. 1987. Skeletal muscle adaptation in racehorses following high intensity interval training. *Equine Exercise Physiology 2.* J. R. Gillespie and N. E. Robinson, eds. Davis, California: ICEEP Publications.

CHAPTER 4

SELECTION

Selection is the most important decision made in the horse business and yet it is a skill in which many are inexperienced and untrained. Most horse owners become proficient in first aid, parasite control, feeding, training, and conditioning but have little knowledge or practice in selecting a horse. The choosing of a horse is a once-in-a-lifetime decision for many and one is stuck with that horse for better or worse. Professionals have an advantage in selecting horses because those who operate breeding, training, or riding programs are constantly observing and evaluating horses and have the opportunity to see how conformation relates to performance.

Selection is choosing the best horse for the intended purpose. It includes consideration of use, temperament, training, age, sex, and breeding, as well as evaluation of the horse's structure or conformation. Judging is analyzing and evaluating performance, conformation and way of going and requires a trained eye along with knowledge of the relationship between form and function.

4.1 General Considerations

Use

The purpose for which a horse is to be used is the single most important factor in selecting a horse. The intended use determines the importance of other considerations such as: general appearance, temperament, training, breed or type, pedigree, age, sex, color, size, conformation, soundness, and way of going. Selecting a potential race horse at one of the yearling sales requires entirely different criteria than selecting the first horse for a young

132

rider. Because they have such a variety of uses a horse unsuitable in one instance may be entirely satisfactory in another.

General Appearance

The general appearance of a horse is less important to many horse people than its training or athletic ability, but all appreciate a horse that is balanced and possesses symmetry (see Figure 4-1). A balanced horse is proportionally as long as it is tall. Thus, the height at the withers, height at the hip, and length of the body should be approximately the same. Willoughby (1975) has collected a tremendous amount of data on body measurements for a number of breeds (Table 4-1). It is interesting to note that the Arabian body length is just slightly shorter ($\frac{1}{2}$ inch) than the height at the withers, whereas

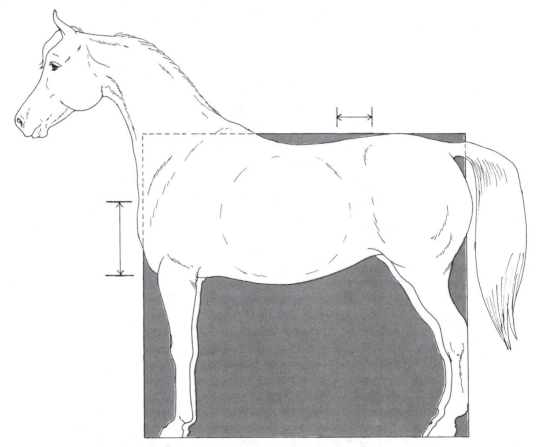

Figure 4-1
Balance and symmetry of a horse.

Table 4-1
Breed comparison of body measurements and proportions

	Arabian	Thoroughbred	Standardbred	Morgan	Quarter Horse	Appaloosa	Draft	Shetland
Height at withers, inches	59.70	63.78	62.7	60.0	59.30	62.0	64.85	39.84
Length of body, inches	59.28	63.78	63.9	62.28	63.25	63.8	70.88	43.23
Height at hip, inches	59.40	63.50	62.5	59.72	59.06	61.75	65.37	40.04
Length of head, inches	23.34	24.52	25.01	23.77	23.43	24.49	27.50	17.65
Length of head as a percentage of withers height	39,19	38.45	39.7	39.6	39.8	39.5	42.4	44.3

the Thoroughbred is exactly as long as it is high. In the Standardbred the body is about an inch longer than the height at the withers; the Morgan and Appaloosa are about two inches longer than high; and the Quarter Horse is nearly 4 inches longer than its height at the withers. A horse low at the withers or high at the hip is said to be "walking downhill" and has a tendency to forge when it moves. The riding-horse breeds are all slightly higher (0.2 to 0.3 inch) at the withers than at the hip, while both the Shetland pony and Draft horse are higher at the hip than withers.

The size of the head varies among breeds, but it should be in proportion with the body. There has been considerable effort to establish a relationship between body parts in the "ideal" horse. A general rule of proportion was proposed by J. Wortley Axe (1905), based on measurements by French hippotomists of the nineteenth century, in which the head was used as the basis of proportion for all other parts (Figure 4-2). The following proportions were proposed:

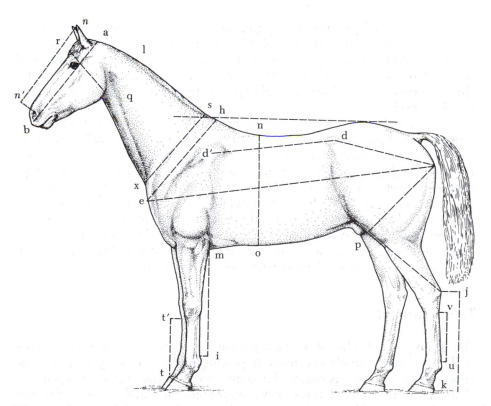

Figure 4-2
Proportions of the horse in profile. (From Goubaux and Barrier)

The length of the head almost exactly equals the distance:

1. From the back to the abdomen—N to O
2. From the top of the withers to the point of the arm—H to E
3. From the superior fold of the stifle-joint to the point of the hock
   —J' to J
4. From the point of the hock to the ground—J to K
5. From the dorsal angle of the scapula to the point of the haunch—
   D' to D

Two and one-half times the head gives:

1. The height of the withers, H, above the ground
2. The height of the top of the croup above the ground
3. Very often the length of the body from the point of the arm to
   that of the buttock—E to F

These laws of proportionality provide general relationships between body parts, but the recent work of Willoughby and others has shown that there are breed differences. For example, the Arabian, Quarter Horse, and Morgan have the shortest heads of the breeds presented in Table 4-1 but the Thoroughbred has the shortest head in proportion to height at the withers or length of body.

The horse is best viewed from a distance so its entire structure can be appraised. It should not possess any obvious faults and the overall impression should be one of attractiveness, correctness, balance, and symmetry. The length of the legs should be in relation to the height and length of the body. The head, neck, and body should be in proportion to each other. All riding horses should be agile, athletic, and tractable. The majority of horses in this country are kept for pleasure. Since the reason for having a pleasure horse is personal enjoyment, owners should be satisfied with their investment. Consequently the color, age, sex, and breed are matters of individual preference that should be considered along with the intended use.

Temperament

The disposition of a horse is of paramount importance. A common mistake made in selecting one's first horse is placing insufficient emphasis on temperament. Experience is required to determine whether a horse is a product of its environment, and abuse or incompetence in handling the horse can result in undesirable and dangerous behavior. A spoiled horse can create problems in its handling, whether the horse be a pleasure, race, or show animal, but fortunately most horses are readily receptive to training and handling and possess a tremendous desire to please. A few horses are mentally deficient; very few are naturally vicious.

Keen observers are capable of determining a horse's frame of mind from its behavior or "body language." Its temperament is generally expressed by the eyes and ears and implemented with the feet or teeth. A mean or angry horse will often pin its ears back, dilate its nostrils, and show white around its eyes. Such a horse is unpredictable, and if truly "mean" it is a potential danger that should not be tolerated. Mean horses are equally dangerous at both ends; they will strike forward, or bite, as well as kick backwards. A horse can inflict as much damage from a bite as it can from a kick.

Training

When selecting a horse, one should consider the extent of its training for its intended use. The less experience one has with horsemanship the better trained the horse should be. No beginner should undertake the training of a young horse without guidance and assistance. On the other hand, most experienced horsepeople prefer an untrained horse because its mind is fresh and can be trained to suit their particular purpose. The training of a horse is a reflection of its disposition, athletic ability, intelligence, and mental capacity.

Breed and Type

There is no one "best" breed of horse; some are better suited for one purpose than another but there is considerable variation in type and ability within all breeds. *Purebred* horses are those registered with a particular breed association (see Chapter 2) while *crossbreds* have at least one registered parent and are usually crosses between two breeds. Horses of unregistered, undistinguished, or unknown ancestry are called *grades*. Many grade horses are outstanding performers and crossbred horses are becoming increasingly popular for specific purposes such as carriage driving, Combined Training, and dressage. The American Horse Council estimates that in the United States at the present time 35 percent of the horses are Quarter Horses, the most popular breed, 16 percent are grades, 12 percent are Arabians, and 10 percent are Thoroughbreds.

A breed is a group of individuals possessing specific characteristics not common to other horses. These characteristics represent "breed type" if they are sufficiently well fixed to be uniformly transmitted. Breed type is an elusive concept since it is based on such divergent characteristics as head shape, head and neck carriage, body structure, color characteristics, or even gaits or way of going. Because type characteristics are often subtle, they require experience and knowledge of particular breeds. For this reason most breeds have all-breed shows where licensed judges evaluate both conformation and performance. In most breed performance classes the animals are judged on type and conformation as well as on their performance. The American Horse Shows Association (in some instances the individual Breed Association) develops the rules and class specifications for horse shows and approves the judges for various breeds.

Breed type, physical characteristics desirable for the breed, is important to the purebred owner and breeder as it can contribute to a horse's value. In the final analysis the selection of a breed should reflect the intended use of the horse. The various breeds of horses are discussed in detail in Chapter 2.

Pedigree and Performance Records

Purebred horses are often selected and priced on the basis of their pedigree. The pedigree is a historical record of ancestry and its value in predicting future performance is only as good as the accuracy of the information on those ancestors. The individual conformation and ability of a horse is most important in selection, but the pedigree can be a useful tool, particularly when a horse is selected at an early age or is being considered as a breeding animal. The best test of a breeding animal is the quality and ability of the offspring it produces, but because this information is often unavailable before purchase or selection, the ancestry or pedigree, is the next best guide to making an enlightened choice. The quality and reputation of the ancestors in the pedigree can either greatly enhance the value of an animal or detract from it. Pedigree fads develop in all breeds from time to time when a particular strain gains temporary popularity, usually based on the performance of the offspring of that pedigree, but good breeders develop their own ideals and stick to them regardless of temporary changes in the popularity of particular individuals, lines, or types within the breed.

Performance records on horses can be particularly useful in choosing a horse for some specific purpose. However, their usefulness depends on the objectivity or subjectivity of the performance involved. For example, which horse is the first to cross the finish line in a race is a fact that can even be verified by a photograph. On the other hand, the winner of a horse show 3-gaited class is an *opinion* made by a judge based on observation and interpretation of what is "ideal." Performances with measurable criteria— such as flat, steeplechase, harness and endurance racing (time), jumping (faults, time, height), and pulling (weight) are more objective than Combined Training, carriage competition (time, faults, and point rating system) or dressage, cutting, and reining (point rating system). Halter classes and performance classes such as gait, pleasure, and driving classes are the most subjective.

Racing offers the classic example of placing value on measurable criteria, and the actual racing records (times, races won, or money earned) of the sire and dam can be the definitive factor in selection. Performance records of other traits can also be useful. Records of showing performance, cutting-horse ability and cow sense are examples of records often considered by horse breeders. Records of reproductive ability are valuable in selecting animals for the breeding herd. In Europe some governments and breed associations have developed comprehensive performance tests for their horses but unfortunately performance testing is seldom used with horses in the United States.

In the final analysis, selection of a horse is a series of enlightened compromises based on all the information that is available. Added to an evaluation of the individual, the pedigree, performance, and offspring all provide valuable information to consider when selecting a horse.

Quality

Quality refers to refinement in the horse and is expressed in the texture of the hair, hide, bones, and joints. Refinement of hair is evidenced by a fine, silky mane and tail. The hair coat should also be fine with a *bloom* or luster. The chin, throat, ears, and legs should be free of excess hair, which other-wise denotes coarseness. Refinement of hide is exemplified by a thin, pliable hide that clearly defines the bones, joints, tendons, muscles, and blood vessels underneath.

The shape and size of the head also expresses quality. It should be proportional to the body, with a triangular shape, pleasing profile, promi-nent eye, and small ears. A large rectangular head, Roman nose, thick muzzle, and large ears indicate coarseness.

There is no visual measure of bone quality as such, so refinement of bone is estimated from the shape and joints of the lower leg. The cannons should appear strong, flat, and clean with good definition of the tendons. Coarse-ness is expressed by swelling, meatiness, and roundness of the cannons and pasterns. The joints of the leg should be well defined, clean-cut, lean, and free from swellings and other unnatural development.

Size

Size in a horse is usually expressed in height at the withers and/or total body weight. The height measurement is in *hands*. One hand represents 4 inches and fractions of a hand are expressed in inches. Thus a horse 62 inches high at the withers would be called 15.2 hands—which is actually 15 hands 2 inches. Height at the withers is used as a show-ring classification of size (ponies, for example, are 13.2 hands and under for show ring purposes) and is often used to describe a horse's size, but it is a poor indicator of a horse's total size or condition. Willoughby has compiled body measurements for various breeds of horses. Table 4-2 from his data demonstrates the differences in height at the withers and total body weight in males of five common breeds. It is interesting to note that the Quarter Horse on the average is the shortest (59.3 in.) but also the heaviest (1178 lb) of these breeds.

The best measure of growth in a horse is weight and the best description of size is a combination of height and weight. However, weight is not routinely used with horses because of the difficulty of accurately obtaining that measurement. Table 4-2 demonstrates how heartgirth more closely follows weight than height. In fact, the correlation between heartgirth and body weight is so high that heartgirth measurements can be used as a remarkably accurate method of estimating weight regardless of breed. Willoughby has developed formulas for predicting body weight from heartgirths in adult horses as follows:

$$\text{(Males) Body weight in lb.} = (.14475 \times \text{heartgirth in inches})^3$$
$$\text{(Females Body weight in lb} = (.14341 \times \text{heartgirth in inches})^3$$

The body weight of immature horses (birth to five years of age) is estimated by the formula:

$$\text{(Colts) Body weight in lb} = (.1387 \times \text{heartgirth in inches} + 0.400)^3$$
$$\text{(Fillies) Body weight in lb} = (.1382 \times \text{heartgirth in inches} + 0.344)^3$$

Other interesting observations from this work are that this weight-prediction formula is accurate for all horse breeds. Secondly, the females in the entire horse family weigh almost as much as the males. Willoughby points out that "horses and their domestic and/or wild relatives are singularly similar in size between sexes, in this respect differing markedly from cattle,

Table 4-2
Average body measurement of adult male horse

Breed	Height (inches) Avg.	Range	Weight (lb) Avg.	Range	Heartgirth (inches) Avg.
Arabian	59.7	(56–63)	933	(800–1050)	67.5
Morgan	60.0	(56–64)	1035	(800–1200)	69.11
Quarter Horse	59.3	(57–61)	1178	(1040–1300)	73.65
Standardbred	63.0	(60–66)	1085	(950–1200)	71.0
Thoroughbred	63.78	(62–68)	1175	(1050–1350)	72.9

and other forms such as deer, goats, sheep, the larger carnivores, anthropoid apes, and, of course, man.''

The size of the horse should be adequate for the desired purpose and appropriate for the task. A rule of thumb is that other things being equal, size is an asset.

4.2   Basic Conformation

The horse is an athlete and its conformation determines its ability to perform. When all the horse's parts (Figure 4-3) are so constructed and proportioned one to another that the horse is perfectly adapted to its work, then it has good conformation (Figure 4-1). Type is a personal preference, but certain conformation characteristics are common to all types. Thus, the following description of desirable conformation applies to all horses regardless of breed.

Head and Neck

The front end is not only important to the appearance of the horse, it also plays a major role in determining the horse's balance, while the length, shape, and attachment of the neck at the withers and shoulder affect nimbleness. A supple horse uses its head and neck for balance and stability. Freedom of motion in the head and neck are associated with freedom of stride. When the withers are prominent, they provide optimum space for the attachment of a graceful neck and enhance a desirable head carriage. For a horse to be well balanced, the neck should be long and lean with the head in proportion to the body.

Head   The head shape and size are indications of refinement, as well as breed type, and should be proportional to the body with a pleasing profile, prominent eye, and small ears (Figure 4-4). A large head, Roman nose, thick muzzle, and large ears indicate coarseness. The head should be "finely chiseled" with good definition of the bony framework. The skin should be thin and the underlying blood vessels clearly defined. The mane should be fine and silky and the chin and jaw should be free from excess long, coarse hair. Such a head gives an appearance of intelligence and refinement and is much more acceptable than a heavy, dull, plain head. The head is important because it is the sensory center for the horse and its structure indicates much about disposition and intelligence. The head should be triangular as viewed

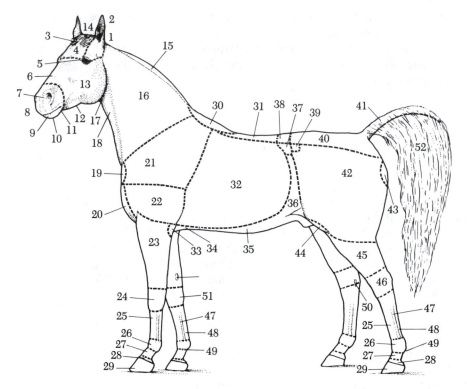

Figure 4-3
Parts of the horse

1 Occipital crest
2 Ear
3 Forelock
4 Forehead
5 Eye
6 Nose
7 Nostril
8 Muzzle
9 Lower lip
10 Chin
11 Chin groove
12 Branches of jaw
13 Jowl or cheek
14 Poll
15 Crest
16 Neck
17 Throatlatch
18 Jugular groove

19 Point of shoulder
20 Chest
21 Shoulder
22 Upper arm
23 Forearm
24 Knee
25 Cannon
26 Fetlock joint
27 Pastern
28 Coronet
29 Hoof
30 Withers
31 Back
32 Ribs
33 Elbow
34 Brisket
35 Belly
36 Flank

37 Coupling
38 Loin
39 Point of hip
40 Croup
41 Dock
42 Thigh
43 Point of buttock
44 Stifle
45 Gaskin
46 Hock
47 Suspensory ligament
48 Flexor tendon
49 Fetlock
50 Chestnut or callosity
51 Trapezium
52 Tail

Figure 4-4
Conformation of head and neck: (*a*) long, meaty head and thick throatlatch; (*b*) Arabian head with a desirable open throatlatch; (*c*) thick throatlatch causing excess pressure when poll is flexed; (*d*) horse with open throatlatch flexed at poll; (*e*) where to determine width between jaws to evaluate proper space for trachea.

from the side; it should have large, powerful jaws and adequate brain capacity; and it should taper to large nostrils capable of great dilation. The profile of the face varies considerably, but a straight or slightly dished face is generally preferred to an arched face or Roman nose. As viewed from the front, there should be width of forehead between the eyes, and they should be placed between the poll and nostrils in such a position to make the head proportionate. Long, narrow heads indicate plainness and are undesirable.

The size of the head varies among breeds, but it should be in proportion with the body. A long, large head or a short, small head can be unbalanced with the body and give the undesirable impression of draftiness or poniness in the horse. Disproportionate heads are often accompanied by plainness in the other features of the head.

The eyes and ears give expression to the horse and indicate its disposition. The eyes should be alert, brilliant, friendly, and widely spaced. They should be prominent without giving the bulging or popeyed appearance called *bovine eyes*. Similarly, they should be large and clear with a deep hazel color. Small *pig eyes* placed close together on the head limit the field of vision, are unattractive, and give an impression of laziness and stubbornness.

Blindness is a serious unsoundness that renders the horse unsafe and significantly reduces its market value. Any cloudiness or discoloration in the eye may indicate a sight problem and should be avoided, as should a partially closed eye or one that is secreting excessively. The eyes of most horses are dark, but occasionally blue eyes are seen, often associated with white spots in the Pinto. Blue eyes have been traditionally discriminated against, but there is no evidence to support the theory that they are in any way inferior or weaker. The deep coloration of the eye is usually all that is visible, but there is a white sclera that encircles the eye. Sometimes this white ring is visible, and gives a horse a wild-eyes, appearance that is undesirable in some breeds. On the other hand, the Appaloosa has mottled skin around the eye, and the visible white sclera is considered a breed characteristic.

The ears should be small, slender, alert, and delicately formed. Their position or set on the head can have as profound an effect on the appearance of the horse as can their size and shape. The ears of mares are often a little larger and not as well shaped as those of stallions. Long, thick, and heavy ears (mule ears) and ears carried horizontally to the side (lop ears) are undesirable. An alert, pricked, mobile ear indicates awareness, whereas an inactive, droopy ear reflects dullness. An overactive ear may suggest a nervous disposition or difficulty in perceiving sight or sound.

The horse must have adequate exchange of air for breathing during exertion. The nostrils should be large but thin and delicate and capable of great dilation. A large nostril is indicative of capacity in other parts of the respiratory system. The nostrils will flare after exercise but should return quickly to normal following rest. The normal discharge if present is colorless and odorless.

The mouth of the horse should be such that the incisor teeth and the lips meet evenly. The lips should be pliable yet quite muscular and capable of pretension. They aid the horse in collecting food and moving it into the mouth. If the lower jaw of the horse recedes so that the upper incisors are prominent and buck-toothed, the horse is said to be *parrot-mouthed*. The opposite condition, in which the lower jaw protrudes in front of the upper jaw, is known as *undershot jaw* or *monkey mouth*. Both conditions can interfere with eating, particularly grazing, and since they may be inherited traits, they are seriously discriminated against in the show ring and breeding herd. The dental pattern and characteristics of the horse's teeth are discussed in Chapter 3.

Neck   An appearance of elegance and grace is provided by a long, lean neck that is functionally flexible to enhance balance and length of stride. There should be a gentle curve from the poll to the withers on top and a straight underline merging high on the body with a well-defined breast area beneath it. This type of neck provides long muscles with a maximum of suppleness and mobility.

The head should attach to the neck in such a manner as to provide ample movement and flexion without impairment of the air passages (Figure 4-4). Therefore, a clean, trim, well-defined throat latch capable of great flexion is desired. A short, thick neck is often correlated with a thick, unyielding throat latch. There should be adequate width between the lower jaws to provide ample space for the windpipe. The head should attach to the end of the neck and not appear to be embedded into it. The attachment of the head to the neck has a direct relationship to beauty and usefulness.

The neck of the saddle horse should be supple and mobile for the best performance. These characteristics are associated with adequate length, whereas short, thick necks are often associated with lack of suppleness, balance, and mobility. There are breed differences in head carriage that reflect differences in the length and attachment of the neck. The extremes are exemplified by the high-headed American Saddlehorse on the one hand and the level-headed American Quarter Horse on the other. It is interesting to note how their way of going is as different as their head and neck carriage. In some breeds, a slight arch or crest on the top of the neck is pleasing and desirable, but an excessive crest, thick upper neck, or *broken crest* (lop neck) are undesirable because they can interfere with flexibility. The stallion should naturally carry more crest than the mare. A thick, "cresty" neck in the mare is usually associated with a lack of femininity. The underline of the neck should be straight and come high out of the shoulder region. A concave neck accompanied by a depression in front of the withers and often a thickened, rounded underline is termed *ewe neck*. Such necks usually result in high-headed horses that have minimal flexion at the poll and are therefore limited athletically. Ewe neck is awkward and unsightly.

Forequarters

The conformation of the forequarters is critical in terms of freedom and length of stride and the longevity of a horse's usefulness. The forequarters (see Figure 4-5) provide propulsion in front, serve as a base of support, and contain shock-absorbing mechanisms that alleviate the concussion of motion. The majority of the weight (60 to 65 percent) is carried on the front legs, and consequently most unsoundnesses from concussion and trauma occur in the front legs of riding horses. Length of stride, smoothness of gait, soundness of legs, and power of propulsion depend upon the architecture of the forequarters.

The two most critical aspects of ideal conformation of the forelimbs are the slope and angles of the bones, to ensure absorption of concussion, and the straightness and trueness of the limbs, so that no one segment receives unusual or abusive wear. Concussion in the forequarters is absorbed by the unique muscular attachment of the forelimb to the body; the sloping shoulder and consequently the angle formed between the scapula and humerus; the angle formed between the humerus and forearm; the small bones and surrounding bursa of the carpus; the sloping, springly pastern with its unique suspensory apparatus; and the expansion and absorption mechanism of the hoof.

Shoulder   The shoulder of the horse should be long, sloping, and muscular, and should extend well into the back. The scapula or shoulder blade is the primary point of attachment of the forelimbs to the rest of the body. The longer the shoulder the greater the area for attachment of the many muscles that tie the forelimb to the vertebral column with a muscular sling that supports the animal's weight, providing an excellent shock-absorbing mechanism. The shoulder should slope well into the back. This decreases the angle between the scapula and humerus, and therefore reduces concussion. A sloping shoulder also provides for free forward motion of the limb by allowing the humerus to move forward and the forearm to extend, thereby allowing maximum length of stride. Long, sloping shoulders are associated with prominent withers and a deep chest and such horses invariably develop a long, free, sweeping stride; those with short, straight shoulders and rounded withers have action that is cramped and less elastic and concussion is increased. A short, straight shoulder is often associated with a short, straight pastern that further shortens the stride and increases concussion.

Arm   The arm is one of the most overlooked regions in the forelimb. The humerus or arm extends from the point of the shoulder, its articulation with the scapula, to the elbow joint, its junction with the forearm. Since the scapula and the humerus are not outwardly distinguishable one from the other, the arm is often overlooked as a part of the shoulder. However, the length, slope, and plane of the arm have a profound effect on the stride, concussion, and set of the legs and foot. To allow the maximum extension of the forearm (length of stride), the arm should be moderately long, well muscled, and fairly upright. An excessively short arm, with its accompanying short muscles, will not advance the forearm sufficiently and the stride will be shortened. The angles created by the point of the shoulder and the elbow provide excellent shock absorption. When the shoulder is straight and the arm is long and horizontal, a horse will tend to "stand over itself." While a long, horizontal arm does not increase concussion, it does cause excessive use of the shoulder muscles so that fatigue sets in and the stride will shorten.

The important point is that the length of the arm is relative to the length of the shoulder and forearm. Kays (1969), in his outstanding discussion of the relation of form to function, concludes that "a long shoulder, a short arm, plus a long forearm makes possible maximum extension of stride and speed."

The plane of the arm is important in determining the set of the feet on the ground. The arm should be in a parallel plane with the spinal column. If the elbow is set in too close to the body, the feet will toe out. The elbow should be clear of the body, but if it is inclined outward too much, the horse will toe in and will stand pigeon-toed. The length of the arm determines whether the legs are set forward or back under the body. The legs should be set well forward.

Foreleg   The forelegs are the hardest part of the horse to keep sound, as they are constantly subjected to the most weight and concussion. It is therefore extremely important that the forearm, knee, cannon, fetlock, pastern, and foot be upright and plumb when viewed from all directions. The forearm is formed by the fusion of two bones, the radius and the ulna, and extends from the elbow to the knee. It should be long, wide, and well directed. The length of the forearm in large part determines the length of stride. Width is an indication of muscling.

The knee is a vital junction between the forearm and the cannon because it must be capable of bearing weight and supporting the body (Figures 4-5 through 4-8). Some racehorse trainers consider the knees one of the most critical areas of all. It should be large, broad, flat, wide, clean and capable of great flexion. A large, broad knee provides a maximum articulating surface for the joint. Width provides room for the tendons which extend and flex the leg. A flat, clean knee is free from thickness, swelling, or puffiness that might indicate injury or deterioration. The knee should be straight from both front and side views—thick, wide, deep, and squarely placed on the leg. A horse that is sprung forward in its knees is termed *buck-kneed*, but this is a much less serious fault than being *calf-kneed*, or back at the knees (Figure 4-5). A calf-kneed horse is considered weak because tremendous additional strain is placed on the tendons, ligaments, and bones. Such a horse is a prime candidate for unsoundness when heavy work is required. The knee should also be straight as viewed from the front (Figure 4-6). A break inward or *knock-kneed* condition and an outward *bow-legged* condition are both considered defects. Sometimes the cannon is not centered in the knee as viewed from the front; this offset knee or *bench knee* (see Figure 4-7) is considered a congenital weakness. Such a condition places additional stress on the medial splint bones and predisposes the development of front leg splints. William Haughton (1968), the famous harness horse driver, tested the knee by picking up the foot and folding the leg back until the foot touches the elbow. "If they don't fold up, if they're tight and can't get up to the elbow, then

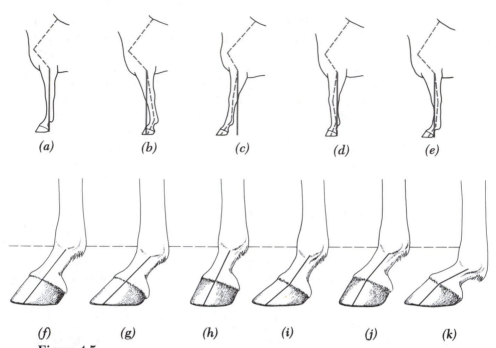

Figure 4-5
Front-leg conformation viewed from the side: (*a*) ideal; (*b*) camped under;
(*c*) camped out; (*d*) buck-kneed; (*e*) calf-kneed; (*f*) proper angle of pastern;
(*g*) pastern with too much slope; (*h*) straight pastern; (*i*) heels too low; (*j*) heels
too long; (*k*) weak pastern.

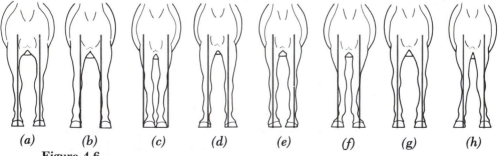

Figure 4-6
Front-leg conformation viewed from the front: (*a*) ideal; (*b*) base wide; (*c*) base
narrow; (*d*) toes out; (*e*) pigeon-toed; (*f*) base narrow, toes out; (*g*) bowlegged;
(*h*) knock kneed.

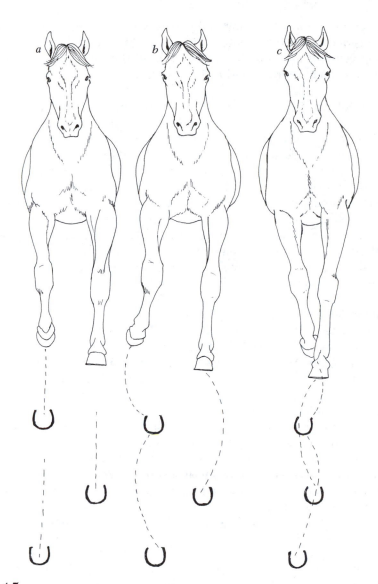

Figure 4-7
Defects in way of going: (*a*) normal way of going; (*b*) paddling, and (*c*) winging inward.

trouble is quite apt to develop. If you can't make the fold, it means the knee isn't functioning properly, its conformation is preventing it from folding completely."

Cannon   The cannon should be short and flat as viewed from the side, and should have tight, fluted, well-defined tendons set well back to give the appearance of abundant support below the knees. When viewed from the front the cannon should be centered in a straight, wide, clean knee. Round-appearing cannons and tendons tied in behind the knee are undesirable as they indicate small tendons and lack of support.

Foot   The fetlock joint, located as it is between the cannon and pastern bones, connects the leg with the foot. It is capable of tremendous flexion and its elasticity from the suspensory ligament disperses a great deal of concussion. The ideal fetlock should be wide, thick, well directed, and free from blemishes. Roughened hair, nicks, and scars on the fetlock are evidence that a horse may interfere (hit part of one leg with another leg) when in motion, and the joint should be examined carefully to see that it is strong, clean, and free from stiffness. Both the slope and length of the pastern are important. The pastern, with its two bones, three articulating surfaces, and suspensory ligament sling, is capable of tremendous flexion and shock absorption. Too long and sloping a pastern (coon foot) causes weakness because it puts undue strain on the tendons, sesamoid bones, and suspensory ligament. On the other hand, a short, upright pastern increases concussion and trauma to foot and fetlock; lameness and unsoundness can result.

The old adage "no foot, no horse" has basis in fact because no matter how fine the conformation of the horse, if the feet will not support it, the animal is worthless. Horses vary greatly in size, shape, and strength of feet, and many farriers feel that horse breeders pay too little attention to foot structure.

The foot should be sized in proportion to the horse. It should be more rounded in front than behind, and wide and deep at the heel. The hooves should be clean, straight, and free from rings and cracks. The specific anatomy and care of the foot are discussed in Chapters 2 and 19, but it should be recognized that the foot itself serves as a shock absorber by yielding at the sole and the pedal joint and expanding and cushioning at the frog and the plantar cushion upon impact.

The set of the forelegs is important for the normal functioning of the limbs (Figure 4-6). The legs should be square under the corners in parallel planes. When viewed from the front, the legs should be straight; the feet should be flat on the ground and should point directly ahead. A line dropped from the point of the shoulder should bisect the forearm, knee, cannon, fetlock, pastern, and foot equally. When viewed from the side, the forearm, knee, and cannon should be in straight line with the shoulder, the pastern, and the slope of the foot.

Defects in the set of the legs take many forms. The feet may be base wide or base narrow, they may toe in or toe out, or they may be a combination of these conditions.Figure 4-6 illustrates the types of conformational defects that can be observed from the front and Figure 4-5 does so from the side.

When the horse is in motion, these conformational faults are usually manifest in faulty flight of the foot (Figure 4-7) which places undue strain on various parts of the limb. A toed-in condition usually accompanies the base narrow condition; the foot has a tendency to "paddle" to the outside when in motion. The toed-out condition is the more serious, particularly when accompanied by a narrow chest or a base narrow condition, because the foot "wings" in and has a greater chance of striking the supporting leg.

The ideal horse stands square and has legs that are straight and feet that move true in flight with no deviation from a straight line.

Body

The body of the horse supports the weight of the rider, contains the vital organs (heart, lungs, digestive tract) and transmits the propulsion of the rear quarters. Hence, the back and loin should be short, straight, strong, and muscular and the chest should be deep and wide. The ribs should be long and well sprung to provide vital capacity for the heart and lungs.

Withers   The withers are the high point of the horse's back, located at the base of the neck, between the shoulder blades. The height of a horse is measured at the withers because this distance is the tallest constant part of the horse. The withers should be prominent and capable of holding a saddle or harness. They should be muscular and well defined at the top, and should extend well into the back. The role the withers play in a horse's performance is often overlooked. Prominent and muscular withers not only provide a maximum surface for the muscles and ligaments that support the head and neck, they also enable the shoulder to have more length and slope, so the entire front action is enhanced. Short, bulky, thick, inflexible necks and low, coarse withers mean horses that are heavy-headed, hard to control, short-strided, and restricted in their action. Horses with low, round, thick withers often have rolling gaits and heavy front ends, and move poorly. These flat, mutton withers are particularly objectionable in a riding horse because the saddle does not stay in place and the horse may be predisposed to forge. When the withers are prominent, the ligaments and muscles that attach the neck to the thorax are much freer to move, and the horse exhibits greater flexibility, coordination, and energy in its movement. High, sloping withers are usually associated with long, sloping shoulders, and the increased length of the muscles in the front end results in a lighter, freer action. It should be emphasized that prominent withers should be accompanied by muscling because thin, overprominent withers are often rubbed by the saddle and result in stiffness and soreness.

Thorax   The thorax is bounded by the back on the top, the ribs on the side, and the sternum at the base. When viewed from the front, the chest should be wide and deep. A narrow chest indicates lack of muscling and vital capacity. An excessively wide chest forces the legs out so that the gait may be rolling and labored. The thorax, when viewed from the side, should be deep and strong in the heartgirth. This region contains the vital lungs and heart and must be well developed for optimal performance. The chest should be deep, with long and well-sprung ribs that project far backward. The rib cage provides a base of attachment for the muscles of the forelimb as well as protection for the vital organs. When the ribs are well arched and project backward, it is possible for the horse to have a long, deep chest and still have a short, straight, strong back. Short, flat, straight ribs decrease the vital capacity of the horse and correspondingly reduce its athletic potential.

The back carries the weight of the rider and must be short, straight, strong, and muscular. The back extends from the withers to the last rib or loin region. A concave or sagging back is termed *swayback* and is undesirable because it denotes weakness. Many long-backed horses become "easy in their topline" (swaybacked) with age. A convex, or *roach back*, is also undesirable because it lacks flexibility of movement, results in leg interference and shortness of stride, and is uncomfortable, inefficient, and unsightly. The loin, which connects the thorax with the powerful propulsion muscles of the hind limb, is sometimes called the *coupling*. It transmits power to the forequarters and therefore must be short, wide, strong, and heavily muscled. A horse weak in its coupling and shallow in the flank may be termed *hound gutted* or wasp-waisted. Such horses lack drive in the hindquarters and make undesirable mounts. A fit horse is often acceptably tight in its middle and cut up in its flank because of its outstanding condition. On the other hand, shallow-bodied horses are undesirable, and depth of flank and strength of loin are the desired norm.

Hindquarters

The main role of the hindquarters is to provide the force for propulsion, and everything about their structure should reflect speed, power, endurance, and athletic ability. They should be long and well muscled; the legs should move in parallel planes, and the hocks should be clean and well placed to allow maximum efficiency of forward motion. The hips should be smooth, should be level when viewed from behind, and should extend well forward. Uneven or knocked-down hips often result in lameness with work. The hips should show definition, but excessive prominence is undesirable because it indicates lack of condition, strength, and endurance. The croup or rump should be long, uniform in width, muscular, and evenly turned over the top (Figure 4-8). The length and width of the croup are important conforma-

Figure 4-8
Conformation of croup: (*a*) level croup of Arabian; (*b*) proper slope for most breeds; (*c*) steep croup.

tional considerations because long muscles are associated with speed and endurance, and width of muscling is related to strength and power. The length of the croup is measured from the hip to the buttocks; the slope is the inclination of this line to horizontal. The desired slope of the croup is determined by breed preferences and intended use. Long-distance runners should have long, level croups, whereas horses that run short distances and are mobile do well with slightly sloping croups. In either case, the extremes should be avoided, and the intermediate long, muscular, rounding yet fairly level croup is preferred. The short, steep croup is faulted because it is often associated with sickle hocks and places undue strain on the hind legs. A croup that slopes and tapers from the hips to the buttocks is *goose-rumped*. A deeply creased croup is preferred because it indicates a well-muscled horse.

Hind Leg   The hind leg propels the horse forward and should have long, heavy muscling through the thigh, stifle, and gaskin. The muscles of the thigh are the most massive and powerful in the horse's body. The femur (thigh bone) should be relatively short and should be inclined forward, downward, and slightly outward. The location of the femur should be such that the legs are set well under the horse with the stifle slightly outward so that there will be a full range of movement for the hind leg. If the femur is carried too far backward, the legs are carried too far to the rear; if it is carried too far forward, the legs are brought too much under the body. The outward inclination of this region is necessary so that the stifle can move freely without striking the belly; at the same time, this outward inclination directs the hocks inward under the body where they can work closely together. The stifle should be muscled to the point of being the widest point in the hindquarter. The gaskin (tibia) should be long and well muscled. A long gaskin increases the length from the hip to the hock, a distance long associated with speed and desirability in form. A long gaskin ensures a maximum range of action and provides a maximum area for attachment of the drive muscles of the hindquarters. A short gaskin decreases the length of stride and is therefore undesirable. The gaskin should have well-defined muscling; it should be broad, wide, and deep toward the hock.

The hock joint is the hardest-working joint in the horse's body. It is the pivot of action that propels the horse forward simultaneously with the contraction of the powerful hindquarter muscles. The hock should be clean, well defined, deep, strong, wide, and flat across, and not rough, puffy, rounded, or fleshy. The lower hock should attach strongly to a short cannon with flat, well-defined tendons that are set well back to give the hock strong support.

The set or angle of the hock when viewed from the side should be neither too straight not too acute (see Figure 4-9). The line connecting the back of the hock and the fetlock should be straight and vertical. A horse's hocks are crooked when there is excessive angulation. Such a horse is said to be "set in its hocks" or *sickle-hocked*, the most common conformational fault of the hind limbs when viewed from the side. The stride of the horse set in its hocks is reduced and the horse will often stand too far under, placing extensive strain on the plantar ligament at the rear of the hock. If this area becomes inflamed, a curb may result, and therefore sickle hocks are termed "curby" conformation. A hock that is too straight receives excessive concussion and trauma.

When hocks are viewed from the rear, they should be set relatively close together under the horse with the cannons parallel (Figure 4-10). If the points of the hock turn inward, the horse is *cow-hocked*; if the horse is also base wide, it is likely to interfere behind (Figure 4-11). If the points of the

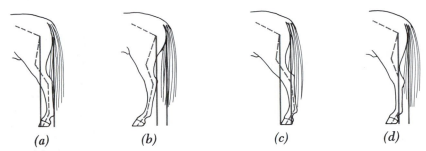

Figure 4-9
Hind-leg conformation viewed from side: (*a*) proper leg set; (*b*) sickle-hocked; (*c*) camped-out; (*d*) hock lacks proper angle, too straight.

hock turn outward, the horse is said to be *open in its hocks* and predisposed to a rotating action behind that puts great strain on the rear leg. A horse should "use its hocks" or "work off its hocks" when in motion, and this is best accomplished when the hocks move in parallel planes with the cannons and fetlocks with no deviation from a straight line.

Inflammation of the hock often occurs in driving horses and is the most common lameness problem in Standardbred racing. Hock problems will sometimes alter the symmetry of gait enough to create back pain and other musculoskeletal problems. Hock problems and other unsoundnesses are discussed in Section 4.3

The fetlock, pastern, and foot of the hind limb are similar in structure to those of the front limb. The fetlocks should be strong and clean. The pasterns should be strong, well defined, and of medium length. The rear pastern may be shorter and less sloping than the front pastern.

The set of the hindlegs when viewed from the side should be such that a plumb line dropped from the point of the buttocks touches or almost touches

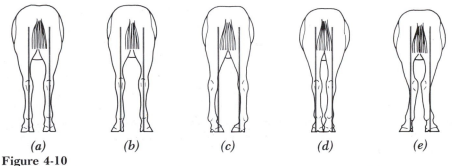

Figure 4-10
Hind-leg conformation viewed from behind: (*a*) ideal; (*b*) stands wide; (*c*) bowlegged; (*d*) stands close; (*e*) cow hocked.

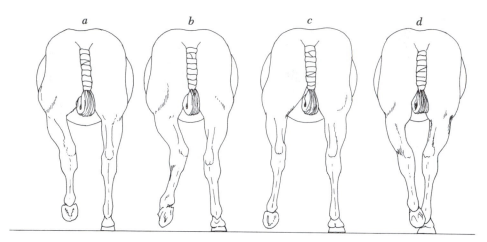

Figure 4-11
Way of going as influenced by conformation of hind legs: (*a*) ideal; (*b*) cowhocked; (*c*) wide; (*d*) close.

the rear border of the hock, runs parallel to the cannon, and strikes the ground 3 to 4 inches behind the heel. When the hind legs are viewed from the rear, a line dropped from the point of the buttocks should fall upon the center of the hock, cannon, pastern, and heel. The toes may turn out slightly so long as the cannons remain parallel.

4.3   Unsoundness and Blemishes

The horse is an athlete, and anything that interferes with its performance reduces its value. Desirable conformation has little value unless the horse is able to perform. Consequently, it is important to be able to recognize and evaluate the common defects that occur in horses. There is no universal classification of defects as unsoundness and blemishes. A splint, for example, can be considered either an unsoundness or a blemish depending on its location and whether it is accompanied by lameness. Determination of the seriousness of a defect requires experience and judgment. Unsoundnesses that are a result of faulty conformation are the most serious because they will continue to recur and may be inherited. Very few horses are completely sound, and most examining veterinarians prefer to use the phrase "service-ably sound for the use intended."

Unsoundnesses are defects in form or function that interfere with the usefulness of the horse. They include defects in conformation, feet, legs, eyes, wind, health, and reproductive functions. They may be congenital or acquired. Horses with congenital defects that result in unsoundnesses should not be used for breeding. Soundness is sometimes classified as *working soundness* or *breeding soundness* since a horse can be used for one purpose even if it is unsound for another. A blemish is an acquired physical defect that does not interfere with the usefulness of the horse but may diminish its value. Some unsoundnesses and blemishes are illustrated in Figures 4-12, 4-13, and 4-14.

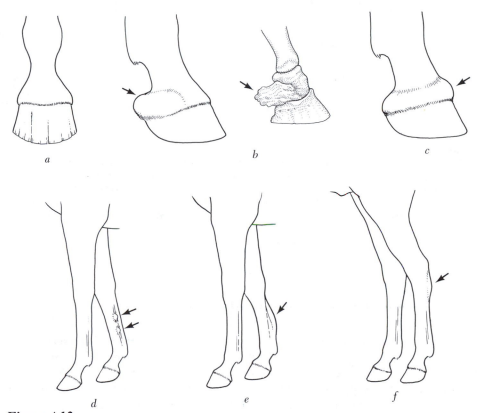

Figure 4-12
Some unsoundnesses and blemishes of the foot, pastern, and cannon. (*a*) *Toe crack*, a split in the front part of the hoof wall (may be partial, complete, high, or low); *quarter crack*, a split in the quarter area of the hoof wall; and *seedy toe*, a separation of the hoof wall near the toe. (*b*) *Side bones*, ossification of the lateral cartilages resulting from injuries that cause calcium to accumulate and harden. (*c*) *Ringbone*, a bony enlargement surrounding the bones of the pastern. (*d*) *Splints tendon*, an extension backward of the flexor tendon, caused by tearing or stretching. (*e*) *Bowed tendon*, an enlargement of the ligament, tendon sheath, or skin below the point of the hock. (*f*) *Curb* (From *4-H Horse Judging Guide*, Cooperative Extension, Cornell University, Ithaca, New York.)

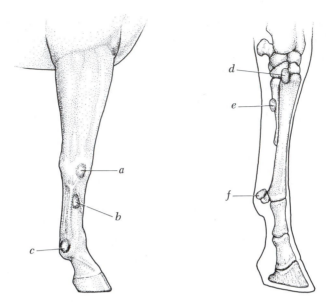

Figure 4-13
Splints of various types and prominent sesanoid bones. Splints *a* and *b* will probably not bother the horse and are not true splints because they are not located on the splint bone: Splint *a* is not affecting the knee and is located on the cannon bone; splint *b* is also located on cannon bone. Splint *d* is located at top of splint bone and involves other knee bones; it will interfere with proper action of knee joint. Splint *e* will rub against the suspensory ligament, which it can damage. The sesanoid bones (*c* and *f*) are enlarged, a condition indicative of sesamoiditis.

Blemishes and Unsoundnesses of the Forelimb

The forelimbs support the major part of the horse's weight and consequently receive the majority of those injuries that result from concussion and trauma. The front legs are particularly at risk in flat racing, which Moyer (1984) at the University of Pennsylvania Veterinary College says "produces the highest number of significant limb injuries per horse per unit of time." Most of the forelimb unsoundnesses occur at the knee or below, and many are accompanied by inflammation of joints, ligaments, and tendons. Injuries that affect the bones and articulation of joints (arthritis or degenerative joint disease) can be generally classified as more serious than injuries to soft tissues such as muscles, tendons, or ligaments.

Splints   A splint is a calcification or bony growth, usually occurring on the inside of the cannon or splint bone area (medial and small metacarpal bones) of the front leg. A splint is usually the result of a tear of the interosseous ligament that binds the splint bones to the cannon, but it can result from any inflammation of the periosteum (periostitis). Splints are a result of trauma but can also have many other causes, such as slipping, running and jumping, getting kicked, or receiving a concussion from hard surfaces. Occasionally a fracture of the splint bones is possible. A developing splint (green splint) can cause pain and lameness, but once a bony callus is formed the splint rarely causes trouble unless the exostosis encroaches on the flexor tendons, suspensory ligament, or carpal joint. In the majority of cases a splint will "set" with rest, and the inflammation resolves itself naturally to a bony blemish. If lameness persists, a more rigorous treatment, such as blistering or surgery, is possible, and is successful approximately half the time. Splints usually occur on the inside of the leg because this area receives the greatest weight. Poor nutrition or faulty conformation, such as over at the knees or bench knees, can be predisposing causes. Occasionally splints occur on the outside of the cannon or on the hind legs. The splint is one of the most common defects of the front limbs and usually occurs in young horses, almost always before 6 years of age.

Sore or Bucked Shins   A bucked shin is an enlargement on the front of the cannon between the knee and the fetlock joints. This enlargement, which

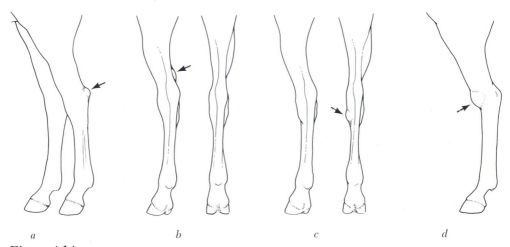

a   *b*   *c*   *d*

Figure 4-14
Some unsoundnesses of the hock area: (*a*) *Capped hock*, an enlargement on the point of the hock, usually caused by bruising. (*b*) *Thoroughpin*, a puffy condition in the hollows of the hock that can be identified by movement of the puff, when pressed, to the opposite side of the leg. (*c*) *Bone spavin* or *Jack spavin*, a bony enlargement that appears on the inside and front of the hock at the point where the base of the hock tapers into the cannon bone. (*d*) *Bog spavin*, any inflammation or swelling of the soft tissue of the hock. (From *4-H Horse Judging Guide*, Cooperative Extension, Cornell University, Ithaca, New York.)

usually occurs in the front limb, is due to trauma to the periosteum, most often caused by concussion. The condition may be confined to soreness, but if periostitis occurs, new bone growth (exostosis) can result in a "bucked" appearance of the shin. This condition often occurs in young horses in rigorous training. Bucked shins can result in lameness that can be corrected by rest and treatment.

Bowed Tendons or Tendonitis   A bowed tendon is an inflammation and enlargement of the flexor tendons at the back of the front cannon. The general cause of bowed tendons is severe strain. In some instances the condition also results from actual rupture of the tendon caused by stress from galloping, jumping, or other strenuous use. Predisposing causes may be calf knees; long, weak pasterns; a long toe and low heel; improper shoeing; being tied in at the knees; and legs that are too fine for the size of the horse. When injury occurs, it is accompanied by soreness to the touch, heat, swelling, and a tendency to flex the knee in order to raise the heel and relieve pressure. In mild cases, the superficial flexor tendon and the suspensory ligament may also be injured. The bowed appearance is a result of formation of fibrinous tissue. The bow may occur anywhere along the cannon and is classified as low, medium, or high depending on its location. Surgical procedures have had only limited success because excessive scar tissue usually develops, and healing is slow because of poor circulation to the region. Treatment requires a long period of convalescence—sometimes more than a year. Prognosis is not good if actual tendon rupture has occurred or if faulty conformation is a predisposing cause.

Sidebones   These are calcifications of the lateral cartilages of the third phalanx or coffin bone. This condition is rarely found in ponies and is less common in the light breeds than in the draft breeds. Lameness may occur during the period of ossification, but after the condition is "set" the animal is usually serviceably sound. The condition is more common in horses that toe in or toe out. Sidebones are considered an unsoundness in a young horse because the premature ossification of the lateral cartilages will result in contracted heels and abnormal foot growth. Rest to eliminate the inflammation and, later, special shoeing to prevent recurrence is the prescribed treatment.

Ringbone   Ringbone is an exostosis of the pastern bone in the form of a raised bony ridge usually parallel to the coronary band. The classification of ringbone as high or low describes the location of the new bone growth, according to whether it occurs on the lower part of the first phalanx above the pastern joint (high) or the lower part of the second phalanx at the coronary band (low). Bone growth in the proximity of the joint (articular periostitis) is much more serious than if there is no joint involvement (periarticular periostitis). The name "ringbone" was given to this condition because the exostosis can encircle or ring the pastern bone, but more commonly the enlargement occurs on the front of the pastern, on the sides, or on the front and sides but not the back. Ringbone commonly occurs on the front pasterns but is also found behind. The usual cause is strain on the ligaments or tendons where they attach to the phalanges. The tearing of the fibers at the point of insertion into the bone causes disturbances of the periosteum, and the resulting inflammation is resolved as exostosis. Faulty conformation that increases concussion is a predisposing cause, but ringbone can also be the result of a direct blow or an injury, such as a wire cut that inflames the periosteum. Lameness, heat, and swelling are results of the periostitis, and treatment is not always effective. Ringbone is a troublesome condition; because of its location, it is subjected to continued trauma and therefore the prognosis is not generally favorable. By the time the exostosis is visible, the condition is chronic and permanent lameness can result; so early diagnosis, treatment, and elimination of the cause are desirable.

Osselets An inflammation of the periosteum on the anterior surface of the fetlock joint may lead to periostitis and subsequent bony outgrowths termed osselets. A sprain or pulling of the joint capsule at its point of insertion in the cannon and long pastern bones initiates the inflammation. Heat will be present, and pain will result in a short, choppy stride and definite lameness. Any conformational fault, such as straight pasterns, that increases concussion is considered contributory. Osselets are easy to detect and a common ailment of racehorses. They usually occur in young horses that are under too much strain from training. As long as no new bone growth occurs on the articulating surfaces of the joint, the prognosis is good, because with rest and treatment the osselets will solidify and become dormant. The ankles of older horses may become enlarged and the flexibility of the joint will thus be impaired.

Sesamoiditis This is an inflammation of the proximal sesamoid bones that is serious because it usually results in chronic lameness. The initial cause is trauma or strain to the fetlock region and injury to the sesamoid bones. Sesamoiditis is another example of inflammation of the bone surface due to tearing of the insertion of a ligament, the suspensory ligament. Swelling and enlargement of the area occur and lameness results. Because the sesamoids provide support for the suspensory ligament, any fracture, exostosis, or roughness of the sesamoids is often a predisposing factor for suspensory lameness.

Suspensory Ligament Unsoundness This type of lameness is common in racehorses. The suspensory ligament attaches to the back of the cannon bone just below the knee, travels downward, and splits above the sesamoid bones into two parts, each attaching to a sesamoid bone. A smaller part continues downward and forward and attaches to the long pastern bone. The suspensory ligament is one of the main supporting structures of the horse's leg and is important in absorbing shock. It is subjected to great stress, and injury can occur to it in any one of a number of sites. Ligament tears high at the point of attachment on the cannon are commonly referred to as check ligament lameness, although it is the suspensory ligament that is affected. Injuries to the middle portion of the ligament usually occur with splints, whereas injuries to the fetlock region of the ligament occur when there is injury to the sesamoid bones. A sprain at the point where the suspensory ligament splits is the most serious because it is extremely painful and because the weight of the horse constantly exerts pressure that prevents proper healing. Suspensory trouble can occur in any of the legs, but is more likely to occur in the front legs. Treatment and prognosis are variable depending on the severity and location of the injury.

Wind Puffs or Wind Galls (Road Puffs or Road Galls)   Wind puffs are soft, puffy, fluid-filled swellings that occur around a joint capsule, tendon sheath or bursa. They are the result of excess synovia and can be found above the knee but usually are on the fetlock and pastern as a result of trauma. Wind puffs occur on both the front and hind legs and are usually a result of heavy work. They rarely cause lameness and are considered common blemishes.

Navicular Disease   Navicular disease is any injury of the navicular bone of the front foot. Faulty conformation and injuries are the most important causes of navicular disease, although nutritional and hormonal imbalances are also possible. A straight pastern and shoulder or a small foot will increase the concussion on the navicular bone, thus forcing it against the flexor tendon and causing excess friction and possible damage. Horses worked repeatedly on hard surfaces are predisposed to the disease, which often affects horses during their prime years (ages 6 to 10). The disease usually begins as an inflammation of the navicular bursa, but it is often complicated by inflammation, ulceration, and partial degeneration of the navicular bone itself, and may progress to exostosis of the bone and calcification of the associated ligaments and cartilage. The term "navicular disease" is also applied to the chipping or fracture of the navicular bone which may or may not be caused by earlier navicular disease damage. An afflicted horse shortens its stride and tends to go up on its toes and to have an increased tendency to stumble. The shoes or foot typically are worn more in the toes than in the heels. When standing, the horse points the toe of the more seriously affected foot, a behavior that has been accepted as typical of a navicular problem. The disease causes varying degrees of lameness, and there is no permanent cure. Corrective shoes that keep the toe short and the heels elevated decrease pressure on the frog and often permit the horse to travel sound. Pain-killers may restore afflicted horses to usefulness for short periods. As a last resort, permanent relief from pain can be accomplished by a posterior digital neurectomy (nerving), but other complications can then arise. A horse that has had a neurectomy is considered unsound even if there are no outward signs of pain or lameness.

Carpitis or Popped Knee   An enlargement of the knee joint as a result of inflammation to the joint capsule, the bones of the carpus, or the associated ligaments is known as carpitis. Carpitis usually results from concussion and trauma, which may cause chip fractures, increased joint fluid, and arthritic modification. The knee is a complex joint composed of 8 small bones connected in 2 rows. Consequently, there are numerous areas where inflammation can occur, and the seriousness of the condition depends on the degree of inflammation, its location, the extent of exostosis, and the amount of articular surface that is affected. Faulty conformation, particularly calf knees and bench knees, can predispose, but trauma from kicks and banging of knees on jumps, against stall walls, and in trailers can also cause carpitis. This condition is considered an unsoundness if lameness or altered function of the knee joint occurs; otherwise it is considered a blemish.

Epiphysitis   Epiphysitis is an inflammation of the epiphysial cartilage plate (growth plate) of the long bones. It almost always involves the front leg and as it is associated with growth it only occurs in young horses. The cause is excessive pressure from either too much weight or concussion prior to maturation. Epiphysitis results in a swelling that is firm and painful and can cause lameness. Prognosis is good with rest until the epiphyses close naturally. Many racehorse trainers now follow a practice of x-raying the "knees" (epiphysial plate at lower end of radius) to see if they are "closed" before they begin strenuous race training.

Capped Elbow or Shoe Boil   A capped elbow is a bursitis or swelling at the point of the elbow and is usually caused when the horse irritates the elbow bursa with the shoe or hoof of the front foot when lying down. It is most commonly found in horses stabled for long periods. The swelling may be extensive, but serious lameness rarely develops. If the elbow is protected by a shoe boot and fibrosis has not developed the results of treatment should be favorable. A scar may be created by a capped elbow and is considered a blemish.

Sweeney   Atrophy of the muscles of the shoulder due to paralysis of the supracapsular nerve is called a sweeney. The condition is usually caused by direct injury to the point of the shoulder and subsequent damage to the nerve. No successful treatment is currently available. Occasionally the nerves degenerate. The horse may be sound with limited use, but often lameness results or the horse becomes lame with extensive use.

Blemishes and Unsoundness of the Hind Limb

The hind limb is the main propulsive force for the horse, and therefore the blemishes and unsoundnesses of this region are primarily a result of strains, sprains, and twists rather than injuries from concussion. This is not to imply that traumatic injuries do not occur behind, as many front-leg unsoundnesses, such as splints, ringbone, and sesamoiditis, also occur occasionally in the hind legs.

Knocked-down Hip   When one hip is lower than the other (viewed from the rear) because of the fracture of the point of the hip on one side, it is termed the hip-down condition. The fracture is usually a result of a direct blow. Horses with a hip down would not make show horses and are a poor risk for racing because they inevitably develop lameness behind and have a crooked, hitching gate.

Stifle Lameness or Gonitis   The stifle is a large, muscular joint that is held together by a number of long ligaments. This structure is subject to a number of different types of inflammation that can affect the patella, ligaments, or joint capsule and that can result in stifle lameness or gonitis. The degree of lameness, the seriousness of the injury, and the prospects for recovery depend on the location, type, and severity of the inflammation. If ligaments have been strained, the prognosis is favorable, but if chronic synovitis or arthritis have occurred, the possibility of recovery is limited.

Stifled or Upward Fixation of the Patella   A particular type of stifle inflammation, in which the patella locks and causes the leg to remain in the extended position, is referred to as the stifled condition. The stifle and the hock are unable to flex and the foot is dragged, but the patella can be released by manipulating the leg forward or backing the horse several steps. A young horse may outgrow this condition, and surgical correction is possible when the condition affects an older animal. The prognosis is favorable so long as gonitis is not severe and arthritis has not developed. There is some evidence that the tendency toward the stifled condition may be inherited.

Stringhalt   Stringhalt is an exaggerated lifting and forward motion of one or both hocks that is spasmodic and involuntary. The cause is not completely understood, but nerve damage to the region may be contributory. Some horses with stringhalt will not exhibit the characteristic flexion after warming up, and the severity of the affliction may be intermittent. Stringhalt is particularly obvious when an animal is backed or turned sharply.

Capped Hock   A capped hock is one of the most common blemishes of the hind limbs. It is a firm enlargement at the point of the hock that reflects an inflammation of the bursa. Capped hock is caused by trauma to the hock, usually as a result of kicking a wall, trailer gate, or some solid object. Extensive fibrosis can occur and a permanent blemish can result, but a capped hock rarely causes serious lameness. A severe injury can cause extensive swelling, but corticoid injections have successfully reduced the inflammation.

Curb   A curb is a hard enlargement on the rear of the cannon immediately below the hock that develops in response to stress. It develops as an inflammation and subsequent thickening of the plantar ligament on the posterior of the hock. Occasionally a curb will also affect the bone, but it is usually confined to the plantar ligament. The condition is associated with faulty conformation, particularly sickle or cow hocks. Kicking or a direct blow may also cause a curb. A curb may result in temporary lameness, but with rest it abates, although the thickened scar tissue remains as a permanent blemish. If periostitis has occurred, then the condition is more serious and a chronic lameness may persist, particularly if there are predisposing conformational faults.

Thoroughpins   A thoroughpin is a soft, fluid-filled enlargement in the hollow on the outside of the hock. The swelling can be pushed freely from the outside to the inside of the hock by palpation. It is caused by strain on the flexor tendon, which causes synovial fluid to escape into the hock hollow. Fortunately, the condition rarely causes lameness and is usually considered a blemish.

Bog Spavin   A soft distension on the inside front portion of the hock joint caused by an inflammation of the synovial membrane of the hock is known as a bog spavin. Faulty conformation (such as straight hocks), strain (resulting from quick stops), and rickets (caused by a nutritional deficiency) may be predisposing causes that result in inflammation of the bursa and an increased production of synovial fluid. A bog spavin, although unsightly, rarely interferes with the usefulness of the horse. In young horses, bog spavins may appear and disappear spontaneously. They may be treated with some success by draining, corticoid therapy, firing, or blistering, but if the cause persists, the condition tends to recur, although it is rarely accompanied by lameness.

Bone Spavin or Jack Spavin   A bone spavin is a bony enlargement on the lower interior surface of the hock joint that may result in limited flexion of the hock. Spavin lameness typically results in a irregularity of gait. Faulty hock conformation, excessive concussion, nutritional deficiencies, and hereditary predisposition are considered causes of the bone spavin, but a traumatic event, such as jumping or vigorous training, is usually required to cause its development. A bone spavin will result in lameness of varying degrees of severity, but prognosis is favorable because as many as two-thirds of those afflicted become serviceably sound, albeit blemished. Lameness may persist as a result of tendon irritation over the point of exostosis, but in some instances an effective treatment is to cut the cunean tendon to relieve tension and pressure over the jack. This operation is relatively safe and simple and often results in immediate relief and soundness for the horse. Horsemen refer to this procedure as "having the jacks cut."

Occult Spavin   Hock lameness without visible exostosis is termed an occult or "blind" spavin. The occult spavin occurs on the articulating surface of the hock joint and is not generally recognized unless accompanied by lameness. Prognosis is unfavorable for blind spavins because they tend to lead to chronic discomfort and further lameness.

Blemishes and Unsoundnesses of the Hooves

Laminitis or Founder   Laminitis is a noninfectious inflammation of the sensitive laminae of one or more hooves. Severe pain can result from circulatory congestion within the foot. A variety of causes have been recognized, including overeating (grain founder), digestive disturbances (enterotoxemia), retained afterbirth (foal founder), lush pastures (grass founder), and concussion (road founder).

When the sensitive laminae become inflamed in a founder attack, the fragile union between the hoof and the laminae weakens, and the pull of the deep flexor tendon may actually create a separation of the laminae and the hoof wall. A foundered horse often has a distorted hoof with characteristic irregular "founder rings," a long toe that curls if neglected, a dished hoof, and a dropped sole caused by the downward rotation of the pedal bone. In severe cases, the pedal bone can protrude through the sole of the foot. There are many treatments for founder, but the prognosis is guarded depending on the extent of alteration of the foot. A new procedure that temporarily replaces the damaged part of the hoof with acrylic resin, combined with therapeutic shoeing, has had encouraging results. If the pedal bone has penetrated the sole, the prognosis is unfavorable (see Chapter 8 for a more complete discussion).

Cracked Hooves or Sand Cracks   Cracked hooves, usually found on the feet of unshod horses, indicate neglect in the care of the foot. They may be called quarter crack, toe crack, or heel crack, depending upon their location on the hoof. Hoof cracks vary in length and depth. When a crack reaches the coronet or the sensitive laminae, lameness usually results. The problem with cracked hooves is that once they begin, the constant pressure from the horse's weight during motion forces the crack to persist. Treatment usually includes special shoeing, a clinch, or special grooving of the hoof. Severe cracks have been successfully repaired with acrylic.

Contracted Heels   Contracted heels is a condition in which the frog is narrow and shrunken and the heels of the foot are pulled together. The foot may become smaller at the ground surface than the coronary band. This condition tends to be a problem in show horses because hoof growth and improper shoeing, which prevents sufficient pressure against the frog, may cause contracted heels. Lameness can eventually result if the contraction is not corrected by special shoeing.

Quittor   A chronic, purulent, inflammatory swelling of the lateral cartilage resulting in intermittent subcoronary abscesses is called quittor. Heat and pain are usually followed by suppurative tracts that periodically heal and reopen. The condition may be caused by a trauma, puncture, bruise, or laceration near the coronary band. Intermittent lameness continues as the condition persists, and permanent lameness can result if the foot is permanently damaged or deformed.

Grease, Grease-heel, or Scratches   An inflammation of the back of the pastern is called grease, grease-heel, or scratches. It leads to a chronic dermititis that results in scabs, skin cracks and eventually granulation clusters. While the cause is unknown, constant moisture, mud, manure and long coarse hair in the region all encourage its onset. It also occurs with race and show horses that perform on tracks or rings that have been treated with chemicals to reduce dust or prevent freezing. Recovery is good depending on treatment and severity of affection. It is interesting that scratches is more common on legs with white markings.

Corns   Corns are caused by a constant irritation to a part of the sole of the foot, usually from poor shoeing practices. After a period of time a lesion develops and the foot must be trimmed to remove the pressure. Prognosis is good if the condition is corrected before permanent damage occurs.

Thrush   Thrush is an infection of the frog of the foot that is quite common in stabled horses. It is caused by an anaerobic organism that causes necrosis of the tissue of the frog and a foul, blackish discharge. Extreme cases can lead to lameness and may require veterinary attention. Generally, when treated early and if proper sanitation is followed, the condition can be easily controlled.

Gravel   Gravel is an infection that penetrates the white line of the sole and travels under the hoof wall between the sensitive and insensitive laminae until it abscesses at the coronet. The term "gravel" arises because a piece of stone is sometimes the causative agent but any wound, crack, bruise, or infection to the area can have similar symptoms. Gravel causes lameness until the abscess erupts or is exposed, after which rapid healing generally occurs.

Seedy Toe   Another problem of the white line of the hoof is seedy toe, a condition where the hoof wall separates at the toe. If not properly managed this condition can become extreme and lameness results. Good hoof-trimming practices and proper first aid will usually correct or control the condition.

Other Unsoundnesses

The majority of the unsoundnesses in the horse result in lameness; however, there are a number of other abnormal conditions that can interfere with the usefulness of the horse. These conditions are briefly mentioned here and are covered more fully in other sections.

Head

Eyes: blindness, cloudy eyes, cataracts, and conjunctivitis (an irritation of the eye)

Mouth: improper meeting of incisors, overshot jaw (parrot mouth), undershot jaw (monkey mouth)

Nostrils: discharge, reflecting respiratory infection

Poll: poll evil, an inflammation of the bursa at the poll that becomes infected

Body

Fistula of withers: an inflammation similar to poll evil except that it affects the bursa at the withers

Saddle and girth sores

Hernias: umbilical hernias occur in both sexes; inguinal or scrotal hernias occur only in the male

Systemic Unsoundnesses

Contagious diseases of any type

Heaves: chronic pulmonary emphysema (a respiratory disorder)

Azoturia: a paralytic metabolic disorder of horses

Roaring: paralysis of the intrinsic muscle of the larynx

Colic: a general term used to describe a variety of digestive disorders

Breeding Unsoundnesses

Genital abnormalities: occur in both sexes

Tipped vulva: "windsuckers"—aspiration of air into vagina that causes chronic infection

Infertility: occurs in both sexes, has a variety of causes

4.4   The Horse in Motion

The nature of movement in the horse is referred to as *way of going*. The *stride* is the repeated limb coordination and placement pattern exhibited by the moving horse. It is the distance and/or time from when a particular foot leaves the ground until that foot again strikes the ground.The cyclical nature of horse motion is due to the fact that successive strides in a gait are similar. There are two phases of the cycle for each limb during the stride, a *stance* phase (weight lifting) when the limb is on the ground and the *swing* phase when the limb is not in contact with the ground (Figure 4-15). The stride can be characterized by a *stride stance phase*, when one or more limbs are on the ground and a *stride suspension phase* when no limbs are in contact with the ground. Depending on the number of limbs on the ground at one time, the stance phase may have periods of single support or *overlap*. The number of suspension phases in a stride differs with the various gaits. The walk and the other slow four-beat gaits (foxtrot, single foot, amble, running walk, and paso) have no suspension phase, while the gallop and canter have one and the trot and pace usually have two.

The speed of a horse is determined by the stride frequency and the stride length (Figure 4-16). The stride of a running horse measures 6.5 to 7.5 meters. It is this extreme length of stride that is largely responsible for the unusual speed of the horse. The racing horse also has a large part of the stride in the suspension phase. A galloping horse has a suspension phase of 20 to 30 percent of the stride, while a trotting or pacing horse has two suspension phases totaling 35 to 43 percent of the duration of the stride. The longer the limbs are off the ground (stride suspension phase) relative to the time on the ground (stride stance phase) and the smaller the overlap, the faster the horse. This was demonstrated in the gait analysis of the famed stretch duel between Secretariat and his stable mate, Riva Ridge, in the Marlboro Cup. Secretariat won because he had a longer stride with greater suspension and reduced overlap time. When Pratt and O'Connor (1976) analyzed the stride it was found that Secretariat had an overlap time of 18.6 percent compared to 27 percent for Riva Ridge.

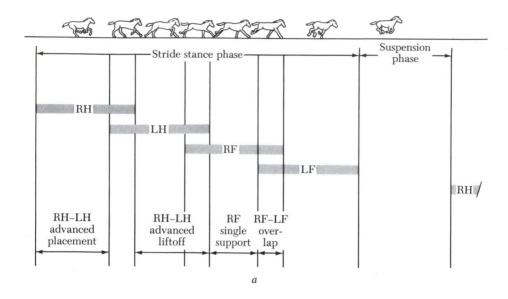

a

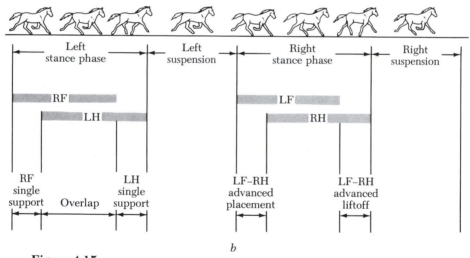

b

Figure 4-15
Diagram showing a limb placement sequence and temporal measurements for the
gallop and trot: (*a*) gallop on the right lead and (*b*) trot. The bars represent the
stance phases of the limbs. LH = left hind limb, RH = right hind limb, LF = left
front limb, and RF = right front limb. (From Veterinary Learning Systems, Inc.)

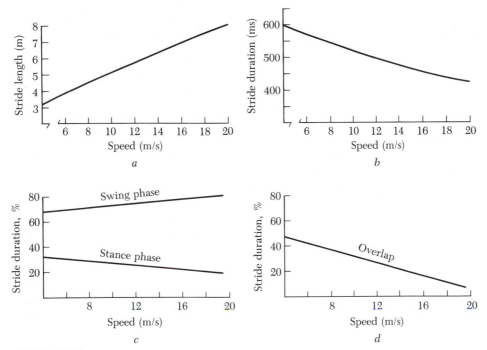

Figure 4-16
Speed-dependent changes in stride parameters of galloping horses: (*a*) stride
length, (*b*) stride frequency, (*c*) stance and swing phase duration for the lead hind
limb expressed as a percentage of stride, and (*d*) overlap duration expressed as a
percentage of stride. (From Adams, 1987.)

Hildebrand (1960) studied how animals run and found that while the
horse and the cheetah both had strides of 23 feet, the cheetah had two
unsupported (suspension phase) periods representing about one half the
stride while the horse had one unsupported period that amounted to about
one quarter of its stride. The cheetah also had a greater stride frequency, 3.5
to 2.5 for the horse. Consequently the cheetah could obtain speeds close to
70 m.p.h. compared to 43 m.p.h. for the horse.

The *gait* describes a specific way of going with a specific sequence of
limb movements that are repeated each stride with a regular cadence. For
example, the walk, trot, and gallop have been classified as natural gaits
because their pattern is distinct and they are commonly occuring gaits.
However, such a classification is an oversimplification because horses have
the ability to perform many gaits naturally. Despite the fact that a number of
gaits are named there are not a finite number of independent and distinct
gaits but rather a transition or continuum from one limb pattern to another.
Unfortunately, there is no standardization of terminology in describing horse

gaits and terms sometimes are used synonymously (i.e., slow gait and step-ping pace; lope and canter; single foot and rack) and at other times describe distinctly different patterns. There have been over 167 symmetrical gaits described for the horse (Hildebrand, 1965). Symmetrical gaits (i.e., walk, trot, and pace) have the limb coordination patterns on one side repeated on the other side half a stride later. In asymmetrical gaits (gallop, canter) the limb pattern on one side does not exactly repeat on the other side.

The recent use of high-speed cinematography, high-speed videography and microcomputers has resulted in increased research on equine locomo-tion, kinematics and biomechanics. To date most of the work on gait analysis has been with racing Thoroughbreds and trotting Standardbreds. There is relatively little information on the other racing breeds and virtually none on the gaited breeds of horse. Gait analysis is being used to predict a horse's racing potential but so far has been limited primarily to assisting in selecting Thoroughbred yearlings. Gait-analysis techniques are also being explored for detecting subtle changes in the stride associated with lameness. The artificial gaits are taught and include the running walk, slow gait, rack, and, in some instances, the pace. The artificial gaits are all modifications of the walk (four beat); the walking horses learn their gait most easily.

The Natural Gaits

The walk (Figure 4-17) is the horse's most useful gait and has been termed the "nearly ideal form of locomotion" and the "mother of all gaits" because so many of the slow four-beat gaits are modifications of the walk. The slowest of the gaits, the walk is characterized by four separate and distinct beats as each foot is placed in a regular 1-2-3-4 cadence. There are variations in the speed and height of the walk depending on the use and type of horse. An English horse should perform the walk with snap and animation while the Western horse will be sure, flat-footed, and ground-covering with less height and action. The dressage tests may call for five different walks: the working walk, collected walk, medium walk, extended walk, and free walk. (Table 4-3) The walk can range from collected to extended with stages in between. The collected walk is slower and shorter-strided than the extended walk. In the collected walk the horse moves forward with neck raised and arched and with the head in a near-vertical position. The hind legs are engaged under the body with good hock action. The steps are elevated and shorter than a normal walk because of increased flexion of the joints. The hind feet touch the ground behind the prints of the front feet. The extended walk by con-trast is ground-covering with the head and neck extended. The rear feet overreach the prints of the front feet. Typically, the head and neck nods up and down twice during each stride but the topline remains level, as there is no lift since there is no suspension phase and the weight is evenly distributed on the stance legs.

Figure 4-17
The walk. Numbers indicate the foot placement sequence in the four-beat gait.
(From Adams, 1987).

The Trot

The trot (Figure 4-18) is a two-beat gait in which the diagonal fore- and hind legs move together. There can be a tremendous variation in the height, length, and speed of the trot. For example, a Standardbred may take rapid, long, ground-covering strides (extended trot) during which the horse will be completely suspended in the air or "floating" twice during each stride. In comparison, a Hackney takes short strides (collected trot) and exhibits much higher action with no period of suspension. There should be a regular 1-2-1-2- beat to the trot.

In dressage riding a collected, working, medium, and extended trot are recognized. With the collected trot the neck is raised and the head carried at the vertical. The hocks are engaged under the body and the steps are

Figure 4-18
The trot. Opposite forelimbs and hind limbs pair to make this a two-beat gait with a period of suspension between each beat. Numbers 1 and 2 show pairing of the limbs. (From Adams, 1987.)

Table 4-3
The five walk gaits in a dressage test

---

1. *Working walk* A regular and unconstrained walk. The horse should walk energetically but calmly, with even and determined steps with distinct marked four equally spaced beats. The rider should maintain a light and steady contact with the horse's mouth.

2. *Collected walk* The horse, remaining "on the bit," moves resolutely forward, with his neck raised and arched. The head approaches the vertical position, the light contact with the mouth being maintained. The hind legs are engaged with good hock action. The pace should remain marching and vigorous, the feet being placed in regular sequence. Each step covers less ground and is higher than at the medium walk, because all the joints bend more markedly. The hind feet touch the ground behind, or at least in, the footprints of the forefeet. In order not to become hurried or irregular, the collected walk is shorter than the medium walk, although showing greater activity.

3. *Medium walk* A free, regular, and unconstrained walk of moderate extension. The horse, remaining "on the bit," walks energetically but calmly, with even and determined steps, the hind feet touching the ground in front of the footprints of the forefeet. The rider maintains a light but steady contact with the mouth.

4. *Extended walk* The horse covers as much ground as possible, without haste and without losing the regularity of his steps, the hind feet touching the ground clearly in front of the footprints of the forefeet. The rider allows the horse to stretch out his head and neck without, however, losing contact with the mouth.

5. *Free walk* The free walk is a pace of relaxation in which the horse is allowed complete freedom to lower and stretch out his head and neck. The hind feet touch the ground clearly in front of the footprints of the forefeet. On a "long rein" the reins are stretched to their utmost. On a "loose rein" the reins are slack.

---

elevated and shorter. The extended trot, on the other hand, is ground-covering as the stride is lengthened, the head and neck are extended, and there is less elevation and action. Many English performance classes may call for two trots—such as the collected and extended trot (Arabian), or pleasure trot and road trot (Morgan). The Western horse does a jog or jog trot, which is slower with less elevation and the rider sitting (not posting).

The pace (Figure 4-19) is a two-beat gait in which the lateral fore- and hind legs move together. Because of the lateral base of support, the pacer tends to throw its body from side to side and thus the pace is an uncomfortable gait to ride. However, the pace is faster than the trot and is a popular gait for harness racing, when speed is preferred to action. There are natural or free-legged pacers, but the majority of horses must be trained to the pace. The pace should have a regular 1-2-1-2 cadence like the trot.

The canter (Figure 4-20) is a fast, three-beat gait done under restraint. Two of the diagonal legs are paired to make one beat while the remaining hind leg and foreleg act independently and are called the leading legs. Consequently, at the canter (and gallop), the horse can be in either the left or right lead. The three beats of the right lead illustrated in Figure 4-20 are: (1) left hind, (2) right hind and left fore, (3) right fore. There is then a moment of suspension and the beat repeats itself in a regular 1-2-3 pause 1-2-3 cadence. When cantering in a straight line, the horse will occasionally

Figure 4-19
The pace. The pace is a two-beat gait in which the lateral forelimbs and hind limbs are paired, with the feet landing as shown. (From Adams, 1987).

Figure 4-20
The canter. The canter is a three-beat gait with the feet landing as shown by the numbers, followed by a period of suspension(s). The nonleading forelimbs and hind limbs are paired to cause one beat of the three beats. (From Adams, 1987).

switch leads to keep from tiring. However, when the horse turns to the left, it will be more stable in the left lead; when the horse turns to the right, the right lead is the "correct" lead. As with the walk and trot, there are variations of the canter ranging from collected to extended. The Western horse does a lope, which is a smooth, slow, straight three-beat canter.

The gallop or run (Figure 4-21) is the fastest gait of the horse. It is an extended canter that results in a four-beat gait; the middle diagonal beat of the canter is extended to two beats because the hind foot hits slightly before the diagonal forefoot. The cadence of the gallop is 1-2-3-4. As with the canter there are left and right leads to the gallop and there is a period of complete suspension in the air. Race horses go at a full gallop or run but many horse-show classes call for a hand gallop which is a long, free, ground-covering, four-beat stride ridden with restraint and under control.

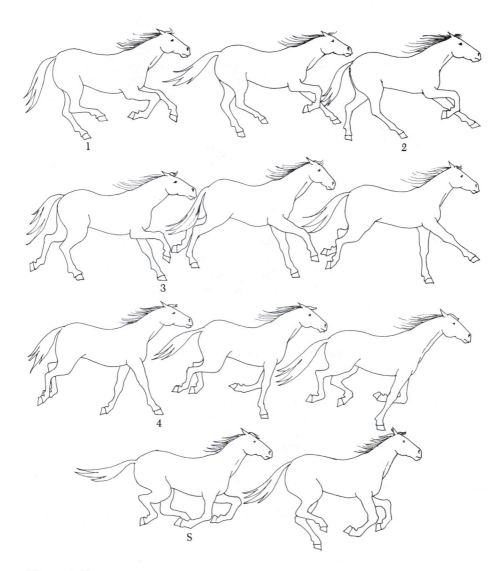

Figure 4-21
The gallop. The gallop is a four-beat gait with a front and opposite hind limb leading and the outer forelimbs and hind limbs paired. This illustration shows a right forelimb and left hind limb lead with the sequence of foot beats indicated by numbers. A period of suspension exists, as shown by S after the lead forefoot leaves the ground. (From Adams, 1987).

The Artificial Gaits

The artificial gaits are a group of related gaits that collectively are four-beat gaits with a sequence similar to that of the walk (left hind, left fore, right hind, right fore). These gaits include the amble, rack, stepping pace, pacing walk, slow gait, running walk, paso, and single foot. Hildebrand (1965) justifiably uses "single-foot" to classify all of them. It is interesting to note that the artificial gaits are confined to the American continent. Two All-American breeds, the Tennessee Walking Horse and the American Saddle Horse originating in Kentucky are the principal "gaited" horses and probably developed from early pacing stock, (Narragansett and Canadian pacers). The Paso Fino is another gaited horse that was developed in South America. There are a number of "gaited" racking, walking, fox-trotting, ambling, and single-footing horses scattered throughout the United States.

The running walk is the fast, four-beat, ground-covering walk characteristic of the Tennessee Walking Horse. It is the smoothest gait of this horse and made this breed popular as a ground-covering riding horse in the South; hence the name *plantation gait*. The running walk is characterized by a smooth, fluid motion; the hind foot overstrides the print of the front, and there is a typical nodding of the head. The running walk should not be confused with the stepping pace. In the running walk the feet on the same side leave the ground at different times. There is an even 1-2-3-4 cadence to the running walk.

The slow gait and the rack are the two artificial, taught gaits performed by the "five-gaited" American Saddle Horse. Since these are slow gaits, extreme action is preferred to ground-covering ability. The slow gait is often called the stepping pace because it resembles the pace. There is lateral movement of the legs; the forefoot and the hind foot leave the ground at the same time. However, the slow gait is not a true pace because as a result of the higher action of the front limb, the hind foot hits the ground before the front foot on the same side. There is a break in cadence in the slow gait; the foot beats $1 \cdot 2 - 3 \cdot 4$. The slow gait is tiring for the horse and uncomfortable for the rider.

The rack is the fastest four-beat gait. It should be performed with animation and in a regular four-beat cadence, 1-2-3-4. The rack is a distinct gait; each foot leaves and strikes the ground at regular intervals. The rack is different and extremely tiring for the horse, but easy on the rider. The rack differs from the running walk in that there is much more up and down motion of the limbs (action) and the hind foot does not overreach the forefoot to such an extent.

The *paso* is a unique four-beat lateral gait. The paso is essentially a broken pace, with more action in front than behind, that is performed at three speeds: classic paso fino, paso corto, and paso largo. Basically the gait changes from collection to speed going from the paso fino to paso largo.

Defects in Way of Going

A horse's gait is examined in terms of length, height, spring, promptness, power, balance, directness, and regularity. Any deviation in the flight of the foot wastes energy and places stress on the limbs. There are a number of conformational faults that can predispose the horse to faulty action. Many defects in way of going can be corrected by proper trimming or shoeing (see Chapter 19).

The most common defects in way of going are deviation in the flight of the foot and interference of one leg with another. There are basically two types of interference: *striding leg interference*, in which one moving leg makes contact with another, and *supporting leg interference*, in which the striding leg strikes a supporting leg (Figures 4-22, 23).

Striding Leg Interference   At the trotting gait, striding leg interference occurs when the hind limb hits the folding forelimb on the same side; at the pacing gait, the interference is between the diagonal fore- and hind legs. Different descriptive terms are used, depending on the point of contact.

Forging is hitting of the sole or shoe of the forefoot with the toe of the hind foot on the same side. Forging usually occurs at the slow trot and causes a characteristic "click, click, click" at this gait that usually disappears at a faster trot.

Cross-firing, a term confined to pacers, is when the hind foot on one side strikes the diagonal forefoot. In other words, cross-firing is forging in the pacer.

Scalping is hitting the hind leg at the coronet with the toe of the forefoot.

Speedycutting is hitting the hind leg at the pastern or fetlock with the toe of the forefoot.

Shin hitting is hitting the cannon or shin of the hind leg with the front toe.

Supporting Leg Interference   At the trotting or pacing gaits, supporting leg interference is the result of hitting one front leg with the other or one hind leg with the other as it strides past the supporting leg. The interference can occur at the coronet, pastern, ankle, shin, knee, or forearm. The term *brushing* is used when contact is slight; the term *striking* is used when interference results in an open wound. Interference is common in race-horses, particularly as they tire during a race, and many are forced to wear protective boots and pads.

Deviation in Flight of the Foot   The flight of the foot should be perfectly straight, but commonly it will be deviated outward or inward. Such deviation from normal can place additional stress on the limbs and cause a myriad of unsoundnesses.

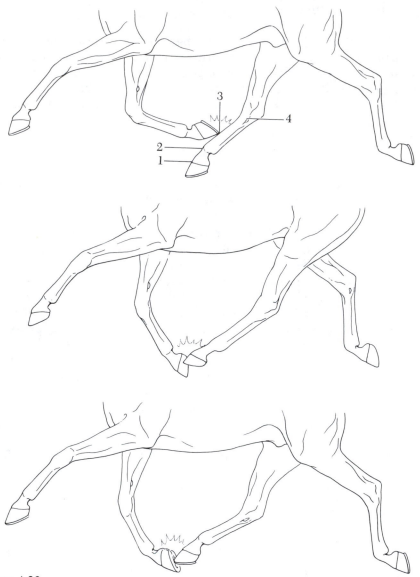

Figure 4-22
Basic interference at the trotting gait (*top*) occurs when the front foot comes in contact with the hind foot on the same side. When the interference occurs at (1), it is known as scalping; at (2) speedy cutting; at (3) shin hitting and at (4) hock hitting. Except for scalping, the interference usually takes place on the inside of the hindleg. At the pacing gait, (*middle*), the basic type of interference between front and hind legs occurs when the hind foot on one side interferes with the opposite front foot and is called cross-firing. Forging (bottom), is a type of interference that occurs at the trot when the hind foot strikes the bottom (sole) of the front foot on the same side. (From Haughton, 1968).

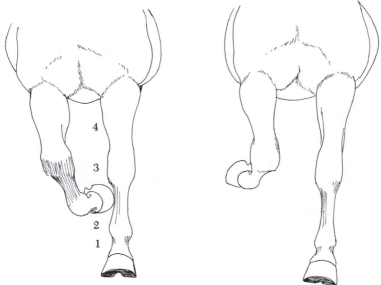

Figure 4-23
Interference occurring when one front leg strikes the opposite front leg in passing: Horses that stand toed-out are more likely to interfere in this manner. (1) ankle hitting, (2) shin hitting, (3) knee hitting, and (4) forearm hitting. (From Haughton, 1968.)

Paddling is throwing the front feet outward while in flight. This action is associated with the toe-in or pigeon-toed position.

Winging out is an exaggerated paddling observed in high-stepping harness horses and saddle horses.

Dishing or winging is throwing the front feet inward while in flight. Dishing is associated with toe-out or splay-foot conformation and is considered serious because it can lead to knee-knocking or other supporting leg interference.

Rolling is a defect of wide-fronted horses in which they roll from side to side as they stride. This is a laboring, unpleasant, inefficient type of action.

Winding or rope walking is the twisting of the striding leg around the supporting leg in such a manner that the horse appears to be tightrope walking.

Trappy is the term used to describe a short, quick, high, and often choppy stride.

Pounding is a hard contact with the ground that causes excessive concussion, often associated with straight shoulders and pasterns and typical of a "heavy-going" horse.

4.5   Vices

A vice is a bad habit that may affect a horse's usefulness, dependability, or health. It is often a reflection of the animal's personality, as vices can be the result of nervousness, viciousness, fear, curiosity, excessive energy, nutritional deficiencies, or boredom. By nature the horse is an athletic animal that is used to roaming and running at will over large areas but with domestication it has become increasingly restricted. "The idle mind is the devil's playground," and the stall is the birthplace of most vices—they are much more common in stabled than pastured horses. Vices that seem to result from boredom and inactivity often disappear with regular exercise. However, a vice is a habit, and once established it is extremely difficult to eliminate. Vices try man's patience and tax his ingenuity while confirming the horse's inventiveness. Vices can be classified as dangerous to the handler, dangerous to the horse, and nuisance habits. Vices dangerous to both the handler and the horse require vigorous correction to prevent tragedy.

Vices Dangerous to Man

Biting and *nipping* are two of the most dangerous vices because a horse can do severe damage to a handler with its powerful jaws. Stallions are particularly apt to bite and should always be watched carefully. Young horses and bored mature horses sometimes get nippy (a much less dangerous habit), but neither vice should be tolerated.

Striking is a natural defense reaction of the horse to fear or confinement, but it is an extremely dangerous habit for the handler. A striking horse can do serious bodily harm to someone in front of it, so it is best to handle the horse from the side and remain alert to its use of its front legs. A horse is most apt to strike when in nose-to-nose contact with another horse. A stallion will often strike when teasing or breeding a mare. A good horseperson anticipates these situations and prevents striking so that it does not develop into a habit or vice.

Rearing is another of the defensive behaviors of the horse and is also dangerous for the handler because the flailing forelegs can come dangerously close to the head. Again, rearing should be anticipated and prevented by firm and drastic handling if necessary. An experienced handler can usually use a lead shank to prevent a horse from striking or rearing. There are situations in which a whip should be used vigorously, as a striking or rearing horse is a potential killer.

Kicking deliberately at a handler is an act of meanness or fear on the part of the horse and must be corrected immediately. A cowkick is a forward and sideward kicking action that can catch a handler at the horse's side. Kicking stall walls or trailer gates is a nuisance habit but creates the possibility of trauma and injury to the legs as well as expensive damage to the facilities. Many capped hocks and curbs had their origin in needless kicking.

Charging is when a horse attacks or savages an unsuspecting attendant in a stall or paddock. This viciousness is most common among stallions but is occasionally observed in mares or geldings. The best correction is prevention by maintaining control of such a horse when in its stall or paddock.

Crowding is when a horse consciously crowds or squeezes the handler against the wall of the stall with its body. Correction requires alertness and anticipation. It is more common with horses housed in straight stalls.

Vices Dangerous to Horses

Cribbing is a habit of force-swallowing gulps of air. It usually requires that the cribber grasp at an object with its incisor teeth and then pulls its neck back in a rigid arch as it swallows air. The condition is considered a habit of boredom but is dangerous to horses because the swallowed air can create gastric upsets or colic. Cribbing is a vice that is frequently copied by other horses.

Wood chewing is a habit that is costly and can be injurious to the horse. Wood chewing is sometimes incorrectly referred to as cribbing but it is a separate vice that does not entail swallowing air. Wood chewers do not usually ingest the wood they chew, but some splinters can cause buccal infections, colic, or excessive tooth wear. Wood chewing is an acquired habit, and an entire herd may pick it up from one horse. Applying creosote to the chewed boards will help decrease the habit but must be repeated annually because with weathering the horse will renew the vice. Wood chewing is a vice rather than a nutrient deficiency, although some horsepeople claim that the horses are "going after the grain in the wood" (see Section 9.3). Chewing rubber fences has caused nylon fiber impactions. Tail chewing is also sometimes a problem among young horses.

Eating bedding, manure, or *dirt* is an unpleasant habit. Dirt and sand eaters are susceptible to colic, and eating of any nonnutritious foreign material can cause digestive disorders (see Section 9.3).

Bolting food is the habit of some horses of eating their grain without adequate chewing. This condition can also result in digestive upsets and decrease the nutrient deficiency of the feed. If several large, round rocks are placed in the feeder, the bolting of the grain will be decreased (see Section 9.3).

Fighting is a perennial problem because some horses are constantly aggressive toward other horses. On the other hand, a group of horses running together will establish a certain hierarchy or "pecking order." An incessant fighter may have to be separated from a herd because it can cause considerable injury to the other horses by biting, kicking, and chasing. Young horses play at fighting but rarely exhibit true aggressive behavior.

Shying is a habit that may reflect poor vision or immaturity and lack of experience. Shying is usually a response to fear and can be dangerous to both horse and handler, as a shying horse is unpredictable and may shy away from one object only to face more serious trouble.

Nuisance Habits

Weaving is a vice of a high-strung, nervous horse. The horse stands in place but weaves its head and neck back and forth as it rocks from side to side. Weaving can be considered a nuisance, but it can have more serious consequences. Some weavers will stress the legs in such a manner that lameness occurs, whereas others will lose weight and become physically exhausted. Weaving is another vice that is easily learned by stablemates.

Stall walking is another vice resulting from nervousness and excessive energy. The horse constantly paces or circles around the stall. This habit can have the same detrimental effect on the horse as weaving.

Pawing is a nuisance and creates additional work for the horseperson. The horse continually paws at the floor and, if possible, digs holes that constantly need repair. Pawing creates the possibility of leg injury.

Mane and/or tail rubbing is considered a vice, but it may be a sign of fungus, lice, or worms. However, some horses rub their tails without any external causes, and this creates a nuisance problem for the owner.

Halter pulling is a vice that is usually a result of poor or inadequate training in tying when young. Some horses can never be cured of the habit of pulling back when tied, and a horse that will not tie is a nuisance that is sometimes dangerous.

REFERENCES

Adams, O. R. 1987. *Lameness in Horses*. Philadelphia: Lea and Febiger.

American Horse Council. 1987. *Economic Impact of the U.S. Horse Industry*. Washington, D. C.: American Horse Council.

Axe, J. Wortley, 1905. *The Horse: Its Treatment in Health and Disease*. London: Gresham Publishing.

Beeman, Marvin, 1973. Conformation: The relationship of form to function. *Quarter Horse J.* (January). 12:82–128.

Beeson, W. M., R. E. Hunsley, and J. E. Nordby. 1970. *Stock Horses, Livestock Judging and Evaluation*. Part 5. Danville, Illinois: The Interstate Printers and Publishers.

Catcott, E. J., and J. F. Smithcors, eds. 1966. *Progress in Equine Practice*. Wheaton, Illinois: American Veterinary Publications.

Churchill, E. A. 1968. Lameness in the standardbred. *Care and Training of the Trotter and Pacer*. Chap. 16. Columbus, Ohio: U. S. Trotting Association.

Cresswell, H., and R. H. Smythe. 1963. Lameness: In *Equine Medicine and Surgery*. 1st ed. Wheaton, Illinois: American Veterinary Publications.

Edwards, Elwyn Hartley. 1980. *A Standard Guide to Horse and Pony Breeds*. New York: McGraw-Hill.

Evans, J. Warren, 1989. *Horses*. 2d ed. New York: W. H. Freeman.

Hanauer, Elsie. 1973. *Disorders of the Horse*. New York: A. S. Barnes.

Haughton, W. R. 1968. Selecting the yearling. *Care and Training of the Trotter and Pacer*. Chap. 2. Columbus, Ohio: U. S. Trotting Association.

Hildebrand, Milton, 1959. Motions of the Running Cheetah and Horse. *J. Mammology* 40:481–95.

Hildebrand, Milton. 1965. *Symmetrical gaits of horses. Science.* 150:701.

Hildebrand, Milton. 1987. The mechanics of horse legs. *American Scientist*. (75)6:500–601.

Hipsley, W. G. 1970. *Judging the Halter and Pleasure Horse in Individual and Team Competition*. Extension Publication No. 65. University of Massachusetts, Amherst.

Kays, D. J. 1969. *The Horse*. Rev. ed. New York: A. S. Barnes.

Leach, D. H., K. Ormrod, and H. M. Clayton. 1984. Standardized terminology for the description and analysis of equine locomotion. *Equine Vet. J.* 16(6):522–28.

Leach, D. H., and Annie I. Dagg. 1983. A review of research on equine locomotion and biomechanics. *Equine Vet. J.* 15(2):93–102.

McCann, J. S., J. C. Heird, C. B. Ramsey, and R. A. Long. 1987. Proportionality of skeletal bone and muscle in horses of different skeletal size and muscle thickness. *Proceedings of the 10th Equine Nutrition and Physiology Symposium, Fort Collins, Colorado*.

Moyer, William 1984. Common injuries of performance horses. *Equine Sport Medicine* (3)3:2.

Moxley, H. F., and B. H. Good. 1955. *The Sound Horse*. Extension Bulletin 330. Michigan State University, Ann Arbor.

Reeves, Richard Stone. 1970. The Perfect Horse. *The Thoroughbred Record* (July 18).

Rooney, J. R. 1972. The Musculoskeletal System. In *Equine Medicine and Surgery*, 2d ed. Wheaton, Illinois: American Veterinary Publications.

Rooney, J. R. 1974. *The Lame Horse: Causes, Symptoms, and Treatment*. Cranbury, New Jersey: A. S. Barnes.

Rossdale, Peter D. 1972. *The Horse*. Arcadia, California: California Thoroughbred Breeders Association.

Stashak, Ted S., 1987. *Adam's Lameness in Horses*. Philadelphia: Lea and Febiger.

Smythe, R. H. 1964. *Horses in Action*. Springfield, Illinois: Charles C Thomas.

Smythe, R. H. 1972. *The Horse, Structure and Movement*. London: J. A. Allen.

Wakeman, D. L. 1965. Selecting and judging light horses. *Light Horse Production in Florida*. Florida Department of Agriculture Bulletin 188:31.

Wentworth, Lady. 1957. *The Swift Runner*. George Allen, Ltd.

Willoughby, David P. 1974. *The Empire of Equus*. Cranbury, New Jersey: A. S. Barnes.

Willoughby, David P., 1975. *Growth and Nutrition in the Horse*. New York: A. S. Barnes.

NUTRITION

CHAPTER 5

DIGESTIVE PHYSIOLOGY

Now, good digestion wait on appetite, and health on both.
   —*Macbeth*, Act III, Scene IV.

Knowledge of horse nutrition, although not as complete as desired nor as complete as for most of the other domestic animals, is increasing. The following is an overview. Readers interested in more detail might wish to consult other books such as *Feeding and Care of the Horse* by Lon D. Lewis, (1982), *Horse Feeding and Nutrition* by T. J. Cunha (1980), and *Horse Nutrition: A Practical Guide* by H. F. Hintz (1983).

5.1   The Digestive Tract

Horses are nonruminant herbivores. They eat fibrous feeds (roughages) but do not have a rumen with bacteria to help utilize the fiber. In ruminants, such as cattle and sheep, bacteria in the rumen produce enzymes such as cellulases that digest the fiber.

Nonruminant herbivores, such as the horse, rabbit, and guinea pig, utilize roughages because they have a relatively large cecum and/or colon that contains the necessary bacterial population for fiber digestion.

Animals such as man, rat, and dog have relatively simple digestive tracts and cannot effectively digest large amounts of roughage. Figure 5-1 is a diagram of the digestive tract of the horse.

Several recent studies have been conducted on equine digestive physiology. An excellent review is provided by Meyer (1982).

189

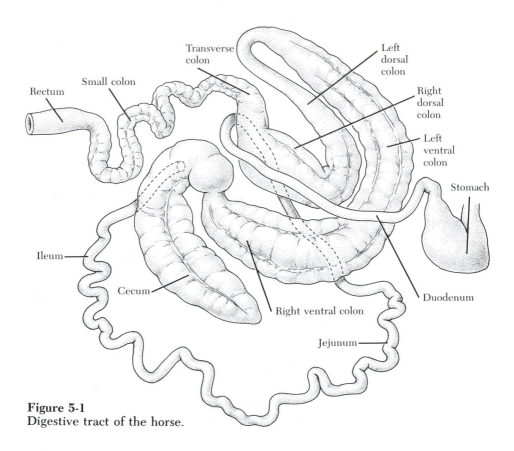

Figure 5-1
Digestive tract of the horse.

Mouth   The mouth is the first part of the digestive system. The teeth should be examined periodically by a veterinarian. (The dental pattern was discussed in Section 3.2.) Horses do not normally have dental cavities comparable to those of humans, but other difficulties can occur. For example, the cheek teeth may develop sharp edges that interfere with chewing or that may even injure the tongue and cheeks. This problem can be corrected by "floating the teeth," that is, filing down the projections with a rasp. Horses with worn or missing teeth should be fed crushed or rolled grains rather than whole grains. Hay can be softened if it is coarsely chopped and then soaked in water. Complete pelleted rations are also useful for horses with poor teeth.

The two main functions of the mouth are to masticate feed and to wet it with saliva. The saliva is produced by three pairs of glands—parotid, submaxillary, and sublingual. The parotid is the largest of these glands and has received most attention from scientists. There is profuse secretion of parotid saliva during eating, but unlike the dog, the horse does not secrete saliva at the sight of food. The quantity of saliva is quite large. Ponies may secrete as much as 12 liters per day (Alexander and Hickson, 1970). The ion composition of parotid saliva is shown in Table 5-1. Excessive loss of saliva can decrease the blood concentrations of chloride and sodium (Stick et al., 1981). Saliva contains a very low concentration of amylase, and this enzyme is probably of minor significance in the process of digestion.

Esophagus   The esophagus in the mature horse is approximately 50 to 60 inches long. Because of the tonus of the muscle of the lower esophagus, vomiting is rare in the horse. The horse is one species in which distension of the stomach can be so severe that it will rupture before vomiting occurs.

Stomach   The stomach of the horse is relatively small. It provides approximately 8 percent of the capacity of the gastrointestinal tract. Alexander (1963) suggested that limited bacterial digestion takes place in the stomach and that the primary product of digestion is lactic acid and not volatile fatty acids. He found the ratio of lactic acid to volatile fatty acids in the stomach of the horse to be 1.7:1, compared with 0.15:1 and 0.12:1 in the pig and rabbit, respectively. Although the horse's stomach is seldom completely empty, emptying of the stomach starts soon after feeding, and there is rapid movement of a meal through the stomach and small intestine into the large intestine. Feed that passes rapidly through the stomach has only limited contact with the gastric secretions. Thus, because of the small capacity of the stomach and the potentially rapid movement of digesta, it is usually recommended that the horse be fed two or more meals daily.

Gastric ulcers can be a problem, particularly in young horses. Signs include depression, colic, and grinding of teeth. The foals frequently lie on their back (Rebhun et al., 1982). The relationship between feeding practices and ulcers is not established. Stress and overuse of anti-inflammatory drugs may be involved in the etiology. Further information on diagnosis and treatment of ulcers can be found in Wilson and Pearson (1985).

Table 5-1
Composition of 24-hour samples of parotid saliva from the horse

Constituent	Concentration (mEq/l)
Na^+	55.0 ± 7.0
K^+	15.0 ± 2.0
Ca^{2+}	13.6 ± 1.7
Cl^-	50.0 ± 13.0
HCO_3^-	52.0 ± 5.6
HPO_4^{2-}	0.25 ± 0.00

SOURCE: Adapted from Alexander and Hickson, 1970.

Small Intestine   The small intestine provides approximately 30 percent of the capacity of the gastrointestinal tract and has about the same capacity as that of the small intestine of a cow of a similar weight. Carbohydrases and proteases, enzymes that digest carbohydrates and protein, respectively, are present in secretions of the intestinal cells and pancreas. The volume of pancreatic juice secretion may be greater than the volume of saliva produced by the parotid glands.

Horses do not have a gallbladder to store bile. Therefore, bile salts that promote emulsification of lipids and are thus important in lipid digestion are secreted constantly into the small intestine. The rate of bile secretion in a horse is estimated to be approximately 300 ml per hour. The contents of the small intestine are quite fluid; they contain only 5 to 8 percent dry matter.

Large Intestine   The large intestine is approximately 25 feet long and is divided into cecum, right ventral colon, left ventral colon, left dorsal colon, right dorsal colon, transverse colon, small colon, and rectum (Figure 5-1). The cecum is a pouch at the junction of the small intestine and colon. The contents of the cecum usually contain 6 to 10 percent dry matter. The colon has the greatest capacity of any segment of the digestive tract and provides 40 to 50 percent of the total capacity of the intestinal tract. The dry matter content of the ingesta increases as the material travels from the cecum to rectum. The ingesta in the first part of the ventral colon, dorsal colon, and rectum contains 12 to 15 percent, 15 to 19 percent, and 19 to 24 percent dry matter, respectively.

The microbial populations of the cecum of the horse and of the rumen of the sheep and cow are qualitatively similar; that is, in general, they consist of the same types of organisms (Smith, 1965). Kern et al. (1973) fed steers and ponies hay with or without oats and found that the numbers of bacteria per volume of cecal ingesta increased when oats were fed to the ponies but not when oats were fed to the steers. Rods, both gram-negative and gram-positive, predominated in the gram-smear counts of both ponies and steers. Celluloytic bacteria numbers per gram of ingesta were similar in the ponies' cecum and steers' rumen whether or not oats were included in the diet.

Kern et al. (1974) compared the microbial population of steers and ponies that were fed timothy hay. The total bacterial, viable bacterial, and ammonia concentrations and pH were higher in the rumen of steers than in the cecum of ponies (Table 5-2).

Further studies are needed to define the actions in the various parts of the large intestine. However, several studies, such as the one by Meyer et al. (1979), have demonstrated that the cecum of the horse can be removed without serious consequences. Horses can also survive when significant amounts of the colon are removed (Ducharme et al. 1985; Sullins et al. 1985).

Table 5-2
Chemical and microbial characteristics of ingesta from intestines of ponies and steers fed timothy hay

Characteristic	Pony gut regions			Steer gut regions			
	Ileum	Cecum	Colon, terminal	Rumen	Ileum	Cecum	Colon, terminal
Ingesta pH	7.4[w]	6.6[w]	6.6[w]	6.9[ax]	7.3[aw]	7.0[aw]	7.2[ax]
Viable bacteria/gm 10^{-7}	36.0[b]	492.0[cw]	363.0[d]	1658.0[ax]	5.4[b]	230.0[c]	12.7[d]
Rods (percent)							
Gram-negative	9.2[a]	63.8[bw]	54.3[b]	33.1[x]	20.3	29.9[x]	25.7
Gram-positive	38.9	6.4	11.2	2.4	0.9	2.7	5.5
Cocci (percent)							
Gram-negative	6.4	33.1	22.6	44.1	33.9	46.9	45.9
Gram-positive	36.2[a]	5.6[a]	11.2[b]	19.4[a]	44.7[c]	22.7[b]	25.3[b]
DNA, μg/g	9.9[b]	8.4[cw]	6.2[c]	51.6[x]	25.5[b]	24.8[bx]	16.0[c]
$NH_3 - N$, mg/100 ml	5.2[w]	2.9[w]	5.4	10.6[x]	15.5[x]	18.2[x]	13.1

[a,b,c,d]Means on the same line (within species), bearing different superscript letters, differ significantly $(P > .01)$.

[w-z]Comparison: fundic vs. abomasum: pyloric vs. abomasum; small intestine vs. small intestine; cecum vs. cecum; colon vs. colon; cecum vs. rumen. Means (between species) on the same line bearing different superscript letters for each comparison differ significantly $(P > .01)$.

SOURCE: Kern et al. 1974.

5.2   Rate of Passage

Approximately 95 percent of the food particles destined to appear in the feces pass through the digestive tract within 65 to 75 hours after ingestion, when horses are fed a hay-grain ration. A slightly faster rate of passage may be expected when only grain or a pelleted diet or fresh grass is fed. The length of the time that food remains in the stomach is variable. Solid particles may reach the cecum within 45 to 60 minutes after eating; however, some particles remain in the stomach for 4 to 7 hours. Fluids may reach the cecum within 30 minutes after ingestion. The stomach empties more slowly in the young foal than in older animals. When barium meals were fed to suckling foals and to weaned foals and the rate of passage followed by radiographing the animals, most of the barium was found to reach the cecum of the weaned foals within 3 hours, whereas after 6 hours there was still barium in the stomach and small intestine of the suckling foal (Alexander, 1946).

The size and digestibility of the feed particles may also influence rate of passage. Inert particles 2 cm $\times$ 2 mm were retained in the intestinal tract much longer than particles 2 mm $\times$ 2 mm (Argenzio et al., 1974).

5.3   Digestibility

Knowledge of the digestive physiology of the horse is essential to the development of sound nutritional practices. There is a need to know not only how the digestive tract of the horse works, but how efficiently it works. One method of measuring efficiency is to determine digestion coefficients for various feeds. Digestibility is calculated according to the following formula:

$$\text{Digestibility} = 100 \times \frac{\text{Nutrient intake} - \text{Nutrient in feces}}{\text{Nutrient intake}}$$

Animals can be housed in metabolism stalls (Figure 5-2) or fitted with collection harnesses (Figure 5-3) so that feces and urine can be collected separately and analyzed. Digestibility can also be determined indirectly by using an indigestible indicator in the ration. By determining the ratio of the concentration of indicator to that of a given nutrient in the feed and the same ratio in the feces, the digestibility of the nutrient can be obtained without measuring either food intake or feces output. The formula is:

$$\text{Digestibility} = 100 \times \left(1 - \frac{\% \text{ indicator in feed}}{\% \text{ indicator in feces}} \times \frac{\% \text{ nutrient in feces}}{\% \text{ nutrient in feed}} \right)$$

Figure 5-2
Metabolism stall used for complete
collection of feces and urine in
nutrition trials.

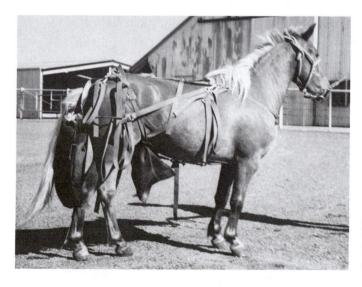

Figure 5-3
The use of harnesses and
bags permits complete
collection of feces and urine.

The values determined by these formulas are really only apparent digestion coefficients because not all the material in feces is residue from the diet. Feces also contain cells sloughed from the lining of the intestine and material that was secreted or excreted into the intestine. The nondietary fraction is called the endogenous fraction. True digestibility or true digestion coefficients are calculated by subtracting the amount of endogenous material from the total amount present in the feces to determine the actual amount of dietary residue. For example, endogenous losses of calcium can be determined by using radioactive calcium, and then the true digestibility of dietary calcium can be calculated (Schryver et al., 1970).

The study of digestive physiology in the horse has been advanced by the use of surgically prepared animals. Fistulas or windows can be placed in various parts of the digestive tract. For example, the pony in Figure 5-4a has a small fistula into the cecum. Horses have been fistulated in the cecum and the dorsal and ventral colon at the same time. Other techniques, such as the reentrant cannula, have also been developed. The pony in Figure 5-5 has a tube in the small intestine that is brought to the outside of the animal and then returned to the small intestine. Such a technique allows measurement of the various nutrients passing from the small intestine into the cecum.

5.4   Site of Digestion

Estimates of the site of digestion and absorption of various nutrients are summarized in Table 5-3.

Protein   The primary site of protein digestion is the small intestine. Dietary protein is broken down by acid hydrolysis and by protein-splitting enzymes (such as trypsin, chymotrypsin, and carboxypeptidases) into amino acids, which are actively absorbed. The proteolytic activity per mg of ingesta in the ileum of the small intestine is 500 times greater than the activity of ingesta in the cecum or colon (Kern et al., 1974). Thus, in the horse the amino acids that are absorbed are supplied by the amino acids in the diet. In contrast, in the ruminant the bacteria of the rumen may alter the amino acid makeup of the diet so that the amino acids that are absorbed are determined by the amino acid content of bacterial protein.

There also appears to be some absorption of amino acids of bacterial origin from the large intestine of the horse. The total contribution of such amino acids to the nutrition of the horse is not known. In one study in which bacterial protein containing N^{15} was placed in the cecum of ponies via a cecal fistula, amino acids containing N^{15} were found in the portal blood of the ponies (Slade et al., 1971). However, other studies have reported that

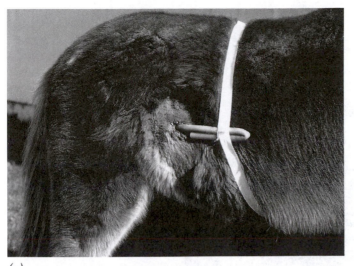

(a)

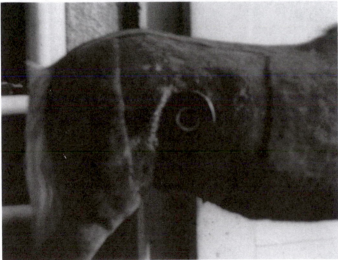

(b)

Figure 5-4
Small *(a)* and large *(b)* fistula or window in the cecum of the pony. Such fistulas are not harmful to the animal and have helped greatly in the study of horse nutrition.

the bacterial amino acids synthesized in the large intestine are not efficiently utilized by the horse (Wysocki and Baker, 1975; Reitnour and Salsbury, 1975).

Young foals can absorb intact proteins such as antibodies up to 36 hours after birth. It is important for the foal to receive the mare's colostrum, which contains antibodies, as soon as possible after birth, because there is no transmission of antibodies across the placenta to the foal.

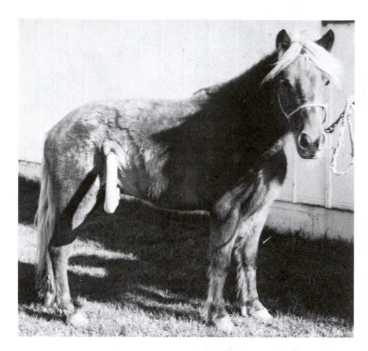

Figure 5-5
Pony with surgical
preparation that allows
collection of ingesta coming
from the small intestine.

Table 5-3
Estimates of site of digestion and net absorption

Dietary fraction	Small intestine (percent)	Cecum and colon (percent)
Protein	60–70	30–40
Soluble carbohydrates	65–75	25–35
Fiber	15–25	78–85
Fats	Primary[a]	—
Calcium	95–99	1–5
Magnesium	90–95	5–10
Phosphorus	20–50	50–80
Vitamins	Primary[a]	—

[a]Estimates of percentage absorbed in various segments are not available, but the small intestine is probably the primary site of absorption of dietary sources of these nutrients.

Carbohydrates   The soluble carbohydrates are digested and absorbed primarily in the small intestine. The carbohydrases of the pancreatic secretions and intestinal mucosa break down the soluble carbohydrates such as starch to glucose and other 5- or 6-carbon sugars, which are readily absorbed. Several carbohydrases, such as sucrase, maltase, trehalase, and β-galactosidase, are present in the small intestine, but activity depends on the age of the horse. Horses at birth have very little maltase or sucrase activity, but the levels gradually increase to a peak at 6 to 7 months. Lactase activity is high at birth but drops off sharply at 3 to 4 years of age (Roberts, 1975). Thus, feeding high levels of sucrase to young foals or feeding high levels of lactose to horses older than 3 years of age could result in digestive upsets and diarrhea because of the enzyme deficiency. Some of the soluble carbohydrates reach the large intestine, where they are fermented by bacteria. The end product of that digestion or fermentation is volatile fatty acids (VFA). The major volatile fatty acids are acetate (2 carbons), propionate (3 carbons), and butyrate (4 carbons). The volatile fatty acids are absorbed from the large intestine and utilized as sources of energy by the tissues of the horse. Propionate can be used by the tissues to synthesize glucose.

The fibrous fraction of the diet is digested primarily in the large intestine. The principal end products of digestion are VFA. The ratios of VFA produced are quite similar to the ratios produced by ruminant microflora (Table 5-4). Increasing the proportion of grain to hay in the diet of both horses and ruminants deceases the relative amount of acetate and increases the relative amount of propionate produced. Such findings also demonstrate that not all of the soluble carbohydrates are digested in the small intestine but that some are digested in the large intestine of the horse. Feeding of pelleted roughages may also increase the percentage or propionate in the cecal fluid, presumably because the rate of passage of pelleted roughages through the digestive tract is faster, and thus more of the soluble carbohydrates reach the lower gut.

Table 5-4
Comparison of ratios of volatile fatty acids in rumen fluid and cecal fluid of cattle and ponies fed rations with varying hay-to-grain ratios (percent)

Hay-to-grain ratio	Species	Acetate	Propionate	Butyrate
1:0	Pony	73	17	8
	Cattle	74	18	8
1:4	Pony	59	25	11
	Cattle	62	22	16

SOURCE: Hintz *et al.* 1971. Also Templeton and Dyer, 1967.

Fat   Dietary lipids are digested and absorbed in the small intestine. The composition of body fat is influenced by the composition of dietary fat because the fatty acids are absorbed from the small intestine before they can be altered by the bacteria in the large intestine. As mentioned earlier, the horse does not have a gallbladder. However, the lack of a gallbladder does not appear to hinder the digestion of fat. Several species, such as rats, deer, camels, elk, whales, porpoises, and pigeons, also do not have gallbladders, but they all secrete bile from the liver. Mature horses can tolerate diets containing high levels of fat. For example, the fat in diets containing 15 to 20 percent tallow, corn oil or peanut oil is 90 to 94 percent digestible.

Vitamins   The fat-soluble vitamins (A, D, E, and K) are apparently absorbed in the small intestine. The horse also absorbs carotene from the small intestine. Carotene, a yellow compound obtained from plant material, can be converted by the tissues of the horse into vitamin A. Not all of the carotene is converted immediately into vitamin A; some is stored in fat deposits. Thus, horses fed diets containing carotene have yellow blood plasma and yellow fat. Cattle, man, and chickens also absorb and store carotene, but many species, such as sheep, rabbits, rats, guinea pigs, water buffalo, swine, camels, and dogs do not store appreciable amounts of carotene in the fat and hence have white fat. However, these animals are also able to convert carotene into vitamin A. It is interesting that two other members of the equine family, the donkey and the zebra, do not absorb carotene and hence have white fat, in contrast to the horse.

Dietary B vitamins are absorbed from the small intestine. Considerable amounts of B vitamins are synthesized by the microflora in the large intestine of the horse (Table 5-5), and some of these vitamins are absorbed (Linerode, 1967). Further studies are needed to quantitate the amounts accurately.

Table 5-5
Vitamin content of samples from hay, cecum, colon, and feces of ponies (micrograms per gram of dry matter)

Vitamin	Source			
	Hay	Cecum	Colon	Feces
Niacin	31	92	143	206
Pantothenate	4	20	31	40
Riboflavin	17	14	15	100
Thiamin	0.4	4	17	9

SOURCE: Adapted from P. A. Linerode, 1967.

Minerals Calcium, phosphorus, magnesium and zinc are absorbed from the small intestine, and other minerals are probably also absorbed from that section of the gut. Calcium is actively absorbed; the mucosal cells of the small intestine contain a calcium-binding protein that facilitates calcium absorption. Vitamin D is necessary for the synthesis of the calcium-binding protein. Thus, when there is a deficiency of vitamin D, calcium absorption is decreased. Phosphorus, but not calcium and magnesium, may also be efficiently absorbed from the large intestine. Alexander and Hickson (1970) have suggested that phosphate may be an important buffer to maintain bacterial activity in the large intestine.

Water The cecum is the primary site of net water absorption, although significant amounts of water are also absorbed from the colon (Argenzio et al., 1974b). The cecum effectively passes drier material into the colon. The amount of water in the digestive tract and feces (that is, the efficiency of water recovery) depends on the diet. The feces usually contain 66 to 76 percent water. The feces of horses fed high-grain diets have a lower water content than the feces of horses fed a hay diet. If a horse is fed only oats, the water content of the feces is 50 percent. Alfalfa hay may cause a temporary increase in the water content. However, Fonnesbeck et al. (1967) found little difference in the fecal water content of horses that were fed and had adjusted to a wide variety of grass and legume hays. Feeding of pelleted feeds increases the fecal water content. Ponies and donkeys may have a slightly drier fecal content than horses fed the same diet. Contrary to popular opinion, feeding horses a wet wheat bran mash does not produce feces with greater water content than feeding a dry wheat bran, assuming, of course, that the horse has access to drinking water.

5.5   Factors Affecting Digestion

Processing of Feeds The method of processing may influence digestion. For example, the pelleting of roughage decreases fiber digestion by approximately 9 to 15 percent. Rolling or breaking the kernel is important in the digestion of small grains, such as wheat and milo. Processing of larger grains, such as corn and oats, does not seem to greatly improve digestibility (Table 5-6). All grains should be cracked, crimped, or rolled for foals and horses with poor teeth. There is no advantage in cooking, fermenting, or predigesting feed for horses with good teeth. Steeping wheat bran in warm water (50°C) for one hour does not affect the digestibility of dry matter, protein, or phosphorus.

Table 5-6
Comparison of digestibility of whole and crimped oats (percent)

Type of oats	Digestibility[a]		
	Dry matter	Crude protein	Neutral detergent fiber
Whole	73.2 ± 4.4	85.6 ± 2.7	36.4 ± 4.0
Crimped	75.8 ± 2.2	84.7 ± 3.0	39.2 ± 4.5

[a]Means are average of six values.
SOURCE: Graves et al., unpublished data.

Level of Intake   The level of intake does not appear to greatly affect the digestibility of all-roughage diets. However, the digestibility of diets containing forage and grain may be decreased with increased dietary intake.

Frequency of Feeding   Frequent feedings (2 or 3 times a day) are usually recommended for horses. Because the horse has a relatively small stomach, severe overeating at one time can produce colic or even a ruptured stomach. Of course, overeating may also produce founder. Therefore, the policy of feeding grain at least 2 times a day seems reasonable. Frequency of feeding, however, does not appear to affect digestibility, at least of complete pelleted diets. For example, when the daily feed of ponies was divided into 1, 2, or 6 feedings, digestibility was found not to be influenced by frequency of feeding, as shown in Table 5-7.

Work   Early studies indicated that activity may have a slight influence on the digestion of the horse. Olsson and Ruudvere (1955) suggested that light exercise might improve digestibility but heavy work may inhibit it. They cited the data of Grandeau and LeClere shown in Table 5-8.

Table 5-7
Effect of feeding frequency on digestibility of complete pelleted horse feed (percent)

Feeding frequency	Digestibility[a]			
	Dry matter	Crude protein	Neutral detergent fiber	Acid detergent fiber
1 time/day	71.5	82.3	45.9	28.2
2 times/day	71.0	80.6	44.6	27.6
6 times/day	72.0	79.5	44.2	28.1

[a]Means are average of 3 values.
SOURCE: Butler and Hintz, 1971.

Table 5-8
Effect of work on digestibility of horse

Amount of activity	Relative digestibility
Rest	100
Walking without load	106
Walking with load	101
Trotting without load	98
Trotting with load	97

SOURCE: Cited by Olsson and Ruudvere, 1955.

Orton et al. (1985) reported that horses trotting for 7 miles per day at a rate of 7 miles per hour digested the dry matter of a diet containing 60 percent chopped oat hay and 40 percent grain mixture at a rate of 73 percent. When the horses were not exercised, the digestibility of dry matter was only 63 percent.

Individuality   The individuality of the horse is sometimes thought to affect feed digestibility. For example, Fonnesbeck et al. (1967) reported that horses differ significantly in their ability to digest crude protein and nitrogen-free extract. This suggests that some of the hard keepers may have an impaired ability to digest feeds.

Associative Effects   No associative effects or interactions were observed when ponies were fed diets containing all hay, 50 percent hay and 50 percent grain, or 20 percent hay and 80 percent grain; that is, the addition of the grain did not appear to influence the digestion of the hay (Hintz et al., 1971). Thompson et al. (1984), however, reported that changing the ratio of oats to hay from 20:80 to 80:20 had an associative effect on protein and cellulose digestibility. Glade (1984) suggested a relationship between fiber digestibility and nitrogen requirements. He suggested nitrogen intakes should be increased when horses are fed poorly fermentable feedstuffs.

Time of Watering   It is often stated that horses should be watered only before feeding because if they are watered after feeding, the digestibility of the diet will be decreased; that is, the water will wash the material out of the digestive tract. There is no evidence to support such a theory. In fact, several studies have demonstrated that time of watering does not affect digestibility. However, watering before feeding is still a good practice, because many horses will not eat unless they are first watered.

5.6   Comparative Digestion

Horses and ponies appear to be quite similar in their ability to digest feed-stuffs. Several studies have shown that the horse can digest protein as efficiently or perhaps slightly more efficiently than ruminants. However, ruminants digest fiber more efficiently than horses. The differences between equine and bovine species vary with the type and amount of fiber (that is, the differences increase as the quality of roughage decreases), but in general, horses are approximately two-thirds as efficient as ruminants in the digestion of fiber. The decreased efficiency of horses is usually attributed to faster rate of passage of digesta. The bacteria in the large intestine of the horse do not have as much time to digest the fiber as the bacteria in the rumen. However, there may be other reasons. Ground roughages in nylon bags were placed in the cecum of ponies and the rumen of cattle via a fistula for the same length of time. A greater percentage of dry matter and fiber disappeared from the bags in the rumen than from the bags in the cecum (Koller, et al. 1978). Comparison of fiber digestion of horses with that of other species is shown in Table 5-9. The horse is much more efficient in the digestion of fiber than the rabbit, even though the rabbit practices coprophagy (eating of feces) and has a large cecum. Some horses, particularly young foals and confined animals, also practice coprophagy. The effect of coprophagy on the digestion of the horse is not known, but it probably is so slight that it can be ignored. As mentioned earlier, the digestion coefficients obtained for horses and for ponies are similar. Further studies comparing other equidae are needed.

Table 5-9
Estimates of relative ability of various species to digest fiber

Species	Relative value	Species	Relative value
Llama	115–120	Mule	66–75
Water buffalo	100–110	Donkey	66–75
Cattle	100	Horse	66–70
Sheep	100	Zebra	66–70
Deer	100	Elephant	66–70
Bison	100	Guinea pig	62–70
Gazelle	100	Pig	50–55
Kangaroo	75–80	Hamster	40–45
Chinchilla	75–80	Rabbit	40–45
Onager	70–75	Vole	40–45
		Tortoise	40–45

Some preliminary reports suggest that the mule, donkey, and onager may be more efficient in the digestion of fiber than the horse.

Ruminants are more efficient than horses in the utilization of phytic acid phosphorus. Bacteria in the rumen produce an enzyme called phytase, which breaks down the phytate molecule and renders the phosphorus available for absorption. Phytase is also present in the intestinal tract of the horse but is not as effective as in the rumen, perhaps because of the faster rate of passage of digesta in the horse.

REFERENCES

Alexander, F. A. 1946. The rate of passage of food residues through the digestive tract of the horse. *J. Comp. Path.* 56:266.

Alexander, F. 1962. The concentration of certain electrolytes in the digestive tract of the horse and pig. *Res. Vet. Sci.* 3:78.

Alexander, F. 1963. Digestion in the horse. In *Progress in Nutrition and Allied Sciences*, D. P. Cuthbertson, ed. Chap. 23, p. 259. London: Oliver and Boyd.

Alexander, F., and J. C. D. Hickson. 1970. The salivary and pancreatic secretions of the horse. In *Physiology of Digestion and Metabolism in the Ruminant*, A. T. Phillipson, ed. Newcastle: Oriel Press.

Argenzio, R. A., J. E. Lowe, D. W. Pickard, and C. E. Stevens. 1974a. Digesta passage and water exchange in the equine large intestine. *Am. J. Physio.* 226:1035.

Argenzio, R. A., M. Southworth, and C. E. Stevens. 1974b. Sites of organic acid production and absorption in the equine gastrointestinal tract. *Am. J. Physiol.* 226:1043.

Argenzio, R. A. 1975. Functions of the equine large intestine and their interrelationship in disease. *Cornell Vet.* 65:303.

Butler, D., and H. F. Hintz. 1971. *Animal Science Memo*, Cornell University.

Cunha, T. J. 1980. *Horse Feeding and Nutrition.* New York, Academic Press.

Ducharme, N. G., F. D. Horney, M. Arighi, J. D. Baird, J. H. Burton, and M. A. Livesey. 1985. Surgical and nutritional implications of extensive colonic resection in ponies and horses. *Proc. 31st Am. Assoc. Equine Pract.*, Toronto, Ontario, p. 505.

Fonnesbeck, P. V., R. K. Lydman, G. W. Vande Noot, and L. D. Symons. 1967. Digestibility of the proximate nutrients of forage by horses. *J. Animal Sci.* 26:1039.

Fonnesbeck, P. V. 1968. Consumption and excretion of water by horses receiving all hay and hay-grain diets. *J. Animal Sci.* 27:1350.

Glade, M. J. 1984. The influence of dietary fiber digestibility on the nitrogen requirements of mature horses. *J. Animal Sci.* 58:638.

Hintz, H. F. 1983. *Horse Nutrition: A Practical Guide.* New York: Arco.

Hintz, H. F., R. A. Argenzio, and H. F. Schryver. 1971. Digestion co-efficients, blood glucose levels, and molar percentage of volatile acids in intestinal fluid of ponies fed varying forage-grain ratios. *J. Animal Sci.* 33:992.

Hintz, H. F., H. F. Schryver, and J. E. Lowe. 1973. Digestion in the horse — a review. *Feedstuffs* (July 2), p. 25.

Kern, D. L., L. L. Slyter, J. M. Weaver, E. C. Leffel, and G. Samuelson. 1973. Pony cecum vs. steer rumen: The effect of oats and hay on the microbial ecosystem. *J. Animal Sci.* 37:463.

Kern, K. L., L. L. Slyter, E. C. Leffel, J. M. Weaver, and R. R. Oltjen. 1974. Ponies vs. steers: Microbial and chemical characteristics of intestinal ingesta. *J. Animal Sci.* 38:559.

Koller, B., H. F. Hintz, J. B. Robertson, and P. J. Von Soest. 1978. Comparative cell wall and dry matter digestion in the cecum of the pony and rumen of the cow using in vitro and nylon bag techniques. *J. Animal Sci.* 47:209.

Lewis, L. D. 1982. *Feeding and Care of the Horse.* Philadelphia. Lea and Febiger.

Linerode, P. A. 1967. Studies on synthesis and absorption of B-complex vitamins in the horse. *Am. Assoc. Equine Pract.* 23:283.

Meyer, H. 1982. Contributions to digestive physiology of the horse. *Zeitschrift Tierphysiol. Tiernahr. Futtermitte.* Supplement 13:1 – 69.

Meyer, H., M. Pferdekamp, and B. Huskamp. 1979. Untersuchungen über die Verdaulichkeit vershiener Futtermittel bei typhylektomietten Ponys. *Dtsc. Tierarzt. Woch.* 86:377.

Olsson, N., and A. Ruudvere. 1955. Nutrition of the horse. *Nutr. Abst. Rev.* 25:1.

Orton, R. K., I. D. Hume, and R. A. Leng. 1985. Effect of exercise and level of dietary protein on digestive function in horses. *Equine Vet. J.* 17:386.

Rebhun, W. C., S. G. Dill, and H. F. Power. 1982. Gastric ulcers in foals. *J.A.V.M.A.* 180:404.

Reitnour, C. M., and R. L. Salsbury. 1975. Effects of oral or cecal administration of protein supplements on equine plasma amino acids. *Brit. Vet. J.* 131:466.

Roberts, M. C. 1975. Carbohydrate digestion and absorption studies in the horse. *Res. Vet. Sci.* 18:64

Robinson, D. W., and L. M. Slade. 1974. Current status of knowledge on the nutrition of equines. *J. Animal Sci.* 39:1045.

Schryver, H. F., P. H. Craig, and H. F. Hintz. 1970. Calcium metabolism in ponies fed varying levels of calcium. *J. Nutr.* 100:955.

Slade, L. M., R. Bishop, J. G. Morris, and D. W. Robinson. 1971. Digestion and absorption of ^{15}N-labeled microbial protein in the large intestine of the horse. *Brit. Vet. J.* 127:xi.

Smith, H. W. 1965. Observations on the flora of the alimentary tract of animals and factors affecting its composition. *J. Path. Bact.* 89:89.

Stick, J. A., N. E. Robinson, and J. D. Krehbiel. 1981. Acid-base and electrolyte alterations associated with salivary loss in the pony. *Am. J. Vet. Res.* 42:733.

Sullins, K. E., T. S. Stashak, S. L. Ralston. 1985. Experimental large intestinal resection in the horse: Technique and nutritional performance. *Proc. 31st Am. Assoc. Equine Pract.*, Toronto, Ontario, p. 497.

Templeton, J. A., and I. A. Dyer. 1967. Diet and supplemental enzyme effects on volatile fatty acids of bovine rumen fluid. *J. Animal Sci.* 26:1374.

Thompson, K. N., S. G. Jackson, and J. P. Baker. 1984. Apparent digestion coefficients and associative effects of varying hay:grain ratios fed to horses. *Nutr. Rpts. Intern.* 30:189.

Tyznik, W. J. 1973. The digestive system of the horse. *Stud Managers Handbook* 9:55. Clovis, California: Agriservices Foundation.

Wysocki, A. A., and J. P. Baker. 1975. Utilization of bacterial protein from the lower gut of the equine. *Proc. 4th Equine Nutr. Phys. Symp.* (January), Pomona, California, p. 21.

Wilson, J. H., and M. M. Pearson. 1985. Serum pepsinogen levels in foals with gastric or duodenal ulcers. *Proc. 31st Am. Assoc. Equine Pract.*, Toronto, Ontario, p. 149

CHAPTER 6

NUTRIENTS

On examining the animal's head, I was particularly struck with the enlarged
and roundish appearance of the facial region. . . . He stepped short, flexed his
limbs with difficulty and apparently with much pain. . . . Bran formed the
greater part of the diet. Thinking that perhaps the kind of food the horses had
been eating had had something to do in producing the disease, I advised that
the bran should be given occasionally, and that some other food should be
made the basis of the support.

—G. Varnell. 1860. *The Veterinarian* 33:493.

6.1 Energy

Several methods are commonly used to determine the energy content of
feeds. The gross energy is determined by igniting the feed in a bomb calo-
rimeter and recording the amount of heat produced. The digestible energy
content is determined in a digestion trial (see Section 5.3), in which the
energy content of the feed and feces is determined. Digestible energy is
gross energy in feed minus energy lost in feces. Metabolizable energy is
determined by correcting digestible energy for energy lost in the gaseous
products of digestion, such as methane, and the energy lost in urine. Net
energy is the term used to refer to the energy that is actually used by the
animal. These methods express the energy content as calories per unit of
weight.

Another method of energy evaluation is to calculate total digestible
nutrients (TDN). This is the sum of all the digestible organic nutrients

(protein, fiber, nitrogen free extract, and fat). The digestible fat content is multiplied by 2.25 because the energy content of fat per unit of weight is 2.25 times that of carbohydrates. TDN is determined in digestion trials and is correlated to digestible energy (DE). Values for TDN can be converted to approximate DE by assuming that 2000 kilocalories is equal to one pound of TDN (4400 kilocalories would equal one kilogram of TDN). TDN and DE are the two most commonly used methods of evaluating the energy content of horse feeds.

Metabolism   The carbohydrate fraction of the diet is the primary source of energy, although protein and fat are also used as energy sources. When high levels of soluble carbohydrates are fed, glucose is the primary product of digestion and is the primary energy source. The importance of volatile fatty acids (VFA), which are the end products of bacteria fermentation, increases as the amount of roughage in the diet increases (see Table 5-4).

The nature of endocrine response to glucose and VFA in the horse differs from the response in ruminants; in horses, VFA does not stimulate the release of insulin. The endocrine response to glucose infusion in horses is more similar to the response obtained in nonruminants such as the dog than to the response obtained in ruminants. Fasted horses are much less sensitive to insulin action than are fed horses (Argenzio and Hintz, 1970; Evans, 1971).

Requirements   The National Research Council (1989) suggests that the maintenance requirements for digestible energy (DE) in kcal/day is equal to $1.4 + 0.03\,W$, where W is weight of the horse in kilograms (as adapted from Pagan and Hintz, 1986a). This equation may overestimate the maintenance requirements of large draft animals because they may have reduced voluntary activity. The requirements for animals with mature weights of more than 600 kg can be estimated by DE (kcal/day) $= 1.82 + 0.0383W - 0.000015W^2$ where W is weight of the horse in kilograms. Energy requirements are given in Tables 6-1A to 6-1G.

Pregnancy does not increase the energy requirements until the last third of the gestation period. At that time, energy intake should be increased approximately 10 percent above maintenance. Intake should be gradually increased throughout the last third of gestation until the level is approximately 20 percent above maintenance. The NRC (1989) estimated that the DE requirements for the ninth, tenth, and eleventh months of gestation can be calculated by multiplying the maintenance requirements by 1.11, 1.13, and 1.20 respectively.

Lactation may greatly increase the energy needs but the amount needed will depend on the quantity and quality of milk produced. The amount of milk produced daily can be expected to be equivalent to 2.5 to 3 percent of body weight during early lactation but there appears to be considerable

Table 6-1A
Daily nutrient requirements of ponies (200-kg mature weight)

Animal	Weight (kg)	Daily gain (kg)	DE (Mcal)	Crude protein (g)	Lysine (g)	Calcium (g)	Phos-phorus (g)	Magne-sium (g)
Mature horses:								
Maintenance	200		7.4	296	10	8	6	3.0
Stallions (breeding season)	200		9.3	370	13	11	8	4.3
Pregnant mares								
9 months			8.2	361	13	16	12	3.9
10 months			8.4	368	13	16	12	4.0
11 months			8.9	391	14	17	13	4.3
Lactating mares								
Foaling to 3 months	200		13.7	688	24	27	18	4.8
3 months to weaning	200		12.2	528	18	18	11	3.7
Working horses								
Light work[a]	200		9.3	370	13	11	8	4.3
Moderate work[b]	200		11.1	444	16	14	10	5.1
Intense work[c]	200		14.8	592	21	18	13	6.8
Growing horses:								
Weanling, 4 months	75	0.40	7.3	365	15	16	9	1.6
Weanling, 6 months								
Moderate growth	95	0.30	7.6	378	16	13	7	1.8
Rapid growth	95	0.40	8.7	433	18	17	9	1.9
Yearling, 12 months								
Moderate growth	140	0.20	8.7	392	17	12	7	2.4
Rapid growth	140	0.30	10.3	462	19	15	8	2.5
Long yearling, 18 months								
Not in training	170	0.10	8.3	375	16	10	6	2.7
In training	170	0.10	11.6	522	22	14	8	3.7
2-year-old								
Not in training	185	0.05	7.9	337	13	9	5	2.8
In training	185	0.05	11.4	485	19	13	7	4.1

Table 6–1B
Daily nutrient requirements of horses (400-kg mature weight)

Animal	Weight (kg)	Daily gain (kg)	DE (Mcal)	Crude protein (g)	Lysine (g)	Calcium (g)	Phosphorus (g)	Magnesium (g)
Mature horses:								
Maintenance	400		13.4	536	19	16	11	6.0
Stallions (breeding season)	400		16.8	670	23	20	15	7.7
Pregnant mares								
9 months			14.9	654	23	28	21	7.1
10 months			15.1	666	23	29	22	7.3
11 months			16.1	708	25	31	23	7.7
Lactating mares								
Foaling to 3 months	400		22.9	1141	40	45	29	8.7
3 months to weaning	400		19.7	839	29	29	18	6.9
Working horses								
Light work[a]	400		16.8	670	23	20	15	7.7
Moderate work[b]	400		20.1	804	28	25	17	9.2
Intense work[c]	400		26.8	1072	38	33	23	12.3
Growing horses:								
Weanling, 4 months	145	0.85	13.5	675	28	33	18	3.2
Weanling, 6 months								
Moderate growth	180	0.55	12.9	643	27	25	14	3.4
Rapid growth	180	0.70	14.5	725	30	30	16	3.6
Yearling, 12 months								
Moderate growth	265	0.40	15.6	700	30	23	13	4.5
Rapid growth	265	0.50	17.1	770	33	27	15	4.6
Long yearling, 18 months								
Not in training	330	0.25	15.9	716	30	21	12	5.3
In training	330	0.25	21.6	970	41	29	16	7.1
2-year-old								
Not in training	365	0.15	15.3	650	26	19	11	5.7
In training	365	0.15	21.5	913	37	27	15	7.9

Table 6–1C
Daily nutrient requirements of horses (500-kg mature weight)

Animal	Weight (kg)	Daily gain (kg)	DE (Mcal)	Crude protein (g)	Lysine (g)	Calcium (g)	Phosphorus (g)	Magnesium (g)
Mature horses:								
Maintenance	500		16.4	656	23	20	14	7.5
Stallions (breeding season)	500		20.5	820	29	25	18	9.4
Pregnant mares								
9 months			18.2	801	28	35	26	8.7
10 months			18.5	815	29	35	27	8.9
11 months			19.7	866	30	37	28	9.4
Lactating mares								
Foaling to 3 months	500		28.3	1427	50	56	36	10.9
3 months to weaning	500		24.3	1048	37	36	22	8.6
Working horses								
Light work[a]	500		20.5	820	29	25	18	9.4
Moderate work[b]	500		24.6	984	34	30	21	11.3
Intense work[c]	500		32.8	1312	46	40	29	15.1
Growing horses:								
Weanling, 4 months	175	0.85	14.4	720	30	34	19	3.7
Weanling, 6 months								
Moderate growth	215	0.65	15.0	750	32	29	16	4.0
Rapid growth	215	0.85	17.2	860	36	36	20	4.3
Yearling, 12 months								
Moderate growth	325	0.50	18.9	851	36	29	16	5.5
Rapid growth	325	0.65	21.3	956	40	34	19	5.7
Long yearling, 18 months								
Not in training	400	0.35	19.8	893	38	27	15	6.4
In training	400	0.35	26.5	1195	50	36	20	8.6
2-year-old								
Not in training	450	0.20	18.8	800	32	24	13	7.0
In training	450	0.20	26.3	1117	45	34	19	9.8

Table 6–1D
Daily nutrient requirements of horses (600-kg mature weight)

Animal	Weight (kg)	Daily gain (kg)	DE (Mcal)	Crude protein (g)	Lysine (g)	Calcium (g)	Phos- phorus (g)	Magne- sium (g)
Mature horses:								
Maintenance	600		19.4	776	27	24	17	9.0
Stallions (breeding season)	600		24.3	970	34	30	21	11.2
Pregnant mares								
9 months			21.5	947	33	41	31	10.3
10 months			21.9	965	34	42	32	10.5
11 months			23.4	1024	36	44	34	11.2
Lactating mares								
Foaling to 3 months	600		33.7	1711	60	67	43	13.1
3 months to weaning	600		28.9	1258	44	43	27	10.4
Working horses								
Light work[a]	600		24.3	970	34	30	21	11.2
Moderate work[b]	600		29.1	1164	41	36	25	13.4
Intense work[c]	600		38.8	1552	54	47	34	17.8
Growing horses:								
Weanling, 4 months	200	1.00	16.5	825	35	40	22	4.3
Weanling, 6 months								
Moderate growth	245	0.75	17.0	850	36	34	19	4.6
Rapid growth	245	0.95	19.2	960	40	40	22	4.9
Yearling, 12 months								
Moderate growth	375	0.65	22.7	1023	43	36	20	6.4
Rapid growth	375	0.80	25.1	1127	48	41	22	6.6
Long yearling, 18 months								
Not in training	475	0.45	23.9	1077	45	33	18	7.7
In training	475	0.45	32.0	1429	60	44	24	10.2
2-year-old								
Not in training	540	0.30	23.5	998	40	31	17	8.5
In training	540	0.30	32.3	1372	55	43	24	11.6

Table 6–1E
Daily nutrient requirements of horses (700-kg mature weight)

Animal	Weight (kg)	Daily gain (kg)	DE (Mcal)	Crude protein (g)	Lysine (g)	Calcium (g)	Phosphorus (g)	Magnesium (g)
Mature horses:								
Maintenance	700		21.3	851	30	28	20	10.5
Stallions (breeding season)	700		26.6	1064	37	32	23	12.2
Pregnant mares								
9 months			23.6	1039	36	45	34	11.3
10 months			24.0	1058	37	46	35	11.5
11 months			25.5	1124	39	49	37	12.3
Lactating mares								
Foaling to 3 months	700		37.9	1997	70	78	51	15.2
3 months to weaning	700		32.4	1468	51	50	31	12.1
Working horses								
Light work[a]	700		26.6	1064	37	32	23	12.2
Moderate work[b]	700		31.9	1277	45	39	28	14.7
Intense work[c]	700		42.6	1702	60	52	37	19.6
Growing horses:								
Weanling, 4 months	225	1.10	19.7	986	41	44	25	4.8
Weanling, 6 months								
Moderate growth	275	0.80	20.0	1001	42	37	20	5.1
Rapid growth	275	1.00	22.2	1111	47	43	24	5.4
Yearling, 12 months								
Moderate growth	420	0.70	26.1	1176	50	39	22	7.2
Rapid growth	420	0.85	28.5	1281	54	44	24	7.4
Long yearling, 18 months								
Not in training	525	0.50	27.0	1215	51	37	20	8.5
In training	525	0.50	36.0	1615	68	49	27	11.3
2-year-old								
Not in training	600	0.35	26.3	1117	45	35	19	9.4
In training	600	0.35	36.0	1529	61	48	27	12.9

Table 6–1F
Daily nutrient requirements of horses (800-kg mature weight)

Animal	Weight (kg)	Daily gain (kg)	DE (Mcal)	Crude protein (g)	Lysine (g)	Calcium (g)	Phosphorus (g)	Magnesium (g)
Mature horses:								
Maintenance	800		22.9	914	32	32	22	12.0
Stallions (breeding season)	800		28.6	1143	40	35	25	13.1
Pregnant mares								
9 months			25.4	1116	39	48	37	12.2
10 months			25.8	1137	40	49	37	12.4
11 months			27.4	1207	42	52	40	13.2
Lactating mares								
Foaling to 3 months	800		41.9	2282	81	90	58	17.4
3 months to weaning	800		35.5	1678	60	58	36	13.8
Working horses								
Light work[a]	800		28.6	1143	40	35	25	13.1
Moderate work[b]	800		34.3	1372	48	42	30	15.8
Intense work[c]	800		45.7	1829	64	56	40	21.0
Growing horses:								
Weanling, 4 months	250	1.20	21.4	1070	45	48	27	5.3
Weanling, 6 months								
Moderate growth	305	0.90	22.0	1100	46	41	23	5.7
Rapid growth	305	1.10	24.2	1210	51	47	26	6.0
Yearling, 12 months								
Moderate growth	460	0.80	28.7	1291	55	44	24	7.9
Rapid growth	460	0.95	31.0	1396	59	49	27	8.1
Long yearling, 18 months								
Not in training	590	0.60	30.2	1361	57	43	24	9.6
In training	590	0.60	39.8	1793	76	56	31	12.6
2-year-old								
Not in training	675	0.40	28.7	1220	49	40	22	10.6
In training	675	0.40	39.1	1662	66	54	30	14.5

Table 6–1G
Daily nutrient requirements of horses (900-kg mature weight)

Animal	Weight (kg)	Daily gain (kg)	DE (Mcal)	Crude protein (g)	Lysine (g)	Calcium (g)	Phos- phorus (g)	Magne- sium (g)
Mature horses:								
Maintenance	900		24.1	966	34	36	25	13.5
Stallions (breeding season)	900		30.2	1207	42	37	26	13.9
Pregnant mares								
9 months			26.8	1179	41	51	39	12.9
10 months			27.3	1200	42	52	39	13.1
11 months			29.0	1275	45	55	42	13.9
Lactating mares								
Foaling to 3 months	900		45.5	2567	89	101	65	19.6
3 months to weaning	900		38.4	1887	66	65	40	15.5
Working horses								
Light work[a]	900		30.2	1207	42	37	26	13.9
Moderate work[b]	900		36.2	1448	51	44	32	16.7
Intense work[c]	900		48.3	1931	68	59	42	22.2

Growing horses:

Weanling, 4 months	275	1.30	23.1	1154	48	53	29	5.8
Weanling, 6 months								
Moderate growth	335	0.95	23.4	1171	49	44	24	6.2
Rapid growth	335	1.15	25.6	1281	54	50	28	6.5
Yearling, 12 months								
Moderate growth	500	0.90	31.2	1404	59	49	27	8.6
Rapid growth	500	1.05	33.5	1509	64	54	30	8.8
Long yearling, 18 months								
Not in training	665	0.70	33.6	1510	64	49	27	10.9
In training	665	0.70	43.9	1975	83	64	35	14.2
2-year-old								
Not in training	760	0.45	31.1	1322	53	45	25	12.0
In training	760	0.45	42.2	1795	72	61	34	16.2

[a]Examples are horses used in Western and English pleasure, bridle path hack, equitation, etc.
[b]Examples are horses used in ranch work, roping, cutting, barrel racing, jumping, etc.
[c]Examples are horses in race training, polo, etc.
SOURCE: National Research Council, 1989.

variation among mares. In general, however, lactation usually increases the energy requirement 50 to 70 percent above maintenance. It is assumed that 792 kcal of DE is required to produce a kilogram of milk.

The amount of energy can be very critical for the performance of the horse. The effects of severe overfeeding or underfeeding are obvious. Guidelines such as those provided by the NRC (1989) are helpful but they are only starting points. Surveys at several Thoroughbred and Standardbred tracks indicate that the racehorses are likely to be fed an amount equivalent to 2.5 to 3.3 percent of their body weight. Thus, there is great variation. The important consideration is the body condition of the horse. Horses should be fed as individuals. More trainers should use scales to determine the proper weight or note changes in weight. European and Asian trainers are much more likely to use scales than are American trainers. Lim (1981) reported that weighing of horses prior to a race is required by the Japanese Racing Association and the weights are displayed on T.V. monitors for the information of the betting public. Body weights at previous races may be included in the racing program. Lim (1981) studied 90 horses in Malaysia. The body weight at which a horse won a race was taken as a standard. It was concluded that the average range of optimal body weight was standard weight plus or minus 16 pounds. The average loss during a race was 9 pounds. Horses that performed well the next week gained the 9 pounds back within 48 hours. Performance was more likely to be adversely affected when the horses were below the optimal weight than when they were above the weight. Estimates of energy requirements for various activities are shown in Tables 6-1A to 6-1G.

The requirements for growing animals depend on the growth rate that is desired. The optimal growth rate is yet to be determined. The National Research Council (1989) estimates of energy requirements of horses growing at moderate or rapid rates are given in Tables 6-1A to 6-1G.

Deficiency   A deficiency of energy in the young horse results in a slow rate of growth and a generally unthrifty appearance. In mature animals there is loss of weight, and poor reproductive and athletic performance.

Toxicity   An excess of energy results in obese horses. The excess fat may decrease performance and reproductive efficiency. Obesity is often considered the most common form of malnutrition in horses in the United States. Also, excess energy makes horses "high" and more difficult to handle. For this reason, grain is not usually fed during initial breaking. Overfeeding of growing horses has been reported to cause a variety of skeletal problems and will be discussed in more detail later.

6.2   Protein

Protein is composed of a chain of smaller units called amino acids, which contain nitrogen. These amino acids are the building blocks the body uses to synthesize body tissues. Some of these amino acids can be synthesized by the body but others, called the essential amino acids, cannot and must be supplied in the diet. The amino acids considered to be essential for most mammals are arginine, histidine, isoleucine, leucine, lysine, methionine, phenylalanine, threonine, tryptophan, and valine.

Metabolism   The small intestine is the primary site of protein digestion in the horse, although significant amounts of nitrogen are absorbed from the large intestine. The nutritional value of nitrogen absorbed from the large intestine is the subject of considerable debate. Some reports indicate that bacteria use nitrogen in the digesta to synthesize amino acids and that the bacterial amino acids are subsequently absorbed from the large intestine, but other reports suggest that the primary products absorbed from the large intestine are not amino acids but ammonia and other non–amino acid nitrogen compounds (see Section 5.4). These latter compounds can be used by the tissues in synthesizing the nonessential amino acids but cannot be used in synthesizing the essential amino acids. The latter theory is the more widely accepted one.

Significant quantities of urea are recycled and hydrolyzed in the gastrointestinal tract of the horse under normal feeding conditions—that is, even when the dietary nitrogen is primarily protein and the protein is not fed at an excessively high level.

Mature horses can use limited amounts of nonprotein nitrogen compounds such as urea in the diet to increase nitrogen retention, but such compounds are not utilized as efficiently as dietary protein and appear to have little practical importance.

Requirements   The quality of the protein (the amino acid content) included in the diet is very important for the growth of young foals. For example, young horses fed a diet in which milk products were the primary protein source grew much faster and more efficiently than horses fed a diet in which linseed meal, and the addition of lysine to be linseed diet greatly improved the growth rate. Weanlings appear to require 0.65 to 0.7 percent lysine in the daily diet. Other amino acids that have been demonstrated to be required for nonruminants such as rat and man are probably also required in the diet of the young horse, but no estimates of the actual requirements are available.

The protein requirement depends on the amino acid content. However, estimates of the total protein required can be made if the diet is assumed to contain a high-quality protein (good mixture of essential amino acids), including an adequate lysine level. Lysine is often the limiting amino acid in vegetable proteins (that is, it limits the growth of the animal because it is not present in adequate amounts). The protein and lysine content of some common feedstuffs is shown in Table 7-4. Soybean meal contains the highest lysine content of the commonly used vegetable proteins. The percentage of protein in the form of lysine is higher in animal products, such as dried skim milk and fish meal, but these products are usually much more expensive than soybean meal. Of course, not only the amount of protein or amino acids, but also their availability or digestibility, is important. The values in the last column in Table 7-4 indicate that most common protein sources are readily digested by the horse. Excessive heating of the protein supplement during processing greatly decreases the digestibility of the protein by the horse.

In general, foals weaned at 4 months should be fed a diet containing 16 percent protein (90 percent dry matter basis). Foals at 6 months of age should be fed at least 14 percent protein, assuming that the diet contains high-quality proteins. Yearlings require 12 percent protein, mature horses require about 8 percent, pregnant mares require 10 to 11 percent, and lactating mares require 11 to 13 percent. Work does not increase the protein requirement when expressed as a fraction of the diet. The increased food intake as a result of the increased energy needs for work compensates for any nitrogen lost in sweat. The National Research Council's estimates of protein requirements, as well as requirements of calcium, phosphorus, and vitamin A are given in Tables 6-1A through 6-1G. Protein concentrations needed are shown in tables 6-2A and 6-2B.

Deficiency   Signs of protein deficiency are general rather than specific. In young animals, poor growth, a high feed-to-gain ratio, and general unthriftiness can result because of protein deficiency. In mature animals, there may be a loss of weight, poor hair coat, decreased reproduction, and decreased hoof growth. Pregnant mares fed protein-deficient diets will have small, weak foals.

Toxicity   Protein toxicity is not a problem. When excessive amounts of protein are fed, the carbon chain of the amino acid is utilized for energy and the nitrogen is excreted via the urine by the kidneys. Excessive ingestion of nitrogen-rich compounds such as urea, however, can be toxic. Urea is converted into ammonia and carbon dioxide by an enzyme, urease, that is produced by the intestinal bacteria. The resulting increase in ammonia absorption can be lethal. One of the clinical effects of ammonia toxicosis is severe central nervous system derangement (Hintz et al., 1970). The earliest signs are aimless wandering and incoordination followed by pressing of the head against fixed objects (Figure 6-1). Once head pressing begins, it is usually continued until the animal falls, and remains at this site until death. Death usually occurs 30 to 90 minutes after the onset of the first signs of nervous system derangement. The first signs of derangement may occur 2 to 10 hours after animals ingest a toxic dose of urea. Although high levels of urea are toxic to horses, horses are more tolerant of dietary urea than cattle. Thus, there is no danger in feeding horses a ration containing urea if the ration is safe for cattle. In the horse, much of the dietary urea is absorbed from the small intestine and excreted via the kidneys. Thus the bacteria in the large intestine do not have an opportunity to convert urea into ammonia and carbon dioxide. Other dangers of feeding horses feeds designed for other species are discussed in Chapter 7.

6.3   Minerals

Calcium

Metabolism   Calcium has many important functions in the body. It is necessary for bone formation (approximately 99 percent of the total body calcium is contained in the skeleton). Calcium is also necessary for normal muscular activity, blood clotting, and enzyme activation. Animals have a regulatory system that tries to ensure that the blood always contains adequate levels of calcium. Whenever the blood calcium level decreases, the parathyroid gland releases a hormone, parathormone, that causes calcium to be released from the bone and thus increases the amount of calcium in the blood. Calcitonin, a hormone produced in the thyroid gland, is activated when the blood calcium level is too high. This hormone decreases the amount of calcium that is removed from the bone. Excessive intakes of calcium result in increased kidney excretion of calcium.

Calcium is absorbed from the small intestine, and the amount absorbed is influenced by many factors. Vitamin D apparently stimulates calcium absorption by inducing the formation of a calcium-binding protein in the cells of the intestine. This calcium-binding protein is important in the absorption of calcium. The level of calcium in the diet also influences the efficiency of calcium absorption, because high levels of dietary calcium decrease the digestibility of calcium. A high level of dietary phosphorus relative to the dietary calcium also decreases calcium absorption. Therefore, the diet should always contain at least as much calcium as phosphorus; that is, the calcium-to-phosphorus ratio should be at least 1 : 1. Lactose, particularly in the diet of the young animal, may increase calcium absorption.

Several tropical grasses such as setaria, buffel, pangola, and kikuyu contain high levels of oxalic acid, which decreases the bioavailability of calcium

Table 6–2A
Nutrient concentrations in total diets for horses and ponies (dry matter basis)

	Digestible energy[a]		Diet proportions		Crude protein (%)	Lysine (%)	Cal-cium (%)	Phos-phorous (%)	Mag-nesium (%)
	(Mcal/kg)	(Mcal/lb)	Conc. (%)	Hay (%)					
Mature horses:									
Maintenance	2.00	0.90	0	100	8.0	0.28	0.24	0.17	0.09
Stallions	2.40	1.10	30	70	9.6	0.34	0.29	0.21	0.11
Pregnant Mares									
9 months	2.25	1.00	20	80	10.0	0.35	0.43	0.32	0.10
10 months	2.25	1.00	20	80	10.0	0.35	0.43	0.32	0.10
11 months	2.40	1.10	30	70	10.6	0.37	0.45	0.34	0.11
Lactating mares									
Foaling to 3 months	2.60	1.20	50	50	13.2	0.46	0.52	0.34	0.10
3 months to weaning	2.45	1.15	35	65	11.0	0.37	0.36	0.22	0.09
Working horses									
Light work[b]	2.45	1.15	35	65	9.8	0.35	0.30	0.22	0.11
Moderate work[c]	2.65	1.20	50	50	10.4	0.37	0.31	0.22	0.11
Intense work[d]	2.85	1.30	65	35	11.4	0.40	0.35	0.25	0.13

Growing horses:

Weanling, 4 months	2.90	1.40	70	30	14.5	0.60	0.68	0.38	0.08
Weanling, 6 months									
Moderate growth	2.90	1.40	70	30	14.5	0.61	0.56	0.31	0.08
Rapid growth	2.90	1.40	70	30	14.5	0.61	0.61	0.34	0.08
Yearling, 12 months									
Moderate growth	2.80	1.30	60	40	12.6	0.53	0.43	0.24	0.08
Rapid growth	2.80	1.30	60	40	12.6	0.53	0.45	0.25	0.08
Long yearling, 18 months									
Not in training	2.50	1.15	45	55	11.3	0.48	0.34	0.19	0.08
In training	2.65	1.20	50	50	12.0	0.50	0.36	0.20	0.09
2-year-old									
Not in training	2.45	1.15	35	65	10.4	0.42	0.31	0.17	0.09
In training	2.65	1.20	50	50	11.3	0.45	0.34	0.20	0.10

[a]Values assume a concentrate feed containing 3.3 Mcal/kg and hay containing 2.00 Mcal/kg of dry matter.
[b]Examples are horses used in Western and English pleasure, bridle path hack, equitation, etc.
[c]Examples are horses used in ranch work, roping, cutting, barrel racing, jumping, etc.
[d]Examples are race training, polo, etc.
SOURCE: National Research Council, 1989.

Table 6–2B
Nutrient concentrations in total diets for horses and ponies (90% dry matter basis)

	Digestible energy[a]		Diet proportions		Crude protein (%)	Lysine (%)	Cal-cium (%)	Phos-phorous (%)	Mag-nesium (%)
	(Mcal/kg)	(Mcal/lb)	Conc. (%)	Hay (%)					
Mature horses:									
Maintenance	1.80	0.80	0	100	7.2	0.25	0.21	0.15	0.08
Stallions	2.15	1.00	30	70	8.6	0.30	0.26	0.19	0.10
Pregnant Mares									
9 months	2.00	0.90	20	80	8.9	0.31	0.39	0.29	0.10
10 months	2.00	0.90	20	80	9.0	0.32	0.38	0.30	0.10
11 months	2.15	1.00	30	70	9.5	0.33	0.41	0.31	0.10
Lactating mares									
Foaling to 3 months	2.35	1.10	50	50	12.0	0.41	0.47	0.30	0.09
3 months to weaning	2.20	1.05	35	65	10.0	0.34	0.33	0.20	0.08
Working horses									
Light work[b]	2.20	1.05	35	65	8.8	0.32	0.27	0.19	0.10
Moderate work[c]	2.40	1.10	50	50	9.4	0.35	0.31	0.22	0.11
Intense work[d]	2.55	1.20	65	35	10.3	0.36	0.31	0.23	0.12

Growing horses:

Weanling, 4 months	2.60	1.25	70	30	13.1	0.54	0.62	0.34	0.07
Weanling, 6 months									
Moderate growth	2.60	1.25	70	30	13.0	0.55	0.50	0.28	0.07
Rapid growth	2.60	1.25	70	30	13.1	0.55	0.55	0.30	0.07
Yearling, 12 months									
Moderate growth	2.50	1.15	60	40	11.3	0.48	0.39	0.21	0.07
Rapid growth	2.50	1.15	60	40	11.3	0.48	0.40	0.22	0.07
Long yearling, 18 months									
Not in training	2.30	1.05	45	55	10.1	0.43	0.31	0.17	0.07
In training	2.40	1.10	50	50	10.8	0.45	0.32	0.18	0.08
2-year-old									
Not in training	2.20	1.00	35	65	9.4	0.38	0.28	0.15	0.08
In training	2.40	1.10	50	50	10.1	0.41	0.31	0.17	0.09

[a] Values assume a concentrate feed containing 3.3 Mcal/kg and hay containing 2.00 Mcal/kg of dry matter.
[b] Examples are horses used in Western and English pleasure, bridle path hack, equitation, etc.
[c] Examples are horses used in ranch work, roping, cutting, barrel racing, jumping, etc.
[d] Examples are race training, polo, etc.
SOURCE: National Research Council, 1989.

Figure 6-1
Head pressing in a pony
caused by brain damage due
to high level of ammonia in
the blood.

by forming calcium oxalate. McKenzie et al. (1981) concluded that calcium deficiency is likely to occur when the total calcium to total oxalate ratio in the feed is less than 0.5. Additional dietary calcium can overcome the effect of oxalic acid.

Requirements   Young animals that are rapidly synthesizing bone and lactating mares have the greatest calcium needs.

Mineral requirements for growth are more difficult to determine than those for maintenance because the criteria of adequacy are not well defined. The rate of body gain, although sometimes used, is not a good criterion because dietary mineral levels that are adequate for optimum weight gain may not be adequate for proper bone formation. In one experiment, ponies fed a low calcium diet for 12 weeks were found to be in negative calcium balance but gained in height and weight at the same rate at those fed higher levels of calcium (Schryver et al., 1974). The criteria for the mineral requirements for growth that were used to obtain the estimated values that follow included mineral retention, bone ash per unit volume of bone, calcium and phosphorus content of bone and of the total body, and specific gravity of bone.

Weanlings may require 150 to 180 mg of calcium per kilogram of body weight per day and yearlings require 100 mg of calcium per kilogram of body weight per day. Thus diets for weanlings that contain 0.6 percent calcium and diets for yearlings that contain 0.4 percent calcium are adequate.

Diets containing 0.25 percent calcium are adequate to maintain mature animals. The mineral requirement of the mare is increased during the last third of pregnancy to provide calcium and phosphorus for mineralization of the developing fetal skeleton. However, the calcium demands of pregnancy are relatively small and can be met by diets containing 0.40 percent calcium.

The mineral requirements of lactation are much greater than those of pregnancy and increase as lactation progresses, reaching a peak at 8 to 12 weeks. During peak lactation, a 450-kilogram mare may produce 20 kg of milk daily containing 800 to 1000 mg of calcium per kg. Lactation calcium requirements can be met by diets containing 0.47 percent calcium (see Tables 6-2A and 6-2B).

Deficiency   Bone is formed by a complicated series of events. A simplified explanation is that cells called osteoblasts produce a protein matrix called osteoid tissue. The tissue is calcified when the proper amounts of calcium, phosphorus, magnesium, and other minerals are present. However, bone is not static; it is continually being re-formed. Calcium and phosphorus are constantly being released by the action of cells called osteocytes and osteoclasts. If the young, growing animal does not have enough calcium to adequately calcify the bone, the bone becomes weak and may be deformed. This condition, *in extremis*, is called rickets. (Rickets occurs only in young animals and can be caused by a deficiency of calcium, phosphorus, or vitamin D.)

Nutritional secondary hyperparathyroidism is a bone disease that may occur in horses if their diet contains a low level of calcium and a high level of phosphorus. A high level of phosphorus interferes with calcium absorption. As discussed earlier in this section, a normal level of blood calcium is required for many functions. Whenever the blood calcium level drops, the parathyroid hormone increases the release of calcium from the bone to maintain the blood level of calcium. Lameness and enlarged facial bones may result (Figure 6-2). The enlarged facial bones result because fibrous connective tissue invades the area from which calcium was resorbed. The bones are lacking in density and sound hollow when tapped. This condition is often called bighead disease, brain disease, or miller's disease. It is called bran disease because feeding high levels of wheat bran, which is very high in phosphorus and low in calcium, produces the condition. The name "miller's disease" was given to the condition because in the past it was common among horses owned by grain millers. The miller often fed his horses the by-products of the milling industry such as wheat bran.

Toxicity   It has been suggested that when animals are fed high levels of calcium for prolonged periods, their bones become very brittle and dense because of an overproduction of the hormone calcitonin (Krook et al., 1969). Whitlock (1970) reported that young horses fed high levels of calcium for long periods had levels of calcitonin that were above normal and evidenced

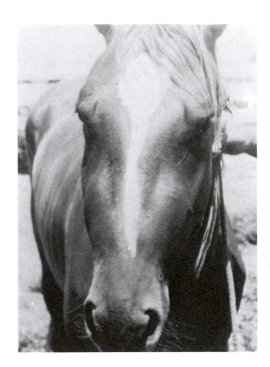

Figure 6-2
"Big head" disease in a horse
caused by excessive phosphorus
and a low level of calcium.
(Photograph courtesy of Texas
A & M University.)

decreased bone resorption. High dietary levels of calcium may reduce the availability of other minerals such as phosphorus, magnesium, and manganese. Therefore, intakes of calcium greatly exceeding the requirements should be avoided.

Phosphorus

Metabolism Approximately 80 percent of the body's phosphorus is contained in the skeleton. Phosphorus is a component of many compounds, such as phosphoproteins. nucleoproteins, phospholipids, and the energy-rich adenosine phosphate compounds. The feces are the primary pathway for phosphorus excretion, but the kidneys also excrete some phosphorus when large amounts of phosphorus are fed. Inorganic phosphate compounds are readily digested by the horse, but some organic compounds, such as phytin phosphorus, are less available. Horses do not utilize phytin phosphorus as efficiently as ruminants but they are more efficient than pigs and chicks, probably because the enzyme phytase, produced by the intestinal bacteria, renders some of the phytin phosphorus available for absorption. Fortunately, the horse can absorb phosphorus from the large intestine, which is the primary site of phytase activity. Although a high-phosphorus–low-calcium ratio greatly hinders calcium digestibility, the opposite effect is much less dramatic. The data shown in Table 6-3 demonstrate that although high levels of calcium inhibit phosphorus absorption, the effect is not so great as the effect of phosphorus on calcium absorption. This may be because much of the calcium is absorbed from the small intestine. Thus, the actual calcium-to-phosphorus ratio of the digesta in the large intestine, a primary site of phosphorus absorption, is much different than the calcium-to-phosphorus ratio in the diet.

Requirements   Dietary requirements for phosphorus are as follows: weanlings require 0.40 percent of the ration, yearlings need 0.25 percent, mature horses need 0.17 percent, pregnant mares require 0.35 percent, and lactating mares need 0.35 percent.

Deficiency   A phosphorus deficiency will result in rickets in the young animal. In a mature animal, deficiency results in osteomalacia (soft bones). Depraved appetite and reduced productive performance are also signs of possible phosphorus deficiency.

Toxicity   The feeding of diets containing high levels of phosphorus, such as wheat bran diets, may be harmful because of the induced calcium deficiency.

Table 6–3
Effect of calcium and phosphorus on digestibility

Percentage in diet		Ca:P Ratio	Estimated true digestibility (percent)	
Ca	P		Ca	P
Trial 1				
0.40	0.20	2.0:1.0	68.1	40.0
0.35	1.19	0.3:1.0	43.5	47.0
Trial 2				
0.15	0.35	0.4:1.0	70.0	53.2
1.50	0.35	4.3:1.0	46.0	47.2
Trial 3				
0.80	0.25	3.2:1.0	75.1	57.4
3.40	0.25	13.6:1.0	52.7	40.7

SOURCE: Schryver, H. F., *et al.* 1974.

However, the situation can be remedied by the addition of calcium to the diet or removal of the source of phosphorus.

Magnesium

Metabolism   Magnesium is an essential constituent of bones and teeth and is required for many body processes as an activator of enzymes. The endogenous losses of magnesium, in contrast to the losses of calcium and phosphorus, are greater in the urine than in the feces. Magnesium digestibility is decreased when excessive amounts of either phosphorus or calcium are present in the diet.

Requirements   Few studies have been conducted on the magnesium requirement, but mature horses apparently require 15 milligrams per kilogram of body weight per day (Meyer, 1960; Hintz and Schryver, 1973).

Deficiency   The signs of magnesium deficiency are hyperirritability, glazed eyes, tetany, and eventual collapse. The condition is often precipitated by some form of stress. Some veterinarians have reported that some horses at sales exhibit these signs and seem to respond to a magnesium-calcium injection. Years ago, it was reported that when mountain ponies or ponies that worked in mines were transported, they suffered from a "transit tetany" that responded to magnesium and calcium injection (Green et al., 1935). Harrington (1971) reported that magnesium deficiency in young foals resulted in calcification of the blood vessels.

Toxicity   There have been no reports of magnesium toxicity in the horse attributed to the diet. However, prolonged breathing of magnesium fumes could be toxic. Studies with other animals demonstrate that high intakes of magnesium cause neurological problems.

Potassium

Metabolism   Potassium is readily absorbed from feeds, and the primary path of excretion is the urine.

Requirements   It has been suggested that the diets of early weaned foals should contain at least 1 percent potassium per day (Stowe, 1971). Other studies indicate that mature horses need approximately 0.4 to 0.6 percent potassium in the diet to maintain positive potassium balance (Hintz and Schryver, 1976).

Deficiency   Reduced appetite is an early sign of potassium deficiency. Weakness and restlessness may also be noted (Meyer et al., 1985).

Toxicity   There have been no reports of potassium toxicity in the horse. A few cases of muscular weakness associated with hyperkalemia (high blood levels of potassium) have been reported in heavily muscled quarter horses (Cox, 1985). The condition does not appear to be the result of excessive intakes of potassium. Dr. Cox suggested that the disease was similar to hyperkalemic periodic paralysis in man which is an inherited disease.

6.4   Trace Nutrients

Trace nutrients are nutrients that are required in small amounts. The requirements are summarized in Table 6-4.

Iodine

Metabolism   Iodine is a component of thyroxine, a product of the thyroid gland that contains 75 percent of the iodine present in the body and controls the metabolic rate. Little research has been conducted on iodine metabolism in the horse, but the element is probably readily absorbed from the small intestine. The urine is the primary path of excretion.

Requirements   The National Research Council (1989) suggests that the iodine requirement is 0.1 ppm.

Deficiency   When mares are fed iodine-deficient diets, their foals may have an enlarged thyroid (goiter) and be stillborn or weak and hairless.

Toxicity   Feeding excessive levels of iodine to the mare (more than 40 mg per day) may result in weak foals with goiters (Figure 6-3) (Baker and Lindsay, 1968; Drew et al., 1975). Such foals often have leg weaknesses and may die shortly after birth. Some types of kelp, a seaweed, contain high levels of iodine. Driscoll et al. (1978) reported cases of iodine toxicity in foals that occurred because the owner fed the mares a supplement at a rate of 12 times the manufacturer's recommendations, which resulted in intakes much greater than 40 mg per day.

Table 6-4
Other minerals and vitamins for horses and ponies (on a dry matter basis)

	Adequate concentrations in total diet				Maximum tolerance levels
	Maintenance	Pregnant and lactating mares	Growing horses	Working horses	
Minerals					
Sodium (%)	0.10	0.10	0.10	0.30	3[a]
Sulfur (%)	0.15	0.15	0.15	0.15	1.25
Iron (mg/kg)	40	50	50	40	1000
Manganese (mg/kg)	40	40	40	40	1000
Copper (mg/kg)	10	10	10	10	800
Zinc (mg/kg)	40	40	40	40	500
Selenium (mg/kg)	0.1	0.1	0.1	0.1	2.0
Iodine (mg/kg)	0.1	0.1	0.1	0.1	5.0
Cobalt (mg/kg)	0.1	0.1	0.1	0.1	10
Vitamins					
Vitamin A (IU/kg)	2000	3000	2000	2000	16,000
Vitamin D (IU/kg)[c]	300	600	800	300	2200
Vitamin E (IU/kg)	50	80	80	80	1000
Vitamin K (mg/kg)	[b]				
Thiamin (mg/kg)	3	3	3	5	3000
Riboflavin (mg/kg)	2	2	2	2	—
Niacin (mg/kg)					
Pantothenic acid (mg/kg)					
Pyridoxine (mg/kg)					
Biotin (mg/kg)					
Folacin (mg/kg)					
Vitamin B_{12} (µg/kg)					
Ascorbic acid (mg/kg)					
Choline (mg/kg)					

[a]As sodium chloride.
[b]Blank space indicates that data are insufficient to determine a requirement or maximum tolerable level.
[c]Recommendations for horses not exposed to sunlight or to artificial light with an emission spectrum of 280–315 mm.
SOURCE: National Research Council, 1989.

Figure 6-3
Enlarged thyroid in foal
caused by feeding excess
iodine to the mare.
(Photograph courtesy of
William Sippel and Florida
Department of Agriculture.)

Iron

Metabolism   Iron is an essential component of the hemoglobin of the red
blood cells. Iron is primarily absorbed from the small intestine. Presumably,
iron absorption in the horse is similar to that reported for other species in
that the efficiency of absorption is determined by the body's need. Iron is
efficiently conserved by the body because the endogenous losses are small.
Loss of blood because of heavy parasite loads or wounds is the primary cause
of iron loss.

Requirements   The National Research Council (1989) suggests that a dietary concentration of 40 ppm of iron daily is adequate for mature animals
but that rapidly growing foals may require 50 ppm.

Deficiency   Anemia is the primary sign of iron deficiency.

Toxicity   Several cases have been reported in which horses died shortly
after being given an intramuscular injection of iron dextran (Wagenaar,
1975).
     Smith et al. (1984) pointed out that vitamin and mineral mixtures are
often given to horses in an attempt to improve performance. Iron is often
included in the mixture. They concluded that excess iron can be harmful.
The excess iron may be stored in various tissues, replace other minerals, and
result in weakened tissues. The condition is called hemochromatosis.
     Smith and coworkers (1984) stated that packed cell volume (concentra-
tion of blood cells) and hemoglobin are commonly used to evaluate iron

status in horses but are relatively insensitive. The horse has a large store of red blood cells in the spleen. When the horse becomes excited, the spleen is contracted, red blood cells released, and the packed cell volume can be increased by as much as 68 percent.

Smith and co-workers evaluated the use of serum ferritin assays. Ferritin is an iron-containing protein. Ferritin was not influenced by excitement of the horses. They concluded that measuring serum ferritin should provide a convenient and accurate method to evaluate iron stores in horses. They stated, "Serum ferritin assays should allow nutritionists, trainers, and veterinarians to determine if horses are iron deficient before instituting iron therapy." They studied 103 horses and found no iron deficiency anemias, but some of the horses appeared to have an overload. They could not determine the cause of the overload because they had neither the history nor the age of the horses. But they suggested that the horses were older and thus accumulated iron over the years or may have been given oral or injected iron in an attempt to increase performance. They concluded that iron deficiency could limit red-blood-cell production but that such a deficiency probably only occurs because of blood loss.

Zinc

Requirements   The daily zinc requirements is probably less than 50 ppm.

Deficiency   The signs of deficiency are decreased zinc blood levels, poor growth, and skin lesions (parakeratosis) (Harrington et al., 1973).

Toxicity   Growing horses fed diets containing 0.54 percent zinc developed anemia, swelling at the epiphyseal region of the long bones, stiffness, and lameness (Willoughby et al., 1973) (Figure 6-4). In some areas the forage may contain high levels of zinc because of fumes from nearby smelters. High levels of zinc may decrease utilization of trace minerals, particularly that of copper.

Selenium

Requirements   The selenium requirement appears to be 0.1 ppm (NRC, 1989). The requirement depends on the vitamin E content of the diet because vitamin E and selenium act synergistically.

Deficiency   Selenium deficiency results in muscular dystrophy in the young foal. The muscles become very pale, hence the name "white muscle disease" (Figure 6-5). The affected foals often die of starvation because they are too

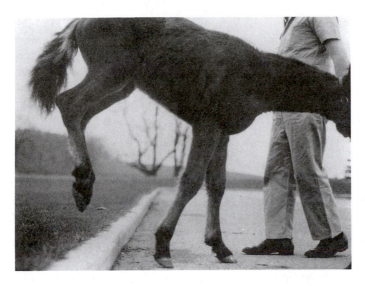

Figure 6-4
Zinc toxicity in this foal caused pain and stiffness, and the animal had difficulty stepping over an 8-inch curb. (Photograph courtesy of R. Willoughby.)

weak to nurse and because the tongue and pharyngeal muscles are weakened. The condition occurs in areas where the soil has a low available selenium content and the mares are fed homegrown feeds.

Roneus and Jonsson (1984) reported the details of 46 foals with muscular dystrophy. Most of the cases were seen within three days of birth. Foals with the acute stage had difficulty standing up, had stiff and swollen muscles, and had tachycardia. Histological changes in the muscles were also described.

Maylin et al. (1980) reported that although only limited amounts of selenium cross the placenta, supplementing mares with 1 mg of selenium per

Figure 6-5
White muscle disease in a foal. Note the striations or pale areas in the intracostal muscles. (Photograph courtesy of R. Whitlock.)

day during gestation and lactation will increase blood selenium levels in foals above those associated with selenium deficiency. Deficiency of selenium causes low blood selenium levels and low levels of the enzyme glutathione peroxidase (Caple et al., 1978). Assay of the enzyme may be a satisfactory method to evaluate selenium status. Dill and Rebhun (1985) concluded that blood selenium levels of less than 0.06 ppm or enzyme levels of less than 15 units suggested selenium deficiency.

Toxicity   Chronic selenium toxicity can result in sloughing of hooves, loss of manes and tails, and eventually death. The condition is called alkali disease. Acute toxicity results in respiratory failure, blindness, and death. The condition is called "blind staggers." Soil in areas of Wyoming, Colorado, South Dakota, and Nebraska has a high selenium content, and plants raised in these areas may be toxic to horses. Commercial mineral mixes may not contain significant amounts of selenium, and thus free-choice feeding of mineral mixes does not produce selenium toxicity. Selenium injections are often used for various muscle disorders, and improper use could result in excess selenium, but few such cases have been reported.

Sodium Chloride (Salt)

Requirements   The requirement for sodium chloride depends on the amount lost in sweat. The National Research Council (1989) suggests that 50 to 60 g of supplemental salt daily will meet the needs of most horses but that additional salt is required in hot climates. Horses given free choice will usually eat to meet their needs. Lactation and sweating increases the salt requirements.

Deficiency   Signs of deficiency include depraved appetite, rough hair coat, and reduced growth. Meyer et al. (1983) reported that ponies deprived of sodium chloride had reduced feed intake, developed a licking habit, and showed signs of dehydration, weight loss, and reduced performance.

Toxicity   There is little danger of excess sodium chloride unless water is not available. However, there have been some reports of fatalities caused by the drinking of salt brine, or the feeding of salt to salt-hungry horses when adequate water was not available. Salt poisoning signs are colic, diarrhea, frequent urination, weakness, staggering, and paralysis of the hind limbs (Friedberger and Frohner, 1908).

Copper

Requirements   The copper requirement is estimated to be 10 ppm (NRC, 1989); however, other reports suggest the requirement may be higher.

Deficiency The activity of the osteoblasts (bone-forming cells) would be decreased and the bones would be thin and weak. Osteochondrosis may also develop (Bridges et al. 1984).

Toxicity The level of copper that is toxic to horses has not been established. Sheep are less tolerant of high dietary levels of copper than any other species that has been studied. Liver damage has been reported in sheep fed diets containing 50 to 60 ppm copper, whereas rats can tolerate diets containing 500 ppm copper. Horses are also more tolerant of excess copper than are sheep. For example, horses were fed 800 ppm copper as $CuCO_3$ for 180 days with no apparent harmful effects (Smith et al., 1975).

Manganese

Requirements The NRC (1989) estimate of 40 ppm is based on studies with cattle.

Deficiency Manganese is required for enzymes that are needed for the formation of cartilage; thus manganese-deficient animals may have shortened and malformed bones. Cowgill et al. (1980) reported that foals from mares fed diets low in manganese had legs so misshapen and joints so enlarged that the foals were unable to flex their limbs. Some bones were quite shortened. The mares were fed an alfalfa hay containing 13 ppm of manganese. The low level was induced by adding large amounts of limestone to the fields to counteract high acidity in the soil caused by sulfur dioxide from a nearby smelter.

Toxicity The toxicity level in horses is not established but excessive levels should be avoided as manganese can interfere with the utilization of other nutrients

Fluorine

Deficiency A fluorine deficiency has not been demonstrated in the horse.

Toxicity Horses are apparently more tolerant of fluorine than are cattle and sheep, but the lesions resulting from toxicity are similar to those observed in other species. Teeth and bones are the parts of the body most severely affected. The severity of the lesion depends on the level of fluoride. Teeth may be mottled and stained. Higher levels cause the enamel to be pitted and off-colored and may also cause severe abrasion and loss of teeth. The bones may become thicker and have abnormal patterns. A "Roman nose" appearance may result because of the thickening of the nasal and maxillary bones (Shupe and Olson, 1971).

Lead

Deficiency   Lead is not an essential nutrient.

Toxicity   An excess of lead in young horses causes pharyngeal and laryngeal paralysis, bone lesions, poor growth, muscular weakness, and anemia (Willoughby et al., 1972). Lead poisoning has been reported in horses located near smelters in Germany, the United States, and Ireland. The pastures contained a high level of lead because of lead emission from the smelters (Burrows and Orchard, 1982). Excessive chewing of wood coated with lead paint can also result in lead poisoning.

6.5   Vitamins

Vitamin A

Requirements   The National Research Council (1989) suggests that diets should provide 30 to 60 IU of retinol or equivalent per kilogram of body weight.

Deficiency   There are many signs of vitamin A deficiency. Anorexia (loss of appetite), night blindness, lacrimation (excessive tearing or watering of the eyes, shown in Figure 6-6), keratinization of the cornea and skin, respiratory difficulties, reproductive failure, convulsive seizures, blindness caused by pinching of the optic nerve, bone lesions, and impaired resistance to disease can result from a deficiency of vitamin A. One of the earliest reports of a probable vitamin A deficiency in the equine is found in the Old Testament in Jeremiah 14:6. The wild asses had trouble seeing and breathing, possibly because a drought had killed all the grass, and the dried grass had little vitamin A value. A deficiency of vitamin A is not likely to happen if good-quality hay or pasture is available.

Toxicity   Donoghue and Kronfeld (1981) fed three ponies diets containing 12,000 µg of retinol (the vitamin A alcohol) per kilogram of body weight per day. The animals appeared unthrifty by 15 weeks with rough hair coats, poor muscle tone, and depression. By week 20 they had lost large amounts of hair and were severely depressed, and death followed. Therefore, overuse of vitamin A supplements must be avoided.

Carotene   Plants do not contain vitamin A but rather a pigment called carotene that body tissues can convert to vitamin A. The efficiency of con-

Figure 6-6
Chronic lacrimation (tearing) in a foal deficient in vitamin A. Note matted hair below the medial canthus of the eye. Narrowing of the nasolacrimal duct, the result of squamous metaplasia, may have caused some of the flowing tears.

version of carotene to vitamin A in horses is not well established but the NRC has estimated that 1 mg of carotene is equivalent to 400 IU of vitamin A.

Carotene is easily oxidized and thus hay stored for prolonged periods may have low levels of carotene.

It has been suggested that carotene has a function in addition to being a vitamin A precursor. Some reports with cattle indicate that the addition of carotene, even in the presence of vitamin A, improves reproductive efficiency. However, the value of carotene for cattle is disputed, and studies in Germany, England, and South Africa failed to find a benefit in horses fed adequate levels of vitamin A (Eitzer and Rapp, 1985).

Vitamin D

Requirements The daily requirement for vitamin D has not been established, but 6.6 IU per kilogram of body weight is known to be adequate. Horses are not likely to need vitamin D supplements when they are fed sun-cured roughages or are exposed to sunlight. Sun-cured hay contains significant amounts of vitamin D, and the ultraviolet rays of sunlight convert the dehydrocholesterol produced by the animal's body into vitamin D.

Deficiency   El Shorafa et al. (1979) reported that ponies deprived of both dietary vitamin D and sunlight for 5 months had reduced growth rate, loss of appetite, and reduced bone ash. However, no problems were detected in ponies given vitamin D or allowed to be in the sunlight.

Toxicity   Levels of 14,000 IU per kilogram of body weight per day result in acute toxicity and calcification of lungs, heart, kidneys, and other organs within 10 days (Figure 6-7). Levels of 3500 IU per kilogram of body weight per day resulted in chronic toxicity. The signs were elevated serum phosphorus, calcification of kidneys, rarefication of bones, severe loss of weight, and death after 3 to 4 months. Levels of 700 IU per kilogram of body weight per day were fed to young ponies for 9 months and no lesions were observed (Hintz et al., 1973).

Harrington and Page (1983) reported an accidental poisoning by vitamin D_3 when the feed manufacturer mistakenly formulated a feed that provided 12,000 IU per kilogram of body weight per day for about 30 days. Signs included limb stiffness, anorexia, weight loss, rib fractures and soft-tissue mineralization. They also reported that horses could tolerate higher levels of vitamin D_2 than vitamin D_3. It was suggested that horses preferentially utilize vitamin D_3.

Vitamin E

Requirements   The requirement for vitamin E has not been thoroughly studied, but foals deficient in vitamin E have been shown to require 27 μg of

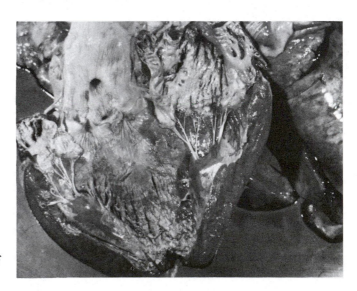

Figure 6-7
Calcification of the heart muscles of a growing foal because of excessive intake of vitamin D.

parenteral tocopherol or 233 μg of oral tocopherol per kilogram of body weight per day to maintain erythrocyte stability (Stowe, 1968). Vitamin E combined with selenium has also been effective in the prevention of white muscle disease. The NRC (1989) suggests that diets should provide 50 to 80 IU per kilogram of feed.

Toxicity   There have been no reports of vitamin E toxicity.

Vitamin K

This vitamin is not considered to be a dietary essential for horses. Vitamin K is produced by bacteria in the intestine, but beware of excess levels of vitamin K. Rebhun et al. (1984) reported that vitamin K_3 injected at rates of 2.2 to 11 mg per kilogram of body weight caused anorexia, colic, stranguria, and hematuria in horses. At necropsy, renal toxicosis characterized by tubular necrosis and dilatation, epithelial degeneration, and interstitial fibrosis was observed. The levels of vitamin K_3 used in the study by Rebhun et al. were those previously recommended by the manufacturers of vitamin K_3 products.

B Vitamins

The National Research Council (1978) has stated that B vitamins need not be added to the rations of most horses, because good-quality hay is an excellent source of B vitamins and significant amounts of B vitamins are produced in the large intestine of the horse. However, the National Research Council (1978) further suggested that additional vitamins might be needed if the horse is under heavy stress, or used for performance such as racing or show, or if very poor quality hay is fed for prolonged periods. Presumably the recommendation for stressed horses was made because the B vitamins play a role in energy metabolism. The greater the energy intake, the greater the B vitamin requirement. However, no information is available as to whether a greater ratio of B vitamins to energy is needed for exercised horses than for horses fed maintenance diets.

Thiamin

Requirements   The requirement for thiamin has not been established, but 3 to 5 mg per kilogram of feed appears to be adequate. Although some thiamin appears to be absorbed from the large intestine (Linerode, 1967), Carroll (1950) demonstrated that horses fed poor-quality hay may develop a thiamin deficiency.

Deficiency Horses fed thiamin-deficient diets exhibited anorexia, nervousness incoordination in the hindquarters, loss of weight, and general weakness (Carroll, 1950). Thiamin deficiency may be induced in horses by the ingestion of plants such as bracken fern and "mare's tail," which contain compounds that impair utilization of thiamin (Figure 6-8).

Toxicity There have been no reports of thiamin toxicity in horses.

Riboflavin

Requirements The riboflavin requirement is not known, but 2 mg of riboflavin per kilogram of feed is adequate (NRC, 1989).

Deficiency Early reports indicated that riboflavin deficiency may produce periodic ophthalmia or moonblindness (recurrent equine uveitis—see Figure 6-9) in the horse (Jones, 1942). However, most veterinarians today do not accept the theory that riboflavin deficiency causes the condition. Recurrent equine uveitis is probably an immunological response to other disease conditions.

Toxicity There have been no reports of riboflavin toxicity in horses.

Vitamin B_{12}

Requirements Mature horses that received 6 mg of vitamin B_{12} per day for 11 months failed to show any signs of B_{12} deficiency (Stillions et al., 1971a).

Figure 6-8
Bracken fern contains an antithiamin compound and may induce thiamin deficiency in horses. Horses seldom eat the plant but will do so when other feeds are not available.

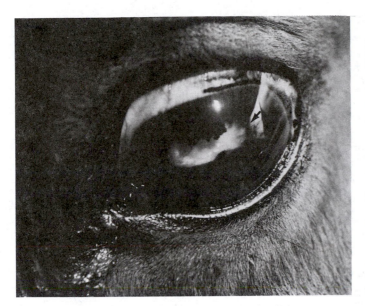

Figure 6-9
Recurrent equine uveitis (periodic ophthalmia, or moonblindness) is often thought to be caused by riboflavin deficiency. However, most recent studies indicate that the condition is an immunological response to other diseases. The cataract (*at arrow*) in the eye is the result of several episodes of recurrent uveitis. Sequelae of this disease can cause severe internal architectural damage, cataracts, and blindness. (Photograph courtesy of R. Riis.)

Significant amounts of B_{12} are produced by the bacteria in the digestive tract (Davies, 1968), and B_{12} can be absorbed from the large intestine (Salminen, 1975).

Deficiency   Vitamin B_{12} deficiency has not been produced experimentally in horses; however, severely debilitated, anemic, heavily parasitized animals appear to respond to B_{12} injections.

Toxicity   No information is presently available on the toxic levels of B_{12} or signs of vitamin B_{12} toxicity.

Biotin

Biotin deficiency in pigs causes dermatitis, cracks in the feet, and lack of coordination. Comben et al. (1984) reported that field studies indicated that daily supplements of 15 mg of biotin improved poor-quality hooves in horses. It was concluded that biotin should be only expected to maintain or to improve the condition of the surface keratin tissues — the periople, hoof walls, sole, and frog and the white-line junctions. Furthermore, several months were needed for the improvement. It was further suggested that 2 mg of biotin daily would maintain the hooves once they were repaired. The NRC (1989) concluded that data are not adequate to establish a requirement.

Other B Vitamins

Few studies have been made of the other B vitamins. Horses fed a low level of niacin excreted more niacin than they consumed, which suggests that niacin is not a dietary essential for horses (National Research Council, 1978). No differences in growth rate were observed when horses consumed 38 mg or 150 mg of panthothenic acid per kilogram of body weight. Thus, diets containing lower levels of pantothenic acid must be fed in order to determine if it is a dietary essential. Low blood levels of folic acid have been reported in horses without access to pasture.

Ascorbic Acid

Requirements   Ascorbic acid (vitamin C) is not considered a dietary essential for horses. Adequate amounts of ascorbic acid are synthesized in the liver of the horse (Stillions et al., 1971b). One early study done with small numbers of animals indicated an increase in the fertility of stallions and mares that were fed vitamin C, but the experiment has never been repeated (Davis and Cole, 1943). Johnson et al. (1973) reported that vitamin C is often used in the treatment of epistaxis (bleeding from the nose) and sometimes seems to alleviate this condition, but the vitamin C levels in the blood of horses with epistaxis were not found to differ from those of horses without epistaxis.

6.6   Water

Metabolism   In addition to drinking water, horses also obtain water contained in feeds and water from metabolic water formed during the oxidation of protein, carbohydrates, and fats within the body. The oxidation of 100 g of glucose, protein, and fats results in 60, 42, and 100 g of metabolic water, respectively. The primary paths of water loss from the body are urine, feces, sweat, and respiration. During periods of water deprivation, the amount of water lost in the feces decreases. Diarrhea results in great water loss and can lead to severe dehydration.

Requirements   Water requirements are difficult to discuss without discussing the many factors that influence the water needs of the horse.

   1. *Temperature*   An increase in the environmental temperature increases the water requirement. Studies in Russia demonstrated that

increasing the temperature from 55° to 70°F increased the water requirement of horses by 15 to 20 percent (Caljuk, 1961).

2. *Activity*   Moderate work may increase the water requirement by 60 to 80 percent above the amount required by the resting animal and hard work may increase it 120 percent. There may be interactions between factors. For example, hard work in a hot environment may increase the water requirement more than one might expect to obtain by simply adding the effect of temperature and the effect of work.

3. *Function*   Mares in the last third of gestation drink approximately 8 to 10 percent more water than non-pregnant mares. Lactating mares may increase their water intake 50 to 70 percent. Of course, the intake will vary according to milk production as milk is approximately 90 percent water.

4. *Food intake and type of food*   Several studies have shown that the total water intake is related to the dry matter intake. Thus, the amount of drinking water needed depends on the water content of the food and the total dry matter intake. The water content of feeds can vary greatly. For example, fresh young grass contains 70 to 80 percent water, whereas hay and most stored grains contain approximately 10 percent water. Several estimates of water requirements are available, but most of them indicate that the horse requires approximately 1 to 2 quarts of water per pound of dry matter consumed. Thus, a 1000-lb horse that consumed 16 pounds of hay per day would need approximately 4 to 8 gallons of water per day.

Other dietary variables also influence water needs. The greater the fiber content, the greater the fecal excretion of water. An increased ash or mineral intake increases the amount of water needed. Water contains varying levels of minerals, depending on the source. In some areas drinking water may supply significant amounts of calcium, copper, magnesium, zinc, and manganese. Of course, water may also contain toxic amounts of minerals such as fluorine, lead, nickel, and selenium. Estimates of water intake are shown in Table 6-5.

Deficiency   It has often been stated that there is no one most important nutrient; that is, all the essential nutrients must be supplied, and a diet is

Table 6–5
Estimated water intake of 1000-lb horse (environmental temperature = 60–70°F)

Activity	Gal/day
Nonworking	4–8
Gestation	7–9
Peak lactation	9–11
Medium work	9–15
Heavy work	12–15

limited if any one of these nutrients is missing or deficient. However, water is the nutrient that man or animal will miss first. For example, ponies deprived of water ate slowly and had decreased their feed intake by the second day. By the third day they were not eating anything, and by the fourth day they were so restless that the experiment was stopped. Each pony lost approximately 40 lb during the experiment, but within 3 days after the end of the experiment, the weight losses were recovered (Turancic and Tvoric, 1971). Donkeys and mules are more tolerant of water deprivation and dehydration than horses (Macfarlane, 1964).

Toxicity   Hot horses or horses that have been deprived of water should not be given large amounts of water at one time or else founder or colic may result.

6.7   Other Equine Species

The requirements previously discussed pertained to the horse. Ponies are generally regarded as "easier keepers" than horses. The efficiency of digestion is similar for ponies and horses, and recent studies have indicated that the maintenance requirements of minerals such as calcium and phosphorus are similar for ponies and horses when expressed on a body weight basis.

The qualitative requirements for nutrients are probably the same for all equidae — horses, ponies, donkeys, mules, zebras, and onagers. Whether the amount of nutrients required per unit of body weight is the same for all is a matter of some controversy. Traditionally, the donkey also has a reputation of being an easy keeper. There is an old saying that thistles and straw are sufficient feed for a donkey. The animal is often fed poor-quality hay yet manages to survive. Unfortunately, few studies have been made on the nutrient requirements of the donkey. Southern farmers claimed that mules required less feed to do the same amount of work as a horse. In fact, it was often stated that "the mule will not require half the feed necessary to sustain a horse. Mules can thrive on hay while horses need grain" (Anon., 1892). Such statements appear to be exaggerations. Lamb (1963) reviewed some early experiments. One experiment in Ohio indicated that mules and horses that were doing the same work ate similar amounts of feed proportionate to their weights. Another experiment in Illinois indicated that the mules ate 0.94 lb of feed per 100 lb of live weight, whereas the horse ate 1.07 lb. Thus, a 1600-lb mule would eat approximately 2 lb less feed per day than the 1600-lb horse. Another experiment cited by Lamb indicated that mules are just as expensive to maintain as horses. The report stated, "it is probably true that the mule is more cheaply maintained but the mule has not been consulted."

Morrison (1957) concluded that mules may require slightly less feed than horses to do a given amount of work. But he also pointed out that although mules will endure more neglect than horses, good care and feed will prove to be profitable. He suggested that the same feeds may be used for both mules and horses and that the same principles apply in adjusting the amount of feed to the size of the animal and to the severity of the work performed. Many zoos have satisfactorily fed rations formulated for domestic horses to wild equidae.

Mules and donkeys are generally considered to be more sensible in eating and less likely to overeat than horses. The mechanism of feed intake control is not known, but several reports suggest that the mule and donkey can be self-fed without the great risk of colic, founder, and obesity that is present when mature horses are self-fed.

REFERENCES

Anon. 1892. In Defense of the Mule. *Weekly Horse World* (November 4), p. 377.

Argenzio, R. A., and H. F. Hintz. 1970. Glucose tolerance and effect of volatile fatty acid on plasma glucose concentration in ponies. *J. Animal Sci.* 30:514.

Baker, H. J., and J. R. Lindsey. 1968. Equine goiter due to excess dietary iodide. *J.A.V.M.A.* 153:1618.

Bridges, C. H., J. E. Womack, and E. D. Harris. 1984. Considerations of copper metabolism in osteochondrosis of suckling foals. *J.A.V.M.A.* 185:173.

Burrows, G. E., and R. E. Borchard. 1982. Experimental lead toxicosis in ponies: Comparison of the effects of smelter effluent—contaminated hay and lead acetate. *Am. J. Vet. Res.* 43:2129.

Caple, I. W., S. J. A. Edwards, W. M. Forsyth, P. Whiteley, R. H. Selth, and L. J. Fulton. 1978. Blood gluthathione peroxidase activity in horses in relation to muscular dystrophy and selenium nutrition. *Aust. Vet. J.* 54:57.

Caljuk, E. A. 1961. Water metabolism and water requirements of horses. *Trudy Vses. Inst. Konevodstra.* 23:295. [As abstracted in *Nutr. Abs. Rev.* 32:574 (1962).]

Carroll, F. D. 1950. B vitamin content in the skeletal muscles of the horse fed a B vitamin low diet. *J. Animal Sci.* 8:290.

Comben, N., R. J. Clark, and D. J. B. Sutherland. 1984. Clinical observations on the response of equine hoof defects to supplementation with biotin. *Vet. Rec.* 115:642.

Cowgill, U. M., S. J. Slater, and J. E. Marburger. 1980. Smelter smoke syndrome in farm animals and deficiency in Oklahoma. *Environ. Pollution* 22:259.

Cox, J. 1985. An episodic weakness in four horses associated with intermittent serum hyperkalemia. *Proc. 31st Am. Assoc. Equine Pract.*, Toronto, Ontario. p. 383.

Cupps, P. T., and C. E. Howell. 1949. The effects of feeding supplemental copper to foals. *J. Animal Sci.* 8:286.

Davies, M. E. 1971. The production of vitamin B_{12} in the horse. *Brit. Vet. J.* 127:34.

Davis, G. K., and C. L. Cole. 1943. The relation of ascorbic acid to breeding performance in horses. *J. Animal Sci.* 26:1030.

Dill, S. G., and W. C. Rebhun. 1985. White muscle disease in foals. *Comp. Cont. Educ. Vet.* 7:S627.

Donoghue, S., and D. S. Kronveld. 1981. Vitamin A nutrition of the equine. *J. Nutr.* 111:365.

Driscoll, J., H. F. Hintz, and H. F. Schryver. 1978. Goiter in foals caused by excessive iodine. *J.A.V.M.A.* 173:858.

Drew, B., W. P. Barber, and D. G. Williams. 1975. The effect of excess dietary iodine on pregnant mares and foals. *Vet. Rec.* 97:93.

Eamens, G. J., J. F. Macadam, and E. A. Laing. 1984. Skeletal abnormalities in young horses associated with zinc toxicity and hypocuprosis. *Aust. Vet. J.* 61:205.

Eitzer, P., and H. J. Rapp. 1985. Zur oralen Anwendung von synthetischem Carotin bei Zuchstuten. *Prak. Tierarzt.* 66:123.

El Shorafa, W. M., J. P. Feaster, E. A. Ott, and R. L. Asquith. 1979. Effect of vitamin D and sunlight in growth and bone development of young ponies. *J. Animal Sci.* 48:882.

Evans, J. W. 1971. Effect of fasting, gestation, lactation and exercise on glucose turnover in horses. *J. Animal Sci.* 33:1001.

Friedberger, F., and E. Frohner. 1908. *Veterinary Pathology.* Trans. by M. H. Hayes. London: Keener and Co.

Green, H. H., W. M. Allcroft, and R. F. Montgomerie. 1935. Hypomagnesaemia in equine transit tetany. *J. Comp. Pathol.* 48:74.

Harrington, D. D., and E. H. Page. 1983. Acute vitamin D_3 toxicosis in horses: case reports and experimental studies of the comparative toxicity of vitamin D_2 and D_3. *J.A.V.M.A.* 182:1358.

Harrington, D. D., C. Marroguin, and V. White. 1971. Experimental magnesium deficiency in horses. *J. Animal Sci.* 33:231.

Harrington, D. D., J. Walsh, and V. White. 1973. Clinical and pathological findings in horses fed zinc deficient diets. *Proc. Third Equine Nutr. and Phys. Symp.* (January), Gainesville, Florida.

Hintz, H. F., J. E. Lowe, A. J. Clifford, and W. J. Visek. 1970. Ammonia intoxication resulting from urea ingestion by ponies. *J.A.V.M.A.* 157:963.

Hintz, H. F., and H. F. Schryver. 1973. Magnesium, calcium and phosphorus metabolism in ponies fed varying levels of magnesium. *J. Animal Sci.* 37:927.

Hintz, H. F., and H. F. Schryver. 1976. Potassium metabolism in ponies. *J. Animal Sci.* 42:637.

Hintz, H. F., H. F. Schryver, and J. E. Lowe. 1971. Comparison of a blend of milk products and linseed meal as protein supplements for young growing horses. *J. Animal Sci.* 33:1274.

Hintz, H. F., H. F. Schryver, and J. E. Lowe. 1973. Effect on vitamin D on calcium and phosphorus metabolism in ponies. *J. Animal Sci.* 37:282.

Johnson, J. H., H. E. Gainer, D. P. Hutcheson, and J. G. Merriam. 1973. Epistaxis. *Proc. Am. Assoc. Equine Prac.*, p. 115. Atlanta, Georgia

Jones, T. C. 1942. Equine periodic ophthalmia. *Am. J. Vet. Res.* 3:45.

Krook, L., L. Lutwak, and K. McEntee. 1969. Dietary calcium, ultimobrachial tumors and osteopetrosis in the bull. *Am. J. Clin. Nutr.* 22:115.

Lamb, R. B. 1963. *The Mule in Southern Agriculture.* Los Angeles: University of California Press.

Lim, A. S. 1981. Body weight and performance. *Proc. 4th Intern. Conf. Drugs in Racehorses,* Melbourne, Australia, p. 93.

Linerode, P. A. 1967. Studies on synthesis and absorption of B-complex vitamins in the horse. *Am. Assoc. Equine Pract.,* New Orleans, Louisiana, p. 283.

Macfarlane, W. V. 1964. Terrestrial animals in dry heat: Ungulates. American Physiological Society, *Handbook of Physiology,* Section 4: *Adaptation to the Environment,* p. 509. Baltimore: Williams & Wilkins.

Maylin, G. Z., D. S. Rubin, and D. H. Lein. 1980. Selenium and vitamin E in horses. *Cornell Vet.* 70:272.

McKenzie, R. A., R. J. W. Gartner, B. J. Blaney, and R. J. Glanville. 1981. Control of nutritional secondary hyperparathyroidism in grazing horses. *Aust. Vet. J.* 57:554.

Meyer, H. 1960. *Magnesiumstoffwechsel, Magneisumbedarf und Magnesiumversorgung.* Hanover: Schaper.

Meyer, H., C. Gurer, and A. Linder. 1985. Influence of potassium intake on potassium metabolism, sweat production and sweat composition. *Proc. Ninth Equine Nutr. Physiol. Soc.,* East Lansing, Michigan, p. 130.

Meyer, H., A. Linder, M. Schmidt, and H. M. Teleb. 1983. Investigations of sodium deficiency in horses. *Proc. Eighth Equine Nutrition Physiol. Soc.,* Lexington, Kentucky, p. 16.

Morrison, F. B. 1957. *Feeds and Feeding.* Ithaca, New York: Publ. by author.

National Research Council. 1978. *Nutrient Requirements of Horses.* Publication No. 6. Washington, D.C.

National Research Council. 1989. *Nutrient Requirements of Horses.* NRC–NAS. Washington, D. C.

Pagan, J. D., and H. F. Hintz. 1986. Equine energetics: Relationship between body weight and energy requirements in horses. *J. Animal Sci.* 83:815.

Rebhun, W. C., B. C. Tennant, S. G. Dill, and J. M. King. 1984. Vitamin K_3–induced renal toxicosis in the horse. *J.A.V.M.A.* 184:1237.

Robinson, D. W., and L. M. Slade. 1974. Current status of knowledge in the nutrition of equines. *J. Animal Sci.* 39:1045.

Roneus, B., and L. Jonsson. 1984. Muscular dystrophy in foals. *Zbl. Vet. Med. A.* 31:441.

Salminen, K. 1975. Cobalt metabolism in horses. *Acta. Vet. Scand.* 16:84.

Schryver, H. F., H. F. Hintz, and J. E. Lowe. 1974. Calcium and phosphorus in the nutrition of the horse. *Cornell Vet.* 64:494

Shupe, J. L., and A. E. Olson. 1971. Clinical aspects of fluorosis in horses. *J. Am. Vet. Med. Assoc.* 159:167.

Smith, J. D., R. M. Jordan, and M. L. Nelson. 1975. Tolerance of ponies to high levels of dietary copper. *J. Animal Sci.* 41:1645.

Smith, J. E., K. Moore, J. E. Cipriano, and P. G. Morris. 1984. Serum ferritin as a measure of stored iron in horses. *J. Nutr.* 114:677.

Stillions, M. C., S. M. Teeter, and W. E. Nelson, 1971a. Utilization of dietary vitamin B_{12} and cobalt by mature horses. *J. Animal Sci.* 32:252

Stillions, M. C., S. M. Teeter, and W. E. Nelson. 1971b. Ascorbic acid requirements of mature horses. *J. Animal Sci.* 32:249.

Stowe, H. D. 1967. Serum selenium and related parameters of naturally and experimentally fed horses. *J. Nutr.* 93:60.

Stowe, H. D. 1968. Alpha-tocopherol requirements for equine erythrocyte stability. *Am. J. Clin. Nutri.* 21:135.

Stowe, H. D. 1971. Effects of potassium in a purified equine diet. *J. Nutr.* 101:629.

Turancic, V., and S. Tvoric. 1971. Effect of water deprivation on horses. *Veterinaria Yugoslavia* 20:179. [As abstracted in *Nutr. Abs. Rev.* 42:1593, 1972.]

Wagenaar, G., 1975. Iron dextran administered to horses. *Tijdscher. Diergeneesk.* 100:562.

Whitlock, R. H. 1970. The effects of high dietary calcium in horses. Ph.D. thesis, Cornell University, Ithaca, New York.

Willoughby, R. A., E. MacDonald, B. J. McSherry, and G. Brown. 1972. Lead and zinc poisoning and the interaction between Pb and Zn poisoning in the foal. *Can. J. Comp. Med.* 36:348.

CHAPTER 7

FEEDS AND FEEDING

Go through the land to all the springs of water and to all the valleys; perhaps
we may find grass and save the horses and mules alive and not lose some of
the animals.

—*1 Kings 18:5*

7.1 Sources of Nutrients

Energy

Grains Any of the grains can be used for energy sources if the characteristics of the grain are considered when balancing the diet. Oats is the grain most preferred by horsemen. It is an excellent feed, but it usually is more expensive per unit of energy than corn. One advantage of oats is that it is a safer feed because it is more difficult to overfeed oats than the other grains. The reasons are that oats are not as digestible and hence are lower in digestible energy (Table 7-1), and the density or weight per volume is less for oats than for the other grains (Table 7-2). Thus, there may be twice as much digestible energy in a quart of corn as in a quart of oats (Figure 7-1). Oats do not have to be processed for horses with normal mouths. Oats also contain slightly more protein and minerals than the other grains.

Corn is often used as an energy source and is usually quite economical. It can be fed whole or cracked or as ear corn but requires closer management

251

Table 7-1
Estimates of energy content of grains and hays (dry matter basis)

Feed	TDN (percent)	Digestible energy (Mcal/kg)
Barley	83	3.66
Corn	88	3.87
Oats	76	3.34
Wheat	87	3.83
Alfalfa, early bloom	56	2.42
Midbloom	52	2.29
Full bloom	49	2.16
Timothy, prebloom	50	2.20
Midbloom	49	2.10
Late bloom	45	1.88

SOURCE: National Research Council, 1978.

Table 7-2
Comparison of weights and volumes of various foods

Feed	Weight of 1 quart (lb)	Volume of 1 pound (qt)
Alfalfa meal	0.6	1.7
Barley, whole	1.5	0.7
Beet pulp, dried	0.6	1.7
Corn, dent, whole	1.7	0.6
Corn, dent, ground	1.5	0.7
Cottonseed meal	1.5	0.7
Linseed meal, old process	1.1	0.9
Linseed meal, new process	0.9	1.1
Molasses, cane	3.0	0.3
Oats	1.0	1.0
Rye, whole	1.7	0.6
Wheat, whole	1.9	0.5
Wheat bran	0.5	2.0

SOURCE: Adapted from Morrison, F. B. 1959.

Figure 7-1
Feed by weight, not volume. There may be twice as much digestible energy in a quart of corn as in a quart of oats because of the greater density and digestibility of corn.

than oats because of the greater energy concentration in corn. Moldly corn should not be fed because it may contain fusarium toxins which cause brain damage or aflatoxins which can cause liver damage.

Barley is commonly fed in the western United States and Canada. The energy concentration of barley is greater than oats but less than corn.

Milo must be ground, crimped, or rolled because the grains are so small and hard that the horse cannot chew and efficiently digest the whole kernel. Wheat must also be processed, but fine grinding should be avoided because of dust problems.

Rye is usually not as palatable as the other grains and should be limited to one-third of the grain mixture. It also must be processed.

Dried brewer's grains are by-products of the brewing industry. In the production of beer, barley is soaked in warm water and allowed to sprout. The germinated kernels form a product called malt. The malt is crushed and heated and the starch is converted enzymatically to sugar. Sugar and other soluble material are removed for further processing. The material remaining after the sugar and soluble material are removed is called brewer's grains. The soluble material is boiled with hops, and yeast is added to convert the sugar to alcohol. The yeast that grows during the fermentation procedure can be harvested and sold as brewer's yeast. Brewer's yeast is an excellent source of B vitamins. Dried brewer's grains are higher in protein but lower in energy than whole grains such as barley and oats. Early research conducted with draft horses indicated that brewer's grains are a wholesome, nutritious, and palatable feed for horses. Because of their high fiber and high protein content, they are useful in the manufacture of complete pelleted feeds for horses.

Molasses Molasses is a good source of energy but it is quite low in protein and phosphorus. Molasses is often added to horse rations to reduce dust and thus to improve palatability.

Vegetables   Fresh vegetables such as carrots and turnips are often used as treats for horses but they are not rich energy sources because they contain almost 90 percent water. Morrison (1959) states that horses and mules may be fed potatoes, cooked or raw, in amounts up to 15 or 20 lb per head daily.

Beet Pulp   Dried beet pulp can be used as a source of energy and roughage for horses. It contains more digestible energy than hay but provides more bulk than grains. Beet pulp contains significant amounts of calcium but has a very low level of phosphorus and B vitamins and contains no carotene or vitamin D.

Citrus Pulp   Citrus pulp is a by-product of the processing of oranges and grapefruit for juice. It consists of the peel and residue from the inside portion. Citrus pulp contains very little protein, phosphorus, or carotene. It contains a high level of calcium because lime is added during the drying process. The palatability of citrus pulp varies with the processing conditions. Care should be taken to feed high-quality pulp. One of the best methods of feeding the dry pulp is to incorporate it in a pellet.

Forages

Hay   Hay provides energy but the energy concentration is much lower than that in grains. Thus, hay cannot supply all the energy needs of animals with a high energy requirement. For example, hard-working horses, lactating mares, and rapidly growing foals may need grain in addition to hay.

The key to simplified horse feeding is to have good-quality hay. Any of the common hays can be fed to horses. The important thing to consider when buying hay is not the kind of hay but rather the nutrition value in relation to the cost. A high quality hay that costs more per ton may be a better buy than cheaper, poor-quality hay. One important consideration when evaluating the quality of the hay is age at harvesting. Young plants contain more digestible energy and nutrients per pound than older plants. As the hay matures, the lignin content increases. Lignin is a structural component and is not digested by horses (Figure 7-2). The effect of the maturity and digestibility of hay is shown in Table 7-1.

Other criteria of good-quality hay are (1) freedom from mold, dust, and weeds; (2) lack of excessive weathering; (3) leafiness and lack of stems; (4) species (legume or grass).

There are two general classes of hay: legume hay and grass or cereal hay. Legume hay contains a higher content of digestible energy, calcium, protein, and vitamin A than grass hay harvested at the same stage of maturity. Alfalfa is the legume hay most commonly fed to horses (Table 7-3). However, many horse owners are prejudiced against feeding it to horses. This prejudice is

Figure 7-2
Mature timothy. Hay made
from such plants contains a
high content of lignin and a
low content of digestible
nutrients.

probably based on the fact that unless good harvesting methods are used, alfalfa hay often is more moldy and dustier than grass. However, in the last 20 years many horsemen have realized the advantages of alfalfa hay, particularly for the broodmare, and thus the use of alfalfa hay has been increasing. Horses should be gradually changed to alfalfa hay, because an abrupt change from grass hay to legume hay could cause digestive upsets. Horses fed alfalfa

Table 7–3
Examples of legume and grass or cereal hays

Legume hay	Grass or cereal
Alfalfa	Barley hay
Birdsfoot trefoil	Bermuda grass
Clover	
Alsike	Bluegrass
Crimson	Bluestem
Red	Bromegrass
Ladino	Fescue
Sweet	Oat hay
Cowpea	Orchard grass
Lespedeza	Prairie grass
Soybeans	Reed canary grass
	Rye grass
	Sudan grass
	Timothy

hay may urinate more and there may be a stronger smell of ammonia in the barn because alfalfa hay contains a higher level of nitrogen than grass hay. The urine will also contain more sediment because of the high calcium content of alfalfa hay; the excess calcium is excreted via the kidneys.

Alfalfa hay, particularly from the Southwest, should be checked for the presence of blister beetles (*Epicauta* spp). The beetles contain a cantharidin that causes colic, fever, degeneration of the lining of the intestinal tract and urinary tract, and eventually death.

Clover and birdsfoot trefoil, when properly harvested, have nutritional characteristics similar to those of alfalfa hay. Moldy red clover hay may cause an extensive amount of saliva excretion (slobbering disease). Moldy sweet clover can induce vitamin K deficiency because of the presence of an antivitamin K factor in the mold. Some producers feel that horses are not particularly fond of birdsfoot trefoil hay and will select other hay when given a choice.

Timothy hay is the favorite of many horsemen. Good-quality timothy hay is an excellent hay for horses, but of course, the grain mixture must contain higher levels of protein and calcium than when legume hay is fed. Other grass hay, such as bromegrass, orchard grass, coastal Bermuda grass, and bluegrass, can also be good horse hays.

Silage   Either legume grass silage or corn silage can be fed to horses and can be a very economical source of nutrients. Silage is not often fed to horses, but Morrison (1959) states that silage can replace from one-third to one-half of the hay usually fed on a dry matter basis. But the silage must be free of mold, because spoiled silage can be poisonous to horses. Ricketts et al., (1984) reported botulism in horses fed big bale silage. They suggested that if big bale silage is used, samples of the material should be carefully selected for a satisfactory aroma, a high dry matter content, a pH of 4.0 to 4.5 and intact plastic wrappings. They concluded that it is difficult to maintain adequate quality control of silage and horse owners should realize that this type of feed inherently carries an element of risk.

Protein

Protein Supplements   Protein supplements that contain a good mixture of essential amino acids are preferred for young growing horses and may also be of benefit to mature horses that require additional protein. Soybean meal contains more lysine than most vegetable proteins and is an excellent protein supplement for horses (Table 7-4).

Two common types of soybean meal contain 44 and 48 percent protein. The 48 percent protein meal is prepared by removing the hulls, which are high in fiber but low in protein; the resulting product is relatively richer in

Table 7 – 4
Protein and lysine content of horse feeds and supplements (percent)

Feeds	Protein	Lysine	Lysine/ protein	Protein digestibility
Alfalfa meal	18	1.20	6.8	75
Alfalfa hay	16	0.96	6.0	74
Clover hay	14	0.90	6.0	75
Corn	9	0.24	2.7	80
Oats	12	0.34	2.8	75
Linseed meal	36	1.20	3.3	70
Soybean meal (44%)	44	3.00	6.8	78
Soybean meal (50%)	50	3.20	6.4	78
Dried skim milk	34	2.60	7.7	85
Timothy hay	7	0.31	4.4	70
Barley	12	0.39	3.3	76
Peanut meal	45	2.30	5.1	75
Beet pulp	9	0.60	6.5	70
Fish meal	6	4.40	7.2	76

SOURCE: Values for protein and lysine content obtained from National Academy of Sciences, 1971. Values for protein digestibility estimated from several sources; major source was Schneider, B. H. 1947.

protein content. Fish meal, dried skim milk, and meat and bone meal are also excellent sources of amino acids but are usually more expensive than soybean meal. Vegetable proteins, such as linseed meal and cottonseed meal, should not be the primary protein source for young horses unless they are fed at higher levels (which may be uneconomical) or combined with additional lysine sources. Cottonseed meal may contain a compound, gossypol, that is toxic to pigs. Studies in California have demonstrated that young horses can be fed rations containing 20 percent cottonseed meal without harmful effects (Moise and Wysocki, 1981). Thus, properly supplemented cottonseed meal seems to be a reasonable protein supplement for young horses. That is, when the total diet contains 0.65 to 0.7 percent lysine.

Urea or other nonprotein nitrogen compounds may be used by the horse in limited amounts under certain conditions, but such compounds are utilized much less efficiently than dietary protein (see Section 5.4).

Legume hays or pellets are also excellent sources of protein for horses.

Minerals

The calcium and phosphorus content and true digestibility of several sources of calcium and phosphorus are shown in Table 7-5. The inorganic sources that are listed are excellent sources of minerals. Grains are poor sources of

Table 7 – 5
Calcium and phosphorus content and true digestibility of some common horse feeds and supplements (percent)

Feed/Supplement	Content Ca	Content P	True digestibility Ca	True digestibility P
Organic				
Beet pulp	0.60	0.10	60	—
Corn	0.02	0.28	—	32
Linseed meal	0.40	0.85	—	30
Milk products	1.30	1.00	79	64
Oats	0.09	0.35	—	40
Wheat bran	0.14	0.15	—	29
Alfalfa hay	1.40	0.20	77	44
Timothy hay	0.32	0.20	70	46
Inorganic				
Bone meal	29.0	14.0	71	58
Dicalcium phosphate	27.0	21.0	74	58
Limestone	35.0	—	69	—
Monosodium phosphate	0.0	22.0	—	58

calcium. The phosphorus in wheat bran is in the form of phytin and is therefore less available than inorganic forms. Legume hays are excellent sources of calcium.

Roughages can be good sources of potassium and, usually, magnesium, whereas grains are low in both potassium and magnesium.

Sodium chloride (salt) should be provided free choice. The trace-mineralized form should be used because it supplies such trace minerals as iodine, copper, manganese, and iron. Either loose or block salt can be used. Waste is minimized when block salt is used, but loose salt permits greater intake.

Of course trace-mineralized salt does not provide any calcium or phosphorus. Furthermore, the use of trace-mineralized salt does not always ensure that the requirements for trace minerals are being met.

Vitamins

Vitamin A The dietary requirement for vitamin A can be met by 5 to 6 mg of carotene per kilogram. Thus, the vitamin A or carotene content of good-quality hay (Table 7-6) is usually adequate for horses, although Fonnesbeck

Table 7–6
Vitamin content of several feeds and feed by-products (mg/kg of dry matter)

Feed/by-product	Carotene	Thiamin	Riboflavin	Niacin	Pantothenic acid
Requirement[a] (mg/kg of dry matter)	6	3	2.2	0.8	2
Alfalfa pellets	100	6	14	40	30
Alfalfa hay	20	3	14	34	18
Timothy hay	14	3	15	30	?
Alfalfa pasture	200	6	11	42	34
Bluegrass pasture	180	5	11	?	?
Dried skim milk	—	4	20	11	36
Brewer's yeast	—	120	34	502	120
Torula yeast	—	7	48	540	73
Corn	3	4	2	20	6
Oats	—	6	1	14	15
Linseed meal	—	9	3	30	12
Soybean meal	—	7	3	25	13
Wheat bran	2	8	3	33	25

SOURCE: Requirements taken from National Research Council. 1978. (Minimum requirements have not been established, but these levels are known to be adequate.) Other values from National Academy of Sciences. 1971. Also from Morrison, F. B. 1959.

and Symons (1967) suggest that horses are not particularly efficient in converting carotene to vitamin A. Hay that has been severely weathered or stored for more than 2 years has a much lower carotene content. Such grains as barley, oats, and wheat do not provide any carotene but yellow corn provides a small amount. Synthetic vitamin A compounds, such as vitamin A palmitate, are inexpensive and stable and can easily be added to feeds containing low levels of vitamin A.

Vitamin D   Sun-cured hay contains approximately 2000 IU per kilogram of feed. A mature horse does not need more than 3300 units per day; thus a vitamin D deficiency does not appear likely when horses receive sun-cured hay. The horses also obtain vitamin D from the action of ultraviolet light on compounds in the skin. Grains and grain by-products contain almost no vitamin D. Supplements that provide high levels of vitamin D include irradiated yeast (9000 to 140,000 IU per gram depending on the type of yeast) and cod-liver oil (85 IU per gram).

Vitamin E   Alfalfa pellets and most good-growing pastures may contain as much as 400 mg of vitamin E per kilogram of dry matter and can be excellent sources of the vitamin for horses. Grains contain low levels of vitamin E (5 to 10 mg per kilogram).

B Vitamins   The B-vitamin content of several feeds is shown in Table 7-6. Brewer's yeast is an economical supplement to use if additional thiamin is needed. Riboflavin deficiency does not occur when the ration contains good-quality hay. Deficiencies of niacin or pantothenic acid are very unlikely to occur in the horse.

The composition of many feeds used in horse rations is shown in Appendix Table 7-1, at the end of this chapter.

7.2   Balancing Rations

Ration balancing is often a tiresome chore and a source of confusion. Several methods can be used but the important principle is that first the horse's requirements must be determined and then the diet must be formulated to satisfy these requirements. Regardless of the method used, a considerable amount of mathematical calculation is necessary unless a computer program is used.

Methods are described in many nutrition books such as *Feeds and Feeding* by Morrison and the National Research Council (1978). Because the methods of calculation are provided in so many other sources it was decided not to include them in this book.

Any of the methods require that the amount of feed intake be known. Thus, it is often difficult to balance rations in this way for horses because the amount of pasture or hay intake may not be known. If the intake of roughage or pasture is not known, a short-cut method can be used.

Short-cut Method   A shortcut to ration balancing is to use guidelines. This method is not as exact as ration balancing but it does have some merit for making preliminary evaluations of diets and rough approximations of reasonable rations. Table 7-7 gives estimates of feed intake and Table 7-8 lists the protein content that is required in the grain mixture of various classes of horse fed legume hay or grass hay. Table 7-9 gives an estimate of the amount of the various protein supplements needed in the grain mixture to provide the necessary level of protein. Table 7-10 lists estimates of the amount of calcium and phosphorus supplements needed for the various classes of horse.

An example of the short-cut method is the formulation of a ration for a 1000-lb lactating mare. From Table 7-7, the mare is estimated to eat approximately 10 to 15 lb of hay and 10 to 15 lb of grain. In this example, the mare is fed timothy hay. According to Table 7-8, the mare's grain mixture should contain 15 to 18 percent protein. In order to provide this level of protein when using soybean meal (50 percent), the grain mixture should contain 15 lb of soybean meal per 100 lb of grain mixture. According to Table 7-10, 10 lb of diacalcium phosphate and 20 lb of limestone should be added per ton of grain mixture when a grass hay is fed to lactating mares. Of course, trace-mineralized salt should be fed free choice. A commercial vitamin could be added if the grass hay was of poor quality.

The kind of grain would depend on its cost and availability. To repeat, obviously the short-cut method is not as precise as calculation method, but it can be helpful in certain situations.

Of course the key to ration balancing is the use of accurate values for feed intake and composition. Forage should be analyzed for protein, fiber, and minerals, and the composition of the grain adjusted accordingly.

Table 7-7
Estimated feed intake (pounds/100 lb of body weight) required for various classes of activity

Activity	Hay	Grain
Maintenance	1½–2	—
Late gestation	1–1½	¼–¾
Lactation	1–1½	1–2
Heavy work	1–1½	¾–1½
Weanlings	¾–1¼	1¾–2

Table 7–8
Estimates of protein needed in grain mix when feeding legume hay or grass hay (percent)[a]

Class	Legume hay	Grass hay
Weanlings	14–16	18–20
Yearlings	12–14	15–18
Mature (maintenance)	8–10	8–10
Gestation	10–12	12–14
Lactation	12–14	15–18

[a]Actual values will vary depending on the protein content of hay and hay:grain ratio used.

7.3   Feeding

The following sections list example diets but it must be remembered that these diets are only *examples*. Many feeds or combinations of feeds can be used. The selection of feeds should be based on nutrient content, cost, availability, and acceptability to the horses. It may not be economical or convenient to use several rations, particularly on small farms. When a single ration is desired, it is usually most economical to formulate that ration for the types of animals with the lowest requirements. The extra nutrients (such as protein and minerals) needed by the other animals (such as weanlings and lactating mares) can be provided by the addition of supplements. Trace-mineralized salt should be fed free choice.

The three classes of horse that are of the greatest concern to the nutritionist are the young foal, the lactating mare, and the hard-working horse. Feeding programs for the various classes are discussed in the following section.

Young Foal   Creep feeding (providing an area where the foal can eat without interference from the mare) can be used in the rearing of foals (Figure 7-3). Feed can be made available within 2 weeks after birth. The foal will not eat significant amounts of feed at this time, but it gradually increases its intake. The creep ration should be highly digestible and palatable. It should contain 16 to 18 percent protein, 0.8 percent calcium, and 0.6 percent phosphorus. Several good creep rations can be obtained commercially, or the ration can be prepared on the farm.

An important consideration in the growth of young foals is the rate of gain desired. Further research is needed to determine the optimal rate of gain for maximum productivity.

Table 7–9
Estimated protein supplement needed to provide various levels of protein in total grain mixture (percent)

Supplement	Pounds of supplement/100 lb total grain mixture							
	10	12	14	16	18	20	22
Dried skim milk (34 percent protein)	None	9	17	24	32	40	50
Linseed meal (36 percent protein)	None	8	16	22	30	39	48
Soybean meal (44 percent protein)	None	6	12	18	24	30	36
Soybean meal (50 percent protein)	None	5	10	15	20	25	30

Note: Grains are assumed to contain approximately 10 percent protein.

263

Table 7–10
Estimated amount of calcium and phosphorus supplement needed per ton of grain mixture

Class	Legume hay	Grass hay
Weanlings	20 lb dicalcium phosphate	20 lb dicalcium phosphate 30 lb limestome
Yearlings	10 lb dicalcium phosphate	10 lb dicalcium phosphate 20 lb limestone
Gestation	10 lb dicalcium phosphate	10 lb dicalcium phosphate 20 lb limestone
Lactation	10 lb dicalcium phosphate	10 lb dicalcium phosphate 20 lb limestone

Note: Amount of supplement can be more accurately calculated when actual intakes of hay and grain are known. Other supplements, such as defluorinated rock phosphate and bone meal, can be used in place of dicalcium phosphate.

Estimates of percent of mature weight attained at various ages are shown in Table 7-11. Information from the older studies is included to demonstrate that horses are expected to grow at a faster rate now. The data in Table 7-11 were used to develop Table 7-12 to provide a rough guide. The data in Table 7-13 demonstrate that horses obtain mature weight at a much earlier age than they obtain mature height.

The relationship between nutrition and skeletal diseases of foals is one of the most perplexing problems on the horse farm. Many controversies surround these diseases—even their names are not universally agreed upon.

Figure 7-3
Some creep feeding can be beneficial for young foals. The foals can eat without interference by the mare.

Table 7–11
Percent of mature weight attained at various ages

Breed	Year	Reference	Age (months)		
			6	12	18
Standardbred	1905	1	36	58	70
Morgan	1945	2	40	61	74
Morgan	1923	3	42	61	75
Grade	1945	2	41	63	74
NRC	1949	4	40	58	70
Average			39.8	60.0	72.6
Thoroughbred	1969	5	48	68	82
Standardbred	1974	6	44	64	79
Arabian	1977	7	46	66	80
Quarter Horse	1961	8	44	63	79
Anglo-Arab	1971	9	45	67	81
Thoroughbred	1979	10	46	67	80
NRC (400 kg)	1978	11	46	66	83
NRC (500 kg)	1978	11	46	65	80
Morgan	1981	12	46	65	81
Standardbred	1981	13	45	67	—
Average			45.7	65.9	80.6

1. Henry. 1912. *Feeds and Feeding.*
2. Dawson, W. M. et al. 1945. *J. Animal Sci.* 4:47.
3. University of Vermont. 1923. *Exp. Stat.*
4. National Research Council. 1949. *Nutrient Requirements of Horses.*
5. Green, D. A. 1969. *Brit. Vet. J.* 124:539.
6. Hintz, H. F. Unpublished data.
7. Reed, R. R. and N. K. Dunn. 1977. *Fifth Equine Nutr. Phys. Symp.* p. 99.
8. Cunningham, K. and S. Fowler. 1961. *La. State Exp. Sta. Bull.* 546.
9. Budzynski, M. E. et al. 1971. *Recz, Nauk. Roln.* Series B 93:21.
10. Hintz, H. F. et al. 1979. *J Animal Sci.* 48:480.
11. National Research Council. 1978. *Nutrient Requirements of Horses.*
12. Balch, D. 1981. University of Vermont, personal communication.
13. Person, B. and R. Ullberg. 1981. *Eq. Vet. J.* 13:254.

Some authors prefer the term *metabolic bone disease* as a collective term. Others such as Dr. Wayne McIlwraith suggest that the term *development orthopedic disease* be used to include *osteochondrosis* (improper maturation of cartilage into bone) *physititis* or *epiphysitis*, (irregular cartilage growth plates) *flexural deformities* (also called *contracted tendons*) and perhaps even *wobbler* (*cervical vertebral malformation*). About the only thing that can be stated with certainty is that the reported incidence of the diseases has increased in the 1970s and 1980s. For example, there was a 20-fold increase

Table 7 – 12
Estimates of body weight at various ages for light horses of various mature body weights

Age (months)	Mature weight (lb)		
	900	1100	1300
2	230	290	330
4	325	400	465
6	415	510	585
8	485	600	700
10	545	680	790
12	600	740	850
14	640	790	900
16	680	840	950
18	710	880	1000

in the number of cases involving skeletal problems of foals at the Ohio State Veterinary College during the period of 1962 to 1984 (Knight et al., 1985).

Many factors such as genetics (Hoppe and Phillipson, 1984; Wagner et al., 1985), trauma, such as running on hard surfaces, and conformation are likely to be involved in the development of the disease.

Several theories have been advanced relating nutrition to development orthopedic diseases. Overfeeding is frequently blamed. Developmental orthopedic diseases are most commonly found in foals that are growing rapidly. Stromberg (1979) reported that feeding high amounts of grain increased the incidence of osteochondrosis in foals. Mayhew et al. (1978)

Table 7 – 13
Percentage of mature height at various ages

Breed	Reference	Age (months)		
		6	12	18
Thoroughbred	1	83	90	95
Arabian	2	84	91	95
Anglo-Arab	3	83	92	95
Quarter Horse	4	83	91	96

1. Hintz et al. 1979. *J. Animal Sci.* 48:480.
2. Reed and Dunn. 1977. *5th Equine Nutr. Physiol. Symp.*
3. Buydzynski et al. 1971. *Recz. Nauk.* Series B. 93:21.
4. Cunningham and Fowler. 1961. *La Agr. Exp. Stat. Bull.* 546.

suggested that some forms of wobblers were related to overnutrition. Kronfeld (1978) suggested that overfeeding of genetically predisposed animals might be responsible for epiphysitis. Rooney (1969) suggested that overweight in the growing horse may displace coffin and fetlock from their normal position. The displacement leads to continuous contractions and shortening of the damper muscle leading to flexural deformities. Glade et al. (1984) reported that a high plane of nutrition influenced the secretion of thyroxin and cortisol, which affected the maturation differentiation in synthetic activities of the growth plate and articular growth center cartilage. Krook and Maylin (1985) suggested that overfeeding calcium to the pregnant mare may lead to hypercalcitonism and abnormal bone development in the fetus and the young foal. The excessive secretion of calcitonin would cause improper maturation of cartilage in the bone and therefore osteochondrosis. They further suggested that the osteochondrosis would predispose animals to fractures at the racetrack.

On the other hand, Knight et al. (1985) suggested that deficiencies of calcium, phosphorus, cooper, and zinc were the primary nutritional causes of skeletal problems. Their suggestion was based on the results of a survey of 19 breeding farms in Ohio and Kentucky. Foals on the farms were evaluated for skeletal problems and given a point score. The degree of correlation between the score and nutrient content of the diet was determined. They concluded that the farms with the fewest problems fed rations that contained at the higher levels of calcium, phosphorus, zinc, and copper.

There is no doubt that copper deficiency can cause osteochondrosis. Several authors have reported cases of osteochondrosis in which foals were raised on farms in which the pasture was contaminated with zinc and they suggested that the zinc induced copper deficiency.

Further experimentation is needed to better determine the mineral requirement of foals and to resolve these controversies. The present recommendation is to use moderation. Excessive energy intake or calcium intake should be avoided but the diet should be evaluated and balanced for calcium, phosphorus, copper, and zinc.

Example of rations that should be adequate for weanlings are shown in Table 7-14.

Yearlings The requirements of protein, calcium, and phosphorus (when expressed as a percentage of the diet) for yearlings are less than those for weanlings because the rate of muscle and bone development has decreased. The yearling ration should contain 12 to 14 percent protein, 0.6 percent calcium, and 0.35 percent phosphorus. Examples of grain mixtures adequate for yearlings are shown in Table 7-14.

Mature Horses The horse that is of least concern to the nutritionist is the non-pregnant, nonlactating horse that is not worked very hard. Of course,

Table 7–14
Examples of rations for yearlings and weanlings fed grass or legume hay[a]

Ingredient	Grass hay		Legume hay	
	Weanling ration (lb/100 lb)	Yearling ration (lb/100 lb)	Weanling ration (lb/100 lb)	Yearling ration (lb/100 lb)
Corn	35	38	42	43
Oats	30	35.5	39	41.5
Soybean meal[b] (44 percent protein)	25	15	11	8
Molasses	6	6	6	6
Limestone[b]	2	1	—	—
Dicalcium phosphate[b]	1	0.5	1	0.5
Trace mineral salt	1	1	1	1

[a]Vitamins may be needed if poor-quality hay is fed. Additional minerals such as selenium, zinc, and copper are needed according to local conditions.
[b]Levels need to be adjusted according to actual analysis of hay.

proper nutrition is just as important for this class of horse, but the absolute amount of the various nutrients it requires is much less. In fact, good pasture or good-quality hay plus trace-mineralized salt can supply all the nutrients needed. Feeding grain to a horse that is seldom worked is not necessary unless the weather is very cold. Cold weather increases energy requirements because the energy is used to keep the animals warm.

Breeding Animals   The effect of nutrition on horse breeding has long been a controversial topic. Some authorities feel that nutrition is very critical, whereas others feel that even if the diet is barely adequate for maintenance, the animals will still be able to reproduce. The body condition of the mare at the time of breeding can be an important factor in reproductive performance. Zimmerman and Green (1978) reported that obese mares were likely to have a lower conception rate than mares in good condition. But too thin is not good. Henneke et al. (1981) reported that mares in good to fat condition had a higher conception rate than mares in thin condition. They suggested a little fat is much better than a little thin. Weight changes can be measured with a scale and weight tape. It is also important to evaluate body condition. Henneke et al. (1981) developed the body score system shown in Table 7-15. They concluded that Quarter Horse mares should have a body score of at least 6 for optimal breeding efficiency.

It is often stated that excess fat can be detrimental to the libido of stallions. Certainly, regulating the body condition of stallions by supervising their diet appears to be a reasonable practice.

Table 7–15
Condition score for horses

Score	
1	*Poor.* Animal extremely emaciated. Spinous processes, ribs, tailhead, and hooks and pins projecting prominently. Bone structure of withers, shoulders, and neck easily noticeable. No fatty tissues can be felt
2	*Very thin.* Animal emaciated. Slight fat covering over base of spinous processes; transverse processes of lumbar vertebrae feel rounder. Spinous processes, ribs, tailhead, and hooks and pins prominent. Withers, shoulders, and neck structures faintly discernable
3	*Thin.* Fat built up about halfway on spinous processes; transverse processes cannot be felt. Slight fat cover over ribs. Spinous processes and ribs easily discernable. Tailhead prominent, but individual vertebrae cannot be visually identified. Hook bones appear rounded, but easily discernable. Pin bones not distinguishable. Withers, shoulders, and neck accentuated
4	*Moderately thin.* Ridge along back. Faint outline of ribs discernable. Tailhead prominence depends on conformation; fat can be felt around it. Hook bones not discernable. Withers, shoulders, and neck not obviously thin
5	*Moderate.* Back level. Ribs cannot be visually distinguished but can be easily felt. Fat around tailhead beginning to feel spongy. Withers appear rounded over spinous processes. Shoulders and neck blend smoothly into body
6	*Moderate to fleshy.* Slight crease down back. Fat over ribs feels spongy. Fat around tailhead feels soft. Fat beginning to be deposited along the sides of the withers, behind the shoulders, and along the sides of the neck
7	*Fleshy.* Crease down back. Individual ribs can be felt, but noticeable filling between ribs with fat. Fat around tailhead is soft. Fat deposited along withers, behind shoulders, and along the neck
8	*Fat.* Prominent crease down back. Difficult to feel ribs. Fat around tailhead very soft. Area along withers filled with fat. Area behind shoulder filled in flush. Noticeable thickening of neck. Fat deposited along inner buttocks
9	*Extremely fat.* Extremely obvious crease down back. Patchy fat appearing over ribs. Bulging fat around tailhead, along withers, behind shoulders and along neck. Fat along inner buttocks may rub together. Flank filled in flush

SOURCE: Henneke, D. R. et al. 1981. *Proc. Equine Nutr. Physiol. Symp.* p. 105.

Very little research has been conducted on the effect of protein on the breeding performance of horses. Studies of other species indicate that very low levels of protein cause cessation of estrus or fetal deaths, but that very high levels of protein during the breeding season are neither beneficial nor harmful.

A severe phosphorus deficiency causes irregularity or cessation of estrus in cattle and rats, and presumably in horses. Deficiencies of many other minerals, such as manganese, iron, and iodine could also result in the mare's failing to settle. However, the requirements for these minerals during the

breeding season do not seem to be much, if any, greater than the requirements for maintenance; that is, there is no evidence that supplementing an
adequate diet with minerals at this time increases the conception rate. One
mineral that may receive greater attention in the future because of its effect
on reproduction in horses is selenium. Selenium is deficient in the diets of
horses fed only homegrown feed in many areas of the United States (see
Section 6.3). Muscle degeneration is frequently observed in young calves,
foals, and lambs raised in selenium-deficient areas. However, studies conducted in New Zealand indicate that selenium deficiency may also cause
reproductive failure in sheep. For example, the percentage of barren ewes in
flocks treated with selenium averaged approximately 8 percent, compared
with 45 percent in untreated flocks.

Stowe (1967) has suggested that combined vitamin A and E supplementation, either by mouth or by injection, significantly improved the reproductive performance of barren mares. In one trial, one group of 9 mares was not
fed a supplement and one group of 9 mares was fed 100,000 IU of vitamin A
and 100 IU of vitamin E per day for 3 months starting approximately one
month before the beginning of the breeding season. Only one of the 9 mares
that were not given a supplement had a live foal, whereas 6 of the 9 mares
that were given the supplement had live foals. Supplements of only vitamin
A or vitamin E were less effective. Similar results were obtained in a subsequent trial in which the vitamins were injected rather than fed. The control
group of 9 mares had 2 live foals and the group of 9 mares injected with
vitamin A and E had 7 live foals. Studies conducted in Europe showed that
mares given vitamin E had a 5 percent increase in conception rate and 7
percent increase in the number of live foals. Several large breeding farms in
the United States have administered a high-level supplementation of vitamin
A and E to mares but the responses have varied. Further studies to determine the effectiveness or perhaps the conditions under which the vitamins
might be effective are necessary before the practice can be routinely recommended. Stallions did not benefit from supplements of vitamin A or E when
fed a "normal" ration (Rich et al., 1981; Ralston and Rich, 1985).

Few experiments have been conducted on the effect of B vitamins on
reproduction. It is usually assumed that unless the utilization of B vitamins
synthesized in the intestine is impaired, or the animals are fed very poor-
quality hay, B-vitamin supplementation is not necessary during the breeding
season. Vitamin C is often given to stallions on breeding farms to improve
semen quality. Davis and Cole (1943) reported that administration of 1 gram
of ascorbic acid per day improved the semen quality and libido of one
Belgian stallion. Further studies are necessary if an accurate assessment is to
be made of the value of such a practice, as other workers have reported that
low fertility does not appear to be associated with a low level of ascorbic acid
in equine semen or serum (Dimock and Errington, 1942).

Pregnant Mares   During the first two-thirds of the gestation period, the fetus is not very large and the nutrient requirements of the mare are not greatly increased, but during the last third of gestation the fetus rapidly increases in size and nutrient requirements are greatly increased. During this time, the intake of bulky feeds such as hay decreases. Thus, more concentrated feeds and grains need to be added to the mare's ration. The mare will still probably eat about 1 lb of hay per 100 lb of body weight and she will need about ¼ to ½ lb of grain per 100 lb of body weight in addition to the hay. Legume hay, such as alfalfa or clover, contains more protein and calcium than grass hays. If a grass hay is fed, the grain mix should contain 14 percent protein, 0.5 percent calcium, and 0.4 percent phosphorus. A ration containing 42 percent oats, 31 percent corn, 10 percent soybean meal, 10 percent wheat bran, 6 percent molasses, and 1 percent limestone would be adequate. If a legume hay is fed, the grain mix does not need to contain any protein or calcium supplements. A mixture of 45 percent oats, 39 percent corn, 10 percent wheat bran, and 6 percent molasses would be adequate.

Lactating Mares   During peak lactation, which usually occurs 8 to 12 weeks after parturition, nutrient requirements are greatly increased (Figure 7-4). A high-producing mare may need as much as 1 to 1½ lb of grain per 100 lb of body weight. Of course, mares vary considerably in their ability to produce milk, and the best rule is to observe the mare and feed accordingly. For example, if she is too thin, increase her grain ration. The grain mix should contain at least 16 percent protein, 0.7 percent calcium, and 0.5 percent phosphorus when a grass hay is fed and 12 percent protein and 0.5 percent phosphorus when a legume hay is fed. A ration containing 36 percent corn, 36 percent oats, 15 percent soybean meal, 5 percent wheat bran, 6 percent molasses, 1.5 percent limestone, and 0.5 percent dicalcium phosphate could be used with the grass hay. A ration containing 42 percent corn, 41.5

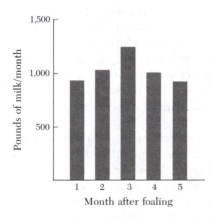

Figure 7-4
Milk production varies greatly among mares, but peak lactation might be expected 8 to 12 weeks after parturition.

percent oats, 6 percent molasses, 5 percent soybean meal, 5 percent wheat bran, and 0.5 percent dicalcium phosphate could be used with the legume hay. Additional vitamins and minerals might be needed, depending on the quality and source of the hay.

Stallions   More information is needed about the effect of nutrition on stallion performance, but most studies indicate that the nutritional requirements for performance are not significantly greater than the requirements for maintenance. The amount of feed necessary varies with the individual, but ¼ to ¾ lb of grain and 1 to 1½ of hay per 100 lb of body weight might be a reasonable amount. The most important rule is to keep the stallion in a reasonable condition. Exercise by riding or lounging not only helps to keep the stallion in good physical condition but also minimizes boredom and keeps him more alert. The stallion should not be allowed to become too thin or run down. In fact, one of the earliest references to the relation between nutrition and reproduction in horses is found in the Bible (Jeremiah 5:8), where it is implied that the libido of well-fed stallions is greater than that of poorly fed stallions. However, the stallion should not be too fat.

The grain mixture described for the pregnant mare should be adequate for the stallion.

As stated previously, these grain ration formulations are only examples. Many different combinations of feeds can be used. The important point is that the nutrients must be present in adequate amounts. Other examples of ration formulations can be found in books by Morrison (1959), Ensminger (1969), Lewis (1982), and Cunha (1980), as well as various experiment station publications.

Hard-Working Horses   Exercise increases the need for energy. Sustained work may increase the energy requirement threefold or more above that of maintenance. Thus, hard-working horses require more feed and, just as important, the feed should contain a high concentration of digestible energy. If diets that consist only of such feed as poor-quality hay, which contains a low amount of digestible energy, are fed, the horse's performance will suffer because of the energy shortage and the horse will probably develop a "hay belly"—that is, it will eat more feed in an attempt to obtain the needed energy but the extra bulk in the diet will distend the intestinal tract. Thus, working horses fed high-quality hay often look trimmer and more attractive than horses fed poor-quality hay.

Working horses fed diets containing 8 percent or more additional fat such as feed-grade animal fat or corn oil are likely to maintain higher blood glucose levels than horses fed conventional diets. Feeding fat may increase stamina and endurance, reduce gut fill, and help maintain body weight because of the increased caloric intake. The caloric requirement, as mentioned earlier, depends on several factors but hard-working horses can be expected to eat amounts of feed equivalent to 2.8 to 3.5 percent of their body weight.

Contrary to popular opinion, muscle is not broken down during work. Some nitrogen is lost in the sweat, but the protein concentration required in the feed of working horses is not greater (in fact, it can be less) than that required for the maintenance of horses in general, because the working horse has a greater total feed intake and thus obtains the necessary total protein intake. For example, Patterson et al. (1985) and Orton et al. (1985) reported that working horses could tolerate diets containing low levels of protein.

Work does not greatly increase the requirement for calcium and phosphorus. As with protein, some calcium and phosphorus are lost in the sweat, but the increased feed intake easily compensates for the slightly increased needs. However, work increases the stress on the skeleton and increases the rate of calcium and phosphorus turnover and bone remodeling. Therefore, a working horse fed a marginal intake of calcium or phosphorus will be more likely to develop lameness than a nonworking horse fed a marginal level of these nutrients. It is particularly important that young horses that are being worked receive adequate minerals, because their bones are still growing.

Further controlled studies are needed to evaluate the requirements of other minerals, but it has been reported that the blood level of electrolytes, such as potassium, sodium, and chloride, may be decreased during extreme prolonged work such as endurance rides (Carlson and Mansmann, 1974). Furthermore, the decreased blood level of potassium may be related to *thumps* (synchronous diaphragmatic flutter) (Mansmann et al., 1974). Thumps, which in some respects is similar to the hiccoughs of man, often occurs in horses during endurance rides. A decreased level of blood calcium may also produce thumps.

There is no doubt that horses require vitamins to perform their best. But are vitamin supplements likely to be of value? Most trainers use vitamin supplements. Winters (1980) interviewed 15 trainers at two Thoroughbred tracks. All trainers interviewed used vitamins. Some thought the vitamins were important while others added them as "insurance." Of course, as mentioned earlier, vitamins, A, D, and K can be toxic if given in large amounts.

If a vitamin supplement is desired, the following amounts per horse per day would not be unreasonable: Vitamin A, 15,000 to 20,000 IU; vitamin E, 1000 IU; thiamin, 20 mg; and folic acid, 20 mg. Vitamin B_{12} is widely used but no benefit has been found in several studies.

In summary, the requirement that is increased most for working horses is the requirement for calories, which can be provided by increasing the grain intake. There may be increases in the protein, vitamin, and mineral requirements; however, the increase in feed intake will generally provide the additional nutrients needed if the feed is of good quality and well balanced.

7.4 Pasture

Pasture can be very important in a horse-feeding program. Good pasture, water, and trace-mineralized salt can provide complete nutrition for many classes of horse, such as mature, nonworking animals, yearlings, and pregnant mares (at least in the first part of the gestation period) (see Figure 7-5). Hard-working horses, lactating mares, and weanlings may require nutrient intakes greater than that provided by pasture. However, several factors, such as pasture management, species of plant, and amount of land determine the amount of nutrition obtained from pastures.

Management   Some features of good pasture management are:

1. Test soil to determine how much lime and fertilizer are needed.
2. Be sure not to overgraze. Too many pastures are simply exercise lots. Rotating pastures is also beneficial to the parasite prevention program.
3. Clipping pastures prevents plant material from becoming too

Figure 7-5
Good-quality pasture, trace mineral salt, and water provide adequate nutrition for most classes of horse. Weanlings, lactating mares, and hard-working horses usually require some additional grain, minerals, and vitamins.

mature, helps maintain a balance of legumes and grasses, and controls weeds.

4. Scatter manure piles.

5. Keep pasture free of mechanical hazards such as wire and nails.

6. Keep pasture free of weeds, particularly poisonous weeds such as nightshade, yellowstar thistle, horse nettle, and ragwort.

7. Keep horses out of the pasture during extreme wet weather to avoid damage to the turf.

Species of Plant   The species of plant used in the pasture depends on local soil and climate conditions. Many different plant species can be used in horse pastures, and the local extension agent can provide appropriate information concerning species and planting and fertilizing procedures.

Bluegrass, bluestem, bromegrass, fescue, orchard grass, and cereal grasses are examples of grasses that are frequently used for horse pastures. In warmer climates, such as that of Florida, pangola grass, bahia grass, carpet grass, or para grass are often used in horse pastures. Legumes used include alfalfa, red clover, white clover, and birdsfoot trefoil. However, red clover does not provide a long-lasting pasture.

Mixtures of grasses and legumes are usually the most satisfactory because they provide a greater total amount of nutrients and longer grazing seasons (Figure 7-6).

For example, Jordan and Marten (1975) reported that reed canarygrass was a good pasture grass for horses early in the grazing season but that in the latter part of the season its palatability greatly decreased because of an increased alkaloid content. Thus, reed canarygrass should not be used as the

Figure 7-6
A mixture of grass and legumes makes an excellent pasture.

only pasture. Kentucky blue grass withstands close grazing and is most productive during hot, dry summer periods. On the other hand, orchard grass grows well under high summer temperatures, has some drought and shade tolerance but is relatively unpalatable at maturity and cannot tolerate close grazing. Birdsfoot trefoil is productive even when grown on poorly drained soil but is more difficult to establish than some of the other legumes. Sorghum-sudangrass hybrids should be avoided because they may produce the cystitis syndrome, which is an irritation of the bladder. Death may occur if kidney infection results.

Tall fescue pasture should be used with care. If consumed by pregnant mares during the eleventh month of gestation it may cause prolonged gestation, thickened placenta, abortion, weak foals, or lack of milk production (Heimann et al., 1981). The toxicity is associated with plants that are infected with the endophyte *Acremonium coenophialum*.

Ryegrass infested with certain molds may cause a neurologic disorder characterized by muscular tremors, locomotor incoordination, and tetany. The condition is called *ryegrass taggers* (Hunt et al., 1983).

Legume pastures containing the mold *Rhizoctonia leguminicola* may cause excessive salivation in horses. The salivation may lead to anorexia and diarrhea. The mold occurs when there is high humidity or rainfall. Sockett et al. (1982) concluded there is no effective treatment for the problem. They suggested that owners should consider cutting the pasture and removing the infested plants. Horses can be returned to the pasture if the plant regrowth is free of the mold. The toxic factor in the mold is an alkaloid called *slaframine*.

Large intakes of alsike pasture may cause photosentization and liver disease in horses (Traub et al., 1982).

Amount of Land

The amount of land required per horse to provide adequate nutrition depends on many factors, such as type or class of horse, rainfall, soil fertility, plant species, shade, and pasture management. One acre of renovated legume grass pasture that was properly managed and enjoyed good growing conditions could provide enough feed for a horse during the grazing season. However, under most common conditions, it is recommended that there be at least 2 to 3 acres of pasture per horse.

7.5   Commercial Feeds

Commercial horse feeds were not used by many horse owners before 1900. However, many expensive tonic feeds were sold. In 1860 Thorley's Feed cost $14 per 100 lb. It was shipped from London and contained beans, barley, flaxseed, Peruvian bark, and quinine tonics. At that time, oats were selling for about $1.40 per 100 lb. The tonic feed in the advertisement shown in Figure 7-7 cost $20 per 100 lb in 1893, when the price of oats was about $1 per 100 lb. In 1901 tonic feeds cost about $15 to $20 per 100 lb, while the ingredients cost about $1.00 to $1.50 per pound.

The tonic feeds usually had linseed meal or grain as a base and contained compounds such as gentian, fennel, fenugreek, charcoal, sulfate, anise, asafoetida, licorice root, mandrake root, walnut bark, iron oxide, and rosin. Exotic names were sometimes used for common materials, such as capsicum for black pepper, frumentum powder for cornmeal, and princess metallic for dried paint. The claims for some tonic feeds were all-inclusive. One tonic claimed to increase endurance, keep colts healthy, make horses fat and their coats glossy, minimize the need for corn and oats in the diet, and cure heaves, coughs, colic, distemper, dropsey, colds, and dyspepsia.

The tonic with the most interesting advertisement was the International Stock Food Tonic, M. W. Savage, the owner of the company, owned one of the most famous horses of all times, Dan Patch (Section 2.4). An advertisement published in 1915 claimed that Dan Patch had been given the tonic every day for more than 14 years. Advertisements for the tonic showing a picture of Dan Patch that was suitable for framing were distributed free of charge, and a Dan Patch Gold Stop Watch was given to everyone who purchased three pails of the tonic.

In the early 1900s many universities warned farmers against buying tonics. In spite of these warnings, tonic feeds were still commonly used in the 1920s and 1930s. Many states eventually made the sale of such feeds illegal.

An early advertisement that did not claim a tonic effect for a feed appeared in the *Breeder's Gazette* in 1886 (Figure 7-8). However, other

Figure 7-7
Advertisement from *The Ohio Practical Farmer* (1893).

Blatchford's Royal Stock Food.

THE MOST COMPLETE FEEDING CAKE
EVER MADE.

A Perfect Milk Substitute or Calf Meal,
—AND—

Unequaled for All Kinds of Young Stock!

INVALUABLE FOR CATTLE, HORSES, SWINE, AND SHEEP.

☞ **For Directions and Testimonials send for PAMPHLET ON FEEDING, issued and mailed free by**

E. W. BLATCHFORD & CO., Sole Manufacturerers, Chicago, Ill.

Figure 7-8
Advertisement from *Breeder's Gazette* (1886).

advertisements for Blatchford's meal indicate that it contained locust beans, wheat, linseed meal, cottonseed meal, and fenugreek. Some high-quality feeds were available before 1900. In 1893 a New Jersey Experiment Station report stated, "A farmer who intelligently exchanges farm products for commercial feeds even at the same price per ton may secure not only an increase in feeding value but also a gain in fertility."

In the late 1890s and early 1900s many states initiated control programs. Massachusetts started analyzing commercial feedstuffs and publishing the results in special bulletins in 1897. Only one horse feed (H. O. Horse Feed) was listed in the 1898 bulletin. The feed contained oat feed and corn. Oat feed was defined as the refuse from factories that manufactured oatmeal for human consumption.

Three horse feeds were listed in the 1903 bulletin. One of them, Blomo, contained dried blood, molasses, and ground corn stalks, and provided 15 percent protein. The five feeds listed in the 1906 bulletin contained mostly oat by-products.

Nine feeds were listed in 1907 and several, such as Molac, Sucrene, and Alfalmo, contained molasses. The molasses was frequently added to increase palatability and to disguise some of the other ingredients.

By 1914, 20 different horse feeds were tested in Massachusetts. It was concluded that "some of them [horse feeds] make very satisfactory grain rations and these, if the feeder is willing to pay the price and does not care to go to the trouble of home mixing, will be found very acceptable providing they do not contain inferior by-products. It is believed, however, that it is possible to home-mix fully as good rations at a lower cost." There were some high-protein feeds but most contained only 9 to 10 percent protein. Some brand names were Iroquois, Algrane, Schumacher Special, O.K., and Very

Best. Feeds containing alfalfa and molasses were Clover Leaf, Kornfalfa Kandy, Peerless, Arab Horse, and June Pasture. Molassine and X-tra Vim contained 75 percent molasses and 25 percent sphagnum moss. The moss was considered to have little value except as a carrier for the molasses.

The 1916 bulletin included such names as Eat-al Horse Feed, King Falfa, Mo-Lene, and Quaker Green Cross. The 1917 bulletin mentioned Alfacorn, King Cotton, Tip-Top, Blue Seal, Peters Re-Peter, Purina O-Molene, Tom Boy, and Good Luck.

In 1921 the bulletin reported, "Mix 500–600 lb of oat feed [oat hulls, shorts, and middlings] with enough corn and hominy meal to form one ton and the result will be a typical horse feed. Oat feed is only slightly superior to mixed hay and should be classed as a roughage." However, the sale of molasses feeds, such as Nu-Life, Badger Prancer, O-Keh, Vim-O-Lene, Your Choice, and Farmer Jones, continued to increase.

By 1928, more than 35 horse feeds were being analyzed by the Massachusetts station. The ingredients were no longer reported in the bulletins, but most of the feeds still contained 9 to 10 percent protein, which suggests that no radical changes were made in formulation. Interesting brand names included Vigor, Bull Brand, Best Ever, Amco Arabs, Domino, and Chelsea.

Although the horse population started to decline about 1915, the number of brands of horse feed (at least according to the Massachusetts sample) continued to increase. In 1935 approximately 50 horse feeds were analyzed. Changes in formulation were evident, as most of the feeds analyzed contained 12 percent or more protein and many contained molasses. In 1933 the Kentucky Department of Agriculture had reported that the horse and mule feeds registered in that state usually consisted of corn, oats, barley, alfalfa, and wheat bran. Molasses was also often included.

In the 1920s some companies added by-products containing high quantities of B vitamins, and vitamin-rich mixtures were added to several feeds in the 1930s. Salt was already being added to rations before 1900.

One of the most common nutritional problems of horses was calcium deficiency, or "big head disease." The first suggestion that the condition was caused by feeding high levels of wheat bran was made in 1860, but the relation between the high phosphorus content of the wheat bran and impaired calcium nutrition was not established until much later. Calcium was first routinely added to commercial horse feeds in the mid-1930s. However, it had been added to general stock rations and fitting rations before that time.

The next important change in feed composition occurred in the early 1960s, when pelleted grains and pelleted complete diets were first extensively manufactured. Of course, pelleting of horse feeds was not a new concept, even in 1950. Pellets or biscuits had been used to feed horses in the Franco-Prussian War of 1870–1871. The Russians fed horses small biscuits composed of oatmeal, pea flour, rye, and ground linseed in 1877. The biscuits were said to provide approximately one-fifth the bulk of oats. The German armies used compressed feeds in World Wars I and II.

The use of pellets has several advantages, such as decreased feed waste, economy of space in storage and transportation, and reduced dust. These advantages appealed to horse owners, and sales of pelleted feeds increased rapidly. Complete pelleted diets also permit the use of properly supplemented by-products and feeds or ingredients that might not normally be accepted by horses. For example, pelleted feeds containing cereal by-products, corn cobs, peanut hulls, or a high level of fat are currently being marketed. Such formulations may become even more important in the future as the competition among feeds and foods increases. Other possible ingredients in pelleted diets include corrugated paper and computer paper.

Disadvantages of the complete pelleted feed may include increased cost and usually result in a dramatic increase in such bad habits as wood chewing and tail biting if the pellets are fed without any other roughage. Thus, hay or other sources of fiber should be made available.

A serious disadvantage of pellets is that some horses may eat them at a very rapid rate. A rapid rate of intake can lead to digestive disorders, such as colic and enterotoxemia, which can be fatal. Therefore, rate of intake should be carefully regulated.

Some horse owners are using extruded feeds, particularly when self-feeding young horses. The extruded feed is pelleted but air is trapped, thus the extruded feed is bulky. Pelleted feeds may weigh twice as much per volume as do extruded feeds. The rate of intake of extruded feeds is much slower than for pelleted feeds. John McKnight of Armstrong Bros. Farms reported that the incidence of digestive disorders such as gastric ruptures greatly decreased when food was changed from pellets to extruded feeds.

Wafers or cubes, particularly of alfalfa, can be fed to horses. The fiber in cubes is not as finely ground as that in pellets, therefore vices observed when only pellets are used may not be seen when cubes are used. For example, Jackson et al. (1984) reported no difference in the amount of wood chewed when horses were fed alfalfa hay or cubes. Wood chewing, however, increased when the weather was wet or cold.

Many companies manufacture horse feeds and most companies have different mixtures for the various classes of horse. There are many factors to consider when determining which feeds to buy. One factor is the reputation of the company. Most companies are reliable and make every effort to consistently produce a high-quality product. Quality control is very important to such companies and they closely monitor their products. Of course, some companies are not so conscientious. The company's reputation can be evaluated by asking other horsemen about their experiences with the company's feed and by checking with the agency that enforces feed laws to determine if the product meets the guarantee on the feed tag. Most feeds meet the guarantee, but there are exceptions.

Another factor is that the feed selected should meet the needs of the horses but should not be provided in excess, which is costly. Most companies clearly indicate the class of horse for which the feed is intended, but the feed tag should be scrutinized to ensure that the feed is adequate. The feed-tag requirements vary from state to state but the minimum amount of information should include minimum crude protein and crude fat and maximum fiber level. Some companies will also provide information about the calcium and phosphorus levels. If information about calcium and phosphorus is not on the tag, the feed dealer might be able to provide it.

The feed should also be selected according to the type of forage that is being fed. The information in Table 7-16 can be used as a rough guide in determining whether the nutrient content of the commercial mixture is adequate for various feeding situations. The table indicates that the advantage of feeding a mixed or legume hay to weanlings is that the grain mixture does not need as much protein. Protein is usually one of the most expensive ingredients in a commercial grain mixture.

The feed tag also lists the ingredients. The wording is often ambiguous so that changes in formulation can be made to take advantage of price changes of the basic ingredients. With such adjustments, the feed can be sold at a lower cost and the nutrient content will still meet the guarantee. However,

Table 7–16

Estimates of nutrient content needed in commercial feeds in order to meet nutrient requirements for various classes of horses

Class of horse	Nutrient (%)	Type of forage feed		
		Legume hay	Mixed hay	Grass hay
Weanlings	Crude protein[a]	14	16	18
	Calcium	0.3[b]	0.6	0.9
	Phosphorus (%)	0.6	0.6	0.6
Yearlings	Crude protein	10	14	16
Mares				
Late gestation	Calcium	0.3[b]	0.3[b]	0.7
Lactation	Phosphorus	0.5	0.5	0.5
Mature horse	Crude protein	8	8	10
	Calcium	—[c]	—[c]	0.2[b]
	Phosphorus	0.3	0.3	0.3

[a]The protein should be of good quality—that is, it should supply the essential amino acids. Soybean meal, milk proteins, and meat meal are examples of protein sources containing a good array of amino acids.
[b]Many mixtures will contain calcium levels greater than 0.3 percent, but the extra calcium is not harmful when an adequate level of phosphorus is provided (but the calcium content of forage should be determined).
[c]The forage will normally provide all the calcium needed.

radical changes may influence the palatability. Of course, factors other than nutrient content determine the value of the feed. The palatability must be satisfactory and the feed must be accepted by the horse. The feed should not be moldy or dusty. Pelleted feeds should not contain a lot of fines (small, dusty particles) or broken pellets.

The services provided by the dealer and the company should also be considered. Will the feed be available when it is needed? Be wary of companies that make outrageous claims and imply that their feed has "magic" qualities.

Commercial feeds designed for other species should be used with caution. The nutrient balance might not be correct and the feed might contain additives that are harmful to horses.

Beef cattle, poultry, or turkey rations may contain monensin, an antibiotic produced by *Streptomyces cinamonensis.* The addition of monensin (rumensin) to beef cattle rations improves feed efficiency by 10 to 15 percent. Monensin is also widely used to control coccidiosis in poultry.

Although very helpful to poultry and cattle, monensin is extremely toxic to horses. The LD_{50} (dose at which it is expected that 50 percent of animals will die) is 2 to 3 mg/kg for horses but 200 mg/kg for chickens and 50 to 80 mg/kg for cattle. Several cases of monensin poisoning in horses caused by the eating of poultry feed have been reported. Whitlock (1978) noted colic, muscular weakness, loss of coordination, and sweating. The horses may die within 12 to 36 hours after the first onset of clinical signs, due to severe damage of the heart muscles.

Muylle (1981) reported that they examined 32 horses with a history of poor performance several months after the ingestion of monensin sodium. Cardiac abnormalities were clearly evident in 8 horses and suspected in 4 others. They concluded that ingestion of monensin by horses can cause sudden death or delayed cardiac circulatory failure.

Amstel and Guthrie (1985) reported cases of salinomycin poisoning in horses. Salinomycin, like monensin, is an ionophore and is used to promote feed efficiency in feed lot cattle. The horse feed was accidentally contaminated with salinomycin at the feed mill. At total of 177 horses were given the feed. Eight died and 120 horses showed clinical signs such as colic, dehydration, ataxia, and muscle spasms. The LD_{50} is only 0.6 mg per kg of body weight for horses. Beef cattle feed containing salinomycin might be expected to provide that amount.

Swine rations can also be dangerous. The antibiotic lincomycin is often added to swine rations to increase rate of gain and feed efficiency. But lincomycin can cause founder, diarrhea, and fatal colitis (inflammation of the colon) in horses. The lincomycin apparently destroys certain bacteria such as those gram positive and gram negative. This allows proliferation of other bacteria producing enterotoxins (Raisbeck and Osweiler, 1981).

7.6 Supplements and Conditioners

There are many supplements and conditioners on the market. When should they be used and of what value are they? Many manufacturers make unrealistic claims and their products are really of little or no value. For example, one company claims that its product will so greatly enhance digestion that there will be almost no fecal material, but in fact, the product has little value as an aid to digestion. Some products are of nutritional value but are sold in a very expensive form. Other supplements are worthwhile only under certain conditions. A vitamin supplement might be very appropriate when animals are under stress such as heavy lactation or hard work and are being fed a very poor-quality hay. But some vitamin supplements are overpriced and have a lot of additional material that is of no known value. Thus, there is no easy answer to questions about the value of supplements and when they should be used. Common sense and experience are the best guides. Do not expect compounds to change the old gray mare into a champion.

7.7 General Management Guidelines

1. *Exercise horses regularly* Horses are athletes and should be given the opportunity to work. Exercise is necessary to keep the muscles in good condition and prevent the horse from becoming too fat.

2. *Make sure that parasite control is adequate* The most common cause of thin horses is a heavy load of parasites. To ensure a good feeding program, parasites must also be adequately controlled. A heavy load of parasites can decrease the efficiency of feed utilization and total feed intake and can prevent weight gains.

3. *Examine teeth regularly* Thin horses often have poor teeth because an animal with poor teeth cannot eat or chew properly. Care of the teeth, including floating (filing or rasping of the sharp edges of the cheek teeth), is important. The diet of older horses with poor or missing teeth can be ground and pelleted to aid digestion.

4. *Feed at regular times* Horses are creatures of habit. They appear to appreciate being fed at regular times. Feeding at regular times may also help decrease some stable vices (see Section 5.5).

5. *Avoid moldy feed* Horses are quite susceptible to moldy feed toxicosis.

6. *Keep the feed manger clean* A clean manger decreases feed waste and helps prevent horses from going off feed. Molds might also develop if the manger is not clean.

7. *Give small, frequent feedings* Small, frequent feedings decrease the chances of gastric distention, founder, or colic. It is recommended that grain be fed at least twice a day if total grain intake exceeds 0.5 percent of the body weight.

8. *Feed by weight, not by volume* There are considerable differences in density among horse feeds. Therefore, when feeding by volume, severe discrepancies may arise. For example, a quart of wheat bran weighs approximately ½ lb, whereas a quart of barley weighs 1½ lb. One quart of corn may provide twice as much digestible energy as one quart of oats because corn weighs more per unit volume and is more digestible.

9. *Make changes in types of feed gradually* Abrupt changes can cause colic or diarrhea, or can cause the horse to stop eating. The most common problem occurs when a low-energy diet is changed to a high-energy diet.

10. *Do not overfeed* Remember the Arab proverb, "Fat and rest are two of the horse's greatest enemies." Excessive fat can reduce performance, reproductive efficiency, and perhaps even longevity.

11. *Make sure that water is frequently available, clean, and fresh* Water should be frequently available—except, of course, when the horse is hot. Automatic water bowls can be very useful because water is thus made available at all times and labor costs are decreased. A small heating unit with thermostat controls prevents water from freezing in the bowl. One unit could be located in such a way that it serves two box stalls. One large unit may serve 8 to 10 horses in a lot. One disadvantage of the automatic waterer is the initial expense; another is that certain horses may play with the waterer and cause water spillage. Unfortunately, horsemen often neglect to keep the automatic waterers clean and functioning properly. When using open tanks in a lot, there should be at least one foot of open water per horse. Excessive water intake by a hot horse may cause serious problems such as colic or founder.

Obviously, very cold or very hot water is not desirable. A range of 45° to 65°F seems reasonable. Horses with damaged teeth sometimes tolerate warm water better than cold water.

REFERENCES

Amstel, S. R., and A. J. Guthrie. 1985. Salinomycin poisoning in horses: Case report. *Proc. 31st Am. Assoc. Equine Prac.*, Toronto, Ontario, p. 373.

Carlson, G. P., and R. A. Mansmann. 1974. Serum electrolyte and plasma protein alterations in horses used in endurance rides. *J.A.V.M.A.* 165:262.

Cunha, T. J. 1980. *Horse Feeding and Nutrition*. New York: Academic Press.

Davis, G. K., and C. L. Cole. 1943. The relation of ascorbic acid to breeding performance in horses. *J. Animal Sci.* 2:53.

Dimock, W. W., and B. J. Errington. 1942. Nutritional diseases of the equine, *N. Am. Vet.* 23:152.

Ensminger, E. M. 1969. *Horses and Horsemanship*. Danville, Illinois: Interstate Printers and Publishers.

Fonnesbeck, P. V., and L. D. Symons. 1967. Utilization of the carotene of hay by horses. *J. Animal Sci.* 26:1030.

Glade, M. J., and T. H. Belling. 1984. Growth plate cartilage metabolism, morphology and biochemical composition in over and under fed horses. *Growth* 48:473.

Glade, M. J., S. Gupta, and T. J. Reimers. 1984. Hormonal Responses to high and low planes of nutrition in weanling Thoroughbreds. *J. Animal Sci.* 59:658.

Heimann, E. D., L. W. Garrett, W. E. Loch, J. S. Morris, and W. H. Pfander. 1981. Reproductive abnormalities in pregnant mares grazing fescue pastures. *Proc. Seventh Equine Nutr. Phys. Symp.* Warrentown, Virginia, p. 62.

Henneke, D. R., G. D. Potter, and J. L. Kreider. 1984. Body condition during pregnancy and lactation and reproductive efficiency of mares. *Theriogenology* 21:897–909.

Hoppe, F., and J. Philipsson. 1985. A genetic study of osteochondrosis dissecans in Swedish horses. *Equine Pract.* 7(7):7.

Hunt, L. D., L. Blythe, and D. W. Holtan. 1983. Ryegrass staggers in ponies fed processed ryegrass straw. *J.A.V.M.A.* 182:285

Jackson, S. A., V. A. Rich, S. L. Ralston, and E. W. Anderson, 1984. Feeding behavior and feed efficiency in groups of horses as a function of feed frequency and the use of alfalfa cubes. *J. Animal Sci.* 59 (Suppl. 1):152.

Jordan, R. M., and G. C. Marten. 1975. Effect of three pasture grasses on yearling pony weight gains and pasture carrying capacity. *J. Animal Sci.* 40:86.

Knight, D. A., A. A. Gabel, S. M. Reed, R. M. Embertson, W. J. Tyznik, and L. R. Bramlage. 1985. Correlation of dietary mineral to incidence and severity of metabolic bone disease in Ohio and Kentucky. *Proc. 31st Am. Assoc. Equine Pract.*, Toronto, Ontario, p. 445.

Kronfeld, D. 1978. Feeding on horse breeding farms. *Proc. 24th Am. Assoc. Equine Pract.*, St. Louis, Missouri, p. 461.

Lewis, L. D. 1982. *Feeding and Care of the Horse*. Philadelphia: Lea and Febiger.

Mansmann, R. A., G. P. Carlson, N. A. White, and D. W. Milne. 1974. Synchronus diaphragmatic flutter in horses. *J.A.V.M.A.* 165:265.

Mayhew, I. G., A. deLahunta, R. H. Whitlock, L. Krook, and J. B. Tasker. 1978. Spinal cord disease in the horse. *Cornell Vet.* 68(Suppl. 6):1.

Merrit, T. L., J. B. Washoko, and R. H. Swain. 1969. *Selection and Management of Forage Species for Horses*. Pennsylvania State University Mimeo. A.S. H-69-1.

Moise, L. L., and A. A. Wysocki. 1981. The effect of cottonseed meal on growth of young horses. *J. Animal Sci.* 53:409.

Morrison, F. B. 1959. *Feeds and Feeding*. 22nd ed. Clinton, Iowa: Morrison.

Muyelle, E. 1981. Delayed monensin sodium toxicity in horses. *Eq. Vet. J.* 13:107.

National Academy of Sciences. 1971. *Atlas of Nutritional Data in United States and Canadian Feed*. Washington, D.C.

Continued on page 294

Appendix Table 7 – 1
Composition of feeds commonly used in horse diets—dry basis (moisture free)

Line no.	Short feed name (Scientific name)	International feed number[a]	Dry matter (%)	DE (Mcal/kg)	TDN (%)	Crude protein (%)	Digestible protein (%)	Lysine (%)	Crude fiber (%)
	Alfalfa								
	(Medicago sativa)								
1	grazed, prebloom	2-00-181	21	2.51	57	21.2	15.6	1.06	22
2	grazed, full bloom	2-00-188	25	2.29	52	16.3	11.4	0.65	33
3	hay, s-c, early bloom	1-00-059	90	2.42	55	17.2	13.4	0.94	31
4	hay, s-c, midbloom	1-00-063	89	2.29	52	16.0	11.6	0.90	32
5	hay, s-c, full bloom	1-00-068	89	2.16	49	15.0	10.1	0.64	34
6	meal, dehy, 15% protein	1-00-022	91	2.42	55	16.3	11.8	0.66	33
7	meal, dehy, 17% protein	1-00-023	92	2.46	56	19.7	13.9	0.96	27
	Bahigrass								
	(Paspalum notatum)								
8	grazed	2-00-464	30	2.11	48	7.9	4.2	—	32
9	hay, s-c	1-00-462	91	1.89	43	5.8	2.5	—	30
	Barley								
	(Hardeum vulgare)								
10	grain	4-00-549	89	3.61	82	13.9	11.4	0.48	6
11	grain, Pacific Coast	4-00-939	90	3.48	79	10.7	7.0	0.35	7
12	hay, s-c	1-00-495	89	1.89	44	8.5	4.7	—	27
13	straw	1-00-498	90	1.63	37	4.0	0.9	—	42
	Beet, sugar								
	(Beta vulgaris, B. saccharifera)								
14	pulp, dehy	4-00-669	91	2.86	65	8.0	5.0	0.66	22
	Bermudagrass								
	(Cynodon dactylon)								
15	grazed	2-00-712	39	2.20	50	9.1	5.2	—	28
16	hay, s-c	1-00-716	91	1.98	45	7.0	4.2	—	34
	Bluegrass, Kentucky								
	(Pao pratensis)								
17	grazed, early	2-00-777	31	2.46	56	17.0	12.4	—	26
18	grazed, posthead	2-00-782	35	2.20	50	11.6	7.4	—	27
19	hay, s-c	1-00-776	90	2.20	50	11.0	5.1	—	30
	Brewers								
20	grains, dehy	5-02-141	92	2.99	68	27.0	20.9	0.95	16
	Brome								
	(Bromus spp)								
21	grazed, vegetative	2-00-892	32	3.00	68	18.3	12.6	—	24
22	hay, s-c, late bloom	1-00-888	90	2.38	54	7.4	5.0	—	40
	Canarygrass, reed								
	(Phalaris arundinacea)								
23	grazed	2-01-113	27	2.38	54	12.0	7.5	—	29
24	hay	1-01-104	91	2.16	49	12.3	7.6	—	33

Line no.	Cell walls (%)	ADF (%)	Cellu- lose (%)	Lignin (%)	Cal- cium (%)	Cop- per (mg/kg)	Iron (mg/kg)	Mag- nesium (%)	Manga- nese (mg/kg)	Phos- phorus (%)	Potas- sium (%)	So- dium (%)	Sul- fur (%)	Zinc (mg/kg)
1	—	—	—	—	2.26	10	200	0.25	28	0.35	2.35	0.20	0.50	18
2	—	—	—	—	1.53	9	330	0.27	25	0.27	2.15	0.15	0.31	15
3	48	38	28	10	1.75	15	200	0.30	32	0.26	2.55	0.15	0.29	17
4	50	40	29	11	1.50	13	180	0.29	29	0.25	1.90	0.14	0.28	17
5	52	42	30	12	1.29	12	170	0.31	27	0.24	1.80	0.14	0.26	17
6	51	41	29	12	1.40	11	330	0.30	31	0.24	2.50	0.10	0.20	22
7	45	35	24	11	1.50	10	400	0.39	31	0.26	2.70	0.10	0.26	22
8	—	—	—	—	0.45	—	60	0.25	—	0.19	1.45	—	—	—
9	—	—	—	—	0.45	—	60	0.19	—	0.22	1.45	—	—	—
10	19	7	—	—	0.05	9	90	0.15	19	0.37	0.45	0.03	0.18	17
11	21	9	—	—	0.05	9	80	0.13	18	0.37	0.58	0.02	0.17	17
12	—	—	—	—	0.21	4	300	0.19	39	0.31	1.49	0.14	0.17	—
13	80	59	37	12	0.24	10	300	0.15	17	0.05	2.01	0.14	0.17	—
14	59	34	—	—	0.75	14	330	0.30	38	0.10	0.20	0.23	0.22	10
15	—	—	—	—	0.49	—	—	0.19	—	0.27	—	—	—	—
16	80	35	23	12	0.40	—	—	0.17	—	0.19	1.57	0.44	—	20
17	—	—	—	—	0.56	10	—	0.20	79	0.40	2.20	—	—	—
18	—	—	—	—	0.46	9	—	0.18	68	0.39	2.01	—	—	—
19	—	—	—	—	0.30	9	260	0.16	93	0.29	1.70	0.14	0.13	—
20	42	23	18	5	0.30	24	270	0.17	42	0.58	0.09	0.28	0.34	30
21	60	31	27	4	0.55	5	100	0.18	—	0.35	2.32	0.02	0.20	—
22	72	44	36	8	0.32	7	100	0.13	106	0.22	2.00	0.02	0.20	—
23	—	—	—	—	0.42	9	150	—	—	0.35	3.64	—	—	—
24	—	—	—	—	0.37	9	150	0.31	106	0.25	1.86	0.39	0.41	—

Appendix Table 7 – 1
Composition of feeds commonly used in horse diets — dry basis (moisture free) — Continued

Line no.	Short feed name (Scientific name)	International feed number[a]	Dry matter (%)	DE (Mcal/kg)	TDN (%)	Crude protein (%)	Digestible protein (%)	Lysine (%)	Crude fiber (%)
	Citrus								
25	pulp wo fines, dehy	4-01-237	90	2.99	68	6.9	3.6	—	14
	Clover, alsike								
	(*Trifolium hybridum*)								
26	hay, s-c	1-01-313	89	2.11	48	14.8	10.1	—	29
	Clover, crimson								
	(*Trifolium incarnatum*)								
27	grazed	2-01-336	17	2.42	55	17.2	12.1	—	27
28	hay, s-c	1-01-328	89	2.16	49	18.0	13.1	—	32
	Clover, ladino								
	(*Trifolium repens*)								
29	hay, s-c	1-01-378	90	2.24	51	21.0	15.6	—	20
	Clover, red								
	(*Trifolium pratense*)								
30	grazed, early bloom	2-01-428	20	2.51	57	21.1	12.0	—	19
31	grazed, late bloom	2-01-429	26	2.42	55	14.5	9.8	—	30
32	hay, s-c	1-01-415	89	2.16	49	14.9	10.0	—	30
	Corn								
	(*Zea mays*)								
33	cobs, ground	1-02-782	90	1.36	31	2.8	0.5	—	36
34	distillers grains, dehy	5-02-842	92	3.08	70	29.8	21.0	0.87	12
35	ears, grnd	4-02-849	87	3.26	74	9.1	5.6	0.20	10
36	grain	4-02-985	88	3.87	88	10.9	8.5	0.30	2
	Cotton								
	(*Gossypium* spp)								
37	hulls	1-01-599	91	1.45	33	4.2	1.1	—	50
	Fescue, meadow								
	(*Festuca elatior*)								
38	grazed	2-01-920	27	2.29	52	11.5	7.3	—	29
39	hay, s-c	2-01-912	88	2.02	46	10.5	5.8	—	33
	Flax								
	(*Linum usitatissimum*)								
40	seeds, meal, solv extd	5-02-048	91	3.04	69	38.9	27.6	1.34	10
	(Linseed meal)								
	Lespedeza								
	(*L. striata, L. stipulacea*)								
41	grazed	2-02-568	31	2.20	50	14.9	10.2	—	38
42	hay, s-c	1-08-591	91	2.07	47	13.9	9.3	—	32
	Linseed — see flax								

Line no.	Cell walls (%)	ADF (%)	Cellulose (%)	Lignin (%)	Calcium (%)	Copper (mg/kg)	Iron (mg/kg)	Magnesium (%)	Manganese (mg/kg)	Phosphorus (%)	Potassium (%)	Sodium (%)	Sulfur (%)	Zinc (mg/kg)
25	23	23	—	—	2.07	6	170	0.16	7	0.13	0.77	0.10	0.07	16
26	—	—	—	—	1.32	6	260	0.41	69	0.29	2.46	0.46	0.17	—
27	—	—	—	—	1.33	—	250	0.29	317	0.32	2.51	0.40	0.28	—
28	—	—	—	—	1.39	—	300	0.29	200	0.20	2.00	0.39	0.28	—
29	36	32	25	7	1.32	9	600	0.29	200	0.24	2.80	0.39	0.18	17
30	—	—	—	—	2.26	—	300	0.51	—	0.38	2.49	0.22	0.17	—
31	—	—	—	—	1.01	—	306	0.43	—	0.27	1.96	0.20	0.17	—
32	56	41	30	10	1.49	11	310	0.45	73	0.25	1.66	0.18	0.17	17
33	89	35	28	7	0.12	7	230	0.07	6	0.04	0.91	—	0.47	—
34	43	—	—	—	0.11	48	200	0.08	20	0.44	0.20	0.10	0.46	35
35	—	—	—	—	0.05	8	80	0.16	6	0.26	0.56	0.05	0.22	18
36	—	—	—	—	0.05	4	30	0.03	6	0.37	0.35	0.01	0.14	21
37	90	71	48	23	0.15	13	150	0.14	10	0.08	0.87	0.02	—	16
38	—	—	—	—	0.60	4	—	0.37	27	0.43	2.34	—	—	—
39	65	43	37	6	0.57	4	—	0.59	24	0.37	1.74	—	—	—
40	—	—	—	—	0.43	28	360	0.67	42	0.90	1.53	0.15	0.44	—
41	—	—	—	—	1.10	—	310	0.29	154	0.28	1.26	0.31	—	—
42	—	—	—	—	1.15	—	330	0.25	184	0.25	1.03	0.30	—	—

Appendix Table 7 – 1
Composition of feeds commonly used in horse diets — dry basis (moisture free) — Continued

Line no.	Short feed name (Scientific name)	International feed number[a]	Dry matter (%)	DE (Mcal/kg)	TDN (%)	Crude protein (%)	Digestible protein (%)	Lysine (%)	Crude fiber (%)
	Milk								
	(*Bos taurus*)								
43	skimmed, dehy	5-01-175	94	4.05	92	36.0	30.3	2.69	0.3
	Molasses								
44	beet, sugar, mn 48% invert	4-00-668	78	3.17	72	8.7	5.3	—	—
45	sugarcane, molasses, dehy	4-04-695	94	3.17	72	9.3	5.8	—	5.0
46	sugarcane, molasses, mn 48% invert	4-04-969	75	3.26	74	4.3	2.0	—	—
	Oats								
	(*Avena sativa*)								
47	grain	4-03-309	89	3.34	76	13.6	10.5	—	12
48	grain, Pacific Coast	4-07-999	91	3.34	77	10.1	6.5	—	12
49	hay, s-c	1-03-280	90	2.07	47	8.9	5.1	—	32
50	straw	1-03-283	92	2.11	40	4.3	2.5	—	40
	Orchardgrass								
	(*Dactylis glomerata*)								
51	grazed	2-03-439	19	2.42	55	18.4	13.2	—	27
52	hay, s-c	1-03-438	89	2.07	47	10.1	6.1	—	36
	Pangolagrass								
	(*Digitara decumbens*)								
53	grazed	2-03-493	19	2.24	51	12.5	8.1	—	29
54	hay, s-c	1-09-459	88	1.98	45	9.6	5.7	—	27
	Prairie								
55	midwest, hay, s-c	1-03-191	90	2.02	46	6.7	3.2	—	33
	Rye								
	(*Secale cereale*)								
56	grain	4-04-047	88	3.52	80	13.8	9.9	0.48	3
	Sorghum								
	(*Sorghum vulgare*)								
57	grain	4-04-383	90	3.52	80	12.6	8.8	0.28	3
	Soybean								
	(*Glycine max*)								
58	hay, s-c	1-04-558	89	2.11	48	15.9	11.0	—	34
59	hulls	1-04-560	92	2.64	60	12.0	7.7	1.61	40
60	seeds	5-04-610	91	4.05	92	43.2	31.7	2.93	6
61	seeds, meal, solv extd	5-04-604	90	3.60	82	50.9	35.7	3.28	7
	Sunflower								
	(*Helianthus* spp)								
62	seeds wo hulls, meal, solv extd	5-04-739	92	3.12	71	50.3	—	1.85	12
	Timothy								
	(*Phleum pratense*)								
63	grazed, midbloom	2-04-905	30	2.15	49	9.6	5.2	—	31

Line no.	Cell walls (%)	ADF (%)	Cellulose (%)	Lignin (%)	Calcium (%)	Copper (mg/kg)	Iron (mg/kg)	Magnesium (%)	Manganese (mg/kg)	Phosphorus (%)	Potassium (%)	Sodium (%)	Sulfur (%)	Zinc (mg/kg)
43	—	—	—	—	1.30	1	10	0.13	2	1.09	1.66	0.05	0.34	68
44	—	—	—	—	0.21	22	100	0.30	6	0.03	6.20	1.52	0.61	18
45	—	—	—	—	0.87	73	240	0.43	52	0.20	3.68	0.19	0.46	33
46	—	—	—	—	1.05	80	250	0.47	57	0.15	3.80	0.22	0.46	30
47	31	17	14	3	0.07	7	80	0.19	43	0.37	0.44	0.18	0.38	33
48	—	—	—	—	0.11	6	90	0.19	42	0.34	0.44	0.16	0.23	—
49	—	36	30	6	0.30	4	400	0.75	120	0.26	1.23	0.17	0.30	—
50	70	47	34	13	0.25	10	200	0.19	37	0.07	2.37	0.40	0.23	—
51	55	31	28	3	0.57	7	170	0.19	40	0.54	3.27	0.04	0.21	17
52	—	—	—	—	0.35	14	110	0.20	40	0.31	3.01	—	0.26	18
53	—	—	—	—	0.45	—	—	0.14	—	0.35	—	—	—	—
54	—	—	—	—	0.37	—	—	0.13	—	0.23	—	—	—	—
55	—	—	—	—	0.41	23	100	0.28	48	0.15	1.01	0.04	—	—
56	—	—	—	—	0.07	8	70	0.14	62	0.36	0.52	0.03	0.17	36
57	—	—	—	—	0.03	11	50	0.20	17	0.33	0.39	0.03	0.16	16
58	—	—	—	—	1.22	9	290	0.79	101	0.28	1.02	0.09	0.24	24
59	67	46	44	2	0.45	18	320	—	14	0.15	1.03	0.05	—	24
60	—	—	—	—	0.28	17	90	0.31	32	0.66	1.77	0.13	0.24	18
61	14	10	8	2	0.13	30	130	0.30	32	0.70	2.19	0.31	0.48	48
62	—	—	—	—	0.41	4	40	0.81	25	1.10	1.10	0.44	—	—
63	—	—	—	—	0.28	11	200	0.15	190	0.25	2.40	0.19	0.13	—

Appendix Table 7–1
Composition of feeds commonly used in horse diets — dry basis (moisture free) — Continued

Line no.	Short feed name (Scientific name)	International feed number[a]	Dry matter (%)	DE (Mcal/kg)	TDN (%)	Crude protein (%)	Digestible protein (%)	Lysine (%)	Crude fiber (%)
64	hay, s-c, pre-head	1-04-881	89	2.20	50	11.5	7.2	—	31
65	hay, s-c, head	1-04-883	88	1.98	45	9.0	4.8	—	32
	Trefoil, birdsfoot								
	(*Lotus corniculatus*)								
66	hay, s-c	1-05-044	91	2.20	50	16.0	12.5	—	30
	Wheat								
	(*Triticum* spp)								
67	bran	4-05-190	89	2.94	67	17.0	14.4	0.68	11
68	grain, hard red winter	4-05-268	89	3.83	87	14.4	10.7	0.42	3
69	grain, soft red winter	4-05-294	89	3.83	87	13.0	9.2	0.57	3
70	grain, soft white winter	4-05-337	89	3.83	87	11.5	7.5	0.35	3
71	hay, s-c	1-05-172	89	1.89	43	8.7	4.9	—	29
72	straw	1-05-175	89	1.50	34	4.2	1.0	—	41
	Yeast								
	(*Saccharomyces cerevisiae*)								
73	brewer's, dehy	7-05-527	93	3.30	75	48.3	32.8	3.33	3

[a]First digit is class of feed: 1, dry forages and roughages; 2, pasture, range plants, and forages fed green; 3 silages; 4, energy feeds; 5, protein supplements; 6, minerals; 7, vitamins; 8, additives.

SOURCE: National Research Council, 1978

Line no.	Cell walls (%)	ADF (%)	Cellulose (%)	Lignin (%)	Calcium (%)	Copper (mg/kg)	Iron (mg/kg)	Magnesium (%)	Manganese (mg/kg)	Phosphorus (%)	Potassium (%)	Sodium (%)	Sulfur (%)	Zinc (mg/kg)
64	64	37	33	4	0.50	6	200	0.15	—	0.25	1.92	0.18	0.13	—
65	70	45	34	11	0.41	5	140	0.16	46	0.19	1.60	0.18	0.13	—
66	44	34	25	9	1.75	9	230	0.51	15	0.22	1.80	0.18	—	77
67	45	12	8	4	0.12	14	190	0.59	130	1.43	1.60	0.04	0.25	120
68	40	—	—	—	0.05	5	40	0.17	44	0.48	0.45	0.03	0.18	43
69	30	—	—	—	0.05	7	30	0.11	36	0.46	0.46	0.02	0.12	48
70	14	4	—	—	0.05	8	40	0.11	40	0.45	0.41	0.02	0.13	30
71	68	41	—	—	0.15	—	200	0.12	40	0.19	1.00	0.28	0.24	—
72	85	54	39	15	0.21	3	200	0.12	40	0.08	1.10	0.14	0.19	—
73	—	—	—	—	0.5	36	100	0.25	6	1.52	1.86	0.08	0.41	42

National Research Council. 1978. *Nutrient Requirements of Horses.* Publ. No. 6. Washington, D.C.

Orton, R. K., I. D. Hume, and R. A. Leng. 1985. Effect of level of dietary protein and exercise on growth rates of horses. *Eq. Vet. J.* 17:381.

Patterson, P. H., C. N. Coon, and I. M. Hughes. 1985. Protein requirements of mature working horses. *J. Animal Sci.* 61:187.

Raisbeck, M. F., and G. D. Osweiler. 1981. Lincomycin associated colitis in horses. *J. Am. Vet. Med. Assoc.* 179:362.

Ralston, S. L., S. A. Jackson, V. A. Rich, and E. L. Squires. 1985. Effect of vitamin A supplementation on the seminal characteristics and sexual behavior of stallions. *Proc. Ninth Equine Nutr. Physiol. Symp.*, East Lansing, Michigan, p. 74.

Rich, G. A., D. E. McGlothlin, L. D. Lewis, E. L. Squires, and B. W. Pickett. 1983. Effect of vitamin E supplementation on stallion seminal characteristics and sexual behavior. *Proc. Eighth Equine Nutr. Physiol. Symp.*, Lexington, Kentucky, p. 85.

Ricketts, S. W. 1984. Thirteen cases of botulism in horses fed big bale silage. *Equine Vet. J.* 16:515.

Rooney, J. R. 1969. *Biomechanics of Lameness in Horses.* Baltimore: Williams and Williams.

Rostkowski, C. M., T. W. Wilson, G. S. Allan, L. J. Deftos, K. W., Benson, F. A. Kalfelz, R. R. Minor, and L. Krook. 1981. Hypercalcitioninism without hypercalcitioninemia. *Cornell Vet.* 71:188.

Schneider, B. H. 1947. *Feeds of the World.* Morgantown: West Virginia University.

Schryver, H. F., and H. F. Hintz. 1972. Calcium and phosphorus requirements of horses: A review. *Feedstuffs* 44:(28)35.

Shelle, J. E., W. D. VanHuss, J. S. Rook, and D. E. Ullrey. 1985. Relationship between selenium and vitamin E nutrition and exercise in horses. *Proc Ninth Equine Nutr. Physiol. Symp.* East Lansing, Michigan, p. 104.

Sockett, D. C., J. C. Baker, and C. M. Stowe. 1982. Slaframine (*Rhizoctonia leguminicola*) intoxication in horses. *J.A.V.M.A.* 181:606.

Stowe, H. D. 1967. Reproductive performance of barren mares following vitamin A and E supplementation. *Proc. Am. Assoc. Equine Pract.*, New Orleans, Louisiana, p. 81.

Stromberg, B. 1979. A review of the salient features of osteochondrosis in the horse. *Equine Vet. J.* 11:211.

Traub, J. L., K. A. Potter, W. Bayly, and S. M. Reed. 1982. Alsike clover poisoning. *Mod. Vet. Pract.* 63:307.

Tyznik, W. J. 1972. Nutrition and Disease. In *Equine Medicine and Surgery.* 2d ed. Wheaton, Illinois: American Veterinary Publications.

Wagner, P. C., S. M. Reed, and G. A. Hegreberg. 1982. Contracted tendons (Flexural deformities) in the young horse. *Comp. Cont. Educ.* 4:5101.

Wagner, P. C., B. D. Grant, B. J. Watrous, L. H. Appell, and L. L. Blythe. 1985. A study of the heritability of cervical vertebral malformations in horses. *Proc. 31st Am. Assoc. Equine Pract.*, Toronto, Ontario, p. 43.

Whitlock, R. H. 1978. Monensin toxicosis in horses. *Proc. 24th Am. Equine Prac.*, St. Louis, Missouri, p. 473.

Zimmerman, R. A., and D. E. Green. 1978. Energy requirements of mares. *Proc 70th Ann. Am. Soc. Animal Sci.*, East Lansing, Michigan, p. 326.

CHAPTER 8

PROBLEMS ASSOCIATED WITH FEEDING

Consarn a hoss anyhow! If they're wuth anythin they're more bother 'n' a teethin baby. Alwas some dun thing ailin' em.
—EDWARD NOYES WESCOTT. *David Harum*. 1898

8.1 Fat Horses and Thin Horses

Many horses are either too fat or too thin. Excess fat is usually a result of overfeeding rather than hormonal problems. There are several conditions that result in overfeeding.

1. Many people enjoy feeding horses used as pets.

2. Many horsemen purposely overfeed show horses or sales horses because fat may help mask undesirable traits. Traditionally, fat horses sell for higher prices than thin horses.

3. Many horsemen think they must greatly increase the feed intake of a mare as soon as she is pronounced in foal. The energy requirement of a pregnant mare increases only during the last third of gestation.

4. Many horses do not receive adequate exercise.

Some of the common causes of thin horses are:

1. *Parasites* In order to have a good feeding program, there should also be a good parasite control program.

295

2. *Milk production*   Some mares produce large amounts of milk and therefore need a considerable amount of energy.

3. *Hard work*   Work can greatly increase the energy requirements of horses. A horse that performs one hour of hard work per day may need twice as much energy as a nonworking horse.

4. *Poor teeth*   Many horses have difficulty eating and chewing. A regular inspection of teeth is a good husbandry practice.

5. *Too much competition*   In group feeding, some horses may be deprived of access to feed by the more aggressive horses.

6. *Malabsorption*   There may be a defect in the gastrointestinal tract, such as lack of enzymes or changes in the intestinal wall, that prevents normal uptake of food.

There is some variation in energy requirements among individual horses, and the old law that "the eye of the master fattens the stock" may be the most important rule. Know the unique traits and needs of your horse and feed it accordingly. Monthly weighings are helpful to monitor weight changes. In fact, there are several reasons for knowing the weight of horses. Many feeding recommendations are based on percentages of the horse's weight. The dosages of drugs, antibiotics, and worming compounds are often based on the weight of the horse.

The best method of determining weight is to use a scale, but when a scale is not available weight can be estimated by using formulas based on measurements such as heart girth and length. For example, Milner and Hewitt (1969) have reported that weight can be estimated by the following formula:

$$\text{Estimated weight (lb)} = \frac{\text{Heart girth}^2 \times \text{Length}}{241.3}$$

In this formula, heart girth is measured in inches and length in inches is measured from point of shoulder to point of hip. Milner and Hewitt measured 108 horses ranging in size from Shetland ponies to draft horses and in age from weanling to mature. The average difference between estimated weights and weights obtained with scales was approximately 6 percent. They compared their results with weight estimates obtained by using tapes supplied by feed companies (Figure 8-1). The tapes also proved to be quite useful; the average difference between estimated and actual weight was less than 10 percent. As mentioned earlier, body condition should be evaluated in as systematic method (see Section 7.3).

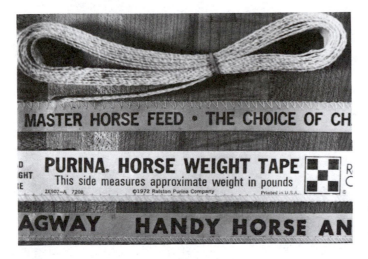

Figure 8-1
Tapes can be useful for estimating the weights of horses.

8.2   Founder

The word "founder" means to fall in or to fall helplessly and go lame. ("Founder" as a noun can also be a synonym for laminitis.) The acutely foundered horse usually does not fall but rather becomes, as the saying has it, "glued to the floor," and if forced to move, does so with great reluctance, placing his forefeet so carefully in a heel-first manner that he may be appropriately described as "walking on eggs." "Acutely" means that the full-blown symptoms appear quite suddenly—for example, when a pony grazing on lush spring pasture is sound one day and foundered the next; or when a horse finds his way into the grain bin that was not closed at night and by the next night he is foundered; or when a mare foals and develops an infection in her uterus because the afterbirth is not expelled completely, and she is foundered 24 to 36 hours later.

The acutely foundered horse is suffering from inflammation of the sensitive laminae of its feet. The sensitive laminae are leaflike structures that are lined up vertically around the region between the hoof wall and coffin bone (Figure 8-2). There are approximately 600 of these laminae on each hoof, and secondary and tertiary laminae branch off from the 600 primary laminae. In most cases, only the front feet are affected by founder; however, all four feet may be affected.

Acute inflammation of the sensitive laminae beneath the hoof wall is the result of an incompletely understood biological phenomenon in which the sensitive laminae become detached form the insensitive laminae. The insen-

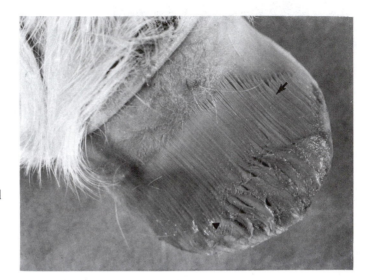

Figure 8-2
Foot preparation (hoof removed), showing sensitive laminae of pony that has had chronic founder. Arrow points to normal laminae. Triangle points to enlarged scarred laminae in area of chronic founder.

sitive laminae make up the inner part of the hoof wall. They interlock (dovetail) with the sensitive laminae, the outer part of the tissue, which cover and are attached to the coffin bone. This interlocking maintains the suspended nature of the coffin bone within the hoof. When a horse places its foot on firm ground, especially when wearing shoes, the hoof wall bears most of the weight and has most contact with the ground. This force of contact is transferred to the laminae and then to the skeletal support, starting with the coffin bone and progressing up the limb. The bones are kept aligned and the force is cushioned by the various ligaments and attached muscles, all of which are elastic in varying degrees. The laminae suspend a large portion of the weight of the horse. It is essential to bear in mind that the laminae are biologically alive and are taking part in the continual hoof growth process, which starts at the hairline.

Degrees of founder range from very mild to severe, resulting in continual pain. The normal position of the bones in the foot is shown in Fig. 8-3. If founder is severe, the coffin bone may drop away from its normal position within the hoof. The greatest drop, or "rotation," as it is usually called, occurs at the toe (Figures 8-4, 8-5). This causes the widened "white line" and, many cases, the "seedy toe" associated with founder. If the coffin bone rotates, the sole has to flatten out also (Figure 8-5). The sole may even become convex (Figure 8-6), or the tip of the coffin bone may actually protrude through the sole, and then the sensitive tissue is exposed to bacterial infection. At the other extreme is the mild "touch" of founder, in which inflammation is evident but no rotation of the coffin bone takes place.

Because of the pain within the laminar area, the horse tries to avoid weight bearing and concussion as much as possible, and thus places its heel

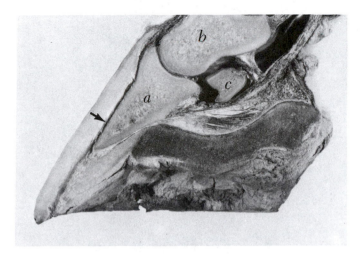

Figure 8-3
Vertical cross section of normal horse foot with long toe in need of trimming. Arrow points to laminar area. Note that front of hoof wall and front of coffin bone (*a*) are parallel. Photo also shows short pastern bone (*b*) and navicular bone (*c*).

to the ground first in a careful deliberate heel-to-toe movement. When the initial inflammation subsides, which happens anywhere from one day to many weeks after the acute founder occurs, the horse is said to be either cured or to have chronic founder. In all cases of chronic founder, there is some rotation of the coffin bone, widening of the "white line," and evidence of abnormal growth rings on the hoof. The sole is flattened and thinner than normal and the front of the hoof wall may have a "dished" concave shape. The distance between growth rings (Figure 8-5) at the toe is less than the distance between the same growth rings at the quarter or heel. The end result is that the toes will curl upward if they are not trimmed regularly.

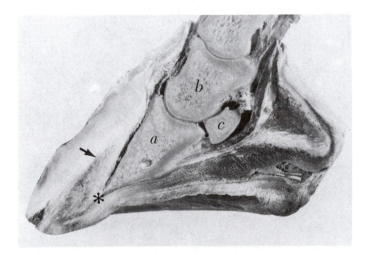

Figure 8-4
Vertical cross section of foot of chronically foundered pony. Note rotation of coffin bone: tip of coffin bone (*asterisk*) should be where point of arrow is. Area between star and arrow is filled with distorted laminae and scar tissue. *a*, *b*, and *c* are the same as in Figure 8-3.

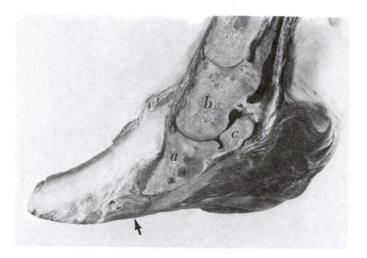

Figure 8-5
Vertical cross section of foot of severe chronically foundered horse. Compare with Figure 8-3: note complete distortion of hoof growth and shriveling up of coffin bone. Arrow points to flat, nearly convex sole. Specimen shows more extensive rotation of coffin bone than is shown in Figure 8-4.

The chronically foundered horse may not show pain but usually moves in a heel-to-toe step unless it is correctively shod or trimmed. Apparently, the rotation of the coffin bone does not change the horse's natural inclination to land flat on the coffin bone. A rotated coffin bone landing flat means that the foot lands noticeably heel first. The defect resulting from the ruptured lamina fills in with scar tissue and always leaves some permanent damage and weakened laminar attachment.

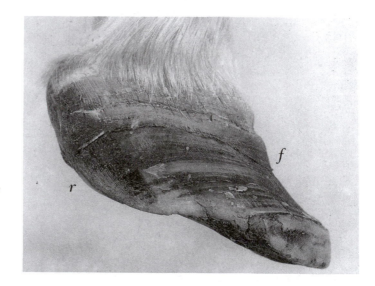

Figure 8-6
Outside of hoof of chronically foundered foot shown in Figure 8-5. Note distortion of distance between growth rings: they are close together at front of hoof (*f*) and much wider apart at heel of foot (*r*). Also note the "dished" concave shape of the front of the hoof wall.

Table 8–1
Incidence of founder

Breed	Total cases, all diseases	Horses foundered	Percent foundered
Appaloosa	565	3	0.5
Thoroughbred	1253	34	2.7
Quarter Horse	1584	44	2.8
Arabs	428	14	3.3
Standardbred	2646	87	3.3
Shetland pony	345	18	5.2
Morgan	251	15	6.0

For reasons that are not well understood, certain breeds and types of horses founder more often than others. Fat ponies and small fat horses that are "easy keepers" are notorious for being targets of founder. Some breeds have comparatively less founder than others. Table 8-1 shows the total number of cases of all diseases for the breeds listed, the total number of founder cases, and the ratio of total cases to founder cases reported by the New York State Veterinary College, Cornell University, Ithaca, New York, from January 1, 1966, to January 1, 1972.

Attaining a body weight that is in keeping with what the feet can hold up is possible by reducing the feed intake over a period of months. This often means a diet of no grain and limited grass hay or limited amounts of low-nutrient hay. Do not overfeed horses and ponies. Excess feed given either by mistake or by neglect over many months causes a large percentage of founder cases. Sudden large increase in grain intake can also cause founder.

Use common sense about riding practices and training practices. Whenever possible, avoid speed on hard surfaces; let animals cool off after exercising them, and continue a parasite control and preventive vaccination program as recommended by your veterinarian.

A retained placenta, reaction to antibiotics, and the use of walnut shavings as bedding can also induce founder. It was usually thought the walnut shavings needed to be fresh in order to induce founder (True and Lowe, 1980), but Ralston and Rich (1983) reported founder in horses bedded with walnut shavings that had been stored for 3 months.

8.3   Heaves

Heaves (chronic obstructive pulmonary disease) is a condition in which the lungs do not work efficiently (Section 16.2). Any management practice that reduces dust may be helpful when horses have heaves. (Horses on pasture seldom have difficulty with heaves.) The use of pelleted feeds, whole clean grains, or wet hay can also be helpful. The hay can be soaked in a water pail 15 to 30 minutes before feeding. The use of hay substitutes such as beet pulp or citrus pulp may also be beneficial. Dusty bedding should be avoided. Hay that had been harvested wetter than normal but treated with propionic-acetic acid to prevent spoilage may have low amounts of dust if the treatment was properly applied.

8.4   Colic

The term "colic" is applied to various conditions of the digestive tract of which pain is the chief symptom (Section 16.2). It is one of the oldest known disorders of horses — Columella described it in the first century after Christ. It has long been considered the most dangerous and costly equine internal disease.

Parasites are often regarded as the primary cause of colic in horses, but diets and feeding practices also can be important factors. Ferraro (1982) stated that "The most important element in the prevention of colic in race horses is consistency of diet and exercise."

Colic is an age-old problem, and many remedies for prevention and treatment have been suggested over the years. The ancient Greeks prescribed 5 drams of myrrh, 6 cotylae of wine, and 3 cotylae of oil for horses with colic. Columella, in A.D 50, reported that the sight of a duck swimming helped cure colic in horses and mules. These treatments did not withstand the test of time, but many practices used by old-time horsemen and veterinarians are still useful.

In 1895, Dr. W. E. Wyman reviewed the causes and treatment of colic. He suggested that the main cause of colic was improper food and water. He recommended that horses be fed at about the same time each day. Grain, if fed in large amounts, should be fed several times daily, and hungry or greedy horses should be allowed to eat rapidly. Smooth stones in the manger were recommended to decrease rate of intake. Diet changes should be made gradually and large intakes of green feeds such as alfalfa, clover, and pea vines should be avoided if horses are unaccustomed to such feeds. Horses fed grain should be allowed to rest 30 minutes before being worked, but if it was necessary to resume work immediately, the work should go leisurely at first and gradually increase in speed. These guidelines are reasonable today. Bacteria in the large intestine need time to adapt to new substrates. In a survey conducted in Florida, veterinarians concluded that rapid change of feeds was one of the most common causes of colic. Excessive intake of grain at one time can cause serious problems, as will be discussed later.

Of course, some practices recommended by Wyman in 1895 seem less reasonable today. For example, it was his practice to thoroughly rub the belly of colicky horses with 2 oz turpentine and 2 oz raw linseed oil. He thought horses should not be watered after feeding because the water would wash the food into the large intestine too quickly and cause colic. This idea is still found in some textbooks today, but research has shown that some water after eating may be helpful. In fact, at the Utah Experiment Station in 1891, J. W. Sanborn watered horses before and after feeding and found no influence on digestibility or digestive upsets. He concluded that horses should be offered water both before and after feeding. We have conducted similar experiments, with similar results. Of course, inadequate water supply can lead to colic because of impaction. Several studies have indicated an increased prevalence of colic caused by impaction in cold climates in the winter. Presumably, the owners were less concerned about supplying water when the weather was cold or the horses were less likely to drink.

Feeding management is important. Tyznik (1982) reported that when a foal is weaned, it may fret for several days and not eat. When it settles down, it is very hungry and overeats grain. This results in overproduction of lactic acid and paralysis of the duodenal valve; the stomach is filled rapidly and does not empty properly. Creep feeding to condition the foal to grain would be helpful. After weaning, the grain intake can be increased gradually, by about 250 g/day, until the desired intake is reached.

Tyznik (1982) also reported problems in stallions put out to graze in hot weather, during which they tend to stand in the shade and not graze. When they are brought into a cool barn, appetite is improved and overeating of grain may result. Feeding hay to the stallions when they first come off pasture and feeding grain later, when they are less hungry, was recommended by Tyznik.

8.5   Eating Problems

Depraved Appetite

Chewing wood, eating dirt, sand, and/or gravel, tail biting, and eating the bark of trees are often observed in horses. The craving for unnatural food may result because of a nutrient deficiency. For example, iron deficiency may cause animals to eat dirt (geophagia). Phosphorus deficiency has been reported to produce pica in several species of animals. Reports from Germany have indicated that pica was observed in horses that were reared on pasture deficient in phosphorus. The pica could sometimes be prevented by feeding phosphorus.

However, most horses seem to chew wood, trees, and hair for reasons other than a nutrient deficiency. One horseman concluded after several years of study that horses chew wood because they like to. Horses fed a complete pelleted diet frequently chew wood and/or eat dirt. Boredom and lack of exercise may also be causes. Horses may develop the habit by imitating other horses.

Chewing of wood can often be alleviated by using hardwoods such as oak, covering the wood with metal, or treating the wood with materials such as creosote. Exercise and accessibility to hay or minerals free choice can also be helpful. Such vices as wood chewing and dirt or stone eating should be discouraged. Replacing wooden fences and mangers is expensive, and excessive ingestion of foreign material may be harmful to the plants, but whether it is harmful, helpful, or of no consequence to the horse depends on the kind of tree and the extent of chewing. For example, the foliage, bark, and seeds of several types of yews are toxic to horses. The bark of the locust tree may be toxic. Leaves of apricot, peach, almond, and wild cherry trees contain a toxic cyanogenic glycoside. Leaves and sprouts of buckeye and oak trees can be lethal. Although in general, little nutritional benefit is derived from bark chewing, in some cases it apparently can be helpful. For example, it has been reported that horses can obtain calcium from the bark of certain trees in Australia. Plains Indians used the inner bark of the round-leaf or sweet bark cottonwood trees as a supplemental feed for horses during the winter.

Excessive chewing of sand or dirt may result in sand impaction of the cecum and colon. Sand impaction is frequently observed when horses are fed on the ground in such areas as the southwestern United States and Florida. Chewing of hair may result in hair balls. Hair balls, although rare, have been reported in the intestines of horses with a habit of licking themselves when shedding.

Coprophagy (eating of feces) is often observed in young foals. Hence, mares should be treated for parasites during late gestation and early lactation to decrease the chances of infesting the foal. Coprophagy is not likely to cause problems (except that it fosters the growth of parasites); in fact, there is some nutritional benefit. Coprophagy is not as common in mature horses as in young foals, but confined horses fed complete pelleted diets are more apt to practice coprophagy than exercised horses with access to hay. Confined horses fed pelleted diets spent approximately 2 percent of their time eating feces, whereas horses fed hay spent less than 1 percent of their time in that manner (Willard et al., 1973).

Fast Eating

Many horses are extremely greedy, rapid eaters. This practice should be discouraged because it may lead to digestive disorders or choking. Spreading the grain in a thin layer, putting large, smooth stones in the bottom of the manger, or feeding several times a day may be helpful. Extruded feeds might also be helpful (see Section 7.3).

Off Feed

What should be done when a horse goes off feed? First of all, try to determine why the horse stopped eating. Is the horse sick? Does it have an elevated temperature? Check the mouth and teeth. A sore tongue or bad teeth are common causes of decreased feed intake. Moldy feed may cause a sudden decrease in feed intake. Perhaps boredom is a cause. Exercise and a change in diet may be helpful. Nutrient deficiencies could also affect intake. Potassium deficiency is possible if roughages are not fed. It has been suggested that horses that go off feed and become "track sour" may respond to thiamin (Tyznik, 1972). Adding two tablespoonfuls of brewer's yeast to the horse's daily ration may be helpful.

Alfalfa hay is usually more palatable than grass hay and may entice the horse to eat. On the other hand, some horses will reject the alfalfa for grass hay. "Sweet feeds," that is, those with molasses, may sometimes increase intake. Treats such as carrots and apples can be added to stimulate the interest of the horse. Bran mashes may be helpful. Often one of the best tonics is green pasture!

Enterotoxemia

Enterotoxemia (overeating disease) is caused by toxins produced by *Clostridium perfingens* Type D bacteria. Swerczek (1976) reported that the condition is most often found in the largest and fastest-growing foals in group-feeding situations. Clinical signs of enterotoxemia are rarely observed. The foals appear healthy at feeding time but may be dead shortly afterward. At necropsy the gastrointestinal tract may be filled with grain and dilated because of gas formation. The cortex of the kidneys can be degenerative and contain areas of hemorrhage. Similar signs are found in lambs with enterotoxemia.

Sound management can help prevent enterotoxemia. Hungry foals should not be given unlimited access to grain. During group feeding, all animals should be given ample space and opportunity to eat. "Boss" foals should not be allowed to crowd out the other foals and eat more than their share. Pelleted feeds containing only grains, with no significant level of fiber, should not be fed free choice. Changes in feed should be made gradually, especially with recently weaned foals brought in from pasture and introduced to grain.

8.6   Metabolic Disorders

Tying-up Disease

Tying-up disease is a muscle disorder that occurs in racehorses and light horses under heavy exercise. It is usually more common in mares than geldings. There is lameness and rigidity of the muscles of the loin. The horse walks as though it has back pain and the muscles feel very hard to the touch. The urine may be coffee-colored because of the myoglobin released from the damaged cells. Some authorities diagnose tying-up as a form of azoturia (Monday morning disease, or black water), a disease that was primarily observed in draft horses not worked on Sunday but given full feed. Azoturia in draft horses also results in rigidity of muscles and release of myoglobin into the urine. In a severe case, the animal will refuse to move, may assume a sitting position, or even fall prostrate on its side. Other authorities suggest that the muscle changes in azoturia are different from those characteristic of tying-up disease.

Reducing the grain intake on days the animals are idle will help prevent tying-up. The relationship of tying-up to nutrition is not clearly defined and much more research needs to be done to clarify it. Many veterinarians feel that injections of vitamin E and selenium are quite helpful in the prevention and cure of tying-up. Others recommend vitamin E, selenium, and thiamin, whereas still others prefer cortisone injections and other methods of treatment. Lack of potassium has also been incriminated (Hodgson, 1985). With rest and proper care, most horses will recover unless there has been extensive kidney damage. Proper training and slow warm up and cool down can also be helpful in the prevention of tying-up.

Sodium bicarbonate (2 percent of the diet) has reported to be of benefit in some cases (Robb and Kronfeld, 1986).

Hyperlipemia

Hyperlipemia is a condition in which there is an unusually high level of fat (triglycerides and cholesterol) in the blood. In fact, the blood appears cloudy. Fat ponies are particularly likely to develop the condition. The condition can be produced in healthy animals by withholding food and it has been reported in animals suffering from equine infectious anemia, equine viral arteritis, and other conditions that caused the animals to stop eating (Schotman and Wagenaar, 1969; Baetz and Pearson, 1972). Some of the changes that occur in blood serum as a result of lack of feed are shown in Table 8-2. Feeding will correct the condition, but animals with severe hyperlipemia may not commence eating even after the original cause for inappetence has been corrected.

Table 8-2
Effect of fasting on triglyceride and cholesterol content of blood serum of ponies

Period	Triglycerides (mg/100 ml)		Cholesterol (mg/100 ml)	
Pretrial	25	25	58	67
After 8 days fasting	196	442	108	193

SOURCE: Numbers in columns 1 and 3 from Baetz and Pearson. 1972. Numbers in columns 2 and 4 from Morris et al. 1972.

Jeffcott and Field (1985) reported that the condition occurs most commonly in fat ponies in late pregnancy and is rarely seen in horses. They suggested that the pony is more likely to mobilize triglycerides than the horse because the pony is less sensitive to insulin.

Jeffcott and Field (1985) concluded that the incidence can be reduced by good husbandry practices such as not allowing animals to get too fat and minimizing stress during the high-risk periods of pregnancy and lactation.

Anemia

Anemia is defined as a deficiency of hemoglobin or number of red blood cells. Hemoglobin is the respiratory pigment of the red blood cells. It combines with and releases oxygen. Thus, the hemoglobin content is of concern to horsemen, particularly racehorse owners, because they feel that the hemoglobin or oxygen-carrying capacity is related to performance. Deficiencies of many nutrients, such as iron, copper, vitamin B_{12}, folic acid, vitamin B_6, vitamin E, and protein, may result in anemia. However, the addition of these nutrients to a balanced ration will not significantly increase the hemoglobin content. For example, an injection of 4 mg of vitamin B_{12} in ponies caused a rapid rise in serum B_{12} but the serum level returned to normal within 5 to 6 days and there was no increase in hemoglobin or hematocrit level. Alexander and Davies (1969) also reported that much of the injected B_{12} was excreted in urine within 1 to 2 days post-injection and there was no increase in hemoglobin content. Carlson (1970) found that a weekly injection of 20 mg of vitamin B_{12} and 100 mg of folic acid did not increase hemoglobin, hematocrit, or red-blood-cell count in polo ponies fed a diet of grass hay and oats. Therefore, an analysis of the feed is needed before supplements are used.

Some reports indicate that heavily parasitized or debilitated animals may respond to vitamin B_{12} therapy but these reports need to be further substantiated (Clifford et al., 1956).

8.7   Stunted Growth and Orphan Foals

If horses are fed a restricted amount of feed during the first year of their life, they might reach full size when mature, but of course the process will take longer, and they cannot be worked as hard or as soon as properly fed horses. Khitenkov (1950) conducted a trial in which groups of foals were fed (1) adequate feed from birth to age 5 years; (2) adequate feed up to 1½ years, insufficient feed from 1½ to 2 years, and then adequate feed; and (3) insufficient feed up to 12 months and then normal feed. The restricted animals grew more slowly, but Khitenkov concluded that under subsequent favorable conditions of feeding and care the horses made up for the delay in growth. Witt and Loshse (1965) conducted a study to determine if poor feeding during winter restricts the growth of horses. The animals were approximately 6 months of age at the start of the trial. One group of horses was fed hay and grain during the winter and the other group was fed only hay. All animals received good feed in the summer. After 3 years the horses that had been fed hay and grain gained an average of 0.5 lb per day during the winter, but horses fed only hay lost 0.51 lb per day during the winter. However, because of the good feed they received during the summer, the horses fed only hay in the winter were able to recover, and after 3 years there were no differences in skeletal size or body measurements between the two groups. Of course, severe diet restriction during the first year of life will result in small, stunted animals.

Animals that have been restricted in growth should be gradually introduced to higher feed intakes. Rapid increases in energy intake could cause severe digestive problems. Furthermore, rapid body-weight gains in previously restricted animals are likely to cause flexural abnormalities of the limbs.

Orphan Foals

Although the rearing of orphan foals has long been a problem, it is much less so now because of recent advances in nutrition. But it is almost impossible to rear an orphan unless it receives colostrum, the first secretions from the mammary gland after the birth of the foal. Colostrum contains antibodies that give temporary immunity to many of the important foalhood diseases. Most foals that do not receive colostrum die. The foal can absorb the antibodies in colostrum only within the first 36 hours after birth. After that time, the structure of the intestinal tract changes and the antibodies can no longer pass through to the blood stream. Therefore, it is critical that the foal receive at least ½ to 1 qt of colostrum the first day. Some farms try to keep a supply of frozen colostrum, obtained from mares with dead foals or high-producing mares, to use if a mare dies or does not produce colostrum. If colostrum is not available, blood can be taken from a mature horse and the serum, which also contains antibodies, separated. Up to ½ pint of serum should be fed to the foal several times throughout the day. Because of potential and serious allergic problems of feeding blood to a newborn foal, a veterinarian should be consulted before this procedure is performed.

Once the foal has received colostrum, several different approaches can be taken in rearing the orphan. Of course, a nurse mare may be used, but nurse mares are often difficult to obtain and it may be difficult to get the foal to nurse or to get the mare to accept the foal. Nurse goats can also be useful. There are several commercial milk replacers on the market. Stowe (1967a) suggests that a milk replacer containing 24 percent protein and 10 percent fat was superior to one containing 24 percent protein and 20 percent fat. Ensminger (1969) suggests that orphan foals can be fed a mixture of one pint of cow's milk, one tablespoon of sugar, and 3 to 5 tablespoonfuls of saturated limewater. He recommends that the foal be fed one-fourth of a pint of this mixture every hour for the first few days of life. Stowe (1967a) reports that bottles and nipple pails are not necessary; the orphan foal can be readily taught to drink from a shallow pan. This is accomplished after a 4- to 6-hour starvation period by offering milk containing an ice cube. The temperature differential between the foal's lips and the cold milk appears to help the foal realize that its lips are in contact with something and he will soon make some sucking or licking movements and readily learn to drink.

8.8   Poisonous Plants

There are a great many plants that can be poisonous to horses. Fortunately, most of them are not very palatable and the toxic principle is often in a low concentration so that the horse has to eat significant amounts of the plants before there is any danger. Nevertheless, the danger of plant poisoning is a real one and steps should be taken to minimize it. Never put hungry horses in a strange pasture, because they are more apt to eat poisonous plants when they are hungry. Be alert for poisonous plants in the pasture. They can be eliminated by digging or by applying chemical weed killers. Fowler (1985) points out that certain plants are poisonous only during a particular season, and it may be necessary to remove horses from pastures during this time. For example, in California, yellow star thistle is a problem especially during the late summer and fall. When the grass is in a short supply, care should be taken to provide animals with supplemental forage or else they may eat the star thistle. Fowler further states that yellow star thistle is one plant for which horses will acquire a taste.

Table 8–3
Some plants that can be poisonous to horses

Plant	Species	Effect
Bracken fern	*Pteridium aquilinium*	Thiamin deficiency[1]
Castor bean	*Ricinus communis*	Severe irritation to intestinal tract[2]
Fiddleneck	*Ansinckia intermedia*	Liver cirrhosis[3]
Golden weed	*Onopis* spp.	Selenium poisoning[4]
Horsetail	*Equisetum* spp.	Thiamin deficiency[5]
Japanese yew	*Taxus cuspidata*	Nervous system damage[6]
Jimsonweed	*Datura* spp.	Alkaloid poisoning[7]
Locoweed	*Astragalus* spp.	Nervous system damage[8]
Oleander	*Nerium oleander*	Digitalis effect[9]
Prince's plume	*Stanleya* spp.	Selenium poisoning[4]
Rattleweed	*Crotalaria spectabilis*	Liver cirrhosis[10]
Red maple	*Acer rubrum*	Hemolytic anemia[11]
Russian knapweed	*Centaurea repens*	Encephalomalacia[12]
Tansy ragwort	*Senecio jacobaea*	Liver cirrhosis[13]
Whitehead	*Sphenosciodium capitallatum*	Photosensitization[14]
Wild cherry	*Prunus* spp.	Cyanide poisoning[15]
Wild onion	*Allium validum*	Hemolytic anemia[16]
Wild tobacco	*Nicotiana trigonophylla*	Paralysis[17]
Woody aster	*Xylorrheza* spp.	Selenium poisoning[4]
Yellow star thistle	*Centaurea solstitialis*	Encephalomalacia[18]

SOURCES:
1. Hadwen, S. 1917. *J.A.V.M.A.* 50:702.
2. McCunn, J. 1947. *Brit. Vet. J.* 103:273.
3. McCulloch, E. C. 1940. *J.A.V.M.A.* 96:5.
4. Trealease, S. F., and O. A. Beath. 1949. *Selenium.* New York: Trealease and Beath.
5. Lott, D. G., et al. 1951. *Can. J. Comp. Med.* 15:274.
6. Lowe, J. E. et al. 1970. *Cornell Vet.* 60:36.
7. Hansen, A. A. 1924. *J.A.V.M.A.* 66:351.
8. James, L. F., et al. 1970. *Am. J. Vet. Res.* 31:663.
9. Wilson, F. 1909. *Ariz. Agr. Exp. Sta. Bul.* 59:14.
10. Cox, D. N., et al. 1958. *J.A.V.M.A.* 133:425.
11. Tennant, B., et al., 1981 *J.A.V.M.A.* 179:143.
12. Young, R. et al. 1970. *Am. J. Vet. Res.* 31:1393.
13. VanEs, L. 1929. *Nebraska Exp. Sta. Bul.* 43:1.
14. Fowler, M. E., et al. 1970. *J.A.V.M.A.* 157:1187.
15. Pijoan, M., 1942. *Am. J. Med. Sci.* 204:550.
16. Pierce, K. R., et al. 1972. *J.A.V.M.A.* 160:323.
17. Burgess, P. S. 1934. *Ariz. Exp. Sta. Bul. Rpt.* 45:44.
18. Cordy, D. R. 1956. *J. Neuropath. Exper. Neurol.* 13:330.

Further information about poisonous plants can be obtained from Kingsbury (1964) and Fowler (1963). Table 8-3 is a partial list of plants that should be avoided.

REFERENCES

Alexander, F., and M. E. Davies. 1969. Studies on vitamin B_{12} in the horse. *Brit. Vet. J.* 125:169.

Baetz, A. L., and J. E. Pearson. 1972. Blood constituent changes in fasted ponies. *Am. J. Vet. Res.* 33:1941.

Carlson, K. A. 1970. *A study of certain hematinics in the horse.* Senior Research Project. New York State Veterinary College, Cornell University, Ithaca, New York.

Clifford, R. J., G. N. Henderson, and J. H. Wilkins. 1956. The effect of feeding penicillin and vitamin B_{12} to mature debilitated horses. *Vet. Rec.* 68:48.

Ensminger, M. E. 1969. *Horses and Horsemanship.* Danville, Illinois: The Interstate Printers and Publishers, Inc.

Ferraro, G. L. 1982. Preventive medicine on the race track. In *Equine Medicine and Surgery.* 3rd ed. R. Mansmann and E. McAllister, ed. *Am. Vet. Publ.* Santa Barbara, California, p. 57.

Fowler, M. E. 1963. Poisonous Plants. In *Equine Medicine and Surgery.* 1st ed. Eaton, Illinois: American Veterinary Publications.

Hodgson, D. R. 1985. Myopathies in the athletic horse. *Proc. Fifth Meeting Assoc. Equine Sports Med.* Reno, Nevada.

Jeffcott, L. B. and J. R. Field. 1985. Current concepts of hyperlipaemia in horses. *Vet. Rec.* 116:461.

Khitenkov, G. G. 1950. The growth of crossbreds on different feeding levels. *Konevodstvo.* 8:15 [Abstracted in *Nutr. Abs. Rev.* 21:751, 1951.]

Kingsbury, J. 1964. *Poisonous Plants of the United States and Canada.* Englewood Cliffs, New Jersey: Prentice-Hall.

Meyer, H., and U. Lemmer. 1973. Mineralstoff- und Spurenelementgehalt in Serum bzw. Plasma des Pferdes. *Dtsch. tierärztl. Wschr.* 80:173.

Milner, J., and D. Hewitt. 1969. Weight of horses: Improved estimates based on girth and length. *Can. Vet. J.* 10:314.

Morris, M. D., D. B. Zilversmit, and H. F. Hintz. 1972. Hyperlipoproteinemia in fasting ponies. *J. Lipid Research* 13:383.

Naylor, J. N. Hyperlipemia. In *Equine Medicine and Surgery.* N. E. Robinson, ed. Philadelphia: Saunders.

Osbaldiston, G. W., and P. R. Griffith. 1972. Serum iron levels in normal and anemic horses. *Can Vet. J.* 13:105.

Ralston, S. L., and V. A. Rich. 1983. Black walnut toxicosis in horses. *J.A.V.M.A.* 183:1095.

Robb, E. J., and D. S. Kronfeld. 1986. Dietary sodium bicarbonate as a treatment for exertional rhabdomyolysis in a horse. *J.A.V.M.A.* 188:602.

Schotman, A. J. H., and G. Wagenaar. 1969. Hyperlipemia in ponies. *Zentralb. Vet.-Med.* A16:1–7.

Schryver, H. F., P. H. Craig, and H. F. Hintz. 1970. Calcium metabolism in ponies fed varying levels of calcium. *J. Nutr.* 100:955.

Stowe, H. D. 1967a. Automated Orphan Foal Feeding. *Proc. Am. Assoc. Equine Pract.,* p. 65.

Stowe, H. D. 1967b. Serum selenium and related parameters of naturally and experimentally fed horses. *J. Nutr.* 93:60.

Swerczek, T. W. 1976. Enterotoxemia. *Proc. Soc. Theriogenology.* Lexington, Kentucky, p. 24.

True, R. G., and J. E. Lowe. 1980. Induced juglone toxicosis in ponies and horses. *Am. J. Vet. Res.* 41:944.

Tyznik, W. J. 1982. The digestive tract of the horse. *Proc. 4th Bain-Fallon Memorial Lecture.* University of Sydney, Australia, p.1

Willard, J., J. C. Willard, and J. P. Baker. 1973. Dietary influence on feeding behavior in ponies. *J. Animal Sci.* 37:227.

Witt, M., and B. Lohse. 1965. Effect of different winter feeding on growth of Fjord horses up to the third year of life. *Zeittschr. Tierzucht. Zucht.* 81:167–199.

Wyman, W. E. 1895. Colic and its prevention. *So. Carolina Ag. Ext. Bull 22.*

REPRODUCTION

CHAPTER 9

ANATOMY AND PHYSIOLOGY OF REPRODUCTION IN THE MARE

Traditionally, the mare is regarded as having a low reproductive efficiency because many mares that have had successful performance records are kept as broodmares without consideration of their existing reproductive problems. Reproduction in the mare is a complex process, and she is unique in several ways, including anatomical site of ovulation, follicular growth patterns, long estrous periods, differentiation of fertilized and nonfertilized ova in the Fallopian tubes, and serum gonadotropin secretion (in pregnant mares). To be a successful horse breeder and to induce the difficult broodmare to conceive and foal, one must have a working knowledge of the anatomy of the reproductive tract and of the physiology of reproduction in the mare.

9.1 Reproductive Tract

The mare's reproductive tract is shown in Figure 9-1. The tract consists of two ovaries, two Fallopian tubes, uterus, cervix, vagina, clitoris, and vulva.

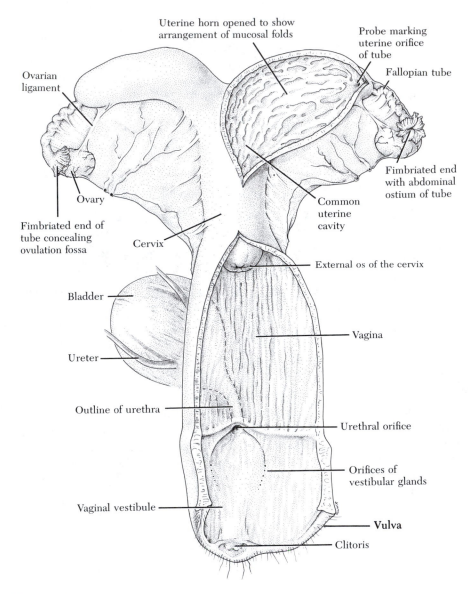

Figure 9-1
Reproductive organs of the mare (*posterior view*) showing vagina and right half of uterus opened. (From Eckstein and Zuckerman, 1956.)

Ovaries

The ovaries function as endocrine and exocrine glands and are the essential organs of reproduction. As endocrine glands, they produce the estrogenic and progestational hormones whose functions are described in Section 9.2. As exocrine glands, they produce and release ova. The anatomical structure of the mare's ovaries has been described by Sisson and Grossman (1953) as bean- or kidney-shaped because of the presence of a definite *ovulation fossa*, the site where ovulations occur (Figure 9-2). The size of the ovaries varies according to the age, breed, size, and reproductive state of the mare. In the mature mare, each ovary weighs approximately 50 to 75 g and is 6 to 7 cm long and 3 to 4 cm wide. Hammond and Wodzicki (1941) have reported that the mature ovary reaches maximum size at 3 to 4 years of age and then decreases in size. Although one ovary is frequently larger than the other, these researchers found that the average weights for each side from several mares were not significantly different. Nishikawa (1959) has studied a Japanese sample and reported that the ovaries of the jenny ass are similar in size and shape to those of the mare, whereas Berliner (1959) has observed that in the United States the ovaries of the jenny ass are larger than mare ovaries.

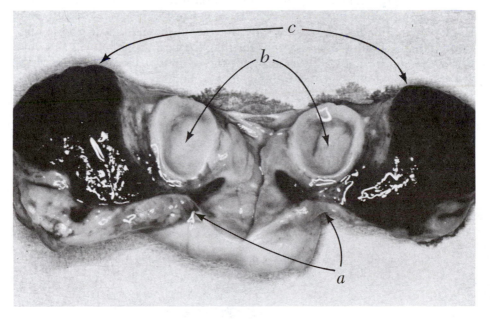

Figure 9-2
Cross section of a mare's ovary: (*a*) ovulation fossa; (*b*) Graafian follicles; (*c*) corpus hemorraghicum. (Photograph courtesy of John Hughes.)

A *serous coat* that is part of the *peritoneum* covers the outer surface of the ovaries except that the *hilus*, where the blood vessels and nerves enter the ovaries, and at the ovulation fossa, which is covered by a layer of primitive germinal epithelium. The inside of the ovaries is composed of connective tissue and follicles and corpora lutea that are in varying stages of development or regression during the breeding season. The *follicles* are not confined to a particular area or outer layer, as they are in other animals, but are distributed throughout the ovary.

The follicles are classified as being either primordial, growing, or Graafian. Primordial follicles consist of an *oocyte* surrounded by a flattened layer of cells known as the *zona pellucida*. As the primordial follicle starts to develop, it becomes surrounded by several layers of granulosa cells and is referred to as the growing follicle. The growing follicle continues to develop and forms the Graafian follicle. The Graafian follicle (Figure 9-3) is surrounded by the granulosa cells and by the *theca folliculi*, which has developed from the connective tissue surrounding the follicle. The theca folliculi differentiates into two layers called the *theca externa* and *theca interna*. The theca externa is a fibrous network that forms a supporting structure for the Graafian follicle. The theca interna, the inside layer of the theca folliculi, contains the vascular supply, and the cells that compose it are responsible for secreting the estrogenic hormones. During development of the Graafian follicle, the granulosa cells continue to increase in number and a cavity or antrum forms within the granulosa cells. At maturity, the cavity or antrum of the Graafian follicle is surrounded by the granulosa membrane and is filled with follicular fluid. Granulosa cells forming the *cumulus oöphorus* surround the ovum and form a stalk that projects the ovum toward the center of the antrum. As the follicle grows and develops, it migrates toward the ovulation fossa. Ovulation (rupture of the follicle) is spontaneous, and extrusion of the

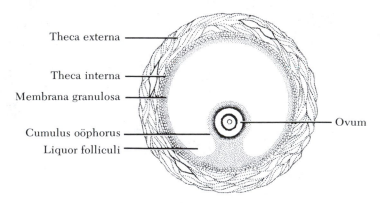

Theca externa

Theca interna

Membrana granulosa

Cumulus oöphorus

Liquor folliculi

Ovum

Figure 9-3
A Graafian follicle. (Diagram courtesy of M. Morris.)

egg occurs through an opening in the ovulation fossa. During each estrous period, there may be several follicles that are increasing in size. Those that fail to reach maturity undergo atresia, a degeneration process.

The development of the *corpus luteum* has been described in detail by Harrison (1946) and by many other investigators. Upon ovulation, the follicle walls collapse and fold inward. A blood clot forms in the cavity as a result of hemorrhage of the thecal capillaries (Figure 9-2). The granulosa cells undergo *lutealization*, and the theca interna cells invade the granulosa wall. The folds of the wall take on the appearance of *trabeculae*, which consist of the theca externa cells and degenerating theca interna cells. The theca externa cells and degenerating theca interna cells invade the lutealizing granulosa cells. The trabeculae are extensively vascularized, and thecal capillaries exist between the luteal cells. By the sixth day, the whole corpus luteum is well vascularized, the theca cells are evenly distributed, the trabeculae are decreased in size and serve as a support for blood vessels, and the theca externa cells begin to pass inward. After approximately 14 days, the corpus luteum decreases in size, and its progestin-synthesizing capabilities decline quite rapidly if the mare is not pregnant. It takes several months for the corpus albicans, a degenerating corpus luteum, to disappear from the ovary.

Blood is supplied to the ovaries via the ovarian artery (Figure 9-4). The ovaries are supplied with an extensive arterial network that forms anastomoses with tributaries of the uterine branch of the ovarian artery. Several veins drain the ovaries and then unite to form the ovarian branch of the ovarian vein. The drainage is such that there is an extensive utero-ovarian plexus. The nerve supply is derived from the renal and aortic plexus of the sympathetic system.

Fallopian Tubes

The Fallopian tubes or oviducts conduct the ova from the ovary to the uterus and are the site of fertilization. They are approximately 30 to 70 mm long and 4 to 8 mm thick and have an inside tubal diameter of 2 to 3 mm. The end adjacent to the ovaries is expanded to form the *ampulla* and the *infundibulum*, which partially surrounds the ovary and directs the ova into the Fallopian tube. The edge of the infundibulum is split so as to form fingerlike projections, or fimbriae. The opening of the oviduct into the peritoneal cavity is called the *ostium abdominale*, and the uterine opening is called the *ostium uterinum*. Three layers of tissue compose the Fallopian tubes: tunica serosa, muscularis (longitudinal and circular), and mucosa.

Some unfertilized eggs are retained in the Fallopian tubes and several of these eggs may be present in the tubules. The fertilized egg is recognized by

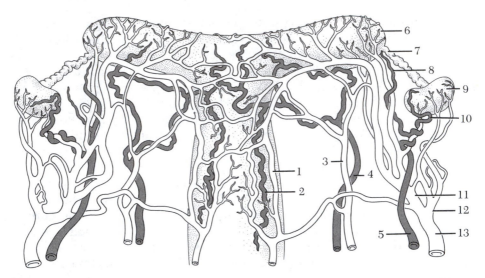

Figure 9-4
Composite diagram of dorsal view of arteries (*cross-marked*) and veins (*shaded*) of
uterus and ovaries of the mare: (1) uterine branch of vaginal vein; (2) uterine
branch of vaginal artery; (3) uterine vein; (4) uterine artery; (5) ovarian artery; (6)
uterine horn; (7) uterine tube; (8) uterine branch of ovarian artery; (9) ovary; (10)
ovarian branch of ovarian artery; (11) uterine branch of ovarian vein; (12) ovarian
branch of ovarian vein; (13) ovarian vein. (From Ginther et al., 1972.)

the Fallopian tubes by an unknown mechanism shortly after conception and
it enters the uterus by the sixth day after conception.

Uterus

The uterus of the mare consists of a body and two horns. The horns, which
are located in the abdominal cavity, are approximately 25 cm in length and
cylindrical. The body of the uterus is located partly in the abdominal cavity
and partly in the pelvic cavity. The body is approximately 18 to 20 cm long
and 10 cm in diameter. The broad ligament of the uterus attaches it to the
abdominal and pelvic walls.

Three coats of tissue similar to the ones forming the oviducts form the
uterine wall. The serous coat is continuous with the broad ligament. The
muscular coat consists of a muscularis longitudinalis, muscularis vascularis,
and muscularis circularis. The muscularis longitudinalis and muscularis vascularis from one inseparable layer of muscle that can be separated from the
muscularis circularis. The third layer of the uterus or mucosa is composed of
a loose connective-tissue network (tunica propria), a deeper layer containing
the long, tubular-shaped uterine glands, and the stratum epithelial.

Uterine blood supply is derived on both sides from three arteries: the uterine branch of the vaginal artery, the uterine artery, and the uterine branch of the ovarian artery. Three veins drain the uterus and correspond to the three arteries. Nerves derived from the uterine and pelvic plexus provide the nerve supply to the uterus.

Cervix

The cervix separates, anatomically and physiologically, the body of the uterus from the vagina. It is approximately 5 to 7.5 cm in length and consists of a powerful, well-developed muscular layer. Extensive longtiudinal folding of the muscle layer leaves no well-defined cervical canal. The lumen is occluded so that it is impermeable to fluids, debris, and bacteria during diestrus and pregnancy. The mucosa of the cervix contains mucosal cells that secrete the mucus found during estrus and pregnancy. The cervix protrudes into the vagina, forming a well-defined angle between the cervix and vagina known as the fornix. One of the cervical folds (frenulum) may extend onto the floor of the vagina.

Vagina

The vagina extends from the cervix to a point just caudal to the external urethral orifice where the vulva starts. It is approximately 15 to 20 cm in length and 10 to 12 cm in diameter. The vagina is divided into the vagina proper and vestibulum vaginae. The hymen separates the two parts and is just anterior to the area where the urethra enters the reproductive tract. The mucous coat contains no glands, is composed of a layer of loose connective tissue, and is covered with a stratified epithelium. Longitudinal and circular muscle fibers are surrounded by a *fibrous adventitia*.

In the mare, the lumen of the vagina is collapsed dorso-ventrally and appears as a transverse slit. It lies on the pelvis so that a seal is formed that helps to prevent movement of infectious contaminants up the reproductive tract.

Blood supply to the vagina is via the internal pudic arteries and drainage is via the internal pudic veins. The nerve supply is from the pelvic plexus of the sympathetic nervous system.

Vulva

The vulva is a terminal portion of the reproductive tract and includes the labia, the reproductive tract caudal to the vagina and clitoris. It hangs over the ischial arch of the pelvis so that the ventral commissure is about 5 cm below the arch. This must be taken into account when a speculum is being passed. The vulva is 10 to 12 cm in length and the external orifice (vulvar cleft) is a vertical slit approximately 12 to 15 cm long. The external urethral orifice is located 10 to 12 cm inside the vulva, just caudal to a prominent transverse fold that forms the hymen.

Clitoris

The clitoris is approximately 5 cm within the vulva. It is the homologue of the male penis. The clitoris is located in the cavity at the ventral commissure of the labia and is enclosed by the prepuce. Following urination or during estrus, the constrictor vulvae muscle layers located in the labia are responsible for eversion of the clitoris. The visible portion of the glans clitoris and the body is about 5 cm long. The clitoral sinus and fossa are areas that harbor the organisms responsible for the veneral disease contagious equine metritis (CEM).

Mammary Gland

The udder is located in the inguinal region and is composed of two gland complexes. The two teats are broad and flat. Each teat is canalized by at least two streak canals that lead from separate teat cisterns and are 5 to 10 mm in length. Each teat cistern is connected to a gland cistern that has a system of ducts leading into it from a secretory gland. The udder is served by the external pudendal artery and is drained by the external pudendal and subcutaneous abdominal veins. The nerve supply is derived from the inguinal nerve.

9.2   Physiology of Reproduction

Estrous Cycle

The average age of the filly at puberty is 12 to 15 months. Russian researchers have reported that crossbred Russian trotter mares reach sexual maturity at 10 to 11 months. Once puberty is reached, the filly begins to come into estrus (heat) in a rhythmic cycle. The estrous cycle is the interval from the onset of estrus until the onset of the next estrus (Figure 9-11). In the mare, it is commonly divided into two periods for practical purposes. The follicular phase or estrus period of the cycle is the period during which the mare shows behavioral signs of estrus, has rapid follicular growth, and finally ovulates. During the estrous period, the mare is interested in and is receptive to a stallion. Most mares in estrus display several characteristic behavioral signs when tested with a stallion (Figure 9-5). Initially, a mare may be agitated but will respond to the stallion and allow him to sniff, nuzzle, and bite her on the neck, flank, and perineal area. Her stance is characterized by a raised tail without switching, hind legs spread apart, and pelvis flexed. The labia of the vulva contract and relax, and there is eversion of the clitoris, which is commonly referred to as "winking." Estrual winking, alone, should not be confused with winking after urination. Mares in estrus usually urinate quite frequently when teased.

Sexual receptivity is influenced by season. The behavioral signs of mares coming out of winter *anestrus* (period of sexual inactivity) are rather variable, that is, some mares have a normal estrous period whereas others are in and out of estrus for several days (Figure 9-6).

The behavioral signs of estrus in the jenny are quite different from those in the mare. Jennies will stand for the jackass with their hind legs spread apart. The head is lowered, the jaws display a chewing motion, and the labia repetitively contract, exposing the clitoris. Jennies urinate frequently, and quite often a slimy mucus is discharged from the vagina.

In both the mare and jenny, the intensity of estrus manifestations increases from weak signs at the start of estrus to a peak at ovulation and then quickly declines to the diestrous state. Hughes et al. (1972b) reported that approximately 50 percent of the mares they studied were in diestrus within 24 hours after ovulation and 80 percent were in diestrus within 48 hours. This finding emphasizes the close relationship between ovulation and the end of sexual receptivity. Occasionally, a mare will ovulate after she has ceased showing signs of estrus.

The diestrous period is the luteal phase of the cycle and begins with ovulation. During *diestrus*, the corpus luteum is formed and secretes progestins. The behavioral response of a diestrous mare to teasing with a stallion varies from a passive, noninterested attitude to complete resentment, often manifested by violent attempts to kick and bite the stallion.

Variations in Estrous Cycle Length

Estrous cycles of the mare are usually 21 to 23 days in length (Table 9-1). A wide range of estrous cycle lengths has been reported in the literature. Cycles vary from 7 to 175 days, but generally, approximately 50 to 60 percent of all mares have a cycle of 21, 22, or 23 days.

There are many causes of prolonged estrous cycles. Season of the year is the most common cause of irregularity. Loy (1970) has classified mares into

(a)

Figure 9-5
Behavioral display of estrus
(heat) by a mare: (*a*)
elevated tail, flexed pelvis,
and acceptance of stallion;
(*b*) eversion of clitoris
(winking).

(b)

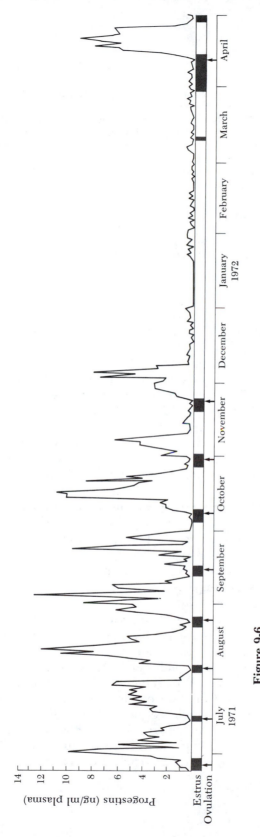

Figure 9-6

Reproductive history of Hi Aggie (■ = estrus; ↑ = ovulation). Notice the winter anestrus December through March. (From Hughes, Stabenfeldt, and Evans. 1972b.)

Table 9-1
Length of estrous cycle and estrous period of mares with 1 or 2 ovulations per cycle

No. of ovulations per estrous cycle	Length of estrous cycle (days)	Length of estrus (days)	Length of diestrus (days)	Corpus luteum lifespan (days)
1	20.7 ± 0.2 (234) (range 13–34)	5.4 ± 3.0 (225) (range 1–24)	15.4 ± 3.1 (201) (range 6–25)	12.6 ± 2.9 (18) (range 8–18)
	20.2 ± 4.5 (76) (range 13–27)	6.1 ± 2.7 (69) (range 1–21)	14.6 ± 3.5 (59) (range 8–25)	12.2 ± 3.3 (17) (range 5–17)

Note: All deviations are standard deviations. Same 11 mares were studied to obtain data for Tables 9-1 through 9-4.
SOURCE: Figures in last column from Stabenfeldt, Hughes and Evans, 1970. All other figures from Hughes, Stabenfeldt, and Evans, 1972b.

three groups according to seasonal variability. There is considerable over-lapping of his classifications because of the variability of mares. In the first category is the polyestrous mare, which cycles regularly throughout the year even though she has some seasonal variations that fall within the normal range for an estrous cycle. The second group includes the seasonally polyes-trous mare, which has a definite breeding season and a definite anestrous period (Figure 9-6). The third classification is the mare that does not cease reproductive activity completely, but does show great irregularity of all characteristics of the estrous cycle during winter and early spring.

Since most mares are seasonally polyestrus, periods of anestrus are noted in most mares that are observed for approximately 2 years. The most common form is winter anestrus. The winter anestrous period is usually 50 to 70 days in length but may be 4 or 5 months long. During the winter anestrous period, follicular development is minimal but some mares may ovulate at irregular intervals. Even though anestrus is primarily a winter phenomenon, cyclic ovulation without estrus has been observed at other times of the year. Other anestrous periods are accompanied by insignificant ovarian activity or the persistent presence of a corpus luteum.

In California, the average estrous cycle length is shortest between April and October (late spring and summer) and longest between November and March (winter). In South Africa, the average estrous cycle is shortest during the summer months of November through February and longest during the winter months of May to August. Winter anestrus is responsible for the longer average length of time from one estrus to the next between November and March (Figure 9-6) (Hughes et al., 1972b). During the winter months, ovulation often occurs without estrus. The estrous cycles can also be interrupted by spontaneously prolongation of the corpus luteum life-span (Figure 9-7). A spontaneously prolonged corpus luteum is a corpus luteum that continues to secrete enough progestins to maintain plasma levels above 3 to 5 ng/ml plasma, which suppresses signs of estrus. The spontaneous prolongation can be due to a failure of the uterus to produce or release the "leutolysin" required to lyse the corpus luteum at the proper stage of the cycle (King and Evans, 1983) or a failure of the corpus luteum to respond. Prolonged cycles are often merely a result of the stud manager's failure to observe estrus. Another frequently encountered problem is that lactating mares that are extremely possessive of their foals may not show estrus. True lactation anestrus occurs in some mares.

Variations in Estrus

The seasonal variability of the estrous cycle, excluding anestrus, seems mainly to be due to variability of the estrous period (Table 9-2). The normal length of estrus is usually 5 to 7 days, but from February to May (late winter and early spring in California), it is 7.6 days. The longest estrous periods are observed when mares are either going into or coming out of winter anestrous. Throughout the year, the diestrous period seems to average 14 to 16 days in length when anestrous periods are excluded.

"Silent heat," the failure to show behavioral signs of estrus, occurs frequently in some mares, although other mares seldom show behavioral signs (Figure 9-8). Usually these mares ovulate and their plasma progestin profile is normal. Approximately 50 percent of mares ovulate at least once without estrus during a 2-year period. The opposite condition, estrus without ovulation, is less frequently observed—it occurred in approximately 3 percent of estrous periods during a 2-year span (Hughes et al., 1972b). Nonovulatory estrus is usually observed immediately preceding the onset of winter anestrus. Split estrous periods, in which the mare is in estrus for a few days, out of estrus for a few days, and then returns to estrus, are observed during approximately 5 percent of the cycles. Split estrous periods average 12 days in length.

Variations in Ovulation Patterns

Diestrous ovulations without signs of estrus occur in most mares some time during a 2-year period (Figure 9-9, Table 9-1). The ovaries are the only part of the reproductive tract that is changed when ovulation occurs during diestrus. If the reproductive history of a mare is not known, the diestrous ovulation may be assumed to occur during a silent heat.

The mare is generally assumed to have one ovulation per estrous period; however, several follicles start to grow during each estrous cycle. Just before estrus, follicles destined to ovulate rapidly increase in size until they are to

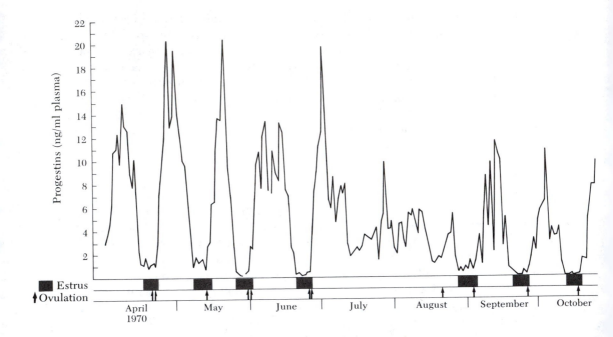

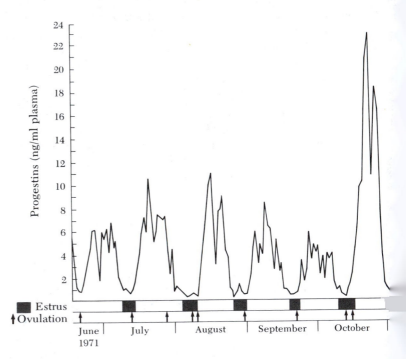

Figure 9-7
Reproductive history of Paisley Star (■■■ = estrus; ↑ = ovulation). Notice the
period of spontaneously prolonged corpus luteum activity during July and August
of 1970. (From Hughes, Stabenfeldt, and Evans. 1972b, p. 119.)

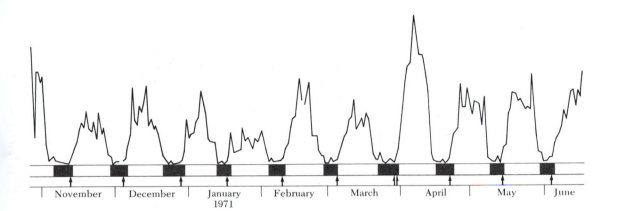

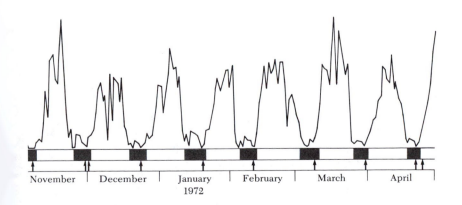

Table 9-2
Effects of seasonal changes on reproduction in the mare

	Jan	Feb	Mar	Apr	May	Jun	Jul	Aug	Sep	Oct	Nov	Dec
Average estrous cycle length (days)	32.3	34.4	26.1	21.9	19.5	21.1	24.6	20.4	20.6	20.5	24.8	30.4
Number of cycles observed	22	20	25	28	39	31	28	27	32	35	25	21
Number of periods of estrus	18	13	26	29	28	28	20	24	27	31	27	22
Average length of estrus (days)	5.06	6.69	8.38	7.76	5.72	4.47	4.70	4.75	4.44	4.53	5.20	7.23
Total ovulations for each month for all 11 mares	20	23	46	45	44	54	48	41	42	44	43	27
Average number of ovulations per month	0.9	1.1	2.1	2.1	2.0	2.5	2.2	1.9	1.9	2.0	2.0	1.2
Total number of diestrous periods without anestrus	16	13	21	27	24	24	21	21	26	27	22	22
Average length of diestrus without anestrus	16.3	16.3	14.2	11.8	14.0	15.8	16.2	15.4	16.1	15.3	16.0	14.3

SOURCE: Adapted from Hughes, Stabenfeldt, and Evans, 1972b, p. 119.

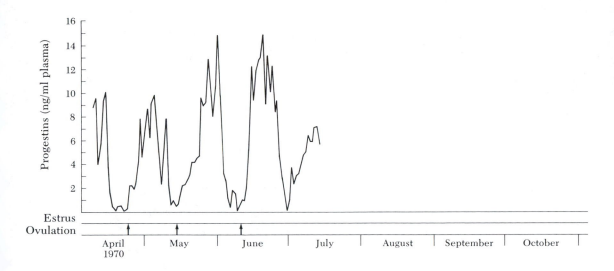

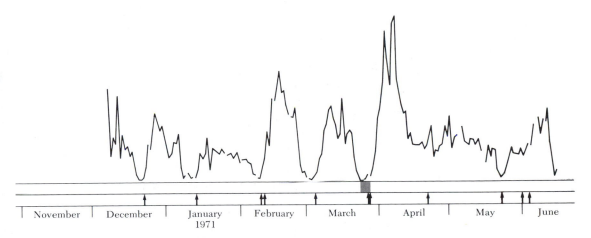

Figure 9-8
Reproductive history of Bashful Boots (▬ = estrus; ↑ = ovulation). Notice the "silent heat" periods and double ovulations. (Graph courtesy of G. H. Stabenfeldt, J. P. Hughes, and J. W. Evans.)

35 to 60 mm long. When a group of Thoroughbred and Quarter Horse mares were observed at the University of California at Davis (Hughes et al., 1972b), twin ovulations occurred approximately 21 percent of the time during a 2-year period (Figures 9-6 through 9-9 and Table 9-3), and several other researchers have reported similar percentages for draft horses, Thoroughbreds and Quarter horses. The percentage of twin conception is lower

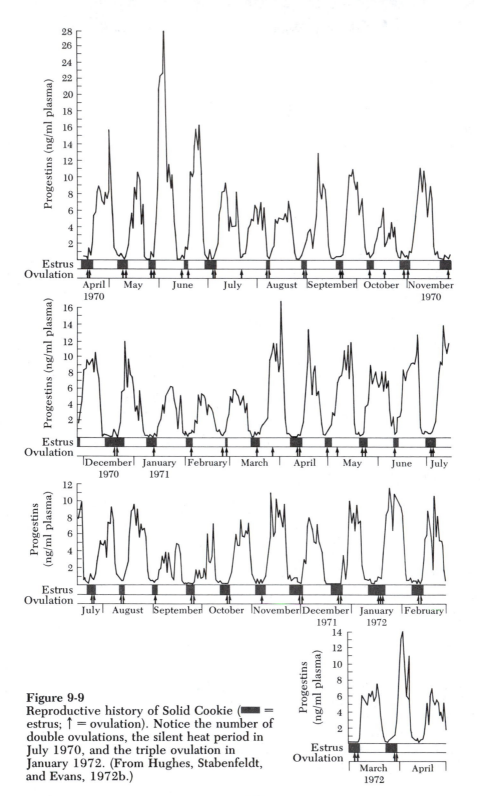

Figure 9-9
Reproductive history of Solid Cookie (■ = estrus; ↑ = ovulation). Notice the number of double ovulations, the silent heat period in July 1970, and the triple ovulation in January 1972. (From Hughes, Stabenfeldt, and Evans, 1972b.)

Table 9-3
Incidence of multiple ovulation by mare and by ovaries, during 2-year period

Mare	No. of triple ovulations	No. of double ovulations	Ovulations				average interval between ovulations (days)
			No. LL	No. RR	No. LR[a]	No. RL[a]	
Aggie Tess	0	7	1	1	2	3	0.7
Bashful Boots	0	9	2	2	1	3	1.5
Double Dark	0	5	2	2	1	0	1.6
Hi Aggie	0	1	0	0	1	0	0.0
Mr. Bruce's Miss	0	2	1	1	0	0	1.0
Paisley Star	0	6	2	1	3	0	1.2
Psychedelic Miss	0	4	0	2	0	2	1.2
Solid Charm	0	4	2	0	1	1	1.0
Solid Cookie	1	25	3	9	10	3	0.7
Solimistic	0	10	2	2	3	3	1.5
Tule Goose	1	11	1	3	6	1	0.5
Total	2	83	16	23	28	16	1.0

[a]First letter indicates ovary in which first ovulation occurred.
SOURCE: Hughes, Stabenfeldt, and Evans, 1972b, p. 119.

333

(1 to 5 percent). The percentage of live foals is 0.5 to 1.5 percent (Berliner, 1959). Some mares (and families) have a tendency to have multiple ovulations, and one mare had 25 twin ovulations in 24 months (Figure 9-9). Multiple ovulations do not appear to affect the duration of estrus, the length of the estrous cycle, or the level of plasma progestins (Table 9-1). A seasonal influence was indicated since few multiple ovulations were observed early in the breeding season (Table 9-4). Double ovulations are rare in pony mares since the incidence is about 2 percent.

Failure of the mare to ovulate during estrus in the breeding season is seldom observed. As previously discussed, it usually occurs during the transition period between winter anestrus and the breeding season. When ovulation fails to occur during the breeding season, the pre-ovulatory follicle becomes quite large and fills with blood (Ginther, 1979). This hemorrhagic follicle may develop normal amounts of luteal tissue or develop none at all. It will gradually recede.

Ovulation is more closely associated with the end of estrus than with the first day of estrus. One may expect ovulation to occur during the last two days of estrus during about 70 percent of the cycles and after the end of estrus for about 15 percent of the cycles. Occasionally, a mare may ovulate prior to showing behavioral signs of estrus; if this occurs, the estrous period is only 1 to 2 days in length (Figure 9-9).

It is not conclusive but the mare appears to ovulate more frequently from the left ovary (Ginther, 1979). The reason for this apparent inequity is unknown.

Table 9-4
Multiple ovulations, by month

Month	No. per month	Percentage per month
January	3	23.5
February	4	22.2
March	10	40.6
April	9	25.7
May	10	32.3
June	10	25.5
July	9	25.0
August	7	21.2
September	8	24.2
October	7	20.0
November	6	17.1
December	4	18.2

SOURCE: Hughes, Stabenfeldt, and Evans. 1972b, p. 119.

Changes in the Reproductive Tract during the Estrous Cycle

The reproductive tract of the mare undergoes several characteristic changes during the estrous cycle. These changes are useful as a guide in determining the optimal time to breed a mare during estrus—especially mares with silent heats. The changes have been discussed by Warszawsky et al. (1972) and are summarized in Table 9-5.

During winter anestrus, the ovaries may be small and inactive, although some mares have considerable ovarian activity. As the mare comes out of anestrus, the ovaries begin to enlarge and soften as the follicles begin to develop. Although several follicles may start to increase in size, usually only one or two reach ovulatory size. The follicle destined to ovulate usually becomes quite prominent just before estrus and starts the rapid growth phase approximately 5 days before ovulation. Generally, maximum follicle

Table 9-5
Changes in mare's reproductive tract during the estrous cycle

Structure	Day of estrous cycle				
	2	4	7	11	17
Anterior portion of vagina (g)	188	216	173	194	230
Cervix (g)	141	168	153	133	164
Uterus (g)	696	674	558	669	849
Uterine horn (left)					
Diameter (mm)	68	55	61	58	62
Length (mm)	151	146	152	154	137
Uterine horn (right)					
Diameter (mm)	58	60	44	61	64
Length (mm)	148	155	144	160	152
Oviduct (Left)					
Weight (g)	5.57	5.72	4.85	6.17	5.20
Length (mm)	147	140	149	162	152
Oviduct (right)					
Weight (g)	7.12	6.83	6.02	6.61	6.61
Length (mm)	154	155	147	166	143
Ovaries	136.7	128.4	122.0	111.7	116.4
Extraluteal tissue (g)	71.9	69.6	70.5	69.0	81.8
Extraluteal fluid (g)	62.7	54.4	42.9	25.6	22.4
Follicles 10–30 mm (no.)	6.3	6.1	7.6	4.6	3.8
Follicles 20–30 mm (no.)	3.4	2.0	1.4	1.7	1.3
Follicles > 30 mm (no.)	1.3	0.9	0.5	0.2	0.1

SOURCE: Adapted from Warzawsky, L. F., et al. 1972, pp. 19–26, 172.

size (40 to 50 mm) is reached the day before ovulation, but it may occur 3 to 4 days before ovulation. Some follicles soften just before ovulation, whereas others do not. Because of the variations in the occurrence and degree of softening, softening of the follicle cannot be used to predict ovulation accurately. Hughes et alobserved that follicles that were 25 to 50 mm in size appeared in addition to the ovulatory follicle in 57.3 percent of the estrous periods of the 11 mares studied. These follicles may continue to increase in size before they regress. Smaller follicles may remain for a couple of cycles and then develop into the ovulatory follicle (Figure 9-10).

Definite changes occur in the cervix during the estrous cycle, and they are quite useful in predicting approaching ovulation. During diestrus, the cervix is covered by a small amount of stocky mucus. The quantity and viscosity of the mucus changes as estrus begins. During estrus, a considerable amount of clear slimy mucus is present. During diestrus, the cervix is tightly constricted and pale pink in color. During estrus, the vascularity increases and the cervix becomes reddish in color. A complete relaxation occurs during estrus to the extent that the cervix may lie on the floor of the vagina and three or more fingers can be passed through it. Palpation reveals the tactile sensitivity and causes muscular contractions of the estrual cervix.

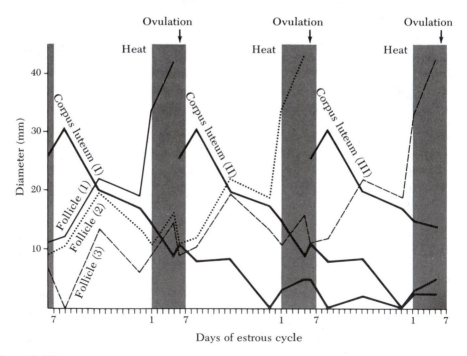

Figure 9-10
Curves of growth of follicle and corpus luteum during the estrous cycle. (From Hammond and Wodzicki, 1941.)

The changes that occur in the vagina are more difficult to interpret than the cervical changes. The most obvious changes that occur in the vagina during estrus are an increase in vascularity, resulting in a change from a pale to a rosy color, and an increase in the size and number of secondary mucosal folds. The walls are dry and sticky during diestrus and become slimy during estrus as a result of changes in cervical mucus secretion.

Hughes et al. (1972b) discussed the changes in uterine form that occur during the estrous cycle. From a practical standpoint, these changes vary in interpretation and their pattern is not sufficiently uniform to be useful in determining the optimal time to breed. The uterus has only slight muscle tone and is turgid during estrus; it may even be termed flaccid. During diestrus, an increase in tone and "tubularity" (contracted state that closely resembles a tube) is observed. However, during late diestrus, tone and tubularity may decrease.

During anestrus, the cervix and uterus lose muscle tone and usually become completely flaccid in most mares that have no follicular activity. In mares that have considerable ovarian activity, the uterus may have some muscle tone and the cervix may be firm and tightly constricted.

9.3   Endocrine Control

Endocrine control of the estrous cycle is not completely understood. Figure 9-11 is a generalized diagram of the sequence of events. The hormones known to control the estrous cycle originate in the pituitary gland, the ovaries, and possibly the uterus.

In the medial basal hypothalamus, neurosecretory cells are responsible for secreting gonadotropin releasing hormone (GnRH). GnRH is a peptide containing 10 amino acids that is released into the blood supply leading to the anterior pituitary gland. In response to GnRH, the pituitary releases luteinizing hormone and follicle stimulating hormone.

The hormones from the pituitary are the *follicle stimulating hormone* (FSH) and the *luteinizing hormone* (LH). They are produced in the basophil cells of the medullary area of the anterior pituitary gland. FSH causes ovarian follicles to grow and to produce increasing amounts of estrogenic substances as they grow. FSH has been purified and its properties characterized by Braselton and McShan (1970) and by Nuti et al. (1972). It is a glycoprotein consisting of 74.9 percent protein, 9.3 percent hexose, 9.1 percent hexosamine, 4.7 percent sialic acid, and 1.3 percent fucose, and it has a molecular weight of approximately 33,200 to 47,900. The amino acid composition is characterized by high levels of cysteine (half-cystine), threonine, lysine, and aspartic and glutamic acids, and a low level of methionine. The NH$_2$-terminal amino acids are phenylalanine and aspartic acids,

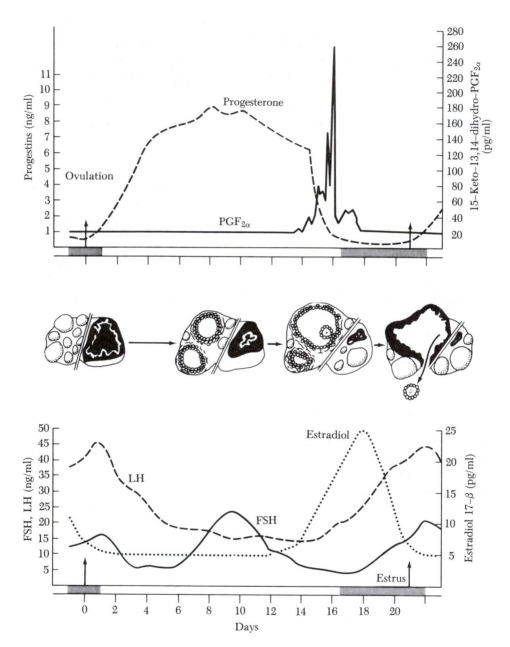

Figure 9-11
The serum concentrations of FSH, LH, estradiol-17β, progesterone and PGF$_{2\alpha}$ and growth of follicles to ovulation during the estrous cycle of the mare. [From research of Evans and Irvine, 1976 (FSH, LH and estradiol-17β); Stabenfeldt, et al., 1972 (progesterone); Neely et al., [1979 (prostaglandin)]. [From Hughes et al. 1978. In *of North America, Large Animal Practice* 2(2):225.]

and the COOH-terminal residues are leucine and glutamic acid. The isoelectric point is pH 4.1

FSH (like LH) is composed of two dissimilar glycoprotein subunits. The alpha subunit is species specific and is common to the various hormones. The beta subunit is hormone specific so that the hormones for specific species have different beta subunits. The pattern of FSH concentration in the blood is quite variable. Some mares have a bimodal curve in which there is one FSH peak during late estrus and early diestrus and a second peak during late diestrus (Evans and Irvine, 1975). Some mares have low concentration during estrus and higher concentrations during diestrus (Ginther, 1979). Others have various combinations of the above (Evans, unpublished).

The luteinizing hormone causes ovulation and initiates the formation of the corpus luteum. Equine LH has a number of LH components and an FSH component. Four LH components have isoelectric points at pH 7.5, 5.9, 6.6, and 7.3, and the FSH component isoelectric point is pH 4.8. The molecular weight is 44,500 to 63,800. LH concentration starts to increase in the plasma 2 to 3 days before ovulation, reaches a peak 1 to 2 days after ovulation, and declines to diestrus levels 6 days after ovulation (Geschwind et al., 1975). FSH and LH work in a synergistic manner in the processes of follicular growth, maturation, and ovulation.

The ovarian hormones that control the estrous cycle are estrogens and progestins. Estradiol-17β is the major ovarian estrogen and is produced by the granulosa cells and by the theca interna of the Graafian follicles. The amount of estradiol-17β in the plasma begins to increase at the onset of estrus, reaches a peak 12 to 27 hours before ovulation, and declines to diestrus levels 5 to 8 days after ovulation (Figure 9-11). Because estrogen reaches a peak before LH peaks, it is possible that the estrogen surge may facilitate the ovulatory surge of LH. Within 24 hours after ovulation, the corpus luteum begins to secrete progestins. By the sixth day after ovulation, maximum levels of plasma progestins have been attained. The corpus luteum remains active for approximately 12 to 14 days, and then rapidly undergoes regression. The roles of the pituitary gland and the uterus in the maintenance and regression of the corpus luteum function are not fully understood. Injection of antibodies against FSH and LH during days 3 to 7 of diestrus causes a significant decrease in the weight of the corpus luteum, indicating that the mare's pituitary contains a substance that is necessary for maintenance of the corpus luteum. Evidence for a so-called uterine luteolysin has been provided by hysterectomy studies conducted by Stabenfeldt et al., (1974), in which functional corpora lutea were maintained for as long as 175 days postovulation after hysterectomy, and by Douglas and Ginther (1979) in which changes in concentration of a hormone, prostaglandin $F_{2\alpha}$, in the uterine veins were determined. Secreted by the uterus, $PGF_{2\alpha}$ is a derivative of prostanoic acid. It is one of six primary types of prostaglandins (E_1 to E_3 and $F_{1\alpha}$ to $F_{3\alpha}$) based on the functions of the cyclopentane ring. Numerical subscripts refer to the number of unsaturations in the side chain and α and β

subscripts refer to the configuration of the substituents in the ring. Prostaglandins have a very short half-life in the blood system since approximately 90 percent is metabolized within 90 seconds of secretion (Pike, 1971). The pattern of $PGF_{2\alpha}$ secretion in the uterine veins (Douglas and Ginther, 1976) and uterine tissue content (King and Evans, unpublished) has been observed to increase at the time of luteolysis (Fig. 9-11). $PGF_{2\alpha}$ and several analogs of it have been synthesized and are used to control the estrous cycle (see Chapter 11).

The interval between the cessation of corpus luteum function and estrus is approximately 3 days. Estrus normally does not occur until the progestin concentration is less than 1 ng/ml in the plasma.

The posterior pituitary gland secretes several octapeptides. Of those secreted, oxytocin plays a role in reproductive function. It is responsible for milk let-down during lactation and for uterine contractions during estrus and parturition. Oxytocin is also used to induce parturition (see Chapter 9).

Placentation and Fetal Development

The fertilized ovum migrates to the uterus approximately 6 days after fertilization. After arriving in the uterus, it may move to the opposite horn or it may remain in the horn on the side ipsilateral to ovulation. Development of the placenta and placentation is a slow process that takes about 150 days to complete. Initially, fetal fluid pressure holds the chorion (outermost membrane) (Figure 9-12) next to the uterine mucosa. At about 40 days, there is an early attachment or interdigitation of microvilli, or fingerlike projections, so the fetus tends to cling to the uterine surface. At about 45 days, macrovilli start to form over the entire surface of the placenta. By day 150, the macrovilli are fully developed exchange units consisting of microvilli and blood vessels. The macrovilli project into crypts in the uterine tissue. The type of attachment is eipitheliochorial, that is, the chorion of the fetus is in contact with the epithelium of the mare's uterus.

In the mare and donkey, the fetal placenta is formed by three membranes (Figure 9-12): chorion, allantois, and amnion. Villi are scattered over the surface of the chorion, and for this reason the attached is frequently said to be "diffuse." The allantois (second membrane) lines the inside of the chorion and is fused with the amnion, thus forming the first water-bag or allantoic cavity. The allantoic cavity is continuous with the bladder by way of the urachus, which passes through the umbilical cord. The amnion is the innermost membrane and forms the second water-bag or amnionic cavity.

The umbilical arteries and veins are located in the connective tissue between the allantois and chorion. The umbilical arteries and their tributaries carry unoxygenated blood and waste products from the fetus to the mare. In general, the blood of the fetus does not mix with maternal blood. However, the two circulations are very close at the junction of the chorion and

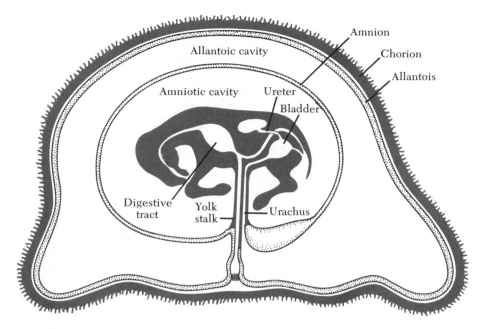

Figure 9-12
Fetus of horse within the placenta. The chorion and allantois make up the
chorioallantois, often called the chorion. (From Frandson, 1974, p. 388. Originally
in Witchi, 1956, *Development of Vertebrates*W. B. Saunders.)

endometrium. At least six layers of tissue separate the maternal and fetal
blood (Figure 9-13): fetal vein, stroma between fetal vein and cytotro-
phohblast, cytotrophoblast, uterine epithelium, uterine stroma, and maternal artery.

The mare has a peculiar specialization, in that *endometrial cups* begin to
form on the thirty-sixth day of pregnancy opposite a transitory, though
well-defined, circumferential thickening of the chorion called the allanto-
chorionic girdle. Allen (1970a) has shown that the endometrial cups of the
mare are composed of fetal cells and maternal blood vessels, connective
tissue, and uterine glands. Development of the endometrial cups can be
divided into five phases: attachment, invasion, phagocytosis, migration, and
cup cell. Their development has been described in detail by Moore et al.
(1975) and is summarized in Figure 9-14. The endometrial cups are described as being a discrete and densely packed mass of large, epithelioid,
decidual-like cells. Upon development of the cups, a central depressed area
filled with coagulum is formed. The accumulation of coagulum, consisting of
degenerate epithelial cells, erythrocytes, polymorphonuclear leucocytes,
and pregnant mare's serum gonadotropin, causes a pouch to form in the
allantochorion. The importance of the secretion of gonadotropins is discussed in the next section.

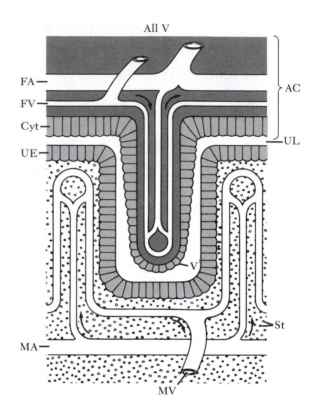

Figure 9-13
Diagrammatic representation of relationship between maternal and fetal tissues in the epitheliochorial type of placenta: FA, fetal artery; FV, fetal vein; AC, allantochorion; UL, uterine lumen; MA, maternal artery; MV, maternal vein; UE, uterine epithelium; Cyt, cytotrophoblast; V, villous; St, stroma; All V, allantoic vesicle. (From Harvey, 1959. p. 433.)

Hippomanes are pieces of amorphous material found in the allantoic fluid during pregnancy. They are apparently formed as a result of material being deposited on a nucleus of cellular debris. The material accumulates in concentric rings as it increases in size and is composed primarily of mucoproteins and calcium phosphate. At parturition, the hippomanes are passed out with the fetus. They are liverlike in texture and are beige, yellowish, or brown in color. They vary in size from 2.5 to several centimeters in diameter.

9.4   Gestation

Length of Gestation Period

Knowledge of the length of gestation and the ability to predict the mare's foaling date are important to the successful management of the pregnant mare. The average gestation period usually ranges from 335 to 340 days.

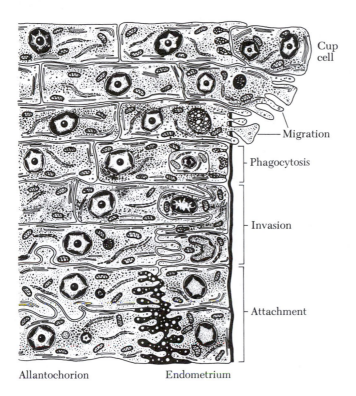

Cup cell

— Migration

- Phagocytosis

- Invasion

- Attachment

Allantochorion   Endometrium

Figure 9-14
A diagrammatic
representation of the
principal events during the
formation of the equine
endometrial cup. The
chorionic girdle becomes
firmly attached to the
underlying endometrium
and the girdle cells invade
and phagocytose the
endometrial epithelium. The
girdle cells migrate down
the length of the
endometrial glands and pass
through the basement
lamina into the endometrial
stroma where they
differentiate into
endometrial cup cells.
(From Moore et al., 1975.)

There is considerable variation in the length of gestation, as evidenced by Paul's (1973) observations of Morgan Horses. The range was 300 to 385 days, the median was 342.5 days, the mean was 339.6 days. Similar observations have been reported for the Thoroughbred by Hintz et al. (1979). They observed a range of 305 to 365 days and a mean of 370.5 ± 7.8 (S.D.) days. From a practical standpoint, foals born in less than 326 days are usually considered premature and those born in less than 319 days show signs of underdevelopment (Ginther, 1979).

Sex of the foal, month of conception, year, number of fetuses, and the individual traits of the mare significantly affect the duration of pregnancy, although size of foal, sire of dam, age of sire, and age of dam apparently have no effect. Colts are usually carried 2 to 7 days longer than fillies, indicating that the sire does exert an influence on gestation length through the genotype. Season of the year seems to have an effect, in that the gestation length is shorter for foals born during warm weather. Mares on a high plane of nutrition foal approximately 4 days before mares on a low plane of nutrition. Therefore, the rate of development of the fetus could be affected by the nutritional plane of the mare. Fetal genotype has an effect; the gestation period of mares carrying a mule fetus (jack × mare) is approximately 10

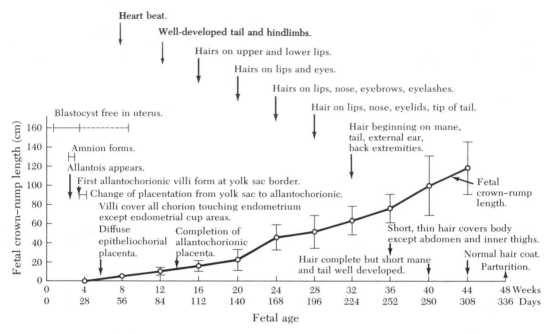

Figure 9-15
Fetal growth. (From Bitteridge and Laing. 1970. p. 98.)

days longer than the gestation period of the same mare when carrying a horse fetus. Figure 9-15 illustrates the growth and development of the fetus.

Hormonal Changes That Occur during Pregnancy

Large quantities of pregnant mare's serum gonadotropin (PMSG), or equine chorionic gonadotropin (eCG) are present in the serum of mares between days 40 and 130 of gestation (Figure 9-16). The rate of secretion and/or absorption is sufficient by 35 to 40 days for it to be detected in the blood. Maximum levels are attained at 55 to 65 days, and the concentration declines until, after 120 to 150 days, it is undetectable in the blood. The amount of PMSG produced by the mare is related to the genotype of the fetus. Mares carrying a mule fetus (jack × mare) produce less PMSG than mares carrying a horse fetus. A jenny carrying a hinny fetus (stallion × jenny) produces more PMSG than the same jenny when carrying a donkey fetus.

PMSG is predominately follicle-stimulating in effect, but it also possesses a luteinizing fraction. The chemical characterization of PMSG has been reviewed by Papkoff (1969). The molecular weight is approximately 28,000. PMSG has a polypeptide content of 30 to 40 percent and a carbohydrate content of 45 percent. It is composed of an α and a β subunit.

Nett et al. (1972) have shown that estrogen starts to increase at approximately 45 days of gestation and increases at a rapid rate beginning at 80 days of gestation. Maximum levels are reached at 200 to 210 days, and then a gradual decline occurs.

Progesterone concentration changes in the blood at different stages of pregnancy are shown in Figure 9-16. Ginther (1979) summarized the changes as follows: days 0 to 8, increases similar to the estrous cycle; days 9 to 27, slight decrease; days 28 to 44, increasing; days 56 to 120, greatest concentration; days 120 to 180, decreasing; days 180 to 300, very low concentration; and the last thirty days, slight increase. Lovell et al. (1975) have observed that a rapid decline in progesterone levels occurs immediately following parturition.

Relaxin is a polypeptide hormone secreted by the placenta during gestation. It is associated with parturition in most species but its role in the mare is not understood. It appears in the blood at approximately 80 days of gestation, rises to maximal concentrations between 176 to 200 days, declines for

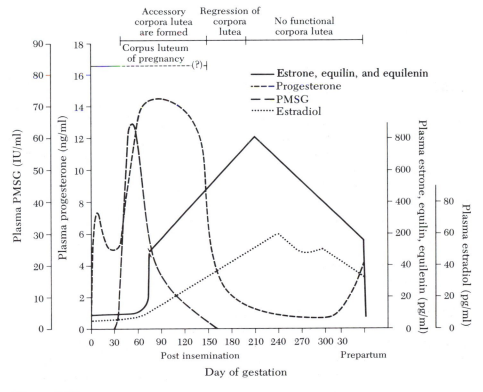

Figure 9-16
Endocrine patterns during gestation of a mare.

about 60 days and then increases steadily from day 240 to foaling (Steward and Stabenfeldt, 1981). After foaling it rapidly disappears from the blood.

Ovarian activity, including formation and ovulation of many Graafian follicles and luteinization of follicles, reaches a maximum during the second and third month of pregnancy. It is during this period that accessory corpora lutea are formed and secrete progestins to maintain pregnancy. Between 150 and 200 days, the corpora lutea undergo degeneration and no large follicles are present. By 140 days, the secretion of progesterone by the corpora lutea is insignificant since removal of the ovaries does not result in abortion.

Hormonal changes that are associated with parturition in the mare are not understood. Two hormones, prostaglandin $F_{2\alpha}$ and oxytocin, probably play a major role in the parturition process. They stimulate myometrial contractions in other species. In the mare, prostaglandin $F_{2\alpha}$ concentration in the blood increases prior to foaling and reaches a peak during the foaling process. Injections of low doses of oxytocin cause an immediate release of prostaglandin $F_{2\alpha}$ when a mare is being induced to foal.

Pregnancy Diagnosis

There are several methods to diagnose pregnancy in the mare, although under normal conditions the first indication of pregnancy is failure to come into estrus 16 to 18 days after being bred. When coupled with a speculum examination of the appearance of the cervix and vagina and a rectal examination for uterine and cervical tone, an accurate presumption of pregnancy can be made. If the mare is not pregnant or is having a silent estrus, the speculum and rectal examination will detect the silent estrus. However, the pseudopregnant condition will not be detected. Ultrasonic echography allows pregnancy diagnosis as early as day 11 to 14 after ovulation with a high degree of accuracy and a fetal heartbeat can be observed at 28 to 30 days.

The most popular method is diagnosis by rectal palpation. This method has a decided advantage in that the result is immediately available. A skilled clinician can detect pregnancy in most mares at 20 to 30 days of gestation and be sure of accuracy after 30 days. The vesicular bulge (fetal membranes) is about the size of a golf ball by day 20 and is the size of a grapefruit by 60 days (Figs. 9-17 to 21). The method for rectal examination is discussed in Chapter 11 and the changes occurring in the ovaries and uterus are given in Table 9-6. After 45 days and until approximately 140 days, the detection of PMSG in the blood is a positive pregnancy diagnosis. Commercially prepared kits to test for PMSG are available to the horse owner who is without veterinary service. The pseudopregnant mare may give a false positive test if

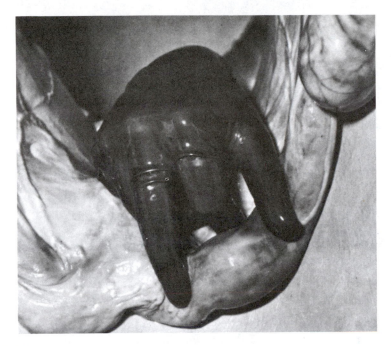

Figure 9-17
Left horn pregnancy, 28 to 30 days. Note the slight enlargement between the
thumb and middle finger. (From R. Zemjanis, 1970.)

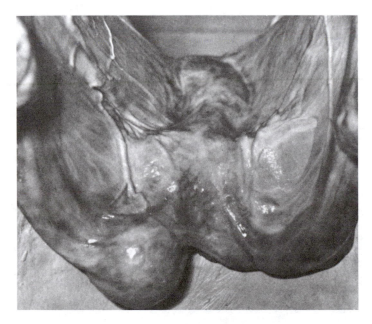

Figure 9-18
Right horn pregnancy, 35 days. (From R. Zemjanis, 1970.)

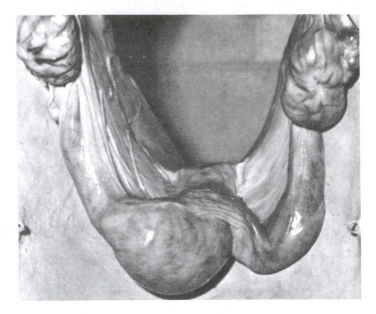

Figure 9-19
Right horn pregnancy, 42 to 45 days. (From R. Zemjanis, 1970.)

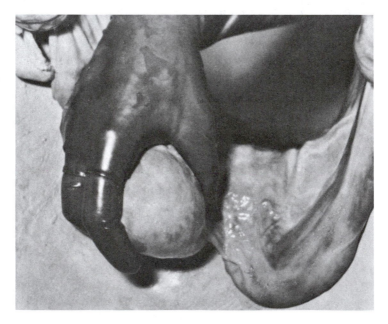

Figure 9-20
Right horn pregnancy, 48 to 50 days. (From R. Zemjanis, 1970.)

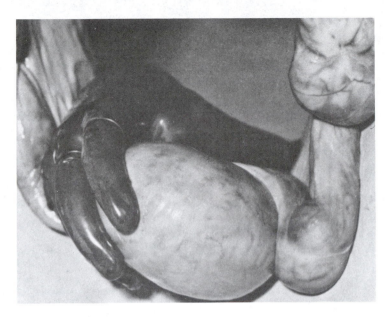

Figure 9-21
Right horn pregnancy, 60 to 65 days. (From R. Zemjanis, 1970.)

she aborted after 36 days. After 90 days,the fetal heartbeat can be determined. The detection of urinary estrogen by chemical means after 120 days of gestation is an accurate method. Fetal movements can often be observed in the mare's flank after the seventh month.

Foal Heat

A mare should come into "foal heat" between 1 and 12 days post-foaling. Even though a wide range (of 1 to 50 days post foaling) has been observed, most mares are in estrus on the seventh to ninth day. The foal heat period may be as long as 12 days, but normally it is approximately 3 days in length.

Table 9-6
Rectal examination for pregnancy

Stage of pregnancy	Ovary	Uterus	Cautions
4 weeks	Position same as in nonpregnant state	2–3 cm bulge Good muscle tone Usually located in horn	Must compare horns and body of uterus. Difficult to find bulge in mares
5 weeks		3–4 cm bulge Can feel fluctuation	with enlarged

continued

Table 9-6
Rectal examination for pregnancy—Continued

Stage of pregnancy	Ovary	Uterus	Cautions
6 weeks		5–7 cm in length and 5 cm in diameter Bulge becoming oval Bulge reaches to junction of horn and body	uterus (i.e. older mares bred on foal heat). May mistake fluctuating pelvic flexure of the colon for pregnant uterus. A filled bladder may be confused with pregnant uterus. Make sure that the hand is cupped under the uterus and each horn is followed to the ovary.
7 weeks		7–8 cm × 6–7 cm Bulge extends into body Fluctuation of fetal fluids Located in pelvic area	
2 months	Moving downward, forward, and medially	12–15 cm × 8–10 cm One-half of body is involved.	
3 months		20–25 cm × 12–16 cm Entire body is involved Starts descent over pelvic brim Can use ballottement to detect fetus	
3–5 months	Positioned further downward, forward, and medially—due to pull of utero-ovarian ligament	Can still palpate fetus Descending over brim of pelvis	
5 to 7 months	Further displacement because increased tension of broad ligaments	Fetus is felt by ballottement Descent continues or may be completed Enlarged uterine artery vibrations can be felt	
7 months to parturition		Easy to palpate fetus Ascent	

Ovulation usually occurs between 9 and 11 days post-foaling. It is a fertile ovulation but care must be taken when the mare is bred during the foal heat (see Section 11.1).

REFERENCES

Allen, W. R. 1970a. Endocrinology of early pregnancy in the mare. *Equine Vet. J.* 2:64.

Allen, W. R. 1970b. *Equine gonadotrophins.* Ph.D. thesis. University of Cambridge, England.

Amoroso, E. C. 1955. Endocrinology of pregnancy. *Br. Med. Bull.* 11:117.

Arthur, G. H., and W. E. Allen. 1972. Clinical observations on reproduction in a pony stud. *Equine Vet. J.* 4:109.

Berliner, V. R. 1959. The estrous cycle of the mare. In *Reproduction in Domestic Animals.* H. H. Cole and P. T. Cupps, eds. Vol. 1 New York: Academic Press.

Bitteridge, K. J., and J. A. Laing. 1970. "The diagnosis of pregnancy." In *Fertility and Infertility in the Domestic Animals.* J. A. Laing, ed. Baltimore: Williams and Williams.

Braselton, W. E., and W. H. McShan. 1970. Purification and properties of follicle-stimulating and luteinizing hormones from horse pituitary glands. *Arch. Bichem. Biophy.* 139:45.

Chevalier, F., and E. Palmer. 1982. Ultrasonic echography in the mare. *J. Reprod. Fert.,* Suppl. 32:423.

Clegg, M. T., J. M. Boda, and H. H. Cole. 1954. The endometrial cups and allanto-chorionic pouches in the mare with emphasis on the source of equine gonadotrophin. *Endocrinol.* 54:448.

Cole, H. H., and P. T. Cupps, eds. 1959. *Reproduction in Domestic Animals.* New York: Academic Press.

Cole, H. H., and H. Goss. 1943. The Source of Equine Gonadotropin. In *Essays in Biology.* No. 107. Berkeley: University of California Press.

Cole, H. H., and G. H. Hart. 1942. Diagnosis of pregnancy in the mare by hormonal means. *J. A. V. M. A.* 101:124.

Dickerson, J. W. T., D. A. T. Southgate, and J. M. King. 1967. The origin and development of the hippomanes in the horse and zebra. 2. The chemical composition of the foetal fluids in hippomanes. *J. Anat.* 101:285.

Douglas, R. H., and O. J. Ginther. 1976. Concentrations of prostaglandins in uterine venus plasma of anesthetized mares during the estrous cycle and early pregnancy. *Prostaglandins* 11:251.

Eckstein, P., and S. Zuckerman. 1956. Morphology of the Reproductive Tract. In *Marshall's Physiology of Reproduction.* A. S. Parkes, ed. 3rd ed. Boston: Little, Brown.

Evans, M. J., and C. H. G. Irvine. 1975. Serum concentration of FSH, LH and progesterone during the oestrous cycle and early pregnancy in the mare. *J. Reprod. Fert.,* Suppl. 23:193.

Frandson, R. D., ed. 1974. *Anatomy and Physiology of Farm Animals.* Philadelphia: Lea & Febiger.

Geschwind, I. I., R. Dewey, J. P. Hughes, J. W. Evans, and G. H. Stabenfeldt. 1975. Circulating luteinizing hormone levels in the mare during oestrous cycle. *J. Reprod. Fert.*, Suppl.23:207-212.

Ginther, O. J. 1979. *Reproductive Biology of the Mare.* Ann Arbor, Michigan: McNaughton and Gunn.

Ginther, O. J., and N. L. First. 1971. Maintenance of the corpus luteum in hysterectomized mares. *Amer. J. Vet. Res.* 32:1687.

Ginther, O. J., M. C. Garcia, E. L. Squires, and W. P. Steffenhagen. 1972. Anatomy of vasculature of uterus and ovaries in the mare. *Amer. J. Vet. Res.* 33:1561.

Hammond, J., and K. Wodzicki. 1941. Anatomical and histological changes during the oestrous cycle in the mare. *Proc. Royal Soc.* 130(B):1.

Harrison, R. J. 1946. The early development of the corpus luteum in the mare. *J. Anat.* 80:160.

Harvey, E. B. 1959. Implantation, development of the fetus, and fetal membranes. In *Reproduction in Domestic Animals.* H. H. Cole and P. T. Cupps, eds. Vol. 1. New York: Academic Press.

Hintz, H. F., R. L. Hintz, D. H. Lein, and L. D. Van Vleck. 1979. Length of gestation periods in thoroughbred mares. *J. Equine Med. and Surgery.* 3:289.

Hughes, J. P., G. H. Stabenfeldt, and J. W. Evans. 1972a. Estrous cycle and ovulation in the mare. *J. A. V. M. A.* 161:1367.

Hughes, J. P., G. H. Stabenfeldt, and J. W. Evans. 1972b. Clinical and endocrine aspects of the estrous cycle of the mare. *Proc. A. A. E. P.*:119.

Jeffcott, L. B., J. G. Atherton, and J. Mingay. 1969. Equine pregnancy diagnosis. *Vet. Record* 84:80.

Jeffcott, L. B., and K. E. Whitwell. 1973. Twinning as a cause of foetal and neonatal loss in the Thoroughbred mare. *J. Comp. Path.* 73:83.

King, J. M. 1967. The origin and development of hippomanes in the horse and zebra. 1. The location, morphology and histology of the hippomanes. *J. Anat.* 101:277.

King, S., and J. W. Evans. 1983. Endometrial prostaglandin $F_{2\alpha}$ synthesizing capability in mares with spontaneously prolonged corpus luteum syndrome. *Physiologist* 26(4):A-31.

Laing, J. A. ed. 1970. *Fertility and Infertility in the Domestic Animals.* Baltimore: Williams and Wilkins.

Lovell, J. D., G. H. Stabenfeldt, J. P. Hughes, and J. W. Evans. 1975. Endocrine patterns of the mare at term. *J. Reprod. Fert.*, Suppl. 23:449-456.

Loy, R. G. 1970. The reproductive cycle of the mare. *Lectures of Stud Managers Courses*:20.

Moore, R. M., W. R. Allen, and D. W. Hamilton. 1975. Origin and histogenesis of the equine endometrial cups. *J. Reprod. Fert.*, Suppl. 23:391.

Neely, D. P., H. Kindahl, G. H. Stabenfeldt, L. E. Edquivist, and J. Hughes. 1979. Release patterns in the mare: physiological, pathophysiological, and therapeutic. *J. Reprod. Fert.*, Suppl. 27:181.

Nett, T. M., D. W. Holtan, and V. L. Estergreen. 1972. Plasma estrogens in pregnant mares. *Proc. Western Sect. of Amer. Soc. Anim. Sci.* 23:509.

Nishikawa, Y. 1959. *Studies on Reproduction in Horses.* Japan Racing Association, Tokyo.

Nuti, L. C., H. J. Grimek, W. E. Braselton, and W. H. McShan. 1972. Chemical properties of equine pituitary follicle-stimulating hormone. *Endocrinol* 91:1418.

Papkoff, H. 1969. Chemistry of the gonadtropins. In *Reprodution in Domestic Animals.* H. H. Cole and P. T. Cupps, eds. New York: Academic Press.

Pattison, M. L., C. L. Chen, and S. L. King. 1972. Determination of LH and estradiol-17β surge with reference to the time of ovulation in mares. *Biol. of Reprod.* 7:136.

Paul, R. R. 1973. Foaling date. *The Morgan Horse* 33:40.

Penida, M. H., O. J. Ginther, and W. H. McShan. 1972. Regression of the corpus luteum in mares treated with an antiserum against equine pituitary fraction. *Amer. J. Vet. Res.* 33:1767.

Pike, J. E. 1971. Prostaglandins. *Sci. Amer.* 225:84.

Rollins, W. C., and C. E. Howell. 1951. Genetic sources of variation in the gestation length of the horse. *J. Animal Sci.* 10:797.

Ropiha, R. T., R. G. Matthews, R. M. Butterworth, F. M. Moss, and W. J. McFadden. 1969. The duration of pregnancy in Thoroughbred mares. *Vet. Rec.* 84:552.

Ryan, R. J., and R. V. Short. 1965. Formation of estradiol-17β by granulosa and theca cells of the equine ovarian follicle. *Endocrinol.* 76:108.

Sisson, S., and J. D. Grossman. 1953. *The Anatomy of Domestic Animals.* Philadelphia: W. B. Saunders.

Stabenfeldt, G. H., J. P. Hughes, and J. W. Evans. 1970. Ovarian activity during the estrous cycle of the mare. *Endocrinol.* 90:1379.

Stabenfeldt, G. H., J. P. Hughes, J. D. Wheat, J. W. Evans, P. C. Kennedy, and P. T. Cupps. 1974. The role of the uterus in ovarian control in the mare. *J. Reprod. Fert.* 37:343.

Stewart, D. R., and G. Stabenfeldt. 1981. Relaxin activity in the pregnant mare. *Biol. Reprod.* 25(2):281.

Turner, C. W. 1952. *The Mammary Gland.* Vol. 1. *The Anatomy of the Udder of Cattle and Domestic Animals.* Columbia, Missouri: Lucas Brothers.

Warszawsky, L. F., W. G. Parker, N. L. First, and O. J. Ginther. 1972. Gross changes of internal genitalia during the estrous cycle in the mare. *Amer. J. Vet. Res.* 33:19.

Zenjanis, R. 1970. *Diagnostic and Therapeutic Techniques in Animal Reproduction.* 2d ed. Baltimore: Williams and Wilkins.

CHAPTER 10

REPRODUCTIVE PHYSIOLOGY OF THE STALLION

The stallion has always been the key to a successful breeding operation. An evaluation of the foaling rate achieved in the Hanoverian breed in Germany between 1815 and 1973 showed that during this period, no improvement of the reproductive rate was achieved (Merkt et al., 1979). Because of the improvement in veterinary service that occurred during this period, the fertility of the Hanoverian stallions may have declined. Today, when some stallions are syndicated for millions of dollars to stand at stud, it is essential that they be managed quite carefully. To develop a management program to ensure continued use of a stallion and to maintain his production of quality semen, the stud manager must have a knowledge of the anatomy and physiology of reproduction in the stallion. This knowledge then serves as the basis for the rest of the management procedures.

10.1 Anatomy of the Reproductive Tract

Figure 10-1 is a diagram of the anatomy of the stallion's reproductive tract. The various parts of the reproductive tract are described in detail in the following sections.

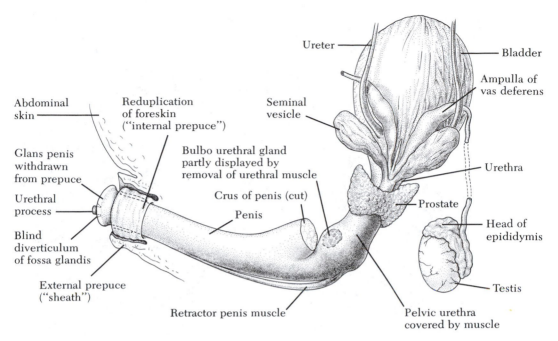

Figure 10-1
Reproductive tract of the stallion (*lateral view*). Bladder and upper urethra are twisted to expose posterior aspects. (From Eckstein and Zuckerman. 1956. p. 43.)

Scrotum

The scrotum of the stallion, a diverticulum of the abdomen, is somewhat asymmetrical, and less pendulous than that of the bull or ram. The scrotum is asymmetrical because one testicle is slightly larger and placed a little further back than the other. (The testicle of a stallion is shown in Figures 10-2 and 10-3.) The testes lie horizontally in the scrotum when the cremaster muscle is relaxed. The scrotum has a smooth appearance when relaxed and is constricted superiorly next to the abdomen. Upon exposure to cold, the external cremaster muscle and the tunica dartos muscle contract so that the scrotum becomes drawn up and wrinkled. The relaxation and contraction of these muscles serve a necessary thermo-regulatory function. The wall of the scrotum is composed of layers that include the skin, the tunica dartos muscle, the scrotum fascia, and the parietal layer of the tunica vaginalis propria. The circulatory system in the scrotum is composed of branches of the external pudic artery, and vein and lymph drainage passes to the superficial inguinal

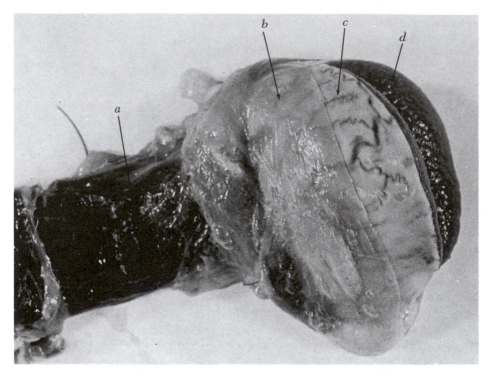

Figure 10-2
Testicle of a stallion showing (*a*) cremaster muscle; (*b*) tunica vaginalis propria; (*c*) tunica albuginea; and (*d*) gland substance.

lymph node. Innervation of the scrotum is derived from the ventral branches of the second and third lumbar nerves.

Testes

The testicles (Figures 10-2 and 10-3) have been described by Sisson and Grossman (1953) as being 10 to 12 cm long, 6 to 8 cm high, and 5 cm wide; each approximately weighs 225 to 300 g; and quite frequently the left one is larger. The tunica vaginalis propria (Figure 10-2), an extension of the visceral layer of the serosa, covers the outer surface of the testicle and envelops the spermatic cord except where the blood vessels and nerves enter the testicle. The tunica albuginea (Figure 10-2) is immediately beneath the tunica vaginalis propria and forms a strong capsule composed of dense white fibrous tissue and smooth muscle fibers. Trabeculae septa of connective tissue project into the gland substance (Figure 10-2) and subdivide the testicular parenchyma into lobules. However, the trabeculae and intralobular septa do not form a distinct mediastinum testis. The septa are interconnected and contain blood vessels and smooth muscle fibers. Each lobule contains seminiferous tubules, interstitial cells (cells of Leydig), and loose connective tissue. The seminiferous tubule is the functional unit of the testis and is composed of three sections. The convoluted tubule in cross-section is composed of connective tissue and a basement membrane that contains layers of spermatogenic cells in various stages of development. Sustentacular cells extend from the basement membrane and are distributed between the spermatogenic cells, thus serving as a source of nutrition to the spermatogenic cells. Each end of the convoluted tubule is connected to a straight tubule which leads toward a rete tubule. The straight seminiferous tubules unite but do not form a rete testis in the center of the testicle as they do in some other species. The seminiferous tubules, after uniting, converge to the

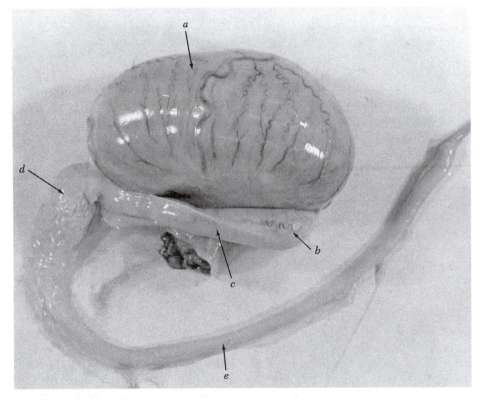

Figure 10-3
Testicle, epididymis, and vas deferens of the stallion: (*a*) testicle; (*b*) head of epididymis; (*c*) body of epididymis; (*d*) tail of epididymis; and (*e*) vas deferens.

attached border of the testicle, where several efferent ducts pierce the albuginea and enter the head of the epididymis.

The spermatic artery, a branch of the posterior aorta, supplies blood to the testicle. The blood leaves the testicle via the spermatic vein, and lymph is drained into the lumbar lymph glands.

Epididymis

At the tunica albuginea, the seminiferous tubules unite into a single duct to form the head of the epididymis (Figure 10-3). The epididymis remains attached to the testicle; the main part of the attached epididymis is known as the body. The tail of the epididymis is located at the lower extremity of the testicle. The epididymis serves a number of functions, such as transporting spermatozoa and other contributions of the testes to semen, concentrating the spermatozoa as a result of water absorption, and providing a place for maturation and storage of spermatozoa.

Vas Deferens

The tail portion of the epididymis gradually merges with the vas deferens (Figure 10-3). Spermatozoa are transported from the tail of the epididymis to the urethra by the vas deferens. A thick muscular wall surrounds a relatively small lumen. Lumen size is constant but the wall thickens considerably to form the ampulla (Figure 10-4) approximately 15 to 20 cm from the entrance to the urethra. The increased size is caused by the presence of numerous glands in the ampullar region. The size of the vas deferens decreases sharply as the duct disappears under the isthmus of the seminal vesicles (Figure 10-4) and opens into the urethra. The opening into the urethra is common with excretory ducts of the seminal vesicles, but there is no ejaculatory duct in the horse as there is in man.

Penis

The penis of the stallion is approximately 50 cm long in the relaxed state; approximately 15 to 20 cm of it lies in the prepuce. During erection, the size increases about twofold. The penis is somewhat cylindrical in shape, but compressed laterally. It is roughly divided into the head, body, and glans. The glans or free end of the penis is bell-shaped, particularly during erection, and the urethral process extends approximately 2.5 cm from the surface of the deep depression or fossa glandis. The size and shape of the donkey penis are similar to those of the stallion.

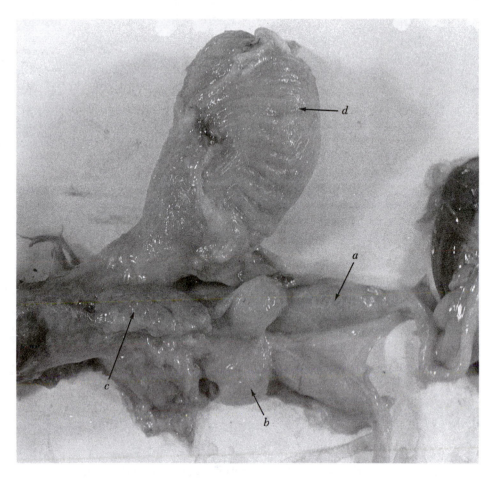

Figure 10-4
Accessory sex glands of the stallion: (*a*) ampulla; (*b*) seminal vesicles; and (*c*)
prostate gland. The bladder is labeled *d*.

Prepuce

The sheath, properly called the prepuce, has been described by Sisson and
Grossman (1953) as a "double invagination of the skin which contains and
covers the free or prescrotum portion of the penis when not erect." The
external prepuce extends from the scrotum to within 5 to 7 cm of the
umbilicus. Before reaching the umbilicus, it is reflected backward and dorsally to form the preputial orifice. The internal prepuce passes backward
from the preputial orifice in such a way that it lines the inside of the external

prepuce. After passing backward for 15 to 20 cm, it is then reflected forward until it approaches the orifice, where it is again reflected backward to form a secondary tubular invagination that contains the relaxed penis. The internal layers of skin contain large sebaceous glands and coil glands or preputal glands whose secretions, together with desquamated epithelial cells, form the fatty smegma.

The circulatory system of the penis is formed from branches of the external pudic arteries and veins, whereas the lymph drainage is to the superficial and lumbar lymph glands. Penile innervation is derived from the pudic, ilio-hypogastric, and ilio-inguinal nerves.

Accessory Sex Glands

Vesicular Glands   The accessory sex glands contribute to stallion semen a fluid of characteristic composition. The vesicular glands (Figure 10-4) are paired glands, each one 15 to 20 cm long and approximately 5 cm in diameter. The long axes are parallel to the vas deferens, and the excretory duct passes under the prostate before it opens into the urethra.

The vesicular glands are responsible for secreting the "gel fraction" to the semen. A major portion of the seminal fluid, as well as fructose, citric acid, and proteins, is secreted by the vesicular glands.

Bulbourethral Glands   The bulbourethral glands are paired, ovoid-shaped, lobulated glands that lie near the ischial arch. Each gland is approximately 4×2.5 cm and has six to eight excretory ducts opening into the urethra behind the prostate ducts. These glands secrete a mucoid secretion.

Prostate Gland   The prostate gland (Figure 10-4) consists of two lateral bulbs (each $3 \times 1.5 \times 0.5$ cm) that are connected by a thin isthmus (2 cm wide). The isthmus lies over the junction of the bladder with the urethra. Approximately 15 to 20 prostatic ducts perforate the urethra.

The thin watery fluid secreted by the prostate gland serves to cleanse the urethra during the ejaculatory process. The fluid is often seen prior to mounting or during the time the stallion's penis is being cleansed prior to breeding.

10.2   Semen Production

Puberty

Hauer et al. (1970) have defined puberty in the stallion as the age at which an ejaculate contains a minimum of 1×10^8 total sperm with 10 percent progressive motility. They found the mean age of puberty in Quarter Horse stallions to be 67 weeks at a weight of 367 kg. Skinner and Bowen (1968) have observed spermatozoa in the ejaculate of Welsh stallions at the age of 11.5 to 14.5 months.

Endocrinology

At present, no detailed studies have been made of the hormone control of the reproductive processes of the stallion. Studies of laboratory animals have developed the following concept of hormone control (Figure 10-5). Gonado-tropin releasing hormone (GnRH) secreted by the hypothalamus is responsible for the secretion of luteinizing hormone (LH) and follicle stimulating hormone (FSH). There may be separate releasing hormones for LH and FSH, but changes in the secretory pattern of GnRH may be responsible for the patterns of LH and FSH. LH stimulates the Leydig cells to synthesize and release testosterone. Testosterone acts via a negative feedback system to the hypothalamus and pituitary to regulate LH secretion. Sertoli cells located in the seminiferous tubules dictate the development and maintenance of sper-

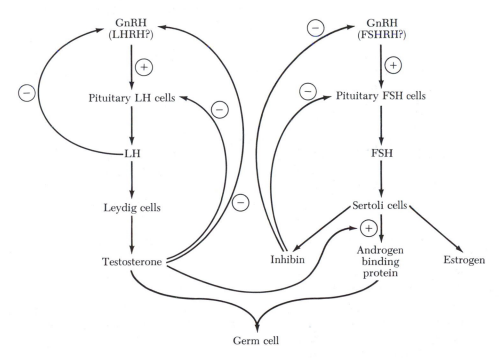

Figure 10-5
Endocrine control of spermatogenesis in stallion. Plus equals positive feedback and minus equals negative feedback.

matogenesis in all mammals. All hormonal regulation of spermatogenesis is probably mediated via the Sertoli cells. These cells are the primary target for FSH. In response to FSH, the Sertoli cells synthesize and secrete inhibin. Though the negative feedback systems at the pituitary and hypothalamus, inhibin controls FSH secretion. FSH and testosterone stimulate the sertoli cells to synthesize and secrete androgen binding protein (ABP). ABP binds testosterone and results in the maintenance of a high testosterone concentration in the seminiferous tubule. Testosterone is responsible for maintaining spermatogenesis and the function of the accessory sex glands. Testosterone reaches the accessory sex glands via the blood system. It may be transported to the epididymis bound to ABP. Prolactin may play a role in the stallion by affecting Leydig cell function during sexual maturation and during photoperiod induced seasonal changes in testicular activity.

Stallion urine contains high levels of estrogen that is produced by the Sertoli cells and interstitial cells. The role of estrogen in the stallion is not presently known.

Spermatogenesis

Spermatogenesis is a process by which primary sex cells in the testis produce spermatozoa. The process starts in the seminiferous tubule and consists of eight stages. This complete series of stages occurs in a given segment (cross-section) of the seminiferous tubule and then the series is repeated in that segment. In other segments of the seminiferous tubule, other cycles are occurring. The process takes 12.2 days and each stage requires a different period of time. Swierstra et al. (1974) has described the stages as follows:

Stage 1   From the complete disappearance of luminal spermatozoa to the onset of elongation of the spermatid nuclei (2.1 days)

Stage 2   From the onset of elongation to the end of elongation of the spermatid nuclei (1.8 days)

Stage 3   From the end of elongation of the spermatid nuclei to the start to the first meiotic division (0.4 days)

Stage 4   From the start of the first to the end of the second meiotic (maturation) division (1.9 days)

Stage 5   From the end of the second meiotic division to the initial appearance of Type B spermatogonia (0.9 days)

Stage 6   From the initial appearance of Type B spermatogonia to the time all bundles of the elongated spermatids have started migration toward the lumen of the seminiferous tubule (1.6 days)

Stage 7   From the time all bundles of elongated spermatids have started their migration toward the tubular lumen until they all reach the lumen (1.5 days)

Stage 8 From the time the elongated spermatids line the lumen to their complete disappearance from the lumen (1.9 days)

The duration of the entire process extends over four consecutive cycles of the seminiferous epithelium. Therefore, the spermatogenic cycle is 49 days in duration. This is an approximation, since the exact scheme for spermatogonial division has not been determined for the stallion.

After development in the testis, the spermatids move through the epididymis. The transport time is 4 to 5 days. Therefore the length of time from initiation of the process until the sperm are ready for ejaculation is 54 days. This must be taken into account when a semen evaluation must be repeated because of poor semen quality. One must wait at least one complete spermatogenic cycle of 54 days. Also, interruption of the spermatogenic process will result in lowered fertility starting approximately 54 days later (for example, after a high fever lasting for 3 to 4 days).

Morphology of the Stallion Spermatozoan

The morphology of the mature stallion spermatozoan has the same basic characteristics as that of other mammalian spermatozoa and has been described in detail by Bielanski and Kaczmarski (1979). The spermatozoan is composed of three principle regions: The head, the neck, and the tail (Figure 10-6).

The head is 6 to 8 micrometers long and 3.3 to 4.6 micrometers wide. It is composed chiefly of a nucleus that contains the genetic material, and is covered anteriorly by the acrosome and posteriorly by the postmucular cap. A plasma membrane surrounds the entire head. The head of the horse spermatozoan is small and slender compared with ovine, bovine, and porcine spermatozoa. The locomotor system, the tail, is connected to the head by the neck. The border between the head and neck is clearly defined by a posterior ring and corresponds to the place of attachment of the plasma membrane with a nuclear envelope. The neck region contains several structures including the proximal centriole, microtubules, the capitulum, and striated columns. The capitulum is located opposite to the thick part of the basal plate. Below the capitulum, the striated columns are situated. They extend to the beginning of the nine dense outer fibers that surround the axial filament. The proximal centriole is located directly below the capitulum. The tail is positioned abaxial to the head so that the head bulges out to one side. The proximal centriole is always located on the side with the greater curvature. The tail of the spermatozoan is composed of three regions: the middle piece, the main piece, and the end of the axial filament or end piece. The middle piece is 8 to 10 micrometers long and 0.5 micrometers wide. It is a source of energy for the spermatozoan. The middle piece starts where the two perpendicular mitochondria in the neck begin spiraling into the mito-

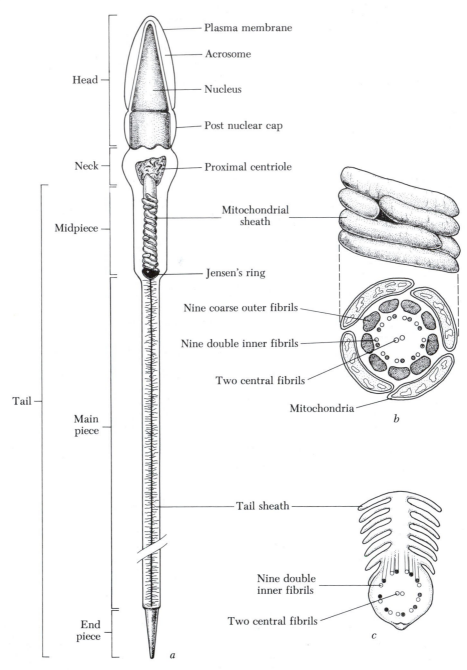

Plasma membrane

Acrosome

Nucleus

Post nuclear cap

Head

Neck

Proximal centriole

Midpiece

Mitochondrial sheath

Jensen's ring

Nine coarse outer fibrils

Nine double inner fibrils

Two central fibrils

Mitochondria

b

Tail

Main piece

Tail sheath

Nine double inner fibrils

Two central fibrils

c

End piece

a

Figure 10-6
Morphology of stallion spermatozoan: (*a*) general view; (*b*) midpiece; (*c*) main piece. [From White, J. G. 1968. In *Reproduction in Farm Animals*. Hafez, E. S. E., Lea and Febriger, eds. Adapted from Wu; 1966. *A. I. Digest*, 14(6):7.]

chondrial helix that surrounds the nine outer, the nine double inner, and two central fibrils. There are about 60 spirals of the mitochondrial helix. Jensen's ring is adjacent to the last spiral.

Nishikawa (1959) has described the spermatozoan of the ass as being 64.1 μm long. The head is approximately 6.9 μm long and 4 μm wide. The middle piece is 9.9 (9.4 to 10.4) μm, and the tail is 47.3 μm long. Therefore, the ass spermatozoan is very similar to the horse spermatozoan except the head is wider and the entire length is longer.

Semen Characteristics

The semen of the stallion, greyish-white in appearance, consists of the seminal plasma or fluid medium and spermatozoa. The whole semen is a series of 8 (range 5 to 10) seminal jets ejaculated in sequence (Tischner et al., 1974). The first 3 jets are milky in color and consistency and contain 80 percent of the total number of spermatozoa ejaculated. The fluid medium of the first 3 jets has a high level of ergothioneine, indicating that the fluid is derived mainly from the ampular glands of the vas deferens. The later jets (range 4 to 10) contain very few sperm cells. They are mucinous in appearance and contain a high concentration of citric acid, indicating that the fluid is derived mainly from the seminal vesicles.

Semen Evaluation

At least four criteria are used to evaluate stallion semen: the volume of semen and the concentration, motility, and morphology of the spermatozoa. The ejaculate volume is usually 50 to 75 ml but may be as great as 150 to 170 ml. At least two factors influence semen volume: season and stallion. The volume of semen is larger during the breeding season than during the winter. The absence of or the amount of gel, which is influenced by stallion variability and season, is an important determinant of volume. The gel-free volume is approximately 40 to 75 ml. The concentration of sperm per milliliter of ejaculate ranges between 30 and 800 million. Pickett et al. (1970) have reported a large variation between stallions, first and second ejaculates, and months of the year for first ejaculates. The mean concentration of spermatozoa in gel-free seminal fluid for first ejaculates was $347.8 \pm 124.9 \times 10^6$/ml, compared with $211.9 \pm 88.1 \times 10^6$/ml for second ejaculates. Consequently, the total number of spermatozoa was influenced by stallion, ejaculation frequency, and season. A mean total number of spermatozoa of $9.3 \pm 3.9 \times 10^9$ was reported for first ejaculates and $4.6 \pm 1.9 \times 10^9$ for second ejaculates. Other reports have reported values ranging from 6.3 to 26×10^9 spermatozoa per ejaculate. The seasonal effect on sperm output and total volume of semen is very marked. The percentage of motile spermatozoa in an ejaculate is usually 60 to 100 percent, and 70 percent is considered quite good. The season of the year does not affect the motility of spermatozoa. Therefore, a stallion is just as fertile during the winter months as during the breeding season; however, he cannot be used to breed as many mares because he does not produce as many sperm. Stallion spermatozoa move in a straight line rather than in a twisting progression.

The percentage of deformed sperm cells incapable of fertilization usually varies between 20 percent and 30 percent, but some stallions have less than 20 percent abnormal spermatozoa. Abnormal types of sperm cells include those with deformed heads, such as small, large, isolated, and polycephaly, and those with deformed tails, such as twisted, abnormally curvy, double tails, and acephaly (Figure 10-7).

Kenny et al. (1971) have observed that the one single seminal factor that best predicted pregnancy was sperm concentration; the best combination of two factors was concentration and volume; the best three factors were concentration, volume, and percentage morphologically normal; and the

Number	Form	Description
N		Normal – right position
		Normal – left position
1		Cytoplasmic droplet in distal position
2		Cytoplasmic droplet in proximal position
3		Protoplasmic droplet in atypical location
4		Single loop of the tail
5		Double loop of the tail
6		Loop of the end part of the tail
7		Spiraling of the tail
8		Tail looped around the head
9		Loose heads
10		Damaged (broken) tail
11		Additional (swollen) acrosome cap
12		Detached cap
13		"Two (or more) headed"
14		"Club" (two spermatozoa joined together)
15		"Dwarf–head"
16		"Gigantic–head"
17		"Pear–head"
18		"Biscuit–head"
19		"Threadlike" midpiece
20		Midpiece divided into fibers and mitochondria
21		Undeveloped spermatozoon

Figure 10-7
Classification of abnormal forms of stallion spermatozoa. (Bielanski et al., 1982. *J. Reprod. Fert.* 32:23.)

best four factors were three plus initial progressive motility after washing and dilution.

A pH determination is useful to help detect incomplete ejaculates and/or the presence of urine in the sample. The average pH for first collections is 7.47 and is 7.60 for a second collection an hour later. If the value for the first collection is higher than the second, an incomplete ejaculation is strongly suspected. Abnormally high pH values may indicate the presence of urine. During the breeding season, the pH values are slightly higher than during the nonbreeding season.

The semen of the jack, *Equus asinus*, has been described by Nishikawa (1959) as being more milky or brownish yellow than stallion semen. The volume per ejaculate averages 50 ml and ranges from 10 to 115 ml. Very little or no gelatinous material is ejaculated. The concentration varies between 95 and 264 10^6/ml and the total number of spermatozoa per ejaculate varies between 8 and 43×10^9. The average total number is 24×10^9.

During the breeding season, most farms make a practice of evaluating the semen every 2 weeks for volume and for the concentration, morphology, and motility of the spermatozoa. If the quality and/or quantity start to decline, the stallion is not used for a few days. The fertility of a stallion is questioned when, upon evaluation of the semen, one or more of the following characteristics exist: the ejaculate volume is below 50 ml, fewer than 50 percent of the sperm cells have normal motility and morphology, and/or concentration of the sperm cells is below 8×10^9 per ml. Before declaring the stallion infertile, several semen collections should be evaluated, because the time interval between semen collections is important. Before a routine semen evaluation is made, the stallion should have one week of sexual rest. Two seminal collections are obtained. A second ejaculate is obtained one hour after the first ejaculate. The second ejaculate should contain approximately one-half the total number of spermatozoa contained in the first. If the difference between the two ejaculates is not approximately 50 percent, one of three conditions may exist: one of the ejaculates may be incomplete; the stallion has low spermatozoan reserves; or spermatozoa have accumulated in the reproductive tract. In such cases, another ejaculate should be obtained the same day as the first two ejaculates and the evaluation process should be repeated one week later to determine the cause. If a stallion has poor-quality semen that may be the result of ill health (fever) or injury, it is advisable to wait at least 60 to 90 days before performing a second semen evaluation. This allows time for one complete spermatogenic cycle to occur. To evaluate the ability of a stallion to produce spermatozoa, daily ejaculates should be obtained for 8 to 10 days after a week of sexual rest. After 6 to 8 days of collection, the number of spermatozoa per ejaculate becomes relatively stable and is a close approximation of the daily sperm production. The number of mares that a stallion can service per season can be determined from the results of this evaluation. Those with poor ability to produce semen are capable of breeding fewer mares per season. Dismount samples, collected when a stallion is dismounting from the mare, are used by some farms to evaluate semen quality, but they are not really indicative of semen quality. The last ejaculate jets normally contain few if any spermatozoa.

When a stallion is used in an artificial insemination program where fresh semen is used or in an A.I. stud where semen is being frozen, it is important to obtain the maximum number of spermatozoa that that stallion is capable of producing. If the stallion is collected too frequently, the concentration of spermatozoa in the semen will decline, as will his libido. Usually, the maximum number of spermatozoa can be obtained when the stallion is collected 3 or 4 times a week. However, the only true test of semen quality is the conception rate.

Factors Affecting Semen Production

Numerous factors influence semen production; the most common have been described by investigators at Colorado State University and are discussed below.

Season of Year   Season of the year has a major influence on seminal characteristics as well as behavior. Semen volume is reduced by approximately 40 percent during the nonbreeding season (Figure 10-8). During the nonbreeding season, ejaculates rarely contain significant amounts of gel, and the average total number of spermatozoa collected per ejaculate is reduced by approximately 45 percent (Figure 10-9). Concentration is also reduced to about 70 percent of the value observed during the actual breeding season. From a practical aspect, the stallion is functioning at approximately 50 percent capacity when required to breed mares three to four months prior to the natural breeding season.

Motility is not affected by season (Figure 10-10) but pH is influenced (Figure 10-11). There is a rise in pH from around 7.35 during the nonbreeding season to 7.6 during the natural breeding season.

Frequency of Ejaculation   When evaluating semen for quality and predicting a stallion's fertility, it is common to make two seminal collections after a week of sexual rest. The second ejaculate is collected one hour after the first. Regardless of season effects, seminal volume and total number of spermatozoa are decreased by the second collection. Generally, seminal volume for the second collection is reduced to about 50 percent of the first collection during the breeding season and is only slightly reduced during the nonbreeding season (Figure 10-8). Total number of spermatozoa per ejaculate for the second collection is usually 50 percent of the value for the first collection (Figure 10-9). If the 50 percent difference is not observed, the

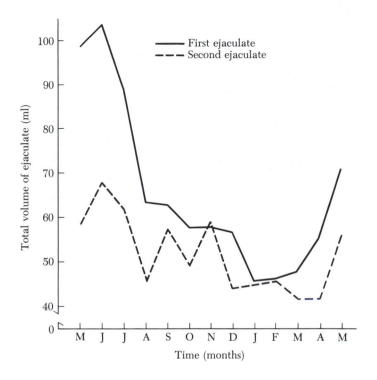

Figure 10-8
Total seminal volume by month for first and second ejaculates. (From Pickett and Voss, 1972.)

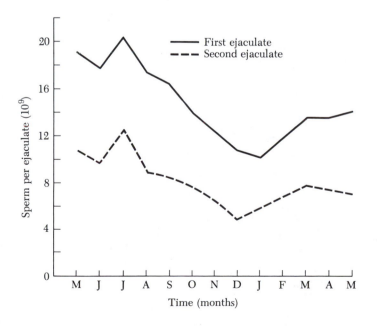

Figure 10-9
Average number of sperm per ejaculate in billions for first and second ejaculates. (From Pickett and Voss, 1972.)

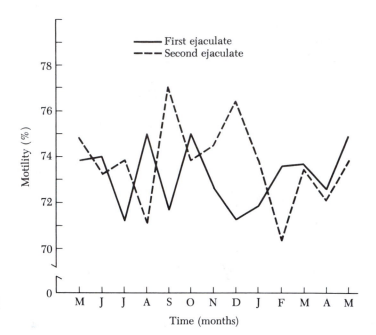

Figure 10-10
Mean monthly variation in motility. (From Pickett and Voss, 1972.)

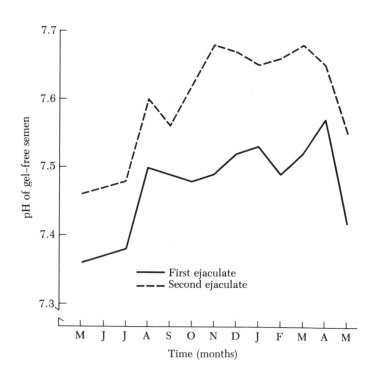

Figure 10-11
pH of gel-free semen by month. (From Pickett and Voss, 1972.)

stallion may have low, high, or no sperm reserves or be immature, or one ejaculation may have been incomplete.

If a stallion is collected several times at hour intervals, gel-free volume and percentage of motile spermatozoa remain relatively constant. Gel volume, total seminal volume, sperm concentration, and spermatozoa per ejaculate decline (Table 10-1). If the stallion maintains his libido, there are usually sufficient spermatozoa to get a mare pregnant to the fifth breeding or semen collection.

Daily sperm output can be determined by collecting a stallion daily for at least eight days (Table 10-2). After six to eight days the number of spermatozoa per ejaculate remains relatively constant and represents the number of sperm that the horse is producing each day. This information can be used to predict approximately the number of mares that can be booked to the stallion if an artificial insemination program is being utilized.

To collect the maximum number of spermatozoa from a stallion during the breeding season, he can be collected three times per week. If collected six times per week (once per day) and rested one day the total number of spermatozoa collected per week is the same as when the stallion is collected three times per week (Table 10-3).

Age Age has minimal influence on gel volume, motility, and concentration. However, gel-free volume, total seminal volume, pH, and total spermatozoa per ejaculate are influenced (Table 10-4). All values are lower for the two- and three-year-old horses except pH, which is increased.

Table 10-1
Seminal characteristics of successive ejaculates

Characteristics	Ejaculate (N = 44)					
	1	2	3	4	5	p^*
Seminal volume (ml)						
Gel	23[a]	6[b]	2[b]	2[b]	1[b]	<0.01
Gel-free	28	23	22	22	22	N.S.
Total	52[a]	29[b]	24[b]	24[b]	23[b]	<0.01
Spermatozoa						
Concentration						
(10^6/ml)	338[a]	182[b]	104[c]	63[c]	51[c]	<0.01
Total (10^9)	8.4[a]	3.8[b]	2.0[c]	1.2[c]	0.9[c]	<0.01
Motility (%)	59	57	63	60	58	N.S.
pH	7.50[a]	7.58[b]	7.65[c]	7.72[d]	7.73[d]	<0.01

Means in the same row with different superscripts differ.
*Based on F-ratio calculated in analysis of variance.
SOURCE: Squires et al., 1979.

Table 10-2
Sperm output (billions) in first 14
days of daily collection

Day of collection	Spermatozoa
1	8.8
2	6.9
3	6.9
4	4.4
5	3.8
6	3.9
7	2.6
8	4.1
9	2.7
10	3.4
11	2.9
12	3.0
13	3.6
14	3.5

SOURCE: Gebauer et al., 1974. *J. A. V. M. A.*
165(8):711.

Table 10-3
The effect of frequency of ejaculation on stallion seminal characteristics

Characteristic	Collections per week		
	1	3	6
Volume			
Semen, gel-free (ml.)	53	56	51
Gel (ml.)	18	8	5
Total (ml.)	72	63	56
Spermatozoa			
Conc. per ml. (10^6)	298	264	170
Total per collection (10^9)	13.5	12.7	7.2
Total per week (10^9)	13.5	38.0	43.1
Motility (%)	56	52	56
Mounts/ejaculate	1.4	1.4	1.4
Reaction time (min.)	2.3	1.6	1.7

1 = 36 ejaculates. 3 = 108 ejaculates. 6 = 216 ejaculates.
SOURCE: Pickett and Voss, 1972.

Table 10-4
Effect of age and ejaculate number on total spermatozoa per ejaculate (10^9) of stallions (number in parentheses)

Ejaculate	Age (years)		
	2–3 (7)	4–6 (16)	9–16 (21)
1	4.5[aA]	9.5[bA]	11.4[bA]
2	2.4[aA]	3.5[abB]	5.5[bB]
3	0.9[aA]	2.6[aB]	2.4[aC]
4	0.6[aA]	1.3[aB]	1.8[aC]
5	0.5[aA]	1.1[aB]	1.2[aC]
Total	8.9[a]	18.0[ab]	22.3[b]

[a,b]Means within a row with different superscripts differ ($P < 0.05$).
[A,B,C]Means within a column with different superscripts differ ($P < 0.05$).
SOURCE: Squire et al., 1979.

Very few stallions are used at stud as 2-year-olds. If they are used, the number of services should be restricted to two per week. As 3-year-olds, they can be used once a day. The number of mares that can be bred to young stallions depends upon the length of the breeding season and whether artificial insemination is used. Approximately six to seven mares can be "hand bred" by the 2-year-old and 10 to 12 can be bred by the 3-year-old. Mature stallions can be used twice daily for short periods if they are rested one day per week. When artificial insemination is used, 10 mares per day may be bred by a stallion that produces good-quality semen.

Testicular Size Various testicular measurements have been used to predict sperm production. Scrotal width is correlated with daily sperm production. A regression equation predicting daily semen production S from scrotal width Y in millimeters is: $S \times 10^9 = -4.8 \times 10^9 + (0.093 \times 10^9)Y$. However, the variability in prediction for a given testicular size reduces the usefulness of the prediction and supplies only a general estimate of a stallion's daily sperm production. Average values are 80 to 120 mm.

Sexual Stimulation The influence of sexual stimulation on seminal characteristics is summarized in Table 10-5. Seminal volume is significantly increased but other parameters (except concentration) are not affected. Concentration is affected because the total number of spermatozoa are diluted in a greater volume of seminal fluid.

Exogenous Steroids Androgens decrease testicular size and inhibit spermatogenesis. Libido has not been affected in closely controlled studies.

Table 10-5
Effect of teasing on stallion seminal characteristics

Characteristic	Treatment	
	Teased	Nonteased
Semen volume (ml)		
Gel	23	8
Gel-free	63	37
Total	86	45
Spermatozoa		
Per ml (millions)	59	118
Per ejaculate (billions)	3.3	3.8
Motility	63	61
pH	7.57	7.59
Sexual behavior		
Reaction time(s)	39	62
Mounts per ejaculate	1.2	1.2

SOURCE: Pickett and Voss, 1972.

Therefore the use of anabolic steroids should be avoided for stallions or potential stallions.

Stallion Infertility

Infertility has many possible causes. A stallion suffering from poor health and/or nutrition, injury, worry, or anxiety will sometimes recover with proper management. Another common problem with the stallion is masturbation. It is difficult to catch some stallions in the act of masturbation, but the presence of dried semen on the abdomen or on the back of the forelegs and a shrinkage of muscles of the loin are good indicators that the stallion has been masturbating. A couple of management practices have been tried to eliminate the vice. Some stallions that are turned outside where they can exercise and see other horses will stop masturbating, but others require that a stallion ring be placed behind the glans penis. The ring is removed before breeding a mare and is subsequently replaced. Care must be taken that a local inflammation does not develop.

Pickett et al. (1976) have implicated mismanagement as the most common cause of abnormal sexual behavior leading to infertility. Overuse of a stallion as a 2- or 3-year-old may lead to the development of poor breeding behavior such as a slow reaction time to breed a mare. A slow reaction time may develop in mature stallions because of overuse during the winter and fall months. Often, the overused young stallion will develop a bad habit of savaging (excessively biting) a mare. Unnecessary roughness in handling a stallion may lead to disinterest in breeding a mare. Disinterest may also result from excessive use of the stallion as a teaser to determine which mares are in estrus.

Breeding Soundness Evaluation

In addition to performing a semen evaluation, a breeding soundness evaluation should be completed on the stallion before purchase or before the start of the breeding season. If possible, the stallion's past breeding records should be examined to determine his past ability to "settle" mares. Pregnancies per service is a very sensitive measure of fertility and is more sensitive than the number of pregnant mares. One must be cautious of the skill of the stallion's manager and breeding techniques since they can be responsible for a poor reproduction record.

The stallion should be in excellent health, as evidenced by a general physical examination and his physical examination and his physical appearance. Regardless of his genetic ability to sire good foals, he must be physically fit and able to perform in the stud. There are several things to consider in a soundness evaluation. Injuries to the hind legs may prevent the stallion from, or cause problems when, breeding mares. Overweight and underweight conditions reduce the potential breeding performance. The condition of the respiratory tract and cardiovascular system may have a direct bearing on the ability of the stallion to survive at stud. Other problems, such as arthritis and melanomas, are common in aged stallions. The digestive tract should be checked for internal parasites. If possible, any history of colic should be ascertained. The reproductive tract should be examined by palpation for any abnormalities. Retention of one or both testicles in the abdominal cavity can be determined by examination of the scrotum. Inguinal hernias are detected by rectal examination.

Libido of the stallion can be evaluated by observing the stallion tease and breed a mare. Ejaculatory disturbances that cause incomplete or inhibited ejaculation resulting from testicular malfunction, physiological inhibition, or psychological inhibition may be observed during the breeding process. Temperament is an important consideration. Stallions that are vicious, stall kickers, or cribbers, or are prone to bite the handler or themselves, are difficult to handle and have a tendency to injure themselves or their handlers. Heritable imperfections, such as overshot and undershot jaws, should be noted.

Artificial Insemination

Artificial insemination (A.I.) has been widely used in the improvement of several species of domestic animals and the mare was reported to the first animal to be artificially inseminated. The technique has been used extensively with mares in the United States, Japan, China, and Russia except for thoroughbreds. Artificial insemination offers many obvious advantages. It permits disease control. (For example, it was practiced in Romania until 1961 to successfully eradicate dourine, a venereal disease, from the horse.) Injured or crippled stallions that have difficulty mounting a mare may be used in an A.I. program. A.I. may be used to prevent injury to the stallion if the mare is nervous or shy at breeding. One of the major advantages is that more mares per season can be bred to the stallion because the stallion is not overused. Mares with physical disabilities that prevent mounting by the stallion may be bred. Caslicked mares, that is, mares whose vulva are sutured to prevent windsucking of contaminants into the reproductive tract, do not have to have the sutures removed for breeding. Mares that do not show behavioral signs of estrus may be bred without risk of injury to the mare to stallion. One objection to A.I. in horses is that if too many mares are bred to a stallion, the market value of his colts may be lowered.

Collection Equipment   Stallion semen is collected for artificial insemination or for evaluation of semen quality by the use of breeder's bag (condom) or artificial vagina. The artificial vagina provides more satisfactory results and consistently less contaminated samples. There are three basic models of artificial vaginas: *Fujihira, Colorado,* and *Missouri.* Each model has design features that determine its use depending on certain circumstances. The Fujihara and Missouri models are lightweight and easy to handle. However, they do not maintain temperature as well as the Colorado model during extreme cold weather. The Colorado model is designed to collect gel-free semen; the other models have to have the collection receptacles modified for collection of gel-free semen. Failure to remove the gel immediately upon collection results in the sperm becoming trapped in the gel, which is very difficult to handle. The Fujihira model (Figure 10-12) has been used quite

Figure 10-12
Fujihira model equine artificial vagina: (*a*) Basic components consisting of: (1) aluminum frame; (2) rubber liner; (3) sponge; and (4) collection bulb. Components are assembled by (*b*) inserting the donut-shaped sponge into the aluminum frame and pushing it to the opposite end; (*c*) inserting the rubber liner; (*d*) tying the ends with string to prevent water leaks; (*e*) filling the unit with water (42°–44°C); (*f*) weighing it to determine amount of water added (unit should weigh approximately 12 pounds); (*g*) attaching a collection bulb and using a sterile lubricant to lubricate the inside of the rubber liner (use a disposable plastic sleeve while smearing the lubricant around).

a

b

c

d

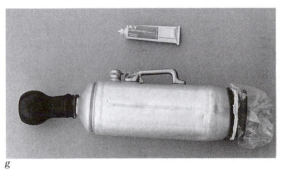

e

f

g

successfully for several years. The artificial vagina is basically a lightweight, rigid frame with a rubber inner liner. Water (temperature of 42 to 44°C at time of collection) is used to fill the space between the liner and the frame and a valve maintains the correct water pressure during collection. A rubber dam is at the end of the vagina to prevent pain if the stallion's penis strikes the end of the artificial vagina. Stimulation at the base of the penis and the temperature of the water are the two main factors responsible for ejaculation. The opening of the artificial vagina is lubricated with a sterile lubricant, and a rubber collection bag is placed over the end of the vagina.

The Missouri model (Figure 10-13) is composed of a combination rubber liner and cone, leather carrying case, and collection bottle. To prepare it for semen collection, the rubber liner is placed in the leather carrying case filled with water (temperature of 42 to 44°C at time of collection) and the collection bottle is attached to cone end of the rubber liner. When ready for collection, the entire apparatus should weigh between 8 and 9 lb (3.6 and 4.1 kg). The Colorado model is more complex in that it consists of several components (Figure 10-14). The inner liner and outer case serve as a water reservoir. A collection bottle with the filter assembly is attached to the cone-shaped end of the outer liner. In cold weather, a protective sleeve slips over the end of the artificial vagina and keeps the collection bottle warm. When ready for collection of a stallion, it weighs 25 lbs (11.4 kg). Sterile disposable AV liners are made for all models of artificial vaginas. They reduce the labor required for cleansing and disinfecting and reduce the risk of cross-infection between stallions.

Collection Procedure   During collection, the stallion mounts a mare that has been properly prepared (see Chapter 11) or a phantom (Figure 10-15). The penis is directed into the artificial vagina, which is held at a fixed position until the stallion begins vigorous thrusts, at which time the collector maintains firm pressure against the glans penis until ejaculation is complete. Ejaculation can be determined visually by flagging of the tail or manually by feeling the pulsations of the urethra at the opening of the artificial vagina.

Some stallions will mount a collection phantom with little or no prior training. To train a stallion to mount the phantom, a gentle mare in estrus is held adjacent to the right side of the phantom. The stallion is allowed to approach the left side of the phantom at an angle of approximately 30 to 45 degrees. As the stallion mounts, the mare is led forward and out of the way. After several teasing sessions, the stallion is tried without the mare. If he is successfully collected, the mare is probably not needed again.

Semen-Handling Procedures   After it has been collected, the semen is held at 100°F in a water bath or incubator. The semen can be used raw, although Hansen (1965) has reported that better results are obtained when the semen is extended with a dilutor before use. Several extenders are frequently used.

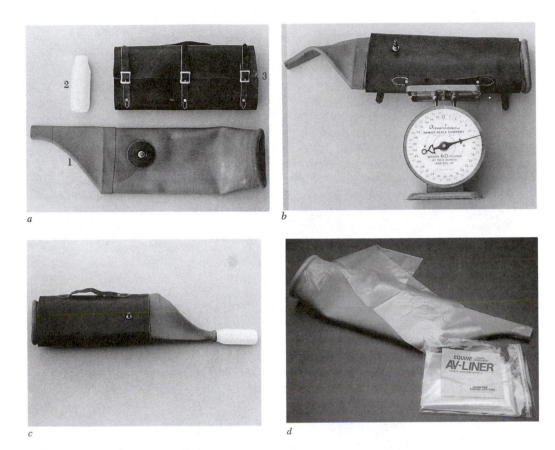

Figure 10-13
Missouri model equine artificial vagina: (*a*) Basic components, consisting of (1) rubber artificial vagina; (2) plastic human baby bottle; and (3) leather "carrying case" or "handle;" (*b*) prepared for use by buckling on the leather handle and filling the AV with warm water (42°–44° C) until the unit weighs approximately 12.5 pounds; (*c*) connecting baby bottle to the end. Immediately prior to collection, a sterile lubricant is used to lubricate the inside of the rubber artificial vagina. (*d*) Disposable liner can be used to improve hygiene and ease of cleaning.

Hughes and Loy (1970) have described skimmed milk and cream-gelatin dilutors. Skimmed milk dilutor is prepared by warming skimmed milk in a double boiler to 95°C for 4 minutes. After the milk has cooled, 1000 units of penicillin, 1 mg of dehydrostreptomycin, and 200 units of Polymyxin B sulfate are added per milliliter of dilutor. Cream-gelatin dilutor is prepared by warming half-and-half cream in a double boiler to 95°C for 2 to 4 minutes. Any scum is removed and the hot half-and-half cream is added to 1.3 grams of Knox gelatin that has been autoclaved with 10 ml of distilled

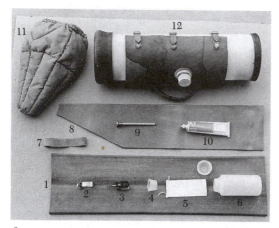

a

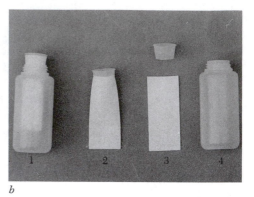

b

c

d

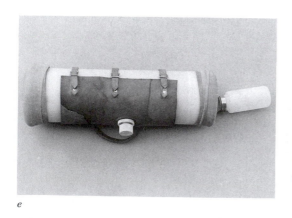

e

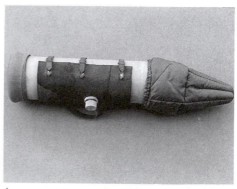

f

Figure 10-14
Colorado model equine artificial model. (*a*) Components consist of straight liner to form: (1) water jacket; (2) hose clamp; (3) screwdriver; (4) filter holder; (5) semen filter; (6) collection bottle and cap; (7) rubber band; (8) cone-shaped liner; (9) thermometer; (10) sterile lubricant; (11) insulated jacket; and (12) frame with leather handle. (*b*) Filters for obtaining gel-free semen. (1) nylon mesh filter; (2) nylon mesh filter inserted into semen collection bottle; (3) in-line milk filter and filter ring; (4) assembly of filter ring, in-line milk filter and semen collection bottle. (*d*)It is filled with water at $42°-44°C$ so it weighs approximately 25 pounds. (*e*) Second step is connecting filter assembly and collection bottle to cone-shaped liner with hose clamp, inserting liner into frame and reflecting end back over large end of the frame. An insulated cone protects the collection bottle from excess sunlight and maintains temperature.

water to a volume of 100 ml. When the mixture cools, antibiotics are added as prescribed for the milk dilutor. The AM extender developed at Texas A&M University is a clear extender and is commercially available. It has advantages when compared with other extenders. Since it is clear, spermatozoa can be easily seen to determine motility after the extended semen has been stored. The AM extender maintains motility over time at a higher rate for most stallions when compared with other extenders. This is advantageous if a stallion's sperm have a tendency to rapidly lose motility. A variety of antibiotics can be added to the extender depending upon the organism(s) that is contaminating the ejaculate or infecting the mare.

The dilutors can be frozen and stored until needed. They are warmed to $10°F$ before use. The semen is diluted $1:1$ or $1:4$ after collection and

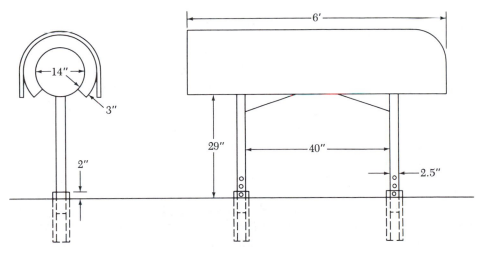

Figure 10-15
Phantom used to collect a stallion.

removal of the gel fraction. For the antibiotics to be effective, the diluted semen should not be used for 20 to 30 minutes. The gel is removed so that the semen can be evenly diluted and handled in a syringe. The gel can be removed by aspiration with a syringe, by straining the semen through 4 to 5 layers of sterile gauze, or by using a filter assembly during collection.

Semen Storage   Semen can be stored in a refrigerator for approximately 24 hours prior to breeding or it can be deep frozen for long-term storage.

If necessary, semen can be stored for about 24 hours by keeping it in a refrigerator at 55°F. Prior to lowering the temperature, the semen is processed by extending it 1 : 1 to 1 : 4 with an extendor such as the AM formula. The extended semen is then placed in a refrigerator for storage. Prior to use, it is warmed to body temperature (100°F). Cooling rate has a significant effect upon sperm motility, morphology, and live-dead staining characteristics. Cooling rates of between 0.2°C and 0.4°C/minute are recommended. Special designed containers such as those made by Hamilton Equine Systems are available for shipping and storing fresh semen (Figure 10-16). They are designed to properly control the rate of cooling and to maintain the final temperature at 4 to 6°C. Using this procedure, fresh semen can be stored for 48 hours and can be shipped to other farms if so desired.

Klug and coworkers in West Germany (1979), Nishikawa (1972), and Amann and Pickett (1984) have described successful methods for processing stallion semen for deep freezing and long-term storage. Conception rates of 63 percent for a single insemination during one estrous period have been achieved. However, marked differences in the freezability have been observed among stallions.

The first step in freezing horse semen is to collect gel-free semen. The ejaculate is evaluated for volume, concentration, and motility. Seminal plasma has unfavorable effects on viability of horse spermatozoa in the liquid and frozen state, so the gel-free ejaculate is diluted with a dilution mixture (Tables 10-6 and 7), incubated for 10 minutes, and centrifuged at 700 to 1000 G. The supernate is removed and the spermatozoa resuspended (Tables 10-6 and 7). Another concentration test is performed to determine the amount of additional diluter required to obtain 5×10^8 actively motile spermatozoa in each insemination dose. Large-volume straws of 1 to 4 ml are used to store the semen.

Insemination of Mare   One method of insemination of the mare has been described by Hughes and Loy (1970). After the perineal area is cleansed with a mild soap solution, a sterile speculum is inserted into the vagina. A sterile Chambers catheter (Figure 10-17) or plastic insemination fusette is then passed through the speculum and into the uterus by way of the cervix. A sterile 10- to 60-cc syringe filled with semen is then attached to the catheter and the semen is deposited into the uterus. Hansen (1965) has

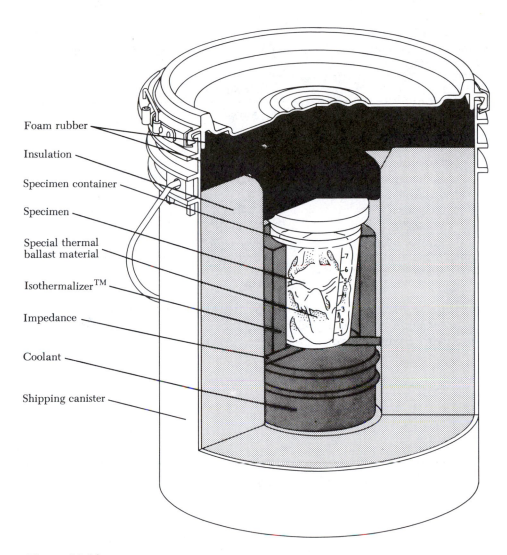

Foam rubber

Insulation

Specimen container

Specimen

Special thermal
ballast material

Isothermalizer™

Impedance

Coolant

Shipping canister

Figure 10-16
Hamilton Equine System's Equitainer™ is a specially designed container to ship
fresh stallion semen. Its design permits the initial cooling rate to be kept within
carefully controlled limits, and prevents the temperature from dropping below the
predesignated minimum temperature, which is typically 4°C. The Equitainer™
maintains the correct refrigerated conditions for up to 40 hours.

Table 10-6
Composition of diluters for Nishikawa and Shinomeija method of freezing stallion semen

	First diluter	Second diluter
Glucose	5.0%	
Lactose	0.3	
Raffinose	0.3	
Sodium citrate	0.15	1st diluter
Sodium phosphate	0.05	+
Potassium–sodium tartrate	0.05	glycerol, 8–10%
Egg yolk	20–5.0	
Penicillin	250 μ/ml	
Streptomycin	250 mcg/ml	

SOURCE: Nishikawa and Shinomeija, 1972.

described an alternate method whereby the operator wears a sterile disposable arm-length glove and guides the catheter through the cervix into the uterus with a finger (Figure 10-18).

10.3   Castration

Age   A castrated horse is referred to as a gelding. Horses chosen for gelding are usually castrated sometime between birth and 2 years of age. Several factors usually determine the age at which horses are gelded. Colts that have

Table 10-7
Composition of diluters for Klug and Martin method of freezing stallion semen

First diluter		Second diluter	
Glucose	6 gm	Lactose solution, 11%	50 ml
Disodium ethylenediamine tetra-acetate	0.370 g	Diluter no. 1	25 ml
Sodium citrate (dihydrate)	0.375 g	Egg yolk	20 ml
Sodium bicarbonate	0.120 g	Glycerin	5 ml
Streptomycin	0.050 g	Orvus-es paste*	0.8 ml
Penicillin	50 units		
Distilled water	100 ml		

*Orvus-Es Paste (OEP434A: Proctor and Gamble, PO Box 599, Cincinnati, Ohio 45201)
SOURCE: Martin, Klug, and Gunzel, 1979.

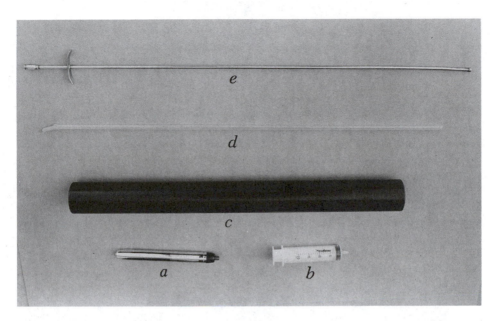

Figure 10-17
Artificial insemination equipment for insemination with aid of a speculum: (*a*)
light; (*b*) syringe; (*c*) speculum; (*d*) plastic insemination fusette; and (*e*) chamber's
catheter.

poor conformation and/or a poor pedigree are gelded as soon as the testicles
descend into the scrotum. The testicles are usually in the scrotum at birth or
arrive there before the tenth month after birth, but occasionally they are not
fully down until the twelfth or fifteenth month after birth. If a horse has good
conformation and a good pedigree and warrants a performance test, he is
kept intact until he fails to meet specific performance criteria. Stallions that
are able to perform but do not sire good foals should also be castrated. There
are too many good stallions available to keep poorly conformed and nonper-
forming stallions or stallions whose progeny are of poor quality.

Advantages of Castration   Gelding a horse has several advantages. Several
geldings may be kept in a paddock, whereas each stallion must be kept by
himself. Geldings are easier to care for, less prone to injury, and easier to
haul because of their attitude. Many people object to working stallions
because they tend to be lazy performers and are not consistent in their
performance.

Procedure for Castration   Because of the possible complications that may
result from castration and the need to anesthetize many of the colts or

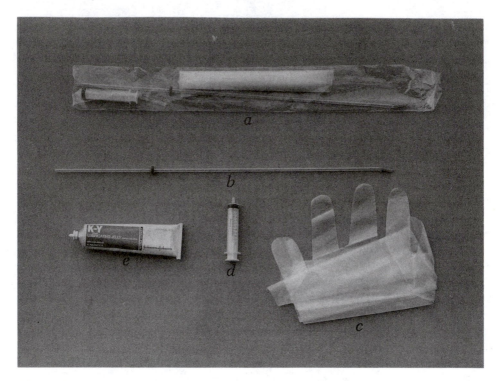

Figure 10-18
Artificial insemination equipment for insemination without aid of a speculum.
Sterile insemination jacket (*a*) consisting of fusette (*b*), syringe (*c*) and glove (*d*).
Sterile lubricant (*e*) is also needed.

stallions to prevent injury, it is customary for a veterinarian to perform the castration. At least three methods of castrating a horse have been described. Lowe and Dougherty (1972) have described the *primary closure method.* An incision is made through the skin and vaginal tunic between the scrotum and superficial inguinal ring on each side of the scrotum. The vessels and ductus deferens are tied by transfixation ligatures. The testicles are removed and all dissected planes are closed with chromic gut. With the *closed technique,* each testicle and spermatic cord still contained within the parietal layer of the tunica vaginalis is freed by blunt dissection from the surrounding tissue well into the inquinal canal. The cord structures are divided by means of an emasculator, which is left in position for approximately one minute. When the *open technique* is used, the tunica vaginalis is freed from the surrounding tissue as in the closed method. Then the tunica vaginalis communis is split with scissors so that all structures that are to be removed can be identified. The testicles, epididymis, and part of the cord are then removed as previously indicated. If the horse has been immunized against tetanus, postoperative care consists of exercise to prevent or control swelling and edema. The stallion libido usually subsides in 4 to 6 months but may last for one year. After castration, it takes several months before the ampullae and vasa deferentia become devoid of sperm and the ejaculate becomes azoospermic (Shideler et al., 1979). However, 7 to 8 days after castration, all spermatozoa obtained in ejaculates are nonmotile and pregnancy is unlikely to result from a mating.

Cryptorchids   An animal with one or both testes undescended into the scrotum is a cryptorchid, more commonly referred to as a ridgeling. The testes may be descended at birth and are usually descended by the age of 10 months. Some colts may be 12 to 15 months of age before descent of the testes. After 15 months of age, colts are considered cryptorchid if the testes have not descended. A positive diagnosis can be made if the vas deferens does not pass through the inguinal canal. These horses are difficult to castrate, and many castrated ridgelings seem to retain their stallion attitude for several months after castration.

10.4   Sexual Behavior

Olfaction is one of the fundamental stimuli of the reproductive responses of the stallion. When the stallion smells the external genitalia of the mare or voided urine, he displays the olfactory or Flehman reflex, in which he extends his neck upward and curls his lip. While exhibiting the reflex he inhales and exhales air in the upper respiratory passages. During the courtship period, the stallion also smells the groin of the mare and bites the mare on the croup and neck. At the first approach to the mare or even upon removal from his paddock to tease mares, the stallion will snort and continues to snort periodically during courtship. Erection in the stallion is usually slow and, in some stallions, may take several minutes. The reaction time for an attempted mount after first visual stimulation is approximately 5 minutes. Upon determination that the mare is in estrus and that erection has occurred, the stallion may mount the mare 2 or 3 times before intromission. Pickett et al. (1970) have observed a seasonal influence on mounts per ejaculate. Fewer mounts (1.5) per ejaculate were observed during the breeding season. The time of copulation varies from a few seconds to several minutes and the ejaculatory reflex lasts from 15 seconds to one minute. The number of intravaginal thrusts necessary to evoke ejaculation is 5 to 10 and their average duration is 11 seconds. (Tischner et al., 1974). Ejaculation can usually be determined by "tail flagging," in which the stallion raises and lowers his tail several times.

10.5   Training a Young Stallion

Training a young stallion to breed mares must be a carefully planned process. The stallion should be completely trained before the breeding season. The mare that is selected should be gentle and in definite estrus. The stallion's approach to the mare should be the same each time after he learns to mount a mare. During the first session, it may take him 30 minutes to an hour before he realizes that he can mount the mare. After a couple of sessions, most stallions know that they are going to breed the mare, and their behavior and approach to the mare should be controlled. During each session, the stallion should become accustomed to the teasing process, to encourage him to let down his penis, and to the hygiene procedure, so that he will stand and not kick the stud manager. With careful training, the stallion will be calm and less likely to injure the mare or the breeding management personnel.

REFERENCES

Amann, R. P., B. L. Thompson, Jr., E. L. Squires, and B. W. Pickett. 1979. Effects of age and frequency of ejaculation on sperm production and extrogonadal sperm reserves in stallions. *J. Reprod. Fert.*, Suppl. 27:1.

Asbury, A. C., and J. P. Hughes. 1964. Use of the artificial vagina for equine semen collection. *J. A. V. M. A.* 8:879.

Berndtson, W. E., J. H. Hoyer, E. L. Squires, and B. W. Pickett. 1979. Influence of exogenous testosterone on sperm production, seminal quality and libido of stallions. *J. Reprod. Fert.*, Suppl. 27:19.

Bielanski, W. 1960. Reproduction in horses. Vol. 1. Stallions. *Instytut Bull.* 116. Kracow, Poland: Instytut Zootechniki.

Bielanski, W., and F. Kaczmarski. 1979. Morphology of Spermatozoa in Semen from Stallions of Normal Fertility. *J. Reprod. Fert.*, Suppl. 27:39.

Douglas-Hamilton, D. H., R. Osol, G. Osol, D. Driscoll, and H. Noble. 1984. A field study of the fertility of transported equine semen. *Theriogenology* 22:291.

Eckstein, P., and S. Zuckerman. 1956. Morphology of the reproductive tract. In *Marshall's Physiology of Reproduction*. 3d ed. A. S. Parkes, ed. Boston: Little, Brown.

Fraser, A. F. 1968. *Reproductive Behavior in Ungulates.* New York: Academic Press.

Hafez, E. S. E. 1968. *Reproductive in Farm Animals.* Philadelphia: Lea and Febiger.

Hansen, J. C. 1965. Artificial insemination. *The Blood Horse* 90:3368.

Hauer, E. P., H. C. Dellgren, S. E. McCraine, and C. K. Vincent. 1970. Puberal characteristics of Quarter Horse stallions. *J. Anim. Sci.* 30:321.

Heinze, C. D. 1966. Methods of equine castration. *J. A. V. M. A.* 148:428.

Hughes, J. P., and R. G. Loy. 1970. Artificial insemination in the equine. A comparison of natural breeding and artificial insemination of mares using semen from six stallions. *The Cornell Vet.* 60:463.

Julian, L. M., and W.S. Tyler. 1959. Anatomy of the Male Reproductive Organs. In *Reproduction in Domestic Animals.* H. H. Cole and P. T. Cupps, eds. Vol. 1. New York: Academic Press.

Kenny, R. M., R. S. Kingston, A. H. Rajamannon, and C. F. Ramberg. 1971. Stallion semen characteristics for predicting fertility. *Proc. Amer. Assoc. Equine Pract.* 17:53.

Lowe, J. E., and R. Dougherty. 1972. Castration of horses and ponies by a primary closure method. *J. A. V. M. A.* 160:183.

Martin, J. C., E. Klug, and A. R. Gunzel. 1979. Centrifugation of stallion semen and its storage in large-volume straws. *J. Reprod. Fert.,* Suppl. 27:49.

Merkt, H., K. O. Jacobs, E. Klug, and E. Aukes. 1979. An analysis of stallion fertility rates (foals born alive) from the breeding documents of the Landgestut Celle over a 158 year period. *J. Reprod. Fert..,* Suppl. 27:73.

Nishikawa, Y. 1959. *Studies on Reproduction in Horses.* Tokyo: Japan Racing Association.

Nishikawa, Y., and S. Shinomeija. 1972. Our experimental results and methods of deep freezing of horse spermatozoa. *Animal Reproduction and AI,* pp. 207–213.

Pickett, B. W. 1974. Evaluation of Stallion Semen. In O. R. Adams, *Lameness in Horses.* Philadelphia: Lea and Febiger.

Pickett, B. W. 1976. Stallion management with special reference to semen collection, evaluation and artificial insemination. *Proc. 1st National Horsemen's Seminar,* pp. 37-47.

Pickett, B. W., L. C. Faulkner, and T. M. Sutherland. 1970. Effect of month and stallion on seminal characteristics and sexual behavior. *J. Anim. Sci.* 31:713.

Pickett, B. W., and J. L. Voss. 1972. Reproductive management of the stallion. *Proc. 18th Conv. A.A.E.P.* pp. 501–531.

Shideler, R. K., E. L. Squires, B. W. Pickett, and E. W. Anderson. 1979. Disappearance of spermatozoa from the ejaculates of geldings. *J. Reprod. Fert.,* Suppl. 27:25.

Sisson, S., and J. D. Grossman. 1953. *The Anatomy of Domestic Animals.* Philadelphia: W. B. Saunders.

Skinner, J. D., and J. Bowen. 1968. Puberty in the Welsh stallion. *J. Reprod. Fert.* 16:133.

Squires, E. L., B. W. Pickett, and R. P. Amann. 1979. Effect of successive ejaculation on stallion seminal characteristics. *J. Reprod. Fert.,* Suppl. 27:7.

Swierstra, E. E., M. R. Gebauer, and B. W. Pickett. 1974. Reproductive physiology of the stallion. I. Spermatogenesis and testis composition. *J. Reprod. Fert.* 40:113.

Thompson, D. L., Jr., B. W. Pickett, E. L. Squires, and R. P. Amman. 1979. Testicular measurements and reproductive characteristics in stallions. *J. Reprod. Fert.,* Suppl. 27:13.

Tischner, M., K. Kosiniak, and W. Bielanski. 1974. Analysis of the pattern of ejaculation in stallions. *J. Reprod. Fert.* 41:329–335.

Wierzbowski, S. 1958. Ejaculatory reflexes in stallions following natural stimulation and the use of the artificial vagina. *Anim. Breed. Abstr.* 26:367.

CHAPTER 11

HORSE-BREEDING PROBLEMS AND PROCEDURES

To be successful, a stallion farm requires skilled and knowledgeable personnel to handle the breeding herd and stallions. Personnel need to understand the basic principles of reproductive physiology and to be able to apply them to practical situations. If they do, many mares that would not ordinarily conceive will do so without too many problems.

11.1 Preparation for Breeding

Teasing

One of the main determinants of a successful breeding season is the skill of the stud manager in handling the teaser while teasing mares and in determining when mares are in estrus. He must be able to recognize the characteristic behavioral signs of estrus and all of their variations. Teasing mares at the proper interval, to determine whether they are in the early phase of estrus, is important. For best results, each mare should be teased once a day, although some breeding farms tease every other day.

Methods Which teasing method is to be used is determined by the number of mares, the physical facilities, and/or the stud manager's preference. If a limited number of mares are to be teased every day, each mare may be brought to the stallion's stall and teased. The stall must be specially constructed so that the stallion can reach the mare with his head and nuzzle her. The retaining wall must be high enough so that he will not attempt to jump out of the stall, and the opening above the wall should be approximately 3½ feet high. The stallion can be taken to each mare in her paddock. The teasing should be done across a specially built teasing wall. Teasing walls permit the stallion and mare to strike at each other without the risk of injuring each other. The teasing wall is approximately 3½ to 4 feet high, 8 feet long, and of solid construction so that there is no opportunity for a foot to get caught between the boards. Many farms tease mares across a regular fence and run a high risk of injury. Some farms take the mare and stallion to a special teasing area that has a teasing wall (Figure 11-1) or rail. The teasing wall is so designed that both the handlers are protected from the mare and stallion but can stay close to the horses to control them. Some farms use a rail made of pipe, but this does not offer protection for the horses and handlers.

When a large number of mares must be teased daily, the teaser stallion may be placed in a "cage" (a specially built stall) inside a large pen. As many as 20 or 30 mares are driven into the pen and teased at one time. Careful records must be kept so that mares that are timid, or are having silent heats, can be removed from the broodmare band and teased individually. To prevent aggressive mares from kicking the other mares or from approaching the

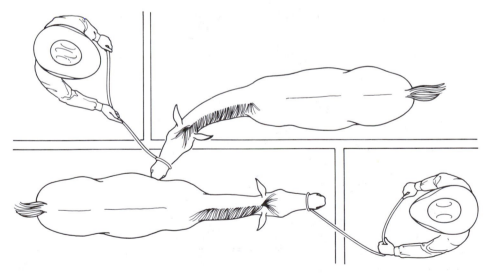

Figure 11-1
Teasing a mare with a stallion across a teasing wall. Both handlers are protected from the kicking and striking of the mare and stallion. (Illustration by M. Morris.)

stallion, a teasing mill can be used (Figure 11-2). The stallion is in a special pen surrounded by 10 pens that hold individual mares.

Pasture teasing is dangerous to both a good teaser stallion and the handler. The teaser stallion should be protected with padding, particularly on his belly. His penis should be covered with a blanket so that he cannot accidently service a mare if he escapes from the handler. Accidental conceptions can also be prevented by using a vasectomized stallion as a teaser. The handler should carry a whip to protect himself from aggressive mares that are out of heat. A pony stallion can be turned loose in a pen or a pasture with mares so long as he is too small to service the mares. If a long drag rope is connected to his halter, he can easily be caught when he has finished teasing the mares.

Teaser   The teaser stallion, by his actions, can often determine the success of detecting mares in estrus. An aggressive individual with a lot of libido that does a lot of squealing seems to work best under most circumstances. Most

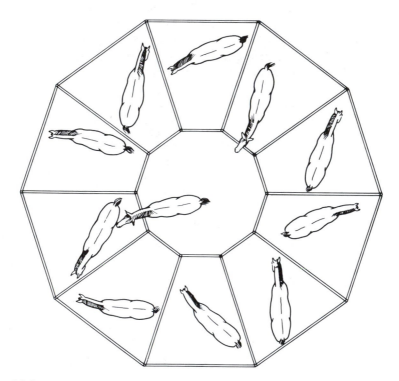

Figure 11-2
"Teasing mill" used by Thane Lancaster to tease several mares simultaneously at the Magic Valley Stallion Station. A stallion is put in the pen in the center. (From *The Quarter Horse Journal* 28 [1975]:202.)

teaser stallions become disinterested in teasing if they are used daily. To help maintain his libido, a stallion can be periodically bred to grade mares or collected with an artificial vagina. When several mares are teased daily, it usually is necessary to use two or three teasers during the breeding season or the stallions lose interest.

Timid mares seldom show estrus to a very aggressive stallion that does a lot of kicking and squealing. A timid mare frequently will display estrus when placed in a stall that is close enough to a stallion so that she can hear and smell but not see him. Maiden mares should be teased with a quiet stallion the first few times that they are in estrus so that they do not become frightened. A common problem for horse owners who have one or two mares to be bred is how to detect when the mare is in estrus, because they have no stallion to tease the mares. Quite often, the mare is delivered to a stud farm so that the stallions can tease the mare. Some mares occasionally show estrus to a strange mare or gelding when they are first introduced. Some horse owners have found that it is economical to buy an inexpensive pony stallion and use him as a teaser until the mare is successfully bred. Vasectomized stallions are useful because a mare will not become pregnant as a result of an accidental breeding.

Hauling Effects   If the owner of a mare keeps the mare at home until she is in estrus and ready to be bred, he sometimes finds that she will not show estrus when she is delivered to the stallion. Some mares are sensitive to strange surroundings and strange people. In such a situation, the stallion owner should have a veterinarian rectally palpate the mare to determine if she is ready to be bred. If she is, she can be artificially bred without risk of injury because she will not accept the stallion.

Maiden, Barren, and Open Mares

The maiden, barren, and open mares should be properly prepared for the breeding season. Because an increasing amount of daylight stimulates a mare to cycle regularly, artificial lights can be used effectively to encourage mares to cycle regularly early in the breeding season. The ability to breed these mares early will help distribute the "book" of a stallion (the mares to be bred to him) and may prevent overusing him during the latter part of the season. All maiden, open, and barren mares should be placed under 16 hours of artificial light per day around the middle of November. Either fluorescent or regular light bulbs can be used and controlled by an electric timer. The average size stall requires approximately 200 watts. The mares may be kept on pasture when a lighted shelter is provided (see section on photo-stimulation).

As the maiden, barren, and open mares come into estrus, they should be given a physical examination by a veterinarian and cultured for uterine infections if warranted. Mares that have uterine infections should be treated as soon as they are diagnosed, to prevent further damage to the uterus. If uterine damage is suspected, uterine biopsies can be obtained and examined microscopically. The biopsy examination is used to predict the mare's ability to conceive and to carry a foal to term.

The vulvas of mares that are aspirating air and contaminants into their reproductive tracts should be sutured (Caslick operation) to prevent further aspiration of air. Many of these mares have a deep-set anus and a vulva that forms a shelf instead of being nearly vertical. Pascoe (1979) has classified mares into three general categories based upon the total length of the vulva from the dorsal commissure to the level of the pelvic brim, and the angle of declination of the vulva (Figures 11-3 and 4). Mares (Group 1) with an effective length of 2 to 3 cm and an index (the effect on length × the angle of declination) of less than 100 seldom had reproductive problems requiring Caslick's operation. Mares (Group 2) with an effective length of 6 to 7 cm and an index of greater than 50 changed as they became older. Relaxation of the pelvic organs and ligaments causes an increase in effective length and angle of declination. Mares (Group 3) with an effective length of 5 to 9 cm and an index of 50 to greater than 200 usually have serious reproductive problems. They require a Caslick operation at a young age and are difficult to manage throughout their reproductive life. Frequently, they have a dished vulvar conformation with the brim of the pelvis almost at the same level as the ventral commissure.

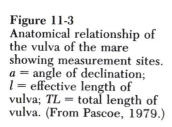

Figure 11-3
Anatomical relationship of the vulva of the mare showing measurement sites. a = angle of declination; l = effective length of vulva; TL = total length of vulva. (From Pascoe, 1979.)

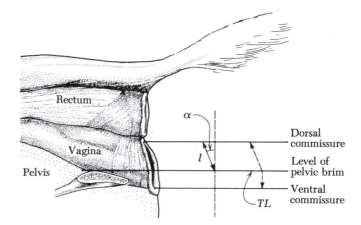

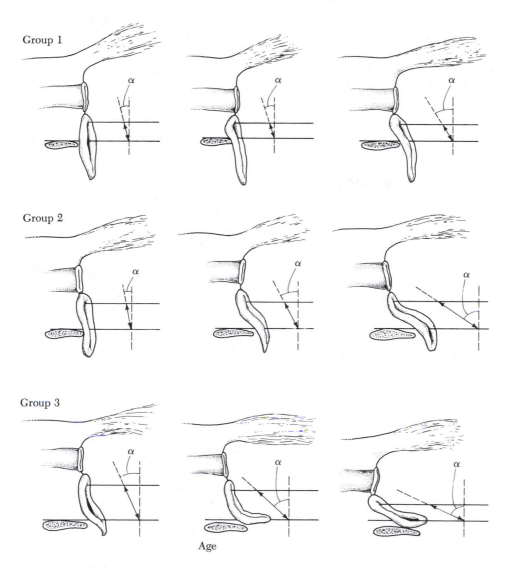

Figure 11-4
Vulval changes in the 3 conformation groups as affected by increasing angles of declination *a* and effective length of vulva *1*. (From Pascoe, 1979.)

Genital Organ Examinations

During the breeding season, all mares are usually teased daily, and as each mare comes into estrus, she is given a genital organ examination by a veterinarian. The examination consists of visual inspection of the cervix and vagina and a palpation per rectum of the uterus and ovaries. The changes in the cervix, vagina, and ovaries are most significant (see Chapter 9) and are correlated with behavioral signs of estrus to determine whether it is the optimal time to breed.

Optimal Breeding Time   The optimal time to breed is just before ovulation. The survival time of the ovum, for breeding purposes, is approximately 12 hours, even though it may be as long as 24 hours. Stallion semen in the reproductive tract of the mare is viable for 48 hours under practical conditions but may be viable for as long as 5 days. If it is not possible to have the mare palpated, breeding should start on day 2 or 3 of estrus, and the mare may be bred every other day until she goes out of heat. This method enables viable sperm to be in the tract while the ovum is viable, even if the mare has a short, 3-day cycle or a longer-than-usual estrus.

Rectal palpation is a practical means of direct examination of the internal genital organs for breeding problems, to detect pregnancy (see Section 9.4) and as an aid to determine the appropriate time for breeding. However, there are certain risks to the person palpating and to the mare being palpated and there are many reasons for inaccuracies. It is not an easy technique to learn and should be done only by a veterinarian, a specially trained equine reproductive physiologist or an individual trained by a veterinarian or equine reproductive physiologist.

The initial step is proper restraint, because most mares object to the procedure. If possible, the mare should be restrained in an examination stock (Figure 11-20). Some mares are still able to kick above the back retaining board and strike the palpater. They may sit down or fall down and injure the palpater's arm before it can be withdrawn from the rectum. If an examination stock is not available, some mares can be restrained with breeding hobbles (Figure 11-18), loaded in a trailer with the butt chain snapped and the door open, held against a wall, or held in a stall. When these methods are used, the palpater stands around a corner or behind something for protection.

After the mare is restrained, her tail is wrapped and pulled to one side to prevent tail hair being introduced into the rectum and to get the tail out of the way.

The palpater wears a glove, usually a vinyl disposable one or rubber, that is lubricated (usually with sodium carboxymethyl cellulose). Lubrication is essential, because the mare's rectum may be dry. Excess friction may result in perforation of the rectum, which creates a serious problem. Many mares will die as a result of peritonitis from the fecal material that enters the abdominal cavity. Lubrication makes it much easier to move the hand around and examine the tract.

To find the tract structures one must be oriented to certain landmarks inside the rectum (Figure 11-5). The first structure that one should recog-

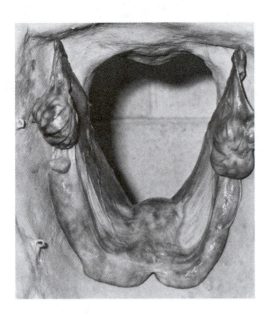

Figure 11-5
Normal nonpregnant uterus of mare,
Anterior view. (From Zemjanis,
1970).

nize is the pelvic girdle. With the hand being held rather flat, it feels sort of mushy since there are structures between the uterus and your hand. The hand is moved further forward, the finger cupped and then retracted to find the junction of the uterine horns with the body. This is a distinct bifurcation and by cupping the fingertips and grasping it at this area, one may follow each horn upward to the respective ovary. For a right-handed palpater, the left ovary is easiest to locate and vice versa for a left-handed palpater. The ovaries are located in front of the upper third of the iliac shaft. They are identified by their oval shape, irregular form, and firm consistency. A more complete description of their size and shape in relation to their reproductive status is given in Section 9.1. To examine the ovary, it may be necessary to flip it from in front of the broad ligament by cupping the first two fingers and rolling it over so that it can be examined freely. Sometimes, this is difficult to accomplish, but failure to do so makes it difficult to accurately feel the follicles or corpora lutea.

After the ovaries are located, one must be able to distinguish follicles from corpora lutea and determine the size of the follicles. Follicles are recognized as a raised smooth surface over a fluctuating cavity. The feel is similar to that of any eyeball. To estimate follicular size, the follicle is compared to width of the palpater's finger or fingers, which have been measured. Immediately after ovulation, an indentation is found at the site of ovulation. The indentation then becomes filled and appears as a soft area. After 24 to 48 hours, the corpus luteum has the feel of fresh liver. After another couple of days, it may feel just like a follicle when it matures. This makes it difficult to be accurate when the reproductive history of the mare is unknown. Palpating on a regular basis during estrus drastically improves accuracy.

Some changes that occur in the uterus during the estrous cycle may be determined by palpation. However, the changes are not very reliable since they vary greatly. At best, they give a general tendency and only serve as a minor aid in determining when to breed a mare. When changes in the ovary and uterus, determined by palpation, are correlated with changes observed in the speculum examination of the cervix and vagina and with teasing, one can usually make an accurate judgment as to when a mare should be bred.

To palpate the cervix, locate the bifurcation of the uterus. The hand is held flat and retracted with a side-to-side waving motion. A muscular structure, the cervix is recognized by its size (6 to 8 cm), rigid muscular feel and cone shape. If it is relaxed, its posterior edge can be flattened with the hand. One cannot grasp the cervix as in the cow.

Most people, when learning how to palpate, have to overcome several problems. After palpating 8 to 10 mares, one should be able to identify the parts of the genital tract, but it takes 500 to 1000 palpations to become reasonably skilled. Upon entering the rectum and sliding the hand forward, it is necessary to remove any fecal balls that are encountered. They can be mistaken for ovaries, and their presence makes it difficult to examine the genital tract. Fecal balls can be distinguished from ovaries by the difference in consistency and freedom of movement. The uterus can be distinguished from a full bladder by blottment. However, the best way to avoid mistaking it for the uterus is to keep retracting the hand over the top of it and trying to pick up the uterus at its bifurcation. The uterus lies posterior and dorsal to the bladder. A full bladder is easy to mistake for a fetus during a pregnancy examination. Real time ultrasonography offers a more accurate method for evaluating the follicular and luteal status of the ovaries since it permits rapid, visual and noninvasive access. Follicles are seen as black non-echogenic areas with relatively smooth outlines. They are non-echogenic because they are fluid filled. Ginther and Pierson (1984a, 1984b) observed that 66 percent of the follicles change from a spherical to a pear-shaped or oblong form on the day preceding ovulation. With ovulation, the preovulatory follicle disappears and an intense echogenic area appears. During the first 2 to 3 days of corpus luteum development, the echogenic area remains intense. The corpus luteum is identifiable throughout its functional life, but the day of diestrus cannot be determined by changes in the ultrasound images.

Often one observation, even without a knowledge of the prior reproductive history of the mare, is sufficient to accurately determine her reproductive status. A skilled operator can determine if the mare has entered the ovulatory season, estimate the stage of the estrous cycle, detect the number of preovulatory-size follicles, detect a failure of ovulation or anovulatory estrus, detect silent estrous ovulations, detect prolonged maintenance of the corpus luteum, detect cystic preovulatory structures and tumors, detect multiple ovulations, and so on. The theory for the treatment is as follows: Two prostaglandin injections are used 15 days apart, since a mare at a stage of the estrous cycle between normal luteolysis and day 5 post-ovulation will not respond to the first prostaglandin injection. However, the mare will be at a stage of diestrus in which she will respond when the second injection is given. HCG is given on day 7 post-prostaglandin treatment to stimulate ovulation, since ovulation normally occurs 7 to 12 days post-treatment. If the mare responds to the hCG treatment, she should do so within 48 hours. Therefore she would be at 6 days post-ovulation when the second prostaglandin injection is given and will respond to it along with the mares which had not responded to the first prostaglandin treatment. Thus all mares are ready to respond to the day 15 treatment. To increase the percentage of mares ovulating within a 2- to 3-day period, the second hCH treatment is given 6 days after the second prostaglandin treatment to stimulate ovulation.

Foal Heat

Care must be taken in order to breed mares successfully during the foal heat. Only those mares that had a normal delivery should be bred. If any of the following conditions exist upon examination by a veterinarian, the mare should not be bred: bruised cervix, lacerations or tears in cervix or vagina, vaginal discharge, placenta retained more than 3 hours, lack of tone in the uterus or vagina, or presence of urine in the vagina. In fact, if there is any question about breeding the mare, it is better to wait. Indiscriminate breeding of mares during the foal heat results in a low (25 percent) conception rate.

Estrous Cycle Manipulations

Several situations commonly occur during the breeding season when the stud manager will need to manipulate the estrous cycle of the mare (the period when the mare comes into heat and ovulates) to increase conception percentages or correct fertility problems. Some of the problems that can be corrected or situations that should be avoided are lactation anestrus, spontaneously prolonged corpus luteum, and overuse of the stallion. Several techniques can be used during the breeding season to manipulate the estrous cycle or some aspect of it.

Human Chorionic Gonadotropin (HCG)   One of the most commonly used techniques is to use human chorionic gonadotropin (HCG) to stimulate follicles to ovulate. This method is particularly useful when a follicle reaches ovulatory size but does not ovulate within a reasonable period. Some farms are using the technique more extensively in that each mare receives an HCG injection 24 hours after the beginning of estrus. The mares are bred 24 hours after the injection, and ovulation usually occurs before 48 hours post-injection. The effectiveness of continued use of HCG has been questioned. Some mares have been reported to not respond to the treatment. Roser et al. (1979) reported that some mares develop an immunological response to HCG (Figure 11-6). However, these mares continue to ovulate, since the anti-HCG antibodies are specific to HCG and do not cross react with luteinizing hormone. The effectiveness of HCG to induce ovulation may be impaired in mares with the immunological response since a large percentage of the injected HCG is neutralized by the antibodies.

Prostaglandins   One of the prostaglandins, $PGF_{2\alpha}$, and some of its analogs, can be used to correct several infertility problems and to manipulate the time of ovulation. Prostaglandins are used because of their luteolytic effect (Figure 11-7). They cause the corpus luteum to undergo involution and cease producing progesterone when they are administered 4 to 12 days or 14 days post-ovulation. (Days 4 and 5 have low success rates.) Plasma progesterone concentration will rapidly decline to below 1 ng/ml and the mare will come into estrus in approximately 2 to 4 days following treatment (Allen and Roswon, 1973). Most mares will ovulate a fertile ovum by 10 to 12 days post-treatment and some will ovulate as soon as 6 days post-treatment. One of the major mare infertility problems is the spontaneously prolonged corpus luteum (Figure 11-8). The corpus luteum does not cease to produce progesterone at approximately 14 days post-ovulation but continues to function for an additional 1 to 3 months. The mare with the spontaneously prolonged corpus luteum is usually normal in all respects except that she has a high level of progesterone in the blood that keeps her out of estrus and from having estrous cycles. Open, barren, or maiden mares that are not pregnant and are not having estrous cycles during the breeding season should be tested for a spontaneously prolonged corpus luteum by analyzing a blood sample for progesterone concentration. The same problem (spontaneously prolonged corpus luteum) in the "pseudo-pregnant" mare makes detection of continued progesterone secretion more confusing and difficult. The pseudo-pregnant mare shown in Figure 11-9 was bred and ceased having estrous cycles. When she was given a pregnancy examination at 45 days, she was found to be open (not pregnant). Either the mare coincidentally developed a spontaneously prolonged corpus luteum, or she may have conceived and reabsorbed the fetus. In both cases, there is an active corpus luteum producing progesterone, which should not be present. When diagnosed, these problems can be corrected by treatment with prostaglandin. The mare will then come into heat and ovulate as previously discussed (Figure 11-10). If a mare conceives and then aborts after 36 days postconception, treatment

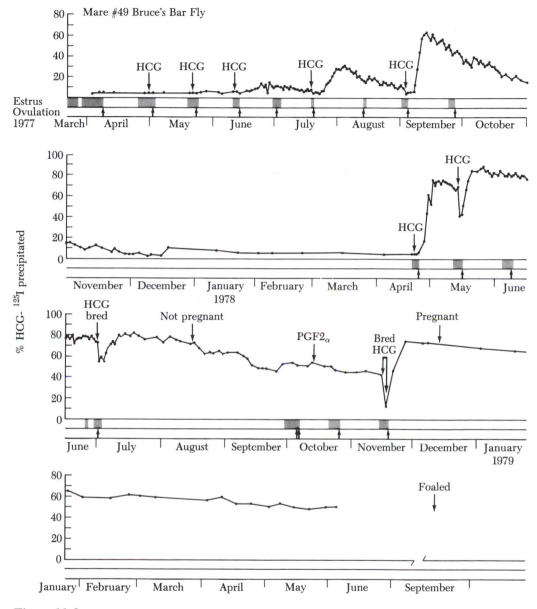

Figure 11-6
Time course of anti-hCG antibody development. Antibody formation occurred after second hCG treatment. ↑ = ovulation; ▬ = estrus. (Roser, Ph.D. Thesis, Univ. of Calif.)

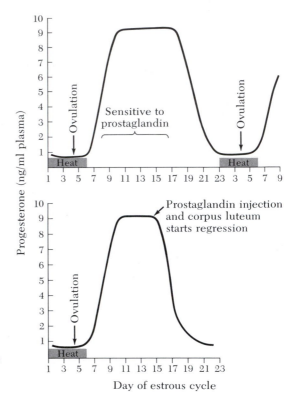

Figure 11-7
Graphs showing days during the estrous cycle when a mare is most sensitive to prostaglandin.

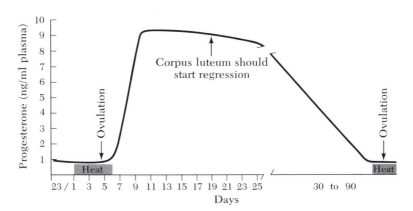

Figure 11-8
Effect of spontaneously prolonged corpus luteum on the estrous cycle.

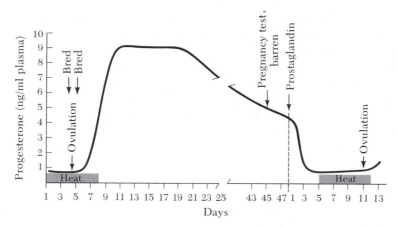

Figure 11-9
Effect of administering prostaglandin treatment to a pseudo-pregnant mare.

with prostaglandins will probably not cause the mare to start cycling again. After 36 days, the endometrial cups are stimulated to form and to secrete pregnant mares serum gonadotropin (PMSG). PMSG suppresses ovarian activity and the mare will not start to cycle again until after PMSG is no longer secreted. Normally, PMSG disappears from the blood at about 120 to 150 days. If the mare is treated with prostaglandins after 36 days of gestation, the corpus luteum present in the ovary that is secreting progesterone will cease

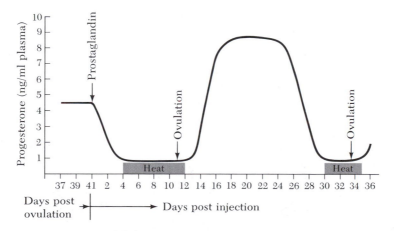

Figure 11-10
Effect of administering prostaglandin treatment to a mare with a prolonged corpus luteum.

to function. Progesterone concentration in the blood will decline but the mare will not come into estrus, because ovarian activity is being suppressed by PMSG. Therefore, the mare remains in anestrus for at least 4 months post-conception. To verify this condition, the presence of PMSG in the blood can be determined.

Another practical use of prostaglandins is to control the length of the estrous cycle (Figure 11-11). It is frequently helpful to breed a mare 10 to 12 days before she is due to come into heat. This is particularly true during the latter half of the breeding season. It is also advantageous to be able to control the time of ovulation of a group of mares that are to be bred artificially with frozen semen, because in such a situation it would be helpful if all the mares came into heat and ovulated at the same time. In a situation where a stallion has a large book of mares and artificial insemination is not permissible, it would be desirable to prevent the stallion's overuse by preventing several mares from ovulating at approximately the same time. To shorten the estrous cycle of an individual mare, prostaglandin administration between days 6 and 12 post-ovulation will cause the corpus luteum to cease producing progesterone and allow the mare to come into heat 3 or 4 days post-injection and to ovulate within 10 to 12 days post-injection. She may ovulate as soon as 2 days post-injection. The ovulation following prostaglandin treatment is as fertile as normal ovulations. All subsequent estrous cycles will be of normal length, and fertility is not decreased.

If it is not possible to tease or palpate a mare to determine when she is to be bred, she can be bred by appointment with the use of prostaglandins. Whatever the day of her estrus cycle on which the treatment schedule is started, the success rate is about 80 percent. The schedule is to inject prostaglandin on days 1 and 15; HCG is injected on days 7 and 21; and the

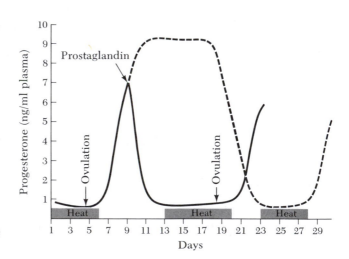

Figure 11-11
Effect of administering prostaglandin to shorten the length of the estrous cycle.

mare is bred on days 20 and 22. This treatment schedule can also be used to synchronize donor and recipient mares in an embryo transfer program.

Prostaglandin treatment may be of benefit to breed mars after they have foaled and before the time of their next normal estrous period. This is particularly true for mares that do not pass the veterinarian's inspection to breed on the foal heat (the first heat period following foaling). Mares are not bred on the foal heat if they have a bruised cervix, lacerations of the cervix or vagina, or a vaginal discharge, or if they retain the placenta more than 3 hours or evidence lack of tone in the uterus or vagina. In addition, there may be other conditions that indicate the inadvisability of breeding on the foal heat. The reproductive tract of the mare may not have had enough time to return to a breedable condition by the time the mare came into heat at 9 to 11 days after foaling, or the mare may have had a very early foal heat ovulation at 7 to 9 days post-foaling. In these situations, the mare can be treated with prostaglandins on the sixth day after her foal heat ovulation, and she will ovulate again within 10 to 12 days or possibly sooner (Figure 11-12). The mare can be bred at least 6 to 8 days sooner than normal and in some cases as many as 10 to 12 days earlier than her next normal estrous period.

Some mares will not cycle during lactation. Usually, such a mare will have her foal heat and then will not come back into heat for a long period. These mares have a persistent corpus luteum as a result of the foal heat ovulation, or they may be extremely possessive of their foals and do not want to approach the stallion. Many of these mares will respond to the prostaglandin treatment in the manner previously discussed.

Prostaglandins can be used to induce abortions at 5 days post-breeding. Prostaglandins cause the corpus luteum to cease producing progesterone, which is required for maintenance of pregnancy. This is useful if a mare is

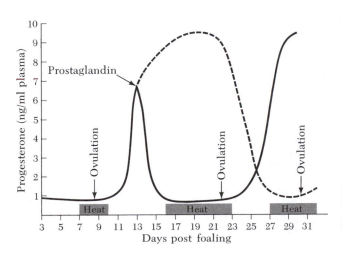

Figure 11-12
Effect of administering prostaglandin to breed the foaling mare before her normal second estrous period.

bred to the wrong stallion by mistake or if the mare had unexpected twin ovulations after she was bred and the stud manager does not want to take a chance that she conceived twins. If treated prior to 36 days post-conception, she should return to estrus.

When using prostalgandins, the stud manager must be aware of the various types of responses in mares following treatment. The chances of successful treatment are lower if prostaglandins are used on the fourth or 5th day post-ovulation. After the 5th day, the success rate is approximately 65 to 80 percent. There are several types of responses: (1) luteolysis as described above; (2) incomplete luteolysis; (3) incomplete luteolysis with subsequent recovery (4) no luteolysis; (5) ovulation within 4 days of treatment; and (6) failure to display estrus (Figure 11-11 and 11-12). Incomplete luteolysis is characterized by a reduction but maintenance of serum progesterone concentrations above 1 ng/ml for several days after treatment. Incomplete luteolysis with subsequent corpus luteum recovery is characterized by a reduction of serum progesterone concentration after treatment, followed by an increase in the concentration. No luteolysis is characterized by increasing or constant serum progesterone concentration despite prostaglandin treatment. The interovulatory periods for these estrous cycles with these responses are close to normal, so the treatment objective is not accomplished. Ovulation occurring within 4 days of treatment occurs after about 30 percent of treatments. In most instances, a follicle of 30 mm or more in size is present on the ovary at the time of treatment. Because of the quick response, the stud manager may not detect the ovulation unless the mare is palpated on a regular basis for a few days post-treatment. Prior to treatment, knowledge of the follicular status of the ovary is extremely important. Failure to display estrus during the ovulatory period occurs after about 25 percent of treatments. This response increases the importance of ovarian palpation

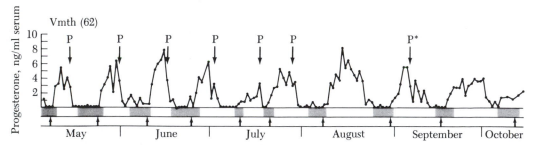

Figure 11-13
Luteolytic responses of a mare treated with prostalene. Successful luteolysis occurred after each prostalene treatment (P). Prostalene injection on 17 July was followed by an ovulation within 4 days. The injection failed to cause complete luteolysis in September. P* = 5 mg of natural $PGF_{2\alpha}$. ▆ = estrus, ↑ = ovulation. (From Keifer et al. 1979.)

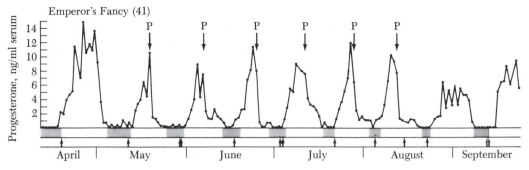

Figure 11-14
Luteolytic responses of a mare treated with prostalene. Incomplete luteolytic responses followed prostalene (P) injections on 6 June, 11 July, 28 July, and 12 August. Ovulation occurred within 4 days of the August treatment. Estrous cycles before and after the prostalene treatment series were normal. ▮ = estrus, ↑ = ovulation. (From Keifer et al., 1979.)

following treatment. Mares that do not display estrus can be successfully bred by artificial insemination.

Oral Progestins   An oral progestin, allyl trenbolone (altrenogest) has been used to successfully treat barren, maiden, and lactating mares which were exhibiting shallow anestrus (active ovaries), prolonged spring estrus, or lac-

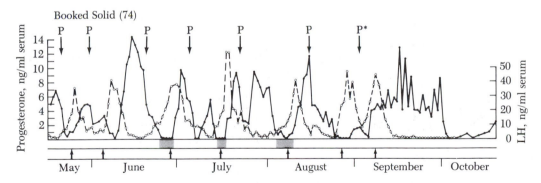

Figure 11-15
Luteolytic responses of a mare treated with prostalene. Incomplete luteolytic responses followed prostalene (P) injections on 31 May and 15 August. Incomplete luteolysis with subsequent luteal recoveries occurred after prostalene injections on 5 July and 22 July. Failure to manifest estrus was observed after prostalene treatments on 21 May, 31 May, and 15 August. Ovulation within 4 days followed the prostalene injection on 21 May. The post-treatment control cycle (September) contained a prolonged luteal phase. P* = an injection of $PGF_{2\alpha}$ (4 mg). ▮ = estrus and ↑ = ovulations. (From Keifer et al. 1979.)

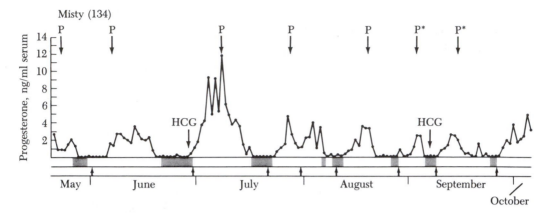

Figure 11-16
Luteolytic responses of a mare treated with prostalene. No luteolysis occurred after prostalene (P) injections on 7 June and 8 July. Prostalene injection on 28 July was followed by an ovulation with estrus within 4 days. Prostalene (4 Mg) was administered on 3 September, and 5 $PGF_{2\alpha}$ (P*) on 15 September. ▬ = estrus and ↑ = ovulations. hCG represents treatment with human chorionic gonadotropin. (From Keifer et al. 1979.)

tation anestrus. The post-treatment response is related to the degree of ovarian activity when mares are treated. The response rate is highest in open, barren, and maiden mares treated during the protracted phase between deep anestrus and normal cyclical activity. It is particularly useful in mares that have been exposed to an artificial-light regimen and are entering the transition period between winter anestrus and normal estrous activity (Allen et al., 1980, and Squires et al., 1978). The condition of true lactational anestrus has been treated successfully with altrenogest, which can also be used to block estrus an ovulation in normal cycling mares (Webel et al., 1975).

After 10 to 15 days oral administration, it is withdrawn and estrus is usually exhibited within 8 days. Most mares show estrus behavior 2 to 5 days after withdrawal. Ovulation occurs within 18 days and most mares will ovulate within 7 to 12 days after treatment.

Gonadotropin Releasing Hormone (GnRH)   Ginther (1979) has reviewed the use of GnRH in the mare. A single injection on the second day of estrus can be used to hasten ovulation. However, daily treatment beginning the second day of estrus is more effective in controlling the time of ovulation and the duration of estrus, since most mares quit displaying estrus within 48 hours after ovulation. More recently, GnRH has been used to stimulate mares to come out of winter anestrus. Several methods of treatment are successful. Initially, three injections were given per day until the mare showed estrus. Most mares would respond within 12 days. Because of the problems arising from the injections, infusion pumps were used, but they had disadvantages. Then, implantable osmotic minipumps were used. Finally, a GnRH analogue implant was developed and is currently utilized. Most mares will ovulate within 3 to 10 days of treatment.

Photostimulation   Early work by Burkhardt (1947) and Nishikawa (1959) demonstrated that the ovarian activity of mares is influenced by photoperiods. During the winter months, ovarian activity is suppressed and the onset of the breeding season can be hastened by artificial long days. Initially, a gradually increasing photoperiod was used to hasten the onset of the breeding season. Then, application of a constant, 16-hour light period per day beginning at 6 A.M. in the winter was shown to be just as effective in stimulating ovarian activity. Later work has demonstrated that a photoinducible phase exists during which a pulse of light can be given to stimulate ovarian activity. The exact time of the photoinducible phase is controversial. Other reports indicate it is 14 to 16 hours after the beginning of dawn and others indicate light must be present 8.0 to 10.5 hours after the beginning of sunset. Either method results in light at the same time relative to the beginning of the photoperiods used in the studies. From a practical standpoint, when mares are exposed to artificial light beginning at 5 to 6 P.M. until 10 P.M., they respond.

For an artificial photoperiod to be successful, a minimum of 2 foot candles are required at the level of the horse's head. The lights may be incandescent or fluorescent. A 200-watt light bulb is sufficient for an average-size stall. Once the artificial photoperiod is started, it should continue until the mare becomes pregnant or until the end of the breeding season. It takes approximately 40 to 100 days for a positive response and approximately 50 percent of the mares will respond by having estrous cycles early in the year. Mares may be kept outside in paddocks.

Pregnant mares that foal in January and February may enter winter anestrus after they foal. Therefore, it is recommended that early-foaling mares be placed under artificial lights the same as barren, maiden, and open mares.

11.2   Infertility

Infertility in the mare can be caused by a number of factors such as poor perineal conformation, tract abnormalities and injuries, and infections.

Poor perineal conformation has been discussed previously (Chapter 9). It is the cause of the pneumovagina or "wind sucker" condition. This is one of the most common causes of infertility. Caslick's operation, suturing the lips of the vulva, will help prevent infectious contaminants from being sucked into the reproductive tract (Figure 11.17).

Numerous reproductive tract abnormalities that cause infertility or lower fertility have been observed. Inside the vagina, scars, abrasions, ulcers, and other defects in the mucosa may cause infertility. Recto-vaginal fistulas are difficult to surgically repair so that the mare is able to carry a foal. Common cervical problems include adhesions, closed cervix, and split cervix (muscles torn, so the cervix cannot form an effective seal). Urine may pool in the cervical area when the anterior part of the tract hangs too low. This may lead to cervicitis and vaginitis. If the urine flows into the uterus, endometritis may develop. Prolonged periods of endometritis usually result in permanent damage to the uterine mucosa and thus infertility due to the lack of prostaglandin $F_{2\alpha}$ secretion by the endometrium. Many of the above problems can be surgically corrected.

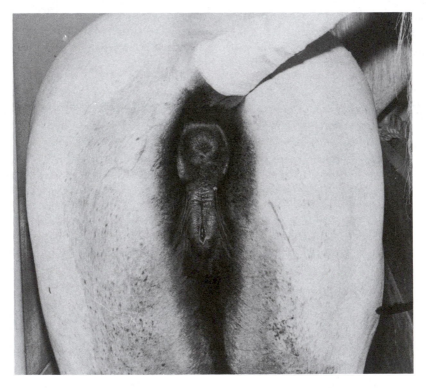

Figure 11-17
Perineal conformation of a mare that aspirates air into reproductive tract. Lips of vulva have been sutured (Caslick's operation) to prevent entry of air. (Photograph by W. Evans.)

The Fallopian tubes may become blocked due to adhesions or tumors. This prevents fertilization from occurring. Ovarian cysts, granulosa cell tumors and underdeveloped ovaries are not uncommon. Ovarian cysts are numerous follicles that fail to ovulate. Underdeveloped ovaries may be due to chromosomal abnormalities. The most common type is failure of the mare to receive one X chromosome (XO genotype); another type is receipt of an extra sex chromosome (XXY genotype).

11.3   Embryo Storage and Transfer

Embryo transfer is the transfer of an embryo (fertilized egg) from one female (donor mare) to another (recipient mare), which carries it to term. Embryo transfer gives a mare an opportunity to produce an increased number of foals each year and during her reproductive life. This method also offers the potential of making faster genetic progress when utilizing genetically superior mares. One of the main advantages it offers is in managing valuable older broodmares who are unable to carry a foal to term. Foals can also be obtained from some subfertile or infertile mares. Generally, the lining of the uterus of these mares has undergone degenerative changes which make it difficult for the embryo to survive. Damages to the uterus can occur during the foaling process so that scar tissue forms. Excess scar tissue decreases the amount of functional uterine tissue. Because the size of the uterus (uterine capacity) is a limiting factor, excess scar tissue can cause an abortion because the embryo receives an inadequate nutrient supply. Another advantage for using embryo transfer in managing older mares is prevention of uterine arterial hemorrhage, which is a common cause of death in old pregnant mares. Collecting embryos from performance and race mares allows them to continue to perform and still produce a foal.

The initial step in the process is to synchronize the ovulations of the donor and recipient mares. These procedures have been previously discussed. The second step is to breed the donor mare, and the third step is to collect and transfer the embryo.

There are two basic techniques of collecting the embryo from the donor and transferring it to the recipient — surgical and nonsurgical. The nonsurgical collection method is to pass a catheter through the cervix into the uterus on the 6th to 8th day post-ovulation, flush with a specialized medium, and collect the medium. The embryo floats out through the catheter into a collection vessel. Surgical collection procedures involve a midventral or a flank incision to expose the uterus. The oviduct ipsilateral to ovulation (on the opposite side) is cannulated and flushed with fluid. The uterine horn is constricted and cannulated so that the flushing media can be collected.

Surgical collection must be performed prior to the sixth day post-ovulation, when the embryo moves into the uterus.

Two methods of surgical transfer, midventral and flank incision, have been used to expose the tip of the uterus horn. After an incision is made in the uterine horn, the embryo is deposited into the lumen of the uterus with an insemination pipette. The nonsurgical method is similar to artificial insemination procedures in that an insemination pipette containing the embryo is passed through the cervix into the body of the uterus. Then, the embryo is slowly expressed from the pipette.

Yamamoto et al. (1982) and others have reported successfully freezing equine embryos at −196°C. Before the technique becomes practical, many details must be investigated and improved.

Micromanipulation of equine embryos is being investigated on a limited scale. Half embryos have been found to be viable, and two pairs of monozygotic twins have resulted from the transfer of "quarter" embryos obtained from blastomeres at the four-cell stage.

11.4   Breeding

Hygiene Procedures

After it has been determined that a mare is ready to be bred, the next consideration is breeding hygiene. Breeding hygiene is important because some infections that cause abortions are the result of introducing bacteria into the mare during breeding.

Mare   The first step is to tease the mare so that she will empty her bladder. The tail should then be wrapped with gauze or some other type of disposable wrapping. (Disposable supplies are used to prevent the spread of bacterial infections.) The mare's external genitalia and the surrounding area should be cleaned with a mild soap such as Ivory or Phisoderm. After the area is clean, the soap should be thoroughly rinsed off because it is spermicidal and causes irritation.

Stallion   The stallion manager must be concerned as to the proper means of hygiene to prevent, or at least reduce, the transmission of infectious pathogens during the breeding process. Washing the stallion's penis prior to breeding was initiated to prevent any dirt, smegma, dead eipithelial skin layers, or other grossly visible particles from contaminating the mare's reproductive tract during service. Also, it has long been thought that washing helps reduce the transfer of real and potential pathogens from stallion to mare. It has been demonstrated that by washing it was possible to alter the bacterial flora present and that the more effective the antiseptic used in the washing, the greater the chance of encouraging the survival of the more resistant and more pathological organisms (Bowen et al., 1982).

Washing with plain water has the least effect on changing the natural flora. Washing with a mild soap, Ivory, may result in an increase in potential pathogens on the penis, especially the coliform organisms. However, semen samples following repeated use of Ivory soap are rarely contaminated. An antiseptic surgical scrub such as Betadine may produce a marked alteration in the bacterial flora by removing or killing nonpathogenic bacteria, allowing the more resistant pathogen types to proliferate. Betadine has encouraged the growth of *Pseudomones aeruginosa* and *Klebsiella* spp. Both organisms cause uterine infection and infertility.

Normal hygiene procedures are to cleanse the stallion's penis prior to the breeding season. A mild soap such as Ivory can be used. During the breeding season the penis should be washed with water and dried with a disposable towel prior to breeding. After breeding, the penis may be rinsed with water. It may be necessary to use a mild soap solution to periodically cleanse the penis of smegma and obvious debris during the breeding season. To prevent the spread of bacteria or other deleterious microorganisms from one stallion to another, a separate wash bucket should be kept for each stallion (or use disposable vinyl liners [plastic bags] in the wash bucket).

Safety

Stallion Most stallions are valuable and should be protected from injury during the breeding process. A stallion can be protected from mare-induced injuries to some extent by tying up one front leg of the mare, but the mare does not then have a strong base of support. Another method is to use a set of breeding hobbles and, if necessary, a twitch (Figure 11-18). The mare can still move around when hobbled but is unable to kick. It is advisable to walk the mare around before the stallion mounts so that she knows she is hobbled. One must be careful to prevent the stallion from getting his foot tangled in the hobbles. A quick release snap or knot should be part of the hobbles. Excessive biting of a mare can be prevented by muzzling the stallion or using a heavy neck drape. This is particularly important when the mare is to be shown at halter or performance classes.

Maiden Mares In breeding young maiden mares, there is a chance of causing injury to the vagina or cervix. Many of the lacerations, tears, or bruises can be prevented by using a breeding roll to restrict entry by the stallion (Figure 11-19). The padded roll is 4 to 6 inches in diameter and approximately 18 inches long. A handle on one end allows easy insertion between

Figure 11-18
Mare prepared for breeding. The mare is hobbled with a set of breeding hobbles
that prevent her from kicking the stallion. The areas around her genital organs
have been washed and her tail has been wrapped with sterile gauze to help
prevent transmission of genital diseases. (Photograph by W. Evans.)

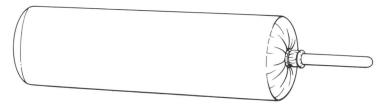

Figure 11-19
Breeding roll used to protect maiden mares from excessive penetration of the
stallion during mating. (Illustration by M. Morris.)

the mare's rump and the stallion just above the penis, as the stallion mounts. In keeping with strict rules of hygiene, the roll should be covered with a disposable cover, such as a plastic examination glove. The maiden should also be examined by a veterinarian and the hymen opened if necessary.

Size Differences   Wide differences in the physical size of mares and stallions may cause difficulties during breeding. Because of the size difference, the stallion may repeatedly mount and then dismount, thus increasing the chance of injury. Short mares that are being bred to tall stallions can be placed on a slight mound in the breeding shed. In the opposite situation, the mare's hind legs can be placed near the back end of a shallow depression. The stallion can then stand on the edge of the depression and service the mare.

Breeding Area

When mares are hand bred, the area that is used for a breeding shed or paddock should be surfaced with a dust-free material and protected from wind or air drafts. The mare and stallion should be washed outside the breeding area to prevent the surface from becoming slick. Sufficient space is necessary to separate the mare from the stallion if she should "explode" unexpectedly because of the hobbles or the approach of the stallion. A palpation chute (Figure 11-20), used to safely restrain mares and/or foals during breeding preparations, such as genital organ examinations, Caslick operations, artificial insemination, and hygiene procedures, should be located adjacent to the breeding area for convenience. A properly designed chute allows a mare to keep her head inside a small foal holding pen, thus preventing many disturbances among mares with foals at side. The chute should also be designed so that it can be opened immediately to free a horse.

11.5   Foaling

Pregnant Mare Care

The care of a pregnant mare for several weeks before foaling warrants special consideration. If possible, it is wise to keep the mare in an area where she can get an adequate amount of exercise to keep her muscles in tone. Mares that are used to being ridden can be ridden until they foal. During the last month, the exercise should not be strenuous. The feet and teeth should also receive proper attention, and good nutritional and health programs should be maintained.

Figure 11-20
Palpation chute used during female
genital organ examinations — side (a)
and front (b) views. The chute is
also used to restrain a horse for
other purposes.

Approximately one month before the expected foaling date, the mare
should be checked to determine whether she has been sutured. If she has
been, the sutures should be removed. Early removal will prevent an unexpected parturition in which the vulva is torn.

When the mare approaches her time to foal, she may be brought into a foaling stall or she may be placed in a large, clean, grassy paddock. Many mares foal on grass and seldom have any complications. Because horses are expensive and the expected foal is the result of a carefully planned mating, most mares are kept in a clean 12 × 14 ft foaling stall for approximately one week before foaling. The stall should be well lighted and free of any projections, such as nails or broken boards. The choice of bedding is important. Straw that is free of dust and cut in long lengths is preferred. Shavings and similar materials will stick to the wet nostrils of a newborn foal and may suffocate it.

Signs of Parturition

The signs of parturition are not always reliable. Some mares show no signs and merely lie down and have their foal. The general indicators are as follows: Approximately 2 to 6 weeks before foaling, the udder becomes distended. A marked shrinkage of the muscles in the croup area occurs 7 to 10 days before foaling as a result of a general relaxation of the muscles and ligaments in the area of the pelvis. There is a tendency for the mare to leave a band of mares and want to be alone just before foaling. The teats fill out their nipples 4 to 6 days before foaling and wax builds up on the end of the teats 2 to 4 days before foaling. Immediately before foaling (less than one day, but the number of hours varies), the wax may fall off and milk starts to drip.

The concentration of calcium increases in mammary secretions just prior to foaling. By determining the calcium concentration with a test kit for total water hardness (Merkt, Total Hardness Test #10025), it is possible to predict when foaling will occur within 4 days. The test strips have four color zones that change color when exposed to a given concentration of calcium. When three or four zones change color, foaling will occur within 4 days. If no more than two zones change color, foaling will not occur within the next 24 hours. Using can decrease the number of nights a mare is observed for foaling (foalwatch), decrease the number of nights she stays in a foaling stall, and cut foaling costs.

Parturition

Parturition is usually described as a three-stage process (Figure 11-21). During the first stage, the muscles in the uterine wall are undergoing rythmic contractions and the cervix is dilating. During this stage the mare may act quite nervous, have cramps, break out in a sweat, and walk around the stall in an anxious manner. The first stage ends when the first water bag (allanto-chorionic membrane) breaks and 2 to 5 gallons of water are expelled. This stage of foaling may last for several hours or only a few minutes. Second-stage labor is expulsion of the foal. The duration of this stage is about 20 minutes but may vary from 10 minutes to 1 hour. This stage is character-

Figure 11-21
Delivery of a normal foal. Stage 1: (*a*) mare rolling; (*b*) mare looking at her side; and (*c*) mare sweating and kicking at her belly. Stage 2: (*d*) first appearance of the placenta; (*e*) appearance of the forelegs; (*f*) body of foal passing through the vulva; (*g*) forelegs tearing placenta. Stage 3: (*h*) placenta torn; (*i*) mare getting up; (*j*) mare up and navel cord broken.

(*a*)

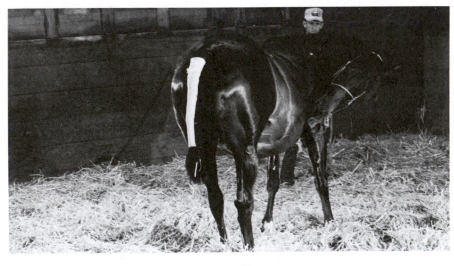

(*b*)

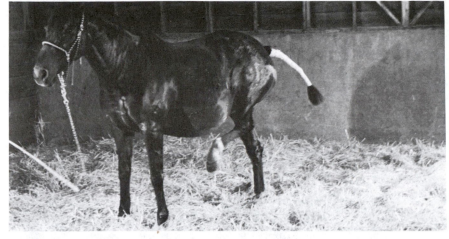

(c)

(d)

(e)

Figure 11-21 (*continued*)

(f)

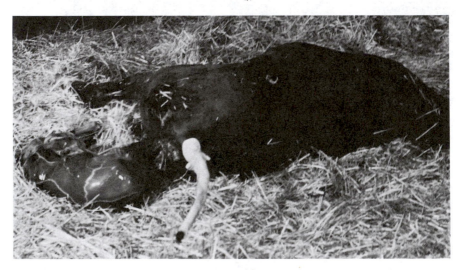

(g)

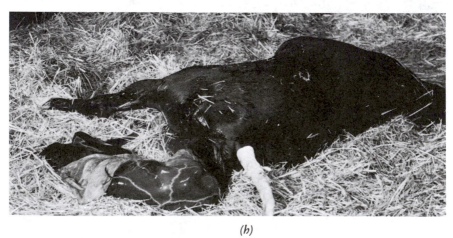

(h)

(i)

(j)

ized by the mare making strong propulsive efforts to supplement the uterine contractions. Most mares are lying on their side during stage two but they may get up and down or roll. Third-stage labor is expulsion of the afterbirth. During the first part of this stage, the mare may continue to lie on her side and rest for a few minutes. When she gets up, the navel cord breaks about 1½ inches from the foal. Within 1 to 2 hours of delivery, the placenta should separate from the wall of the uterus. When the afterbirth has passed, stage three is completed.

Induction of Parturition Under appropriate conditions, the mare can be successfully induced to foal. If the technique is to be utilized, the gestation length must be at least 340 days. Colostrum should be present in the mammary gland, the cervix relaxed, and the foal in the proper position. Early induction results in weak, immature foals and an inadequate supply of colostrum and milk. Although induction of parturition is not a standard technique, it is recommended for certain situations when an attendant is required. When neonatal isoerythrolysis (jaundice foal) is suspected, one must keep the foal from nursing; induction insures that someone is present at foaling. It is also useful for mares with injuries, pre-parturient colic, prolonged gestation period, or colostrum loss.

There are several methods used to induce parturition. Use of oxytocin has been more reliable than prostalgandins, analogs of prostaglandins, and/or dexamethasone. The average interval from injection of oxytocin to completion of foaling is about 30 minutes. Normally, the mares have no dystocia, don't retain fetal membranes, and produce ample colostrum and milk.

Foaling Problems

The normal position of the foal during delivery is shown in Figure 11-22. The forelegs are extended and the head and neck rest on them. The forelegs appear first, and the muzzle does not become visible until the legs are out approximately to the knees. The most difficult part of the foaling process is passage of the shoulders. The body is passed quite easily until the hips pass

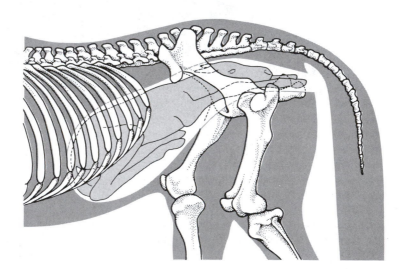

Figure 11-22
Normal position of a foal during foaling. The forelegs are extended and the head and neck rest on the forelegs. The hindlegs are extended backward as the body passes outward.

the mare's pelvis. As the hips pass, a slight delay in the rate of delivery usually occurs. When the foaling process starts, it is advisable to leave the mare alone unless she has trouble. Some indications of trouble are: one foreleg out to the knee, two forelegs out but no head, head out without forelegs, or prolonged labor without any parts visible. At the first indication of trouble, an experienced person should be notified. A foaling attendant who does not know what to do should not attempt to correct the difficulty, but should keep the mare up and moving until help arrives.

Dystocia is any foaling problem that prevents delivery of the foal by the mare's efforts alone. There are two types of dystocia, fetal and maternal. Maternal dystocias include atonic (failure of muscles to contract) uterus, uterine torsion, and pelvic problems. Fetal dystocias include malpositions (Figure 11-23) and/or malformations of the fetus. Dystocias are serious because they usually result in death of the fetus and/or mare. The genital tract of the mare is very sensitive to trauma and the uterus can be easily perforated. Lesions in the lining of the uterus are common after dystocia as is severe bruising of the uterine wall. Cervical lesions are inevitable after prolonged obstetrical help and are often followed by the development of adhesions, which can result in permanent infertility. Vaginal lesions can become serious, particularly if the wall is perforated.

In cases of malposition, the foal may be manipulated to correct the position. Sometimes traction will correct minor cases of dystocia. Certain types of dystocia may require caesarean section or embryotomy to remove the foal—i.e., uterine torsion, transverse presentations (back of foal against cervix), grossly deformed foals, dead foals, ankylosis and deformed legs, and wry necks.

The afterbirth should be passed by the mare within a couple of hours, and it is usually passed 10 to 15 minutes post-foaling. If it is not, a veterinarian should be notified so that a treatment can be given to aid in its passage. Manual removal will damage the uterus, and part of the placenta may break off. Retention of the placenta or any part of it may lead to founder. However, some mares will retain a portion of the placenta for 24 hours without ill effects. To be sure that the entire placenta has been passed, it should be spread out on the floor or ground and visually examined to make sure that no part of it has been retained (Figure 11-24).

As soon as the placenta has passed, the mare that previously had her vulva sutured should be resutured as soon as possible.

A few mares may display symptoms of colic within a few minutes of foaling. These symptoms last for only a few minutes. The abdominal pain is due to contraction of the uterine muscles. A mild sedative used for colic may be used for severe cases.

Mares may encounter other problems that are serious and may result in death if not detected immediately. Internal or external hemorrhage caused by lacerations during foaling are serious. When major blood vessels rupture

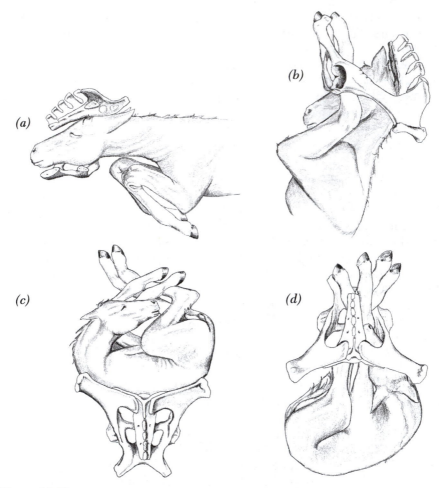

Figure 11-23
Malpositions of foal: (*a*) flexed capri (knees); (*b*) nape posture; (*c*) transverse
dorsal presentation; (*d*) transverse ventral presentation; (*e*) elbow lock with
dogsitting posture; (*f*) dogsitting posture; (*g*) wry neck or flexed neck; and (*h*)
breech presentation.

and the mare bleeds into the peritoneal cavity, death usually occurs. The
prepubic tendon may rupture and cause death. A recto-vaginal fistula occurs
when the leg is forced through the vagina and rectal wall and then tears out
the shelf that separates the anus and vulva. Reconstructive surgery may be
successful.

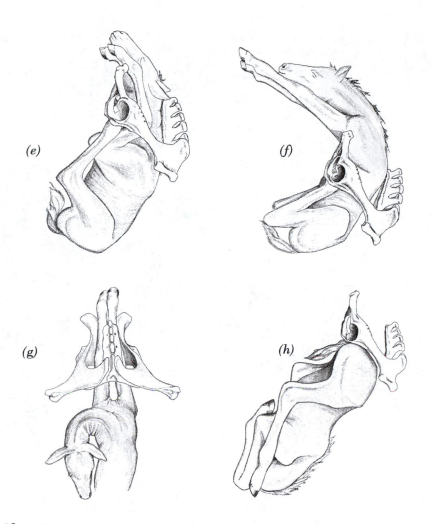

(e)   (f)

(g)   (h)

Abortion

Abortion is the expulsion of the fetus prior to 300 days of gestation. Abortions are classified as infectious or non-infectious.

Infectious abortions fall into two categories—acute and chronic. An acute infectious abortion is caused by an organism that attacks and kills the fetus. Most infectious abortions are of the chronic type and are caused by *Streptococcus zooepidemicus*, *Klebsiella* spp., *E. coli*, *Pseudomonas aeruginosa*, and *Mucor* and *Aspergillus* spp. Equine Herpes Virus I (Rhinopneumonitis virus) abortions can assume epizootic proportions. Abortions due to *Salmonella abortus equi* were very common but are now relatively uncommon.

Non-infectious abortions are caused by many different factors. Preg-

Figure 11-24
Placenta has been spread out on the ground to make sure that it is complete and that part of it has not been left in the mare: (*a*) pregnant horn; (*b*) body; (*c*) point of rupture. (Photograph courtesy of Patricia Barry.)

nancy in the uterine body rather than the uterine horn can lead to fetal development in a limited area where the placenta lacks the ability to supply the required nourishment. There is a direct relationship between endometrial fibrosis and abortion due to inadequate placental exchange. Strangulation of the umbilical cord due to torsion and premature rupture and the subsequent stoppage of blood flow leads to fetal death. Some fetal abnormalities are incompatible with life of the fetus and cause abortion. Trauma sustained by the mare very rarely causes abortion. Twin abortions take place at any time during pregnancy but occur most frequently after the eighth month. These abortions are usually caused by fetal malnutrition. One fetus dies to insufficient surface area for placental attachment, and autolysis leads to death of the other fetus. There are many drugs and plants and some nutritional deficiencies that lead to abortions.

Foal Care

Immediate Care   Upon its delivery, the foal should be examined to make sure that it is breathing. If necessary, the placenta should be removed from its nostrils and artificial respiration should be given immediately. Blowing into the foal's mouth or working its ribs as well as vigorously rubbing its body will stimulate respiration. In some cases, lifting up the foal and gently dropping it may start respiration.

Navel   The navel stump should be dipped in a tincture of 10 percent iodine. The navel cord should not be cut during delivery, but the foal should be allowed to break it to prevent excessive bleeding. The stump should be treated and examined for a few days after delivery. Lack of treatment or the presence of urine on the stump may lead to an infection, which in turn leads to navel ill. The urine on the stump is due to failure of the urachus (Figure 9-12) to close and is referred to as previous or persistent urachus. Silver nitrate or other applications usually result in prompt closure or it can be sutured.

Tetanus   After treatment of the navel stump, a tetanus antitoxin is administered to the foal and mare. To enhance tetanus prevention, the mare can be given a tetanus toxoid booster shot approximately one month before foaling. Sufficient antibodies are formed to protect the mare and the foal (via colostrum). Some farms also administer a prophylactic dose of antibiotics, but the practice is controversial. For antibiotics to be effective, they must be administered for 3 to 5 days.

Nursing   The normal foal should stand and nurse within a few minutes. If the foal is not up and nursing by 2 hours after delivery, it should be helped up and guided to the mare's udder. It is important for the foal to receive the colostrum because it is a laxative and contains antibodies that protect the foal. The foal's digestive tract permits absorption of the colostrum antibodies for approximately 36 hours after birth. It is produced by the mare for a maximum of 48 hours.

To ensure that a foal receives the passive transfer of immunoglobulins, the colostrum can be tested immediately after parturition for its concentration of immunoglobulins with a colostrometer or an agglutination test kit. The foal is then tested at 12 to 18 hours after birth to insure that the immunoglobulins were absorbed. A variety of test kits are available for the foal tests. If the colostrum test is negative, colostrum from a frozen colostrum bank can be given to the foal. If the foal tests negative, it should be given colostrum immediately and retested 12 hours later. If it tests negative again, it can be given a plasma transfusion or other treatment by a veterinarian.

Defecation   Within 4 to 12 hours after birth, the foal should pass the meconium, which is the fetal excrement. If the foal fails to eliminate the meconium or if constipation is evidenced by persistent straining and elevation of the tail, an enema consisting of 1 or 2 quarts of warm, soapy water should be given. The treatment should be repeated until yellow feces appear.

Eyes   Within 1 or 2 days after birth, some foals' eyes will start to water because the eyelids and lashes are turned in (a condition called entropion). The eyelids should be rolled out and an eye ointment rubbed in the eye.

Diarrhea   Diarrhea is a common problem of foals, particularly when the mare has her foal heat. If it ceases after the mare goes out of heat, no treatment is necessary. Persistent diarrhea should be treated by a veterinarian. Quite often a digestive disturbance occurs as a result of consumption of too much milk. Reducing the feed consumption of the mare or muzzling the foal for a few hours will eliminate the condition. A squirting type of diarrhea can dehydrate a foal and lead to death within a matter of hours.

Orphan Foals

Occasionally, a mare will die during foaling or shortly thereafter. Some mares may reject their foal or fail to produce milk. If possible, the foal should be transferred to another mare. This is difficult unless a "nurse mare" is kept for foster mother qualities. Oil of linseed or whiskey poured over the foal will disguise the foal's odor so that another mare will let it suckle. On some farms, the use of a milk goat has been successful. If these methods are unsuccessful, the foal will have to be bottle fed (see Section 8.7) or be bucket fed. Young foals can be taught to drink milk from a bucket by withholding feed and water while keeping the foal isolated for 5 hours. After 5 hours, the foal is allowed to suckle one's finger as it is immersed in a bucket of milk. Most foals will continue to drink the milk and they should have ad libitum access to milk until they are ready for weaning, at 3 to 4 months of age. A grain mixture and hay should be available to the foal. It will start consuming a little of each in a few days.

Foal Problems

Limb weaknesses and deformity are quite common in foals and may cause difficulties in rising and moving about early in life. These conditions include knuckling over at fetlock joints, overextension of the fetlock joints, and contracted forelegs. Most of these conditions respond to exercise but corrective splinting may be required. Convulsive syndrome, also called barker, dummies or wanderers, is a nervous disorder resulting from circulatory dysfunction. The brain undergoes a period when it fails to receive sufficient oxygen. Premature separation of the placenta from the uterus or premature separation of the umbilical cord are two common causes.

A large percentage of foal deaths can be attributed to septicemias (blood stream infections) during the first couple weeks of life. When the septicemia is present at birth, the foal may be semi-comatose and is called a sleeper. *Actinobacillus equui* bacteria cause septicemia immediately after birth. *Streptococci* infections are similar but occur later at four to six months. Naval-ill or joint-ill is a septicemia that settles in the joints.

Failure to receive passive immunity occurs in a significant percentage of foals. The failure may result from: (a) failure to nurse within 24 hours of birth, (b) premature lactation by the mare, (c) failure to absorb antibodies or (d) low immunoglobulin content of the colostrum. If given colostrum at birth or protected with antibodies until its immune systems function, the foal suffers few, if any, consequences. Other types of immunity failure include combined immunodeficiency disease, due to a deficiency of B- and T- lymphocytes, which is found in Arabian foals. Due to lack of an immune system, these foals die when they are about 5 months old when the passive immunity is lost. Agammaglobulinemia is a deficiency of the B-lymphocytes. Selective deficiency of IgM, a specific immunoglobulin, has been observed.

Trauma during foaling may result in broken ribs, a ruptured diaphragm, or a ruptured bladder. Other problems encountered include a closed rectum, congenital heart defects, and jaw deformities.

Hernias are defects in the body wall that permit some of the abdominal viscera to protrude. They usually occur at the navel or through the inguinal ring. Both conditions may correct themselves if they are not too serious. Otherwise, surgical correction is required before the blood supply to a protruding piece of intestine is cut off.

Neonatal isoerythrolysis (jaundice foal, isohemolytic icterus, or NI) is caused by an incompatibility of blood groups between mare and foal (see Section 13.3). Antibodies to the foal's red blood cells are formed by the mare and secreted in the colostrum. After the foal nurses and absorbs the colostrum, its red blood cells are destroyed. The foal becomes anemic and dies unless it is treated with blood transfusions. The problem can be avoided by preventing the foal from receiving the colostrum. If a problem is suspected, a simple test is to place some colostrum on a microscope slide along with a small drop of the foals blood. If the foal's red blood cells clump together or *hemolyze*, the colostrum is not safe and the foal should not be allowed to suckle the mare until the colostrum is no longer produced.

Weaning

The stress of weaning is reduced if the foal is fed grain and hay beginning a few days after birth. At 4 to 6 months, the foal is receiving a small percentage of its daily nutrient intake from the mare. If possible, it is desirable to wean two foals at the same time. By placing them in the same box stall or paddock, they seem to fret less than foals weaned by themselves. The mare and foal should be separated so that they are unable to see or hear each other. After a week, the weaning process should be completed.

Once the mare and foal have been separated, the mare's feed should be reduced for a few days so that the "drying up" process takes less time. The mare should not be milked out, because the udder will fill up and get tight. If the mare is too uncomfortable, an oil preparation containing spirits of camphor can be rubbed on the udder a couple of times a day for 4 or 5 days. After approximately one week, the udder should get soft and flabby. Some horsemen milk out the ½ to 1 cup of liquid left in the udder at the end of a week.

11.6   Business Aspects of Horse Breeding

Stallion Contracts

Since 1960, light-horse production has become a lucrative business. It is not uncommon for mares to be shipped from coast to coast to be bred. Because the two parties who participate in breeding the mare seldom know or live close to each other, stallion owners are using contracts that specifically state the conditions under which the mare will be bred. The purpose of the contract is to prevent misunderstandings and hard feelings, and if it does not, it does not fulfill its intended purpose.

Most people are not familiar with legal terminology and the contract conditions for mating, and thus do not fully understand the obligations of the stallion and mare owners. The contract should be simply worded so that it is easy to understand and so that both parties know their obligations. All the terms and conditions should be clearly spelled out so there is no opportunity for misunderstandings or hard feelings. Several items should be included in a basic contract, but a contract should be written for a specific farm since circumstances are different for each farm.

The stallion and the mare should be clearly identified. The identification form should include the stallion's registration number and the name and address of the owner or lessee. The owner of the foal needs this information to complete the registration application. The stallion owner's name and address do not have to appear on the Breeder's Certificate since the manager of the stallion may sign it. If the manager of the stallion signs the Breeder's Certificate, it may be difficult to obtain the stallion owner's name and address at a later time. To complete the Stallion Breeding Report, it is necessary to know the mare's registration number and the recorded owner of the mare at the time of service. Following is a typical example of wording to obtain this information:

> "The Horse Breeding Farm, Route 2 Box 89, College Station, Texas 77840, is willing to breed your mare, _____, Reg. No. _____, by _____ and foaled in 19 ____, during the breeding season of February 1, 1988 to June 15, 1988, to the stallion, Mr. Who #97,545, owned by and standing at the Horse Breeding Farm, Route 2 Box 89, College Station, TX, for $500.00 upon the following terms and conditions."

Other significant information is included in the terms and conditions. The age of the mare is given so that the stallion manager will know if he needs to give the mare any special consideration, particularly if she is barren. A definite breeding season is stated. The limited season is important to the stallion owner so that the horse can be shown or used for other purposes during the remainder of the year. To help prevent overuse of a stallion, some owners will limit the number of times that a mare will be bred. Many mares are problem mares and simply will not conceive regardless of how many times they are bred, even though they appear to be normal. The location of the breeding farm where the stallion will stand for this breeding season is given so that the mare owner knows exactly where to deliver the mare. The stud fee and the terms and conditions of breeding are also stated.

In many instances, payment of a booking fee or deposit is required to validate the contract. The terms of payment should be outlined in the contract. The fee may be due upon return of the contract at the time the mare is booked, or it may be due at a later date. The wording should clearly state whether the money is refundable. If it is, it is a booking deposit. If the mare does not conceive and the money is not refundable, it is a booking fee. High booking fees are a source of irritation to many mare owners, particularly if the mare goes home barren and the owner has encountered other excessive expenses for board, examination, vanning, trimming, and so on. Sometimes it is difficult for the stallion owner to decide the correct booking deposit or fee for a stallion because he wants to discourage mare owners from changing their mind, forfeiting the deposit, and breeding their mares to another stallion.

Payment of the amount of the stud fee that is due after the booking fee or deposit has been paid should be outlined in detail. There are as many ways of handling it as there are stallion owners. It is quite common for Quarter Horse stud fees to be paid when the mare is returned home. Other farms are more lenient in obtaining payment. They may specify a given date for payment, such as October 1, or the fee may be due when the foal is born. The stallion owner usually makes certain guarantees concerning pregnancy or birth that must be fulfilled before payment is due or a refund is given. A stallion owner's obligations may be considered completely fulfilled if the mare is positively in foal at 45 days. In this author's opinion, the stallion owner has performed his responsibilities when the mare conceives and should not be financially responsible for later events leading to abortion that are beyond his control. A live foal guarantee means different things to different people, so the stallion owner's definition must be fully spelled out. The existence of a previously live foal is verified by a veterinarian if a piece of the dead foal's lung floats in water. With a live foal guarantee, the mare owner may have to make a critical decision at birth if the foal does not breathe immediately after presentation. Before artificial respiration is given or the foal is stimulated to breathe by other means, the owner or attendant must rapidly evaluate the foal for conformation and appearance to see if an attempt to save it is economical. In most cases, other feelings override economic decisions and an attempt is made to save the foal. It is common to see the conditions specified of a live foal that stands and nurses or that stands and nurses without assistance. These conditions usually mean that there is a reasonable chance for a healthy foal. Most farms require notification and/or certification by a licensed veterinarian within a short time of failure of live birth. If this is not forthcoming, the mare owner is held for payment of the stud fee. This is the reason many farms send a bill for the stud fee approximately 11 months after the last service when the payment is due if a live foal is guaranteed. Other farms that have previously collected the stud fee require the certification before the fee is refunded. It is to the mare owner's benefit that the stud fee be returned within a given time of certification of live birth failure. Most stallion owners actually reserve an option to refund the fee or rebreed the mare only during the following season. This option allows the stallion to be sold without obligations to breed him with certain mares, to change the breeding fee, and/or to change the location of the stallion.

Another important condition of the contract may be that a Certificate of Service is not to be issued until the stud fee and all other charges that are due under the contract have been paid. This encourages the mare owners to pay all bills, for if they do not, they cannot register the foal.

In addition to the stud fee, all other expenses that are to be paid by the mare owner should be specified. The daily rates for boarding the mare, or

straight hair and one of which is homozygous for curly hair, that is, $Cc^{cr}Ss \times Cc^{cr}ss$.

The expected genotypic results for the "color" locus are $[(1/2)(C) + (1/2)(c^{cr})] \times [(1/2)(C) + (1/2)(c^{cr})] = (1/4)(CC) + (1/2)(Cc^{cr}) + (1/4)(c^{cr}c^{cr})$. The expected phenotypic results are 1/4 chestnut + 1/2 palomino + 1/4 cremello.

The expected results for the "hair" locus are $[(1/2)(S) + (1/2)(s)] \times [(1)(s)] = (1/2)(Ss) + (1/2(ss)$. The expected phenotypic results are 1/2 with straight hair and 1/2 with curly hair.

The expected joint frequencies for the 6 genotypes are $[(1/4)(CC) + (1/2)(Cc^{cr}) + (1/4)(c^{cr}c^{cr})] \times [(1/2)(Ss) + (1/2(ss)] = (1/8)(CCSs) + (1/8)(CCss) + (1/4)(Cc^{cr}Ss) + (1/4)(Cc^{cr}ss) + (1/8)(c^{cr}c^{cr}Ss) + (1/8)(c^{cr}c^{cr}ss)$.

The expected frequencies of the joint phenotypes can be found by adding the frequencies of genotypes that look alike. In this example, the phenotypic frequencies are the same as the genotypic frequencies, but in other examples, this would not be true. For example, if the second animal were heterozygous for the curly hair gene, there would be 9 possible genotypes but only 6 phenotypes. Another way to find the joint phenotypic frequencies is to multiply the phenotypic results for each locus together. For example, (1/4 cremello + 1/2 palomino + 1/4 chestnut) × (1/2 straight hair + 1/2 curly hair) = 1/8 cremello straight hair + 1/8 cremello curly hair 1/4 palomino straight hair + 1/4 palomino curly hair + 1/8 chestnut straight hair + 1/8 chestnut curly hair.

Epistasis

The joint phenotypic frequencies will be different if the genes at one locus affect the expression of genes at another locus, a form of gene action called *epistasis*. The rule is to obtain the genotypic frequencies as before, then examine these to determine the phenotypes. An example is the gene for white, W. (There is no true albino gene in horses. Horses with the W gene have colored eyes.) The W gene masks any other colors that may be present. In addition, when homozygous for W, the fetus is resorbed before birth so that no WW animals are born, which results in an apparent lowered conception rate. A homozygote for ww will express the color determined by the other loci.

Suppose that a white stallion that is heterozygous for black is mated to a white mare that is also heterozygous for black (both WwEe). The results at the "white" locus are, for genotypic frequencies, $[(1/2)(W) + (1/2)(w)] \times [(1/2)(W) + (1/2)(w)] = (1/4)(WW) + (1/2)(Ww) + (1/4)(ww)$, and for the phenotypic frequencies, 1/4 no foal + 1/2 white + 1/4 color. The results at the "black and chestnut" locus are, for genotypic frequencies, $[(1/2)(E) + (1/2)(e)] \times [(1/2)(E) + (1/2)(e)] = (1/4)(EE) + (1/2)(Ee) + (1/4)(ee)$, and for phenotypic frequencies, 3/4 black + 1/4 chestnut. The joint genotypic re-sults are [(1/4)(*WW*) + (1/2)(*Ww*) + (1/4)(*ww*)] × [(1/4(*EE*) +(1/2)(*Ee*) + (1/4) (*ee*)] = (1/16)(*WWEE*) + (1/8)(*WwEE*) + (1/16)(*wwEE*) + (1/8)(*WWEe*) + (1/4)(*WwEe*) + (1/8)*wwEe*) + (1/16)(*WWee*) + (1/8)(*Wwee*) + (1/16)(*wwee*). The joint phenotypic frequencies are [(1/16)(*WWEE*) +(1/8)(*WWEe*) + (1/16)(*WWee*)] = 1/4 no foal; [(1/8)(*WwEE*) + (1/4) (*WwEe*) + (1/8)(*Wwee*)] = 1/2 white; [(1/16)(*wwEE*) + (1/8)(*wwEe*)] = 3/16 black; and (1/16)(*wwee*) = 1/16 chestnut. If only live foals are counted, 2/3 are white, 1/4 are black, and 1/12 are chestnut, since 1/4 of the matings will not produce a live foal to be counted.

A gene that masks the effect of another gene at a different locus is said to be *epistatic* to the masked gene; that is, in the example just described, the gene for white is epistatic to the genes for black and chestnut.

Linked Genes

The simple formula for finding genotypic results with respect to several loci will not work if the loci are linked. Linkage or a tendency for genes at certain loci to be inherited together occurs when the loci are near each other on the same chromosome. The phenomenon of crossing over (exchange of genetic material on corresponding parts) between the members of a pair of paired chromosomes in the production of germ cells does, however, result in some recombination of genes from the linked loci. The frequency of recombina-tion of genes depends on the distance between loci. Since horses have only 32 pairs of chromosomes, many genes must be linked on the same chromo-somes. Linkages of genes located on five chromosomes have been found. Most of these involve blood factors as described in Chapter 13. The close linkage between the locus for roan versus non-roan and the locus for black versus chestnut, however, has significance for breeders of roan horses.

12.4   Selection for Dominant and Recessive Genes

The major problem in selection for simply inherited traits is selection against recessive genes. Dominant genes can obviously be selected against very easily because all animals with the dominant gene can be readily identified.

Sex-linked recessive genes can also be eliminated quite quickly. All males as well as all homozygous females with the gene can be identified and culled. Unless female carriers are preferred to the homozygous normal females, the result is that the frequency of the sex-linked recessive gene will decrease by 1/2 each generation. Similarly, the frequency of affected males drops by 1/2 each generation, and no affected females are born—as long as no affected males are used in the breeding program. The decrease may be a little more rapid if mares that have affected foals are also culled.

Eliminating non-sex-linked recessive genes is more difficult, because heterozygous carriers cannot be readily identified. Even then, simply eliminating all affected individuals (those homozygous for the recessive gene) can be quite effective. If q is the frequency of the recessive gene, then the frequency of affected animals is q^2. After n generations of selection, the frequency of the gene is $[q/(1 + nq)]$ and the frequency of affected animals is $[q/(1 + nq)]^2$. This formula works as long as the phenotypically normal carrier animals are not preferred to the homozygous normal animals.

Elimination of recessive genes can be increased by testing for carriers—especially males—and then using only those males that have a high chance of not being a carrier. Testing does require time and effort.

The best method of testing to identify carrier males is through test mating by artificial insemination (A.I.) to a random sample of mares. Because A.I. has not been widely practiced with horses, this method will not be discussed here.

The problem of detecting carriers is obvious. There is no way to tell the difference between a "normal" horse with two desirable genes and a "normal" horse that carries one undesirable gene in addition to the one desirable gene. A carrier, however, will transmit the undesirable gene to half of his foals.

A carrier stallion is identified if he sires at least one undesirable foal, since any affected foal must have received one undesirable gene from the stallion as well as one from its dam. If the carrier stallion is mated to a group of mares, the expected proportion of normal and undesirable foals can be determined if the proportion of ova carrying the desirable and undesirable genes is known. This proportion can be determined for the testing methods described in the remainder of this section.

Mating to Homozygous Recessive Mares

The chance of obtaining a normal foal by mating a carrier stallion to a homozygous recessive mare is 1/2, as shown in Figure 12-8. Thus, the chance of obtaining all normal foals by mating a carrier stallion to n homozygous recessive mares is $(1/2)^n$, since the type of foal in each birth is independent of the type of foal in other births. This probability is the chance of not obtaining any undesirable foals, and thus is the chance of *not* detecting a carrier from the foals produced by n matings to homozygous recessive mares. The chance of detection plus the chance of nondetection must be 1. Therefore, the chance of detecting a carrier is $1 - (1/2)^n$. See Table 12-5 for a listing of chances of detection for different numbers of matings.

Table 12-5
Chances of detecting a carrier stallion from various types of matings

| Number of foals n | Detects only one recessive | | Detects all recessives |
	Mating to homozygous recessive mares $1 - (1/2)^n$	Mating to known carrier mares $1 - (3/4)^n$	Mating to own daughters $1 - (7/8)^n$
1	.50	.25	.12
2	.75	.44	.23
3	.88	.58	.33
4	.94	.68	.41
5	.97	.76	.49
6	.98	.82	.55
7	.99	.87	.61
8	~1.00	.90	.66
9		.92	.70
10		.94	.74
15		.99	.87
20		~1.00	.93
50			~1.00

Mating to Known Carrier Mares

Another method is to mate the stallion to known carrier mares — of the type Ee. The chance of obtaining all normal foals out of n foals will be $(3/4)^n$. The chance of detection will be $1 - (3/4)^n$, as indicated in Table 12-5.

Mating to Own Daughters

A third method is to mate a stallion to his own daughters. This method provides an equal chance of detecting all recessive genes that the stallion may be carrying. For a specific recessive gene, mating a stallion to daughters of a known carrier stallion is equivalent to mating to his own daughters.

In calculating the chance of detection by mating a stallion to his own daughters, both genes of each of his original mates are assumed to be normal. The result, as shown in Figure 12-14, is that half the daughters will be carriers if the stallion is a carrier. The chance of the undesirable trait appearing in a foal when a carrier stallion is mated to a daughter is 1/8, and the chance of obtaining n normal foals from n such matings is $(7/8)^n$. The chance of detecting a carrier stallion from matings to his own daughters is $1 - (7/8)^n$, as shown in Table 12-5.

Step 1: Original mating

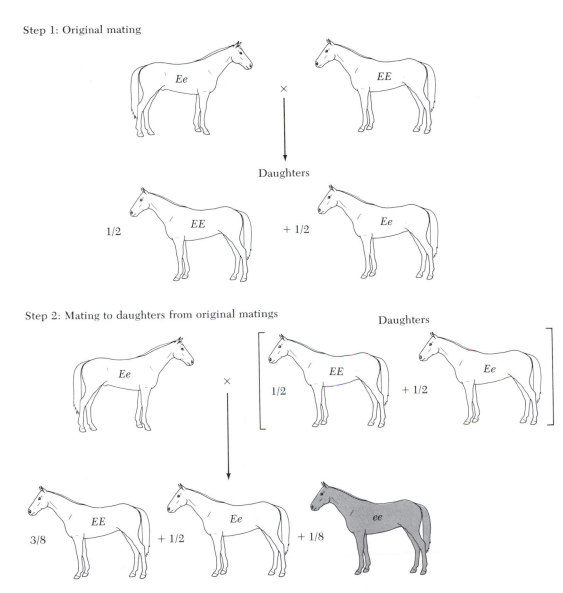

Figure 12-14
Detecting a carrier stallion from matings to his own daughters.

Noncarrier Cannot be Proved

A noncarrier stallion can never produce a foal with an undesirable trait caused by a recessive gene. Thus, any stallion with an undesirable foal is immediately labeled a carrier. A stallion, on the other hand, can never be completed proved to be a noncarrier. There will always be some small chance of nondetection no matter how many normal foals he sires.

The principles discussed in this chapter will be applied in Chapter 13 in the discussion of parentage tests and color inheritance and in the listings of simply inherited lethal and semilethal traits.

REFERENCES

Anderson, W. S. 1939. Fertile mare mules. *J. Heredity* 30:549.

Archer, R. K. 1961. True haemophilia (haemophilia A) in a Thoroughbred foal. *Vet. Record* 73:338.

Auerbach, C. 1961. *The Science of Genetics.* New York: Harper & Row.

Baldwin, J. T., Jr. 1957. An early report of a fertile female mule. *J. Heredity* 39:84.

Bartlett, O. 1963. Zeehorses and zebronkeys. *Animals* 2:394.

Benirschke, K., L. E. Brownhill, and M. M. Beath. 1962. Somatic chromosomes of the horse, the donkey, and their hybrids, the mule and hinny. *J. Reprod. Fert.* 4:319.

Benirschke, K., R. J. Low, L. E. Brownhill, L. B. Caday, and J. DeVenecia-Fernandez. 1964. Chromosome studies of a donkey–Grevy zebra hybrid. *Chromosoma* 15:1.

Benirschke, K., R. J. Low, M. M. Sullivan, and R. M. Carter. 1964. Chromosome study of an alleged fertile mare mule. *J. Heredity* 55:31.

Benirschke, K., and N. Malouf. 1957. Chromosome studies of equidae. In *Equus.* H. Dathe, ed. Vol. 1, No. 2. Berlin: Tierpark.

Blakeslee, L. H., R. S. Hudson, and R. H. Hunt. 1943. Curly coat of horses. *J. Heredity* 34:115.

Bonadonna, T. 1966. The interspecific and reciprocal crossing horse-donkey and the fertility of resulting hybrids. *World Review of Animal Production* 3:46.

Cradock-Watson, J. E. 1967. Immunological similarity of horse, donkey, and mule haemoglobins. *Nature* 215:630.

Craft, W. A. 1938. The sex ratio in mules and other hybrid mammals. *Quart. Rev. Biol.* 13:19.

Gray, A. P. 1972. *Mammalian Hybrids.* Slough, England: Commonwealth Agricultural Bureaus, Farnham Royal.

Groth, A. H. 1928. A fertile mare mule. *J. Heredity* 19:413.

Heck, H. 1952. The breeding of the Tarpan. *Oryx* 1:338.

Hsu, T. C., and K. Benirschke. 1967. *An Atlas of Mammalian Chromosomes.* New York: Springer-Verlag.

Hutt, F. B. 1964. *Animal Genetics.* New York: Ronald Press.

Johansson, I., and J. Rendel. 1968. *Genetics and Animal Breeding.* New York: Freeman.

Jones, W. E., and R. Bogart. 1971. *Genetics of the Horse.* East Lansing, Michigan: Caballus.

King, J. M., R. V. Short, O. E. Mutton, and J. L. Hamerton. 1965. The reproductive physiology of male zebra-horse and zebra-donkey hybrids. *J. Reprod. Fert.* 9:391.

Lasley, J. F. 1972. *Genetics of Livestock Improvement.* Englewood Cliffs, New Jersey: Prentice-Hall.

Lloyd-Jones, O. 1916. Mules that breed. *J. Heredity* 7:494.

Lydekker, R. 1912. *The Horse and Its Relatives.* New York: Macmillan.

Makino, S. 1943. Notes on the cytological feature of male sterility in the mule. *Experentia* 11:224.

Martin, G. H., Jr. 1962. The zebronkey: A riddle of science? *Univ. Philippines Vet.* 7:23.

Petzsch, H. 1962. Birth of a zebroid from a ♀ ass mated with a ♂ Grant's zebra. *Saugetierk. Mitt.* 10:61.

Rice, V. A., F. N. Andrews, E. J. Warwick, and J. E. Legates. 1970. *Breeding and Improvement of Farm Animals.* New York: McGraw-Hill.

Roberts, E. 1929. Zebra-horse cross. *J. Heredity* 20:12.

Ryder, O. A., L. G. Chemnik, A. T. Bowling, and K. Bernirschke. 1985. Male mule foal qualifies as the offspring of a female mule and jack donkey. *J. Heredity* 76:379.

Sangor, V. L., R. E. Mairs, and A. L. Trapp. 1964. Hemophilia in a foal. *J. A. V. M. A.* 142:259.

Smith, H. H. 1939. Fertile mule from Arizona. *J. Heredity* 30:548.

Trujillo, J. M., C. Stenius, L. Christian, and S. Ohno. 1962. Chromosomes of the horse, the donkey, and the mule. *Chromosoma* 13:243.

Trujillo, J. M., Betty Walden, Peggy O'Neil, and H. B. Anstall. 1965. Sex-linkage of glucose-6-phosphate dehydrogenase in the horse and donkey. *Science* 148:1603.

Trujillo, J. M., Betty Walden, Peggy O'Neil, and H. B. Anstall. 1967. Inheritance and sub-unit composition of haemoglobin in the horse, donkey, and their hybrids. *Nature* 213:88.

CHAPTER 13

COLOR INHERITANCE, PARENTAGE TESTING, GENETIC LETHALS

Most breeders are interested in one or more of three kinds of genetic effects that are simply inherited. All breeders are concerned about lethal or semi-lethal traits. Lethal traits caused by a single dominant or recessive gene can be eliminated by procedures described in Section 12.4. Some lethal or semilethal traits will be described briefly in Section 13.3. Color inheritance, which is of special interest to the color breeds such as Appaloosas, Pintos, Paints, Buckskins, and Palominos, will be discussed in Section 13.1. Tests of parentage have become increasingly important with the possible use of embryo transfer and artificial insemination. Simply inherited traits can be used successfully in many cases of disputed parentage — not to prove parentage, but to exclude some animals as possible parents. The most useful traits for parentage tests, in addition to the obvious color traits, are genetically controlled types of blood proteins, enzymes, red-blood-cell antigens and lymphocyte antigens. Some of these have been shown to be linked to each other and to color traits. The potential of these traits for parentage exclusion will be described in Section 13.2.

13.1 Color Inheritance

The color of a horse has little to do with its performance, although some colors may contribute to the value of the horse if those colors or patterns are currently popular. Many breeders, such as those breeding Standardbreds or

Thoroughbreds or Quarter Horses, have only a slight preference for the color of their horses. Others may want certain color patterns, such as a particular type or types of an Appaloosa, a Paint or Pinto, a Buckskin, a Palomino, or an American White Horse. Color is a primary means of identification and also may be a first indicator of questionable parentage. More likely the value of color for most people is as an enjoyable pastime in guessing the color of the foal from a particular mating or in trying to fathom the apparent complexities of color inheritance. In this section the odds for and against the "color" breeders will, in part, be explained. Such information may help others to understand the genetics of certain basic colors.

The Colors

Before discussing the genetics of coat color, a brief description of the basic colors and patterns may be needed because very few people, even those closely associated with horses, will agree on the names of all colors. All horses carry the genetic makeup (genotype) either for black or chestnut (the horse world's word for shades of red or brown). Other genes will modify or may even mask the expression of the genes for black or chestnut, but for this discussion all horses will be considered to be genetically either black or chestnut. White markings on the head and legs will be ignored in discussion of the colors.

The basic colors are summarized in Table 13-1. The insert following page 496 illustrates the colors. The pictures do not necessarily capture the true colors of the horses or even the exact color for the supposed genotype, but they do give a reasonably accurate idea of the colors described. Colors may appear to be different in sunlight than in shade, and may seem to change at different times of the year; they may also be slightly modified by the diet or other environmental factors.

The Blacks   The basic modifications of black are bay, seal brown, buckskin, dun, grulla, and perlino. Except for the perlino all the blacks have black points—lower legs, mane, and tail. The double dilution to perlino leaves only a trace of rust at the points.

A true *black* horse cannot have any red or brown hairs. *Seal browns* are black with some brown hair with the amount ranging from a few brown hairs only on the muzzle to large brown (not red) areas over various regions of the body. Brown hair is most apparent on the muzzle, the flanks, behind the shoulders, and along the gaskin. *Bays* show an even greater modification of black—the body is some shade of red and only the points are black. Black and dark seal brown can be diluted to *grulla*—ranging from a slate gray appearance to a muted light brown appearance of the body with a black back (dorsal) stripe. The head is usually darker than the body. Dilution refers to

Table 13-1
Brief description of basic color patterns (excluding white markings on head and legs)

	"Blacks"				"Chestnuts"		
Color	Body	Lower legs	Mane and tail	Color	Body	Lower legs	Mane and tail
Black	Black	Black	Black	Chestnut	Brown or red	Brown or red	Darker or lighter brown or red
Bay	Red	Black	Black	Sorrel	Reddish yellow	Reddish yellow	Darker or lighter
Seal brown	Black with brown areas	Black	Black	Red dun[a]	Yellow-red	Darker than body	Dark red
Grulla[a]	Dilute black	Black	Black	Palomino	Gold	Slightly darker	White
Buckskin[b]	Yellow-red	Black	Black	Cremello	Off-white	Off-white	White
Perlino	Off-white	Rust	Rust				

[a]Has dark back (dorsal) stripe,
[b]With a dorsal stripe would be called dun.

individual hairs and is not a mixture of white and colored hairs as is gray or roan. Bays and light seal browns are diluted to *buckskin*. The diluted red body may range from a light yellow, almost cream color, to a more reddish yellow. Again, the points are black. If a dorsal stripe is present, the pattern is often called *dun* rather than buckskin. Bays can be doubly diluted to *perlino*. The body is off-white (pearl), the skin is pink, and the eyes are nearly pink and appear washed out. The black points are diluted to rust, with the shade ranging from very noticeable to almost unnoticeable. Perlinos are not albinos. White markings, if present, can be seen against the off-white pearl color.

The Chestnuts   Chestnuts never have black points, although sometimes the mane and tail may be so dark red or brown that the appearance is black. The lack of black on the legs of chestnuts will always allow a clear-cut distinction between blacks and chestnuts. Confusion exists from region to region of the country and from light-horse to draft-horse people about the terms "chestnut" and "sorrel." *Sorrel* for this discussion will be defined as the lightest shade of chestnut—yellow with only a hint of red. All other chestnut types, ranging from liver chestnut to light red chestnut, will be called chestnut.

The *red dun* requires special mention. The body color is diluted chestnut, yellow-red, although the shade may sometimes be confused with light red shades of chestnut. The legs are a darker shade of chestnut than the body. The mane and tail may be quite red, and sometimes the head is dark red and a dark red dorsal stripe runs down the back.

The *palomino* dilution of chestnut is well known. The body is some shade of gold and the mane and tail are white or nearly white. Doubling the palomino dilution of chestnut results in the *cremello*. The body is off-white (creme), the mane and tail are white, and the eyes are washed out — almost pink. Probably because of the eyes, the cremellos and perlinos are sometimes called albinos, although they are not true albinos because some pigment is present.

White Patterns   The addition of white hairs to hair of the base colors can greatly change the overall appearance. Some of these modifications are listed in Table 13-2. The greatest change is when white hairs replace all of the colored hairs as in the *dominant white* horse. Dominant white horses are not albinos although they have pink skin; they have normally colored eyes and occasionally, small pigmented spots.

Gray and roan are often confused. Both are mixtures of white and colored hair. Both can occur with all nonwhite colors. *Graying* is progressive, however. A foal that will turn gray is born with a colored coat. After the first shedding, a few white hairs will be found interspersed with the colored hairs. Each shedding results in a larger and larger proportion of white hairs, until the horse may appear to be white. Gray-white is easily distinguished from dominant white by the dark skin pigment, especially on the head of a

Table 13-2
Brief description of white patterns

Color	Description
Dominant white	White, occasionally small, coin-sized pigmented areas, pink skin, colored eyes
Roan	Body a constant mixture of white and colored hairs; legs, neck, head are solid color
Gray	All areas are a mixture of colored and progressively more white hairs
Tobiano	White spotting on any color (see Figure 13.10)
Overo	White spotting on any color (see Figure 13.12)
Splashed white	White spotting on any color (see Figure 13.13)
Sabino	White spotting on any color
Flaxen	Yellow to white mane and tail on "chestnuts" only

gray. Often, graying is first noticed on the head, in contrast to the *roan* which usually has a solid-colored head (except for white markings).

The pattern of white hairs on the roan does not change with age, although due to shedding and growth of the winter coat the amount of white may appear to change in spring or fall. Roaning is usually confined to the body and is often greatest over the loin and hips. The legs and head are solid colored. Usually, the neck is also solid.

Two terms associated with spotted horses, *piebald* and *skewbald*, refer not to the patterns of spotting but to the color of the colored hair. Piebald is black with white spotting and skewbald is any color except black with white spotting. There are, however, at least two usual patterns of white spotting, each with extreme variations in the amount of white.

The *tobiano* pattern has been described as what would be expected from pouring a bucket (or barrel) of white paint over the back of a solid-colored horse. The head is usually normally marked. The edges of the white areas are usually smooth. White extends over the back. The legs are predominantly white.

The *overo* pattern is almost, but not quite, the opposite of tobiano. The head is often heavily marked with white — bald faces are common. White does not usually extend over the back. The edges of the white areas are quite irregular. The white areas are predominantly on the body or neck and often are enclosed by colored areas. The legs are often a streaked mixture of colored and white areas.

Two other spotting patterns — *splashed white* and *sabino* — which may be inherited independently of tobiano and overo resemble the overo pattern and are discussed under major modifying loci.

Flaxen (yellow to almost white) manes and tails are not really an addition of white hairs to the body coat but are useful in identification. Flaxen manes and tails occur only on chestnuts and *not* on blacks. Expression is variable — sometimes the mane only, sometimes the tail only, and sometimes both. Lighter shades of chestnut are more likely to have flaxen manes and tails. Flaxen manes and tails are more rare and usually are darker on horses with darker shades of chestnut.

Color inheritance in the horse is complex and is not completely understood. Many of the inferences in this section are based on studies by Castle and his associates. When uncertainties exist, a question will be raised and a speculation as to its answer will be given.

Major Color Loci   Genes from four or five loci interact to cause most of the major color patterns. The effects of these genes are illustrated in Figure 13-1 and will be discussed in this section. Several other genes mask, spot, gray, or roan some of these colors. In addition, modifier genes that have relatively small effects probably cause much of the variation in the basic colors.

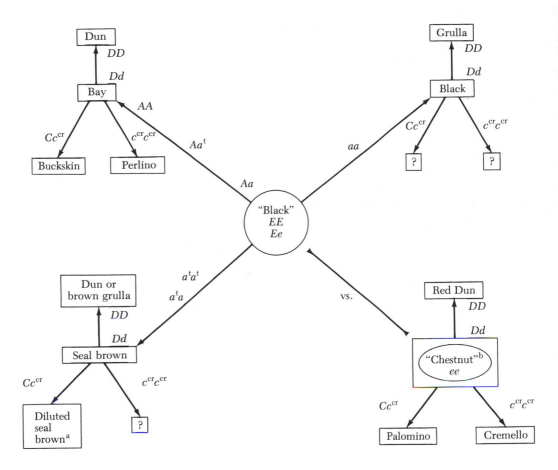

Figure 13-1
Chart showing probable colors from combinations of four major color loci. The capital letters represent dominant alleles for 2 alleles at a locus; for 3 alleles at a locus, the alleles are listed in order of dominance. The dilution gene c^{cr} is incompletely dominant. Joint effects of D and c^{cr} are not known although probably the dilution is not enhanced and a dorsal stripe occurs because of the D gene.

[a] Brown hair diluted to light brown but black not diluted.

[b] Effects of A, a^t, and a with ee are not known.

The C Locus Nearly all mammals except the horse carry a recessive gene for the true pink-eyed and pink-skinned albino. Such a gene has not been found in the horse. Several genetic pathways result in a white or near-white horse, but none give the true albino. Thus, all horses are thought to be homozygous for the gene C, which allows other color genes to express themselves. In the remainder of this book, the *CC* designation will be assumed and not written in discussion of color genotypes except for another allele, discussed later, assigned to this locus, that results in dilution of red (brown) color.

The *B* and/or *E* Loci   Geneticists do not all agree on what to call the gene that causes a horse to be basically chestnut ("chestnut" is really red or brown). In many species as well as the horse, a dominant gene, *B*, results in the formation of black eumelanin, a pigment causing black coat color and a dark skin. In most species, but perhaps not in the horse, a recessive gene, *b*, results in chocolate brown eumelanin, causing chocolate coat color and brown or pinkish skin. Some geneticists claim the *bb* genotype results in the red or brown chestnut color of horses. The skin color, however, is dark in most, if not all, chestnuts. In most species a gene at the *E* locus, when homozygous, (*ee*), prevents formation of eumelanin and codes for formation of red (brown) pigment called phaeomelanin. The dominant allele at the *E* locus, *E*, allows expression of eumelanin unless modified by genes at still other loci. Horses of the genotypes *BBee* or *Bbee* would be basically chestnut but with dark skin. Because most, if not all, horses appear to be homozygous, *BB*, the chestnut caused by the *ee* genotype essentially is recessive to black, *EE* or *Ee*. Thus for all practical purposes whether the genotypes causing basically black or basically chestnut colors are:

Hypothesis I:		Hypothesis II:
$\left.\begin{matrix} BB \\ Bb \end{matrix}\right\}$ black	OR	$\left.\begin{matrix} BBEE \\ BBEe \end{matrix}\right\}$ black
bb } chestnut		*BBee* } chestnut

is not important. Sponenberg and Beaver (1983) state that most breeds have only the *B* allele at the *B* locus with some exceptions being the North American Spanish Horse, the American Quarter Horse, and possibly the Morgan horse. The frequency of the *b* allele is undoubtedly very low.

For convenience the single locus nomenclature will be used to distinguish between basically black and basically chestnut coat colors, with all horses being assumed to be *BB*. Thus, *EE* or *Ee* is black, and *ee* is chestnut.

The next section describes a locus, *A*, with genes that restrict the expression of the black pigment so that *Aa* or *Aa^t* with *EE* or *Ee* results in bay, *a^{t}a^t* or *a^{t}a* with *EE* or *Ee* results in seal brown and *aa* with *EE* or *Ee* is black. According to such a theory, however, all homozygous black matings should breed true; that is, *aaEE* × *aaEE* should give only *aaEE* and heterozygous black matings, *aaEe* × *aaEe* would give, on the average, ¾ black and ¼ chestnut offspring. However, a small number of bay foals always seem to result when large numbers of black by black matings are made. Errors in distinguishing between dark brown and black cannot account for the discrepancy. Therefore, Castle postulated that another allele, E^D, extends the dark pigment so much that it actually masks the effect of the bay pattern gene, A. He called this the gene for dominant black and described the color as jet black, which does not fade in sunlight as the ordinary black may do. The foal coat is also thought to be black, whereas the first coat of the recessive black is often more of a mouse gray color. The existence of a similar gene in other mammals helped lead him to this conclusion. Thus, E^D is thought to extend the dark color even if A or a^t is present (that is, E^D is epistatic to A and a^t). A jet black could be E^D_aa or $E^D\_A\_$ or $E^D\_a^ta^t$ or $E^D\_a^ta$, where the blank indicates that the paired gene has no effect. There is no direct evidence that the gene E^D exists in the horse.

As expected for a recessive color such as chestnut, chestnuts mated together must always breed true. The *chestnut rule* is that mating a chestnut stallion to a chestnut mare can produce only a chestnut foal.

Since "blacks" may carry the gene for chestnut, "blacks" mated together may produce "chestnut" as well as "black" foals.

The A Locus   Most breeders know that bay is dominant to black. The reason is that genes at another locus, the A locus, determine whether the black pigment is expressed fully or is restricted to mane, tail, and legs. There are probably three or four alleles at the A locus. In order of decreasing dominance, they can be named A^+, A, a^t, and a. The A^+ allele creates the wild-type coat, which resembles the protective coloration of most wild animals, as exemplified by the Przewalski's horse shown in Figure 13-2. The net appearance is somewhat that of diluted bay but with a dorsal stripe. There is some dispute about the existence of this allele, but the A^+ allele really does not concern the breeder of modern horses. The other three alleles modify the black pigment, as shown in Table 13-3. The A gene restricts the black areas to the legs and to the mane and tail. The result is the bay horse, with reddish (bay) body, black legs when not hidden by white stockings, and black mane and tail. The a^t allele, in the a^ta^t or a^ta genotypes, does not restrict the black as much, so the result is the dark brown or seal brown horse, which is sometimes mistaken for a black horse. Again, the points will be black. The very dark seal brown horse can be distinguished from a true black horse only by the presence of brown hair around the muzzle or on the flanks. The picture in the color insert that illustrates the dark seal brown is somewhat lighter than would normally be seen, so that some brown is visible. Lighter seal brown horses show more brown areas, particularly at the flank and behind the shoulder. This brown does not have a reddish appearance, although distinguishing between dark bay and light seal brown may sometimes be difficult due to minor modifier genes that may lighten or darken the basic color. The most recessive allele, a, when homozygous, does not affect the

Figure 13-2
The Przewalski's horse shows the restrictive effects of the wild-type allele at the *A* locus on the genotype for formation of black pigment.

Table 13-3
Effect of restriction genes on genetically black and chestnut colors

Genotype at A locus	Black EE or Ee	Chestnut ee
AA *Aa*ᵗ *Aa*	Bay	Chestnut (various shades)
aᵗaᵗ *aᵗa*	Seal Brown	Chestnut (various shades)
aa	Black	Chestnut, shade is unknown, possibly sorrel

distribution of the black pigment, and thus a uniformly black horse will result. Such a black horse may fade from exposure to the sun so that the body color may seem to be blackish-bay.

Sponenberg and Beaver (1983), however, state that a dominant gene at another locus is responsible for the seal brown pattern. They have assigned the symbol P, for the pangaré effect, to the gene which results in light phaeomelanic areas on the muzzle, over the eyes, on the flanks and on the inside of the legs. The effect of this gene is most obvious on blacks, where it creates the typical seal brown pattern, and on sorrels, where it creates the blonde sorrel pattern seen in some Belgians. The pangaré or a similar effect is common on many relatives of the horse, including the donkey and the onager. Data to confirm the existence of the pangaré gene have not been published. Thus in Figure 13-1, the a^t gene with ee is listed as causing the seal brown pattern.

The effects of genes of the A locus on the chestnut genotype are not known, although the aa pair may result in sorrel. McCann (1916) found that sorrels breed true as would $eeaa$ genotypes mated to each other. His studies, however, did not show a clear pattern for the relationship between sorrel and the darker shades of chestnut.

The C Locus, Again   As shown in Table 13-4, two kinds of dilution of color are known. The c^{cr} gene is incompletely dominant and appears to dilute only red or light brown hair but not black hair. The Cc^{cr} genotype results in a dilution of the basic coat color. The $c^{cr}c^{cr}$ genotype dilutes the color even more, so that such horses are sometimes called albinos—although they are

Table 13-4
Effects of dilution genes on genetically black, bay, and chestnut colors

Locus	Genotype	Basic color		
		Black[a]	Bay[b]	Chestnut
D "Dominant dilution"	DD Dd	Grulla	Dun	Red dun
	dd	Black	Bay	Chestnut
C "Palomino dilution"	CC	Black	Bay	Chestnut
	Cc^{cr}	Black	Buckskin	Palomino
	$c^{cr}c^{cr}$	Black[c]	Perlino	Cremello

[a]Also dark seal brown,
[b]Also light seal brown,
[c]Gene action not known.

not true albinos because they have colored eyes and some pigment. The *CC* genotype has no effect on the basic color. The dilution acts by causing a portion of the hair shaft to lack pigment (Gremmel, 1939).

Castle (1960) presented evidence that the *Cc*ᶜʳ genotype has no effect if the genotype at the *A* locus is *aa*. Similarly, Singleton and Bond (1966) suggested that *c*ᶜʳ has no effect on either *a*ᵗ genotypes or *a* genotypes. Thus, *c*ᶜʳ would cause dilution only if the *A* allele is present. Adalsteinsson (1974), however, has shown that the *A* allele is not necessary but that the dilution effect is only on red or flaxen color and not on black.

The *Cc*ᶜʳ genotype may turn the dark chestnut horse into a sooty (not clear) dark palomino with light mane and tail. The light chestnut horse would become a clear, light palomino. The *c*ᶜʳ*c*ᶜʳ genotype would turn both the dark and light chestnut horses into what is called cremello (creamy white, near white, blue-eyed) with light mane and tail. The cremello horse is sometimes incorrectly called an albino.

If the basic color is bay, the heterozygous dilution genotype, *Cc*ᶜʳ, would dilute the red body color to buckskin. The body colors of buckskins are similar to those of palominos, but the manes, tails, and legs are black. Some buckskins also have a black dorsal stripe and some zebra striping on the legs and shoulders, although these characteristics may be due to the presence of other genes at other loci. The *c*ᶜʳ*c*ᶜʳ genotype dilutes the bay color to what is called *perlino*, which also is sometimes incorrectly called albino. The tails and manes are darker (a rusty color) than the body color and darker than those of the cremellos. Some aged gray horses also have a white body color and reddish mane and tail but can easily be distinguished by the dark pigment around the muzzle.

The *D* Locus   Adalsteinsson (1974, 1978) and Van Vleck and Davitt (1977) presented evidence for a second kind of dilution that is dominant and operates on all colors. The genotypes *DD* and *Dd* will dilute black and dark seal brown to grulla, bay and light seal brown to dun (almost certainly with a dorsal stripe), and chestnut to red dun. A dorsal (back stripe) is typically associated with the *D* gene.

Major Modifying Loci

Loci for masking, graying, roaning, and spotting have major effects that affect coat color. These loci are listed in Table 13-5.

The *W* Locus   A dominant gene, *W*, for white color, has mixed merit for breeders of white horses. The gene masks all other color patterns. Many geneticists had suspected that the homozygous genotype, *WW*, was lethal. In 1969, Pulos and Hutt confirmed this suspicion. They concluded that

Table 13-5
Loci of genes that mask or cause variation in basic color patterns

Locus and allele	Effects
W	Dominant clear white; masks all colors; brown, hazel, or blue eyes; lethal before birth when homozygous.
W^{ap}	Dominance for white blanket on rump of Appaloosas; may cause leopard pattern as homozygote with proper modifiers; allele proposed by Jones and Bogart (1971).
w	Recessive for normal color.
G	Dominant progressive gray; frequency of white hairs mingling with basic color increases with age; first coat has no white hairs; may become white when aged; GG may gray faster than Gg.
g	Recessive; no graying.
Rn	Dominant roan; mixing of white hairs with basic color; no change with age; homozygote is lethal since there have been many reports of no roans breeding true. There has been one report, however, of two true-breeding roans.
Rn^{ap}	Intermediate dominance; may cause a roaning over hips in Appaloosas; allele proposed by Jones and Bogart (1971).
rn	Recessive; no roaning.
T	Dominant for white spotting (Tobiano pattern); spotting independent of other basic color; homozygote may have more white.
t	Recessive; allows normal expression of genes at other loci.
o	Recessive for white spotting (Overo pattern).
o^{e}	Recessive for Overo spotting, but when homozygous will cause lethal white foal syndrome; allele proposed by Jones and Bogart (1971).
O	Dominant; allows normal expression of genes at other loci.
F	Dominant for mane and tail similar to body color.
f	Recessive allele; usual expression is flaxen mane and tail but only on chestnuts and sorrels.
S	Dominant silver dappling in Shetlands; strong dilution of black to dark cream or light chocolate; almost to dappled chestnut; red is only slightly diluted but black points of bay diluted to white so that appearance is similar to palomino.
s	Recessive; normal allele.
Sl	Gene for silver or varnish roan; may be at same locus as Rn; modifiers can result in slight silvering to completely white.
Sl^{ap}	Homozygous: may cause leopard pattern with proper modifiers. Heterozygous: frosty hip blanket or overall frosty roan depending on modifiers. Alleles proposed by Jones and Bogart (1971).
blo	Recessive for blotchiness; may be associated with Appaloosa patterns, allele proposed by Jones and Bogart (1971).

because no abortions or dead foals were observed, the lethal effect must occur very early in pregnancy. The result is that the apparent conception rate for matings of white horses is reduced. Matings of $Ww \times Ww$ would be expected to result in ¼ WW, ½ Ww, and ¼ ww genotypes but the WW genotype is not born, so two-thirds of the foals born are white (Ww) and one-third colored (ww), rather than the three-to-one ratio that would be expected from crosses of horses heterozygous for a dominant gene (Figure 13-3). The allele w is recessive and the ww genotype allows expression of other color loci. Dominant white horses have colored eyes (blue, brown, or hazel) and usually have small spots of skin pigment, so they are not true albinos.

Another gene in the W series has been hypothesized by Jones and Bogart (1971) as being responsible for one of the Appaloosa color patterns. The W^{ap} gene may cause the white blanket on the rump with the extent of the blanket depending on modifier genes. This gene may be similar to that proposed by Miller in his study of Appaloosa inheritance, which will be discussed later in this section.

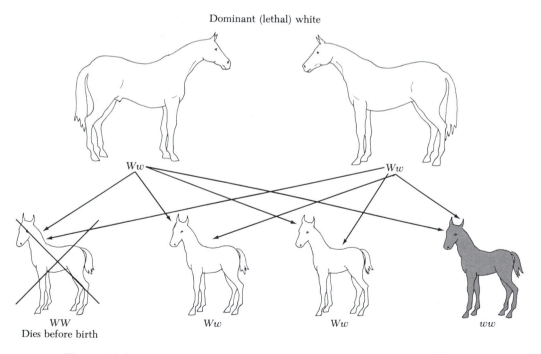

Dominant (lethal) white

Ww Ww

WW
Dies before birth

Ww Ww ww

Figure 13-3
The dominant white gene masks the effects of other color genes but is lethal when homozygous.

The *G* Locus   The gene for progressive graying is designated as *G*. The *g* gene, when homozygous, *gg*, allows regular color expression. The graying action begins after the first coat, which shows the basic color, is shed. Each succeeding coat has an increasingly larger number of white hairs mingled with hairs of the base color. Homozygous *GG* horses may exhibit graying earlier in life and to a greater degree than heterozygous (*Gg*) horses, and both may appear white when aged (Figure 13-4). Most so-called "white" horses are really gray horses whose hair has become completely white. The skin pigment will, however, be dark in places where the hair was originally colored and around the muzzle, in contrast to the dominant white horse, which has a pink muzzle. Although dappling may occur with most colors, it is especially noticeable on gray horses. Progressive graying can occur with and without dappling. Some gray horses retain a dark mane and tail for some time, whereas the mane and tail of others turn white before the body (Figure 13-5). Many gray horses also develop many very small brownish or dark spots as they age, which result in a "flea-bitten" appearance.

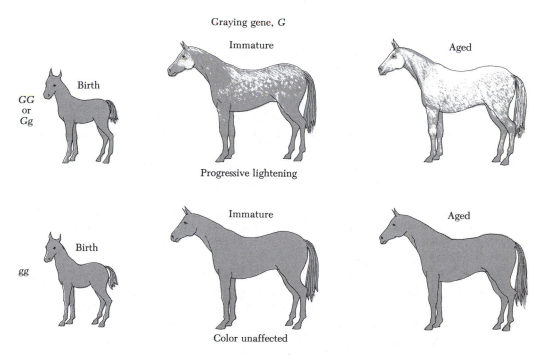

Figure 13-4
Horses carrying at least one graying gene will become progressively whiter with each shedding.

(a)   *(b)*

Figure 13-5
The manes and tails of gray horses may turn white earlier or later than the body:
(*a*), X-Star Special, a 5-year-old Arabian stallion with white mane and tail; (*b*),
Zane Grey's Silver Tip, a Missouri Fox Trotting stallion with darker mane and tail.
(Photograph (*a*) courtesy of Sue Anne Yarborough, Jasper, Arkansas; photograph
(*b*) courtesy of Ed Edwards, West Plains, Missouri.)

The *Rn* Locus   The gene for roaning, *Rn*, is easily confused with the graying
gene because it also results in white hairs mingled with the basic color hairs.
The white hairs, however, are present in the first coat and do not increase in
frequency with age, although horses carrying both the roaning and graying
genes are born with white hairs and would become progressively gray. The
appearance of roans (and grays) depends on the base color (see Figure 13-6).
A chestnut base results in a chestnut roan or rose gray. A bay when roan may
be called a red roan. A sorrel may be a strawberry roan. Actually, the most
accurate terminology is obtained by combining the base color and roan or
gray, that is, bay roan, gray on bay, and so on. When present with genes
distinguishing a color breed, the roan gene often detracts from the color
pattern. Roaning can hide some Appaloosa patterns and can make paint and
palomino patterns indistinct. Some breeders believe that the homozygote
RnRn is lethal because there have been many reports that roans never breed
true. Hintz and Van Vleck (1979), from an analysis of registrations of Belgian
horses, found a 2 : 1 ratio of roan to solid foals from roan to roan matings and
a 1 : 1 ratio from roan to solid matings. Smith (1979, personal communication) found similar results from an analysis of 1940 and 1941 registrations of
Tennessee Walking Horses of 63 matings of roan and solid stallions to mares
that had roan sires and dams. These results are expected if the homozygote,
RnRn, is lethal so that roan horses are always *Rnrn* (Figure 13-7). However,
Singleton (1969) cites a 1918 report of Babcock and Clausen concerning two

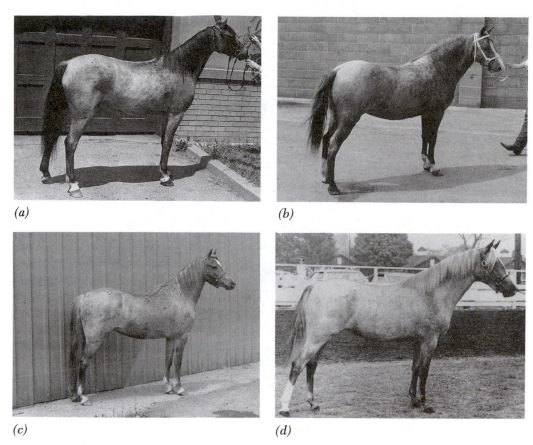

(a)　　　　　　　　　　　　(b)

(c)　　　　　　　　　　　　(d)

Figure 13-6
Roan ponies showing the characteristic solid color (lack of interspersed white hairs) around the head and neck and on the legs. Notice the difference of the intensity of roaning; (a), roan on a bay background (note black points); (b), roan on a chestnut background (note chestnut points); (c), roan on a light chestnut background; (d), roan on a chestnut background with light mane and tail (GlanNant Rhyme, a champion Welsh Pony). (Photographs (a), (b), and (c) courtesy of Harold A. Willman, Ithaca, New York. Photograph (d) by Tarrance Photos, courtesy of Mrs. Karl D. Butler, Ithaca, New York.)

stallions, a bay roan that sired 256 bay roan foals and a red roan that sired 230 bay roan and 24 black roan foals. These stallions were also described by Wentworth in 1913. Since both were listed as red roan stallions and because there is always the chance of confusing roans with grays, it is possible that they were actually gray on a bay background and, if homozygous for G, would breed true. Many modifier genes probably affect the percentage of white hairs in the roan because the variation in the amount of white is great.

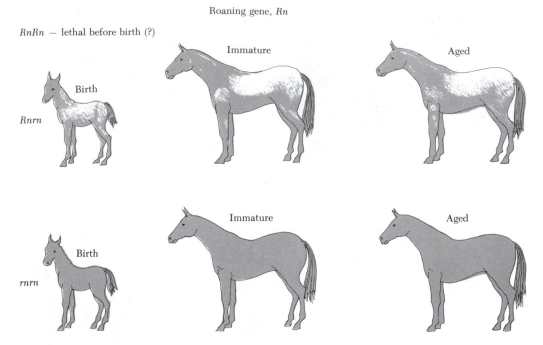

Roaning gene, *Rn*

RnRn — lethal before birth (?)

Immature   Aged

Birth

Rnrn

Immature   Aged

Birth

rnrn

Figure 13-7
The roan horse is born with white hairs interspersed with colored hairs. The
amount of white does not change with age.

Most roans are much darker around the head, neck, and legs than on the
body (Figure 13-6), and this characteristic helps distinguish roan from gray.
The amount of roan appears to vary in different seasons because of the
length of hair and the amount of undercoat that shows.

Jones and Bogart (1971) have speculated also that in the series of alleles
at the *Rn* locus there is an intermediate dominant gene *Rn*ᵃᵖ that may cause
roaning on the hips of Appaloosas.

The *S* Locus   Shetlands have a dominant gene that is responsible for silver
dappling — black is diluted to a dark cream or light chocolate, almost to a
dappled chestnut; the mane and tail are white or nearly white (Figure 13-8).
Although the red body color of the bay is not diluted much, if at all, by the *S*
gene, the black points are diluted nearly to white. Thus the *S* gene together
with the genotype for bay results in the anomaly, a bay horse without black
mane and tail. Usually enough black hairs are mixed with the white hairs to
suggest that the silver dapple gene is responsible, especially if the color is on
a pony. The effect of *S* on a chestnut genotype may not be noticeable or may

lighten the mane and tail. Homozygotes, *SS*, may be diluted to a greater extent than heterozygotes. The combination of *S* and the graying gene may produce a white or gray-white pony that turns white by 1 or 2 years of age. The foal coat may even contain white hairs. Such white ponies usually will show pigment on the nose or around the eyes. This color pattern also is found in Australian pony and stock horses and is called taffy rather than silver dapple.

The *Sl* Locus   There have been some reports of a progressive silvering gene, but the reports do not clearly distinguish between its effects and those of *G* and *Rn* genes. The gene is named *Sl* and results in a silver or varnish roan that varies from slight silvering to complete white. An allele at this locus, *Sl*^ap^, is thought by Jones and Bogart to cause, when homozygous, a leopard-

Figure 13-8
Hilltop's Little Bummer, a silver dappled "chestnut" pony. Although the body is dappled, there is a "mask" on the face and the mane and tail are white. Although the color may sometimes be called chestnut, the genotype is for black but is modified by the silver dappling gene. (Photograph by Frank's Photo House, courtesy of Jim and Loretta Murphy, Huron, South Dakota.)

spotted Appaloosa if the proper modifiers are present. The heterozygous genotype $Sl^{ap}sl$ may give a frosty hip blanket or overall frosty roan depending on modifiers present, according to Jones and Bogart (1971).

The T Locus   At least two forms of white spotting are thought to exist. One is dominant and the other is recessive. Spotting is independent of other color patterns and presumably may occur on any color except, of course, white. The dominant gene for white spotting is designated *T* for the tobiano pattern (Figure 13-9). The tobiano pattern has been described as similar to that obtained by pouring white paint over a solid-color horse (Figure 13-10). The head is marked similarly to that of a solid-color horse. Horses that are *tt* will not be spotted unless spotting is caused by genes at another locus. The homozygote *TT* may have more white than the heterozygote. The amount of white also is probably determined by modifier genes.

The O Locus   The overo pattern of recessive spotting is due to the *oo* genotype (Figure 13-11). The overo pattern is in some respects the complement of the tobiano pattern. A horse exhibiting this pattern is likely to have a colored back and underside; white spotting is mostly confined to the middle of the body as shown in Figure 13-12. The dominant allele, *O*, codes for normal coloring. Spotting in horses that carry *O* will be due to genes at another locus. As these spotting genes are at two different loci, they may in combination account for much variation in spotting patterns, but variation in amount of spotting must also be due to many modifier genes. Because two overo parents occasionally have solid-colored foals, there is some reason to doubt if the overo pattern is due to genes at a single locus. A number of modifier genes, however, may create the variation in amount of white spotting. With no modifiers for white transmitted along with the overo allele, *o*, a solid foal may be genetically *oo* but not show the overo pattern. The only test would be to mate to another overo with many modifiers.

Tobiano spotting

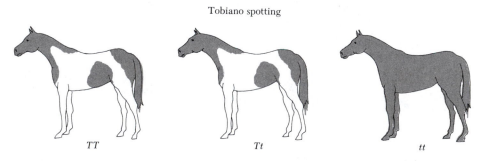

TT   Tt   tt

Figure 13-9
The tobiano pattern is apparently dominant and is characterized by white over the back and up the legs, and by a normally marked head.

Figure 13-10
Tobiano pattern on Cherokee War
Chief, a Paint Horse champion.
(Photograph by Marge Spence,
courtesy of the American Paint
Horse Association, Fort Worth,
Texas.)

Overo spotting

OO   Oo   oo

Figure 13-11
The overo pattern apparently requires a pair of recessive genes and is
characterized by color over the back and on the legs and by much white on the head.

Figure 13-12
Overo pattern on Flash Thru Bars, an International Pinto Horse Halter Champion. (Photograph by Marge Spence, courtesy of Dorothy Lawrence, Sperry, Oklahoma.)

Two other types of white spotting have been postulated by Sponenberg and Beaver (1983); splashed white and sabino. The splashed white pattern (Figure 13-13) is said to be recessive and somewhat resembles the overo pattern in that white does not cover the back except with extreme splashing. The legs, however, are usually white without streaking and the edges between the colored and white areas are clearly defined rather than irregular. The amount of white varies but always seems to rise from the bottom side of

Moderate splashed white   Medium splashed white   Extensive splashed white

Figure 13-13
Splashed white is thought to be due to a dominant gene with the amount of white determined by many modifier genes. Note the extensive white on the head as compared to the Tobiano pattern.

the horse. Minimum white would be a small splash on the belly, often accompanied by a considerable amount of white on the legs. The gene for splashed white may be responsible for the crop-out phenomenon (a small splash near the underline) seen infrequently in matings of solid-colored horses. Splashed white with a large amount of white might be confused with the tobiano pattern if the white goes over the back. In that case, however, the head is likely to be nearly all white except for the ears.

The sabino pattern also seems to arise from the underline and front and rear flanks but is quite irregular and can be confused with extensive, although blotchy, roaning on the legs and lower body. Sponenberg and Beaver (1983) reference Wiersema as having determined that the controlling gene is dominant.

Locus for Flaxen Mane and Tail   There is some confusion as to the reason for the whitish mane and tail of the Palomino. Some speculate that a recessive gene called flaxen is responsible. Others believe that the optical density of the basic brown pigment is diluted so much in the mane and tail that it appears to be lighter than the body color. The flaxen mane and tail also appears on chestnut horses, particularly Belgians and Shetlands. Figure 13-14 shows chestnut horses with and without the light mane and tail. Wentworth (1913) presented evidence that indicates that the gene for flaxen mane and tail is recessive, at least on chestnut backgrounds. Very likely it has no effect on a genetically black horse.

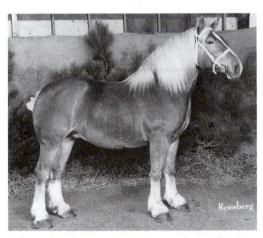
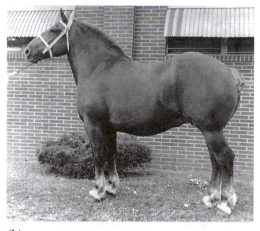

(a)   (b)

Figure 13-14
Two Belgian mares: (a), Trixie du Marais, with white mane and tail; (b), Sunny Lane Nanette, with chestnut mane and tail. (Photograph (a) by Remsburg, courtesy of Donald J. Wack, Zelienople, Pennsylvania; photograph (b) courtesy of Leo J. Fox, David City, Nebraska.)

The *blo* Locus   Blotchiness in the color pattern may be the result of the homozygous genotype *bloblo* for blotchiness (Jones and Bogart, 1971). This gene may act as a major modifier of the Appaloosa patterns.

The Mottled Locus   The mottled or parti-colored skin, which appears particularly on the genitalia and muzzle of the Appaloosa, may be due to a dominant gene *M* (Figure 13-15).

Loci for Markings   Many genes may be responsible for white stockings, stars on the forehead, the blaze face, and so on. The gene for white snip on the nose is thought to be recessive by some geneticists and dominant by others. Similarly, it is not clear whether the blaze or white stripe on the front of the face or the white forehead star are due to recessive or dominant genes. Karner (1980) in an unpublished study has found that the amount of white markings appears to be inherited as a quantitative trait with a high degree of relationship between the amount of white on the legs and on the head. Some

Figure 13-15
Co-Regent, an Appaloosa stallion, showing the effects of the mottling gene around the face. (Photograph by DiGinio Photography, courtesy of Lynn and Jack Nankivil, Sahaptin Farm, Winona, Minnesota.)

interesting sidelights are that white stockings are more common on the rear legs than on the front legs, and that the amount of white is usually greater on the rear legs than on the front legs.

Appaloosa Loci   The inheritance of the many color patterns of the Appaloosa is not completely clear. Miller, in an undated report, studied 9,955 parents and progeny of Appaloosa breeding, which led him to propose the following model for the expression of the white blanket over the rump and croup and for the expression of small, dark spots in the area of the blanket: A dominant gene Ap is necessary for expression of the Appaloosa blanket and dark spots. This gene seems to correspond to the W^{ap} gene proposed by Jones and Bogart (1971). Another gene, wb, is necessary for the blanket and still another gene, sp, is necessary for the dark spots. The gene effects of these two loci are sex influenced, so that the blanket and dark spots can occur in homozygous and heterozygous males but only in homozygous females, as shown in Table 13-6. Modifiers for both the blanket and spot characteristics determine the extent of the blanket and spots. Enough blanket modifiers may result in what is essentially a white horse. Spots can occur without the blanket, and the blanket can occur without spots. (Figures 13-16 and 13-17).

Crew and Buchanan-Smith (1930) proposed that leopard spotting consisting of dark spots on white background (Figure 13-18) is the result of a

Table 13-6
A genetic model for the inheritance of the white blanket and dark spots characteristic of Appaloosas

Genotypes	In the presence of $Ap(ApAp$ or $Apap)$	
	Male	Female
	Blanket characteristic	
$WbWb$	Solid	Solid
$Wbwb$	Blanket	Solid
$wbwb$	Blanket	Blanket
	Spotting characteristic	
$SpSp$	No dark spots	No dark spots
$Spsp$	Dark spots	No dark spots
$spsp$	Dark spots	Dark spots

In the absence of $Ap(apap)$

$apap$ results in no blanket and no dark spots, although Wb and Sp may be present.

SOURCE: Adapted from Miller, R. W., *Appaloosa Coat Color Inheritance.*

 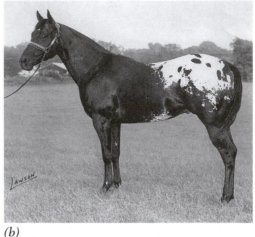

(a)   (b)

Figure 13-16
Appaloosas showing a white blanket without spots (a) and with spots (b): *a*, Lace Hankie, an Appaloosa mare; *b*, Prince Jet Band, an Appaloosa stallion. (Photograph (*a*) courtesy of Lon and Val Whitson, Livingston, California; photograph (*b*) courtesy of Joan McGloon, Woodbury, Connecticut.)

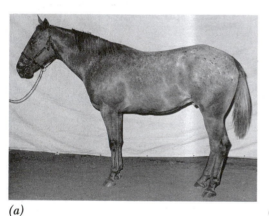 

(a)   (b)

(c)

Figure 13-17
Appaloosas showing dark spots without blanket on a chestnut background (*a*), dark spots on a light background (*b*), and dark and white spots on a chestnut background (*c*): (*a*), Joker's Red Baron, an Appaloosa gelding; (*b*), Red Hill Coco Bars, an Appaloosa mare; (*c*), Jocay's Hi Johnny, an Appaloosa stallion. (Photograph (*a*) by Norman R. Olson, courtesy of Marilyn S. Olson, Libuse, Louisiana; photograph (*b*) by CORA, courtesy of Cora Morris, Marshall, Arkansas; photograph (*c*) courtesy of Carolyn Ann Young, Jocay Stables, Indianapolis, Indiana.)

THE COLORS OF THE HORSE

a. **Black.** Black all over except white markings; no brown hairs on muzzle or flanks. Rapid Chick, Quarter Horse stallion. Photograph by Morgan Studio, courtesy of John Bishop, Carthage, Missouri.

b. **Grulla.** Dilution of black or seal brown hair, not a mixture of black and white hair; black points and usually dorsal stripe. Poco Frosty Blue, Quarter Horse stallion. Photograph courtesy of Emerson Davitt, Quarter Dale Farm, Rensselaer, New York.

c. **Dark Bay.** Dark brown with red on body, black on points—mane, tail, legs. Pute Cee Bonanza. Quarter Horse stallion. Photograph courtesy of W. M. Daniel, Ellijay, Georgia.

d. **Dun (buckskin †).** Dilution of dark bay or seal brown; sooty yellow to yellow-red body, black points and usually dorsal stripe. Apple Cash, American Buckskin stallion. Owned by K. G. Ormiston, Rialto, California. Photograph by Bill McNabb, Jr., courtesy of American Buckskin Registry Association, Inc., Anderson, California.

e. **Light Bay.** Light red to sandy red body, black points. Flaigor, Arabian stallion. Photograph by Polly Knoll, courtesy of Nodoroma Farms, Ocala, Florida.

f. **Buckskin (dun †).** Dilution of ligher shades of bay; clear light yellow to dark cream body, black points, often with dorsal stripe. Hollywood Buck, Quarter Horse. Photograph by Carol Dickinson, courtesy of owner Steve Fleming, Topeka, Kansas.

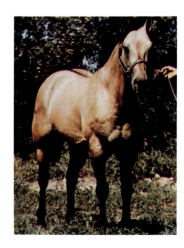

† Buckskin registries refer to darker shades as buckskin, but scientific journals and dictionaries refer to buckskin as a light shade of dun.

g. **Seal Brown.** Often mistaken for black but with brown hairs on flanks or muzzle, black points. *Ribereno, Thoroughbred stallion. Photograph courtesy of W. T. Pascoe III, Newport Beach, California.

h. **Light Seal Brown.** Similar to dark bay but without reddish cast; brown areas noticeable, black points. Deck Jack. Quarter Horse stallion. Photograph by Darol Dickinson, courtesy of Robert L. Kowalewski, East Aurora, New York.

i. **Gray on Bay.** Progressive graying may occur on any color. Heritage Desiree, Arabian mare. Photograph courtesy of Heritage Hills Arabians, Lake Geneva, Wisconsin.

j. **Overo pattern on Bay.** White spotting may occur on any color. Calico Cactus, Pinto stallion. Photograph courtesy of Bob and Holly Welch, Watertown, Minnesota.

k. **Roan on Dark Chestnut.** Roaning is constant from birth, usually with little roaning on legs and head, and may occur on any color. Trotting pony. Photograph by Richard Dobec, Ithaca, New York.

l. **Appaloosa pattern on Sorrel.** Appaloosa patterns may occur with any color. Dial Bright Too, Appaloosa stallion. Photograph by Darol Dickinson, courtesy of Elbert County Land & Cattle Company, Elbert, Colorado.

m. **Dark Red Chestnut.** Similar to dark bay, but mane, tail, and legs are not black. Notice lighter chestnut on lower legs. Bar Bob Panzer, Quarter Horse stallion. Photograph courtesy of Dean Thomas, Dean Thomas Quarter Horses, Utopia, Texas.

n. **Dark Palomino.** Sooty dilution of dark chestnut, similar to dun but notice near white or flaxen mane and tail and no black on legs. Usually does not have dorsal stripe. Monte Que, Jr., Quarter Horse stallion. Photograph by Eisler, courtesy of Mr. and Mrs. Herman R. Ewell, East Earl, Pennsylvania.

o. **Red Chestnut.** Sometimes will be lighter than shown here. Notice chestnut points. Mane and tail may also be lighter. Applevale Donalect, Morgan stallion owned by Mr. and Mrs. Lawrence A. Appley, Hamilton, New York. Photograph by A. C. Drowne, American Morgan Horse Association, Inc., Hamilton, New York.

p. **Palomino.** Clear dilution of chestnut. Some shades of palomino may be even lighter than shown here. Similar to buckskin except has near white mane and tail and has no black on legs. El Taco Dorado, double-registered Palomino and Quarter Horse stallion. Photograph courtesy of Lyman A. Myers, Vinita, Oklahoma.

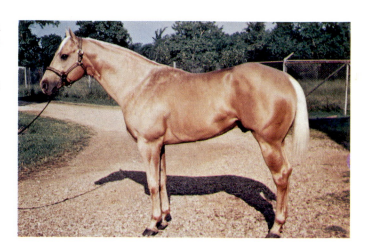

q. **Red Dun.** Basically chestnut. May be dilution of brown chestnut. Light body, darker red legs, often with barring. Usually has dark red dorsal stripe, often with darker mane and tail. Connie Coed, Quarter Horse mare. Photograph courtesy of Mary Aspegren, Mundelein, Illinois.

r. **Light Red Dun.** Body color similar to palomino but has darker red legs and often darker red mane and tail and dorsal stripe. This horse shows leg barring. Raisin Sand, Quarter Horse stallion. Photograph courtesy of James C. O'Daniel, Plain Dealing, Louisiana.

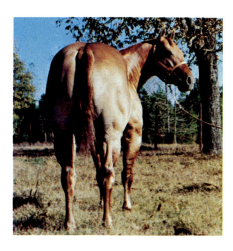

s. **"Black" Chestnut.** May be mistaken for black but will show brown hairs, especially on muzzle, flanks, and lower legs (in contrast to dark seal brown). Is actually dark liver chestnut. Vigilmarch, Morgan stallion. Photograph by Jacobson, courtesy of Herbert V. Kohler, Kohler, Wisconsin.

t. **Liver Chestnut.** Very dark brown. Brown chestnut is very evident on lower legs, flanks, and muzzle. Bennfield's Ace, Morgan stallion. Photograph courtesy of A. C. Drowne, American Morgan Horse Association, Inc., Hamilton, New York.

u. **Sorrel.** Difficult to distinguish from light chestnuts but has more yellowish body showing little or no red. Mane and tail are often same as body, but with flaxen mane and tail, may appear similar to dark palomino. Skip's Pamela, Quarter Horse mare. Photograph courtesy of W. G. Brown, Jr., Alexander, Arkansas.

v. **Dark Chestnut with Flaxen Mane.** Flaxen mane and tail may appear on all except the darkest chestnuts—even red duns—but never on blacks, bays, seal browns, duns, or buckskins. Narindi #28095, Arabian mare. Photograph courtesy of Sue Anna Yarborough, Valley Y Ranch, Jasper, Arkansas.

w. **Perlino.** Double dilution of bay. Off-white or pearl body with rust color on tips of mane and tail and sometimes on lower legs. Also called type B albino—not a true albino. Lady Dawn, American Albino mare. Owned by Jackie Spencer, Marysville, California. Photograph by Stoney Hochstrat, courtesy of Mrs. Lowell A. Boaz, Marysville, California.

x. **Cremello.** Double dilution of chestnut. Off-white or cream body with even lighter mane and tail. Also called Type A albino—not a true albino. This is a foal from the mating of two Palominos; notice white blaze. Some Palominos are this light but will become darker with age. Photograph courtesy of James E. Dinger, University of Maryland, College Park, Maryland.

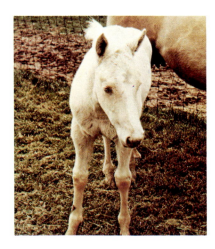

(a)

(b)

(c)

(d)

Figure 13-18
Appaloosa stallions showing the variation in size of spots of the leopard spotting pattern. The size and distribution of the spots depends on modifier genes. (a), Prince's Joker; (b), Joker's John E.; (c), D-R Joker's Playboy; (d), Oxburn's Snafu. (Photograph (a) courtesy of Richard Herrmann, Theilman, Minnesota; photograph (b) courtesy of Mr. and Mrs. E. D. Pollard, Caldwell, Idaho; photograph (c) courtesy of Linda Roberts, South Lyon, Michigan; photograph (d) courtesy of Lynn and Jack Nankivil, Sahaptin Farm, Winona, Minnesota.)

dominant gene. The leopard mule shown in Figure 2-54 is evidence to support this view since only one gene for spotting could have been inherited from the Appaloosa mother.

Sponenberg (1983), who studied leopard spotting in the Austrian Noriker breed, concluded that leopard spotting in that breed is due to an incompletely dominant gene. A pair of the genes results in a few-spotted (nearly white with relatively small and few spots) leopard whereas the heterozygote would have many large spots on a white background. The two types may not always be distinguished from each other because of modifier effects. The typical Appaloosa blanket over the rump, however, is rarely found in the Noriker breed. Norikers usually are either solids or leopards and do not show the typical mottled skin or striped hooves of North American Appaloosas.

Jones and Bogart (1971), however, suggest that leopard spotting can result for the $W^{ap}W^{ap}$ genotype, or from $W^{ap}w^{ap}$ or from $Sl^{ap}Sl^{ap}$ with enough modifiers. According to their proposal, $W^{ap}w^{ap}$ would usually result in the blanket pattern, as would $W^{ap}W^{ap}$ with few modifiers, and $Sl^{ap}Sl^{ap}$ would result in silver roans or horses with frosty blankets over their hips that might, with the proper modifiers, cover virtually the entire body. Appaloosa roaning, mainly over the hips, is thought by Jones and Bogart to be due to a dominant gene Rn^{ap}. Modifier genes cause the expression to vary, but the Rn^{ap} gene is not associated with leopard spotting.

Obviously, many questions about color inheritance in horses are unanswered. Many modifier genes must be responsible for differences in shading of the basic colors. Horse breeders apparently will continue to be surprised by some of the colors of foals that result from some of their matings. Perhaps color is not very important to breeders of working horses, but it is to color-horse breeders, and much more knowledge is needed.

The basic rules of genetics in Chapter 12 can be used to calculate the frequencies of possible genotypes that will result from particular matings. These genotypes can then be used, together with Figure 13-1, the color insert, and the information in this section to predict the frequencies of various colors and color patterns.

13.2   Blood Factors and Genetic Tests for Disputed Parentage

Differences in at least 32 physiological characteristics of blood are known to be the result of alleles at an equal number of loci. More are discovered each year. Eight loci control the red-blood-cell antigen systems listed in Table 13-7. The details of the laboratory procedures for determining these blood

Table 13-7
Genetic loci that control
red-blood-cell antigen systems

Antigen locus symbol	Minimum number of alleles
A	11
C	3
D	15
K	2
P	9
Q	5
T	4
U	2

factors are complicated but have been elucidated by many researchers, including Stormont of the University of California at Davis, and Sandberg at the Agricultural University of Sweden and their colleagues. Blood protein systems controlled by other loci are shown in Table 13-8. The phenotypes of most of these systems are determined by electrophoresis. Many new alleles and systems can be expected to become known as blood analysis techniques become more refined.

In the mid-1970s immunogeneticists discovered that antigens occurring on lymphocytes are controlled genetically. Genes that control graft rejection in animals are known as histocompatibility genes. The controlling loci are closely linked in most species in what is called the major histocompatibility complex (MHC). The products of these loci are found on the surface of lymphocytes and thus in the horse, are referred to as Equine Lymphocyte Antigens (ELA). Although in many species several loci have been characterized, in the ELA system only two systems have been reported: *ELA-A* and *ELA-D*, both with several alleles. Two other loci, *ELY-1* and *ELY-2*, that code for lymphocyte antigens are not linked to the ELA system (Table 13-9). Because these systems are carried on lymphocytes it is possible that particular lymphocyte antigens may be associated with resistance to disease. For example, an allele at an histocompatibility locus is associated with resistance to Marek's disease in chickens. About 90 percent of horses with heaves or founder have been found to carry an antigen at the *ELY-1* locus that has a frequency of only about 50 percent (Lazary et al., 1982). Mottironi et al. (1981), however, did not find any association between ELA antigens and a genetic disease called combined immunodeficiency (CID) in Arabian horses that affects the immune system.

Table 13-8
Genetic loci controlling enzyme and protein markers in the blood of horses

Genetic system	Locus symbol	Minimum number of alleles
Albumin	*Al*	3
Transferrin	*Tf*	9
Xk protein	*Xk*	3
Carboxylesterase	*Es*	5
Pre-albumin	*Pr*	10
Vitamin D binding protein	*Gc*	2
Carbonic anhydrase	*Ca*	3
Catalase	*Cat*	2
Phosphoglucomutase	*PGM*	3
Phosphohexose isomerase	*PHI*	2
6 phosphogluconate dehydrogenase	*PGD*	3
Acid phosphatase	*Ap*	2
Hemoglobin α-chain	*Hb*	2
Cholinesterase	*Ch*	4
Autolytic factor	*Lf*	2
Mitochondrial glutamate oxaloacetate transaminase	GOT_M	2
Soluble malic enzyme	*ME1*	2
Glucose-6-phosphate dehydrogenase	*G6PD*	2
Hypoxanthine-guanine phosphoribosyl transferase	*HGPRT*	2
Phosphoglycerate kinase	*PGK*	2

Table 13-9
Comparison of *ELA-A, ELA-D, ELY-1* and *ELY-2* lymphocyte antigen systems

	ELA-A	*ELA-D*	*ELY-1*	*ELY-2*
Minimum number of alleles	15	6	2	2
Linkage	ELA-D	ELA-A	?	?
	Blood group A	Blood group A		
Tissue Expression:				
lymphocytes	Yes	Yes	Yes	Yes
platelets	Yes	No	No	Yes
red blood cells	No	No	No	Yes

SOURCE: Adapted from summary in Antczak, D. 1984a.

Both red-blood-cell antigens and other blood markers can be used for tests of disputed paternity, partial verification of parentage, and as a permanent record of identification.

At most loci that control the red-blood-cell antigens, the alleles are codominant except for one recessive allele. Most alleles at loci for most of the blood proteins and ELA antigens are codominant. If the blood type of the mother is unknown, only codominant systems are useful in determining possible identities of the father because of the difficulty of determining whether a horse with a dominant phenotype is homozygous or heterozygous.

Within a breed, the number of segregating alleles at each locus is also important. If nearly all horses of a breed have a particular allele at a locus, then that locus will not be very useful in parentage tests. Similarly, the larger the number of blood systems that can be used, the greater the chance of excluding an incorrect parent.

Some frequencies of different alleles for the red-blood-cell antigen loci of Shetlands and Thoroughbreds, as determined by Stormont and Suzuki (1964), are given in Table 13-10. This table shows that, especially for Thoroughbreds, the frequency of the recessive allele for most of the blood antigen systems is relatively high except for the Q system, where the problem is that several genotypes may have the same phenotype. These systems are essentially reduced to a dominant-recessive basis because the frequency of codominant genotypes is relatively low.

The codominant systems, exemplified by the albumin and transferrin loci, are generally more efficient in parentage tests because both genes are identified in the offspring as well as the disputed parent. Frequencies in this system for some Shetlands, Thoroughbreds, and Arabians are shown in Table 13-11. Table 13-11 also shows that a marked similarity exists between Arabians and Thoroughbreds with respect to the transferrin frequencies, which is not surprising because many of the Thoroughbred ancestors were Arabians or were closely related to Arabians. In fact, frequencies of various genes are used to indicate pathways and similarities in the development of breeds.

In the survey taken in the early 1960s (see Table 13-12), 6 alleles were identified in the esterase system, but only two alleles were found in Arabians and 3 were found in Thoroughbreds.

Stormont has estimated the efficiency of excluding the wrong stallion from the correct stallion for the eight blood antigen systems and the albumen and transferrin systems. The efficiency depends, as indicated in Table 13-13, on the number of alleles, their frequencies, and their dominance relationships. The efficiency values for Shetlands and Thoroughbreds are given in Table 13-13. The greater amount of heterozygosity and greater number of alleles in the Shetlands results in more efficient exclusion for them than for Thoroughbreds.

Table 13-10
Frequencies of red-blood-cell alleles in Shetland ponies and Thoroughbreds

Locus	Allele	Frequency in	
		Shetland	Thoroughbreds
A	a^{A_1}	.311	.705
	$a^{A'}$	.285	.029
	a^H	.036	.004
	$a^{A'H}$	.060	.000
	a	.308	.262
C	c^C	.652	.732
D	d^D	.139	.000
	d^J	.122	.150
	d	.739	.850
K	k^K	.180	.064
P	p^P	.342	.206
	$p^{P'}$	.048	.091
	p	.610	.703
Q	q^Q	.152	.508
	q^R	.387	.000
	q^S	.010	.104
	q^{QR}	.131	.076
	q^{RS}	.189	.312
	q	.131	.000
T	t^T	.450	.659
U	u^U	.317	.148

SOURCE: Stormont and Suzuki. 1964.

The *ELA-A* system is especially powerful for parentage testing because of the large number of alleles and the lack of a high frequency for any allele (Table 13-14). The *ELA-A* system together with the red-blood-cell and serum protein systems provides a very efficient procedure for parentage exclusion.

A difficulty with studying blood factors is that each laboratory may have a different name for each locus and each allele at that locus. Since 1974 researchers have agreed to standardize their reagents and naming systems. The names of known alleles were standardized in 1974. Thus, the alleles named in Tables 13-10, 13-11, 13-12 may have different names today than when those studies were made.

Table 13-11
Frequencies of albumen and transferrin alleles in American Shetlands, American Thoroughbreds, South African Arabians, and South African Thoroughbreds

Locus	Allele	American Shetlands ($n = 273$)	American Thoroughbred ($n = 150$)	South African Arabians ($n = 45$)	South African Thoroughbreds ($n = 54$)
Albumen	Alb^A	.387	.214	.620	.278
	Alb^B	.613	.786	.380	.722
Transferrin	Tf^D	.172	.267	.300	.167
	$Tf^{F°}$	.460	.563	.477	.648
	Tf^H	.026	.027	.056	.009
	Tf^M	.031	0	0	0
	Tf^O	.108	.090	.167	.046
	Tf^R	.203	.053	0	.130

°The frequency of Tf^F would be the sum of the frequencies of alleles for subtypes Tf^{F1} and Tf^{F2}.

SOURCE: Osterhoff, Schmid, and Ward-Cox. 1970; and Stormont and Suzuki. 1964.

Table 13-12
Frequencies of esterase alleles in Arabians and Thoroughbreds

Breed	Allele					
	Es^F	Es^1	Es^S	Es^{X1}	Es^{X2}	Es^0
Arabian	0	0	.987	.022	0	0
Thoroughbred	.037	0	.926	0	.037	0

SOURCE: Osterhoff, Schmid, and Ward-Cox. 1970.

Exclusion Principle

The basic rule in parentage tests is that an animal cannot be proven to be a parent of another but can only be proven not to be the parent. There are two rules to keep in mind in dealing with parentage tests. First, a blood-group factor cannot be present in a foal unless it is present in at least one parent.

Table 13-13
Expected frequencies of excluding the incorrect stallion when the true sire and another sire are possible fathers

Locus	Fraction of expected exclusions	
	Shetlands	Thoroughbreds
Blood antigen locus		
A	.220	.038
C	.010	.004
D	.170	.078
K	.081	.049
P	.118	.165
Q	.079	.030
T	.041	.009
U	.069	.078
All above combined	.572	.378
Blood protein locus		
Alb	.181	.138
Tf	.366	.291
Tf and Alb systems combined	.481	.389
All systems combined	.778	.620

SOURCE: Stormont, Suzuki, and Rendel. 1965.

Table 13-14
Estimated frequencies of *ELA-A* antigens

ELA-A allele	Thoroughbreds (60)	Standardbreds (67)
1	.00	.25
2	.16	.01
3	.18	.05
4	.00	.11
5	.18	.07
6	.01	.06
7	.01	.05
8	.00	.10
9	.19	.00
10	.04	.25
Blank[a]	.23	.05

[a]Would include newly defined alleles 11 to 15 plus any not yet defined.
SOURCE: Adapted from Bailey, E. and P. J. Henney. 1984.

Second, if a parent is homozygous for a particular blood-group gene, the foal also must have that allele.

Two examples will help to explain how parentage tests work. Table 13-15 gives the genetic information on a foal and its dam as well as on the supposed sire in this first example. Each locus is examined to see if the results are consistent with genetic principles. At the *A* and *C* loci, the genotype of the disputed sire is in agreement with that of the foal and its dam, but at the *D* locus the dam must contribute either D^d or D^- because her genotype is not completely known. (An allele indicated with a negative superscript is recessive to all others). The foal, however, has D^c and D^d, so that D^d must have come from the dam. Then the D^c gene must come from the foal's sire. The disputed sire does not possess the D^c allele so he is excluded

Table 13-15
Phenotypes of foal, dam, and disputed sire for an example of parentage testing

Horse	System							
	A	*C*	*D*	*K*	*P*	*Q*	Alb	Tf
Foal	A^a	C^a	$D^c D^d$	K^a	P^a	Q^a	B	D
Dam	$A^a A^b$	C^a	D^d	—	—	—	A	DH
Disputed sire	A^a	C^a	D^d	K^a	P^a	Q^a	A	D

as the sire. The albumen locus provides further proof in exactly the same way.

A more difficult problem in disputed parentage occurs when the dam's phenotype is not known. An example that concerns possible sires of a foal is shown in Table 13-16. The presence of recessive genes at loci *A, C, D,* and *P,* as well as the lack of any codominant phenotypes in the foal or disputed sires, makes it impossible to exclude either sire as a possible parent with respect to those loci. The results at the *K* locus are also in agreement for either sire. From inspection of what is known at the *Q* locus, it can be seen that if the dam contributed a Q^b allele, sire II would be excluded and if the dam contributed a Q^c allele, sire I would be excluded. Information on the dam is, however, not known. There is no disagreement for either stallion at the albumen locus. At the codominant transferrin locus, however, there are no genes in common between the foal and sire II, and that sire is, therefore, excluded. Sire I is not excluded and, thus, may be the sire of the foal.

Parentage tests are obviously useful for ensuring proper identification for registered breeds and especially for stallions used artificially — particularly if the use of frozen semen and artificial insemination becomes more practical and widespread. For example, all dairy bulls used artificially in the United States are required to be blood-typed for several different blood systems. Since 1970 in Sweden all foals of the North-Swedish trotter and Swedish trotter breeds are pedigree-checked by blood typing before registration is allowed.

Gene Linkages

Chromosome maps for domestic species are not well developed as compared to those for experimental animals such as mice, rats, and chickens. Nevertheless, chromosome mapping (establishing which loci share chromosomes) in the horse has been spurred by typing of blood factors. Genes at loci that are linked on the same chromosome tend to be inherited together — the closer

Table 13-16
Phenotypes of a foal and two possible sires for an example of parentage testing

Horse	System							
	A	*C*	*D*	*K*	*P*	*Q*	Alb	Tf
Foal	A^a	—	D^c	—	P^b	Q^bQ^c	A	DH
Sire I	A^b	C^a	D^d	—	P^b	Q^c	A	DF
Sire II	A^a	C^a	D^c	—	P^a	Q^b	A	FO

the linkage the less likely that recombination will occur. Linkages and approximate map distances are shown in Figure 13-19 for 17 loci. Four linkage groups corresponding to chromosome segments less than 50 map units in length have been identified for 14 loci. In addition, three loci are known on the X chromosome. The locus for hemophilia A is also likely to be on the X chromosome as in most species. (Although only a few cases of hemophilia have been reported in horses, all have been male, which supports the X-chromosome location.)

Of special interest are the linkages of three loci controlling coat color patterns—roan, tobiano, and chestnut. The roan and tobiano loci are very tightly linked and the roan and chestnut loci are tightly enough linked that if *Rn* and *e* genes were on the same chromosome then they would stay together in 9 out of 10 germ cells. In fact, Sponenberg et al. (1984) reported that a bay Belgian stallion (*ERn/ern*) when mated to chestnut mares (*ern/ern*) produced only one chestnut roan foal (*eRn/ern*) and one bay foal (*Ern/ern*) out of 57 matings, indicating an even closer linkage. The other foals expressed the stallion's *ERn* chromosome (30 bay roans) and the stallion's *ern* chromosome (25 chestnut, nonroan).

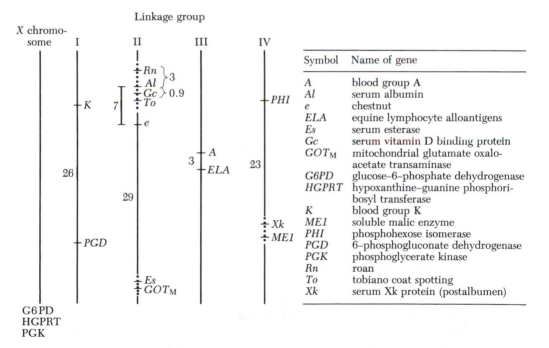

Figure 13-19
Approximate gene map of the horse. A dotted line indicates that the gene order has not been established. Loci on X chromosome have not been ordered. From Andersson, L. 1983.

Pleiotropy is when a gene affects more than one characteristic. For example, the *ELY-1* allele may be associated with a higher than expected frequency of heaves. Similarly the *Aa* and *Qa* antigens appear to be associated (perhaps directly) with neonatal isoerthrolysis. That disease and others with a possible genetic basis are discussed in the next section.

13.3   Simply Inherited Lethals

According to Britton (1962) as many as 20 percent of all foals are born with some form of abnormality, although Rossdale (1972) has estimated a lower frequency of four percent. Some of these are severely handicapped or are born dead. Many other conditions may result in abortion or resorption of the fetus early in gestation, resulting in a lowered foaling rate. Most of these abnormalities either are influenced by the joint action of many genes or are nongenetically controlled accidents in development. Some of the more common lethals and abnormalities will now be discussed, and an indication will be given of whether they may be simply inherited. Trommershausen-Smith (1980) in an excellent review of genetics and disease in the horse has pointed out that evidence for genetic causes of many defects thought to be genetic is not convincing. Any defect that could, no matter how rare, not be explained in any other way for many years was labeled as genetic without adequate pedigree study and especially without experimental matings. Because most congenital defects of the horse are rare, because horses are not as numerous as other domestic animals, and because the horse is expensive and has a long generation interval, extensive genetic studies of the horse are infrequent. Nevertheless, several defects appear to be genetically determined or are associated with a genetically determined characteristic. The first 11 of the following defects are in this category, then 13 defects that may have or have been thought in the past to have a genetic basis are listed.

Combined Immunodeficiency (CID)

Deficiency of B- and T-lymphocytes has been found in Arabian foals, all of which have died by 5 months of age because of lack of resistance to disease. Many die specifically as a result of adenoviral infections of the respiratory system. The evidence is that the genetic basis for the deficiency is an autosomal recessive gene: (1) only Arabian foals have been reported to be affected; (2) full siblings have been affected, but the incidence of such an occurrence is low; (3) the syndrome in the horse is similar to a condition in man that is known to be genetic; (4) abnormalities (lymphopenia and immunoglobulin deficiency) have been detected before signs of infection; (5) affected foals come from normal parents; (6) both males and females are affected, and (7) matings of known carriers result in a 3:1 ratio of normal to affected offspring. Estimates of the frequency of CID in the Arabian population are as high as 2.3 percent, which implies as many as 25 percent of all normal Arabians are carriers of the recessive gene for CID. A physiological test for carriers has not been found but would be much more desirable than progeny testing to find carriers (Section 12.4).

Combined immunodeficiency may be caused with other immunological disorders such as failure of passive transport (FPT, antibodies of the colostrum do not get to the foal's blood system), agammaglobulinemia (B-cell deficiency but normal T-cell function), and selective immunoglobulin M (IgM) deficiency. Table 13-17 summarizes the characteristics of these disorders and indicates how CID can be distinguished from other immunodeficiency disorders.

Neonatal Isoerythrolysis of Foals

Neonatal isoerythrolysis, hemolytic icterus, and hemolytic disease are three names for the destruction of red blood cells of the foal by serum antibodies in the first milk (colostrum) of the mare. This condition is somewhat similar to the phenomenon caused by Rh factor incompatibility in humans. The mare may be stimulated to produce antibodies against the red blood cells of the foal if she somehow receives red cells that are antigenically different from hers. These cells may be received by transfusion and antigen vaccination or even by the passage of the foal's blood through an abnormal placenta to maternal tissues. Apparently, the antibodies that result cannot pass back to the fetal blood, so the foal is born with no harmful effect of the incompatibility. The antibodies may, however, build up in the colostrum of the mare. If the newborn foal nurses, the antibodies in the milk pass directly to the foal's blood and destroy the red blood cells. The foal becomes anemic and sluggish within 12 to 36 hours and generally dies within a few days. A yellowish and continuous discoloration of the urine occurs after 24 to 48 hours. During late pregnancy, the mare's serum can be tested for the presence of such antibodies. Further confirmation of the diagnosis requires an agglutination test using the dam's serum and red blood cells of the sire. Fortunately, isoerythrolysis is rare, so the passage of the incompatible blood antigens of the foal to the blood of the mare (sensitivization) does not occur often.

The blood factors most commonly associated with neonatal isoerythrolysis are antigens coded for by Aa and Qa (Section 13.2). Not all mares having incompatible alleles at the A and Q loci, however, produce antibodies against Aa and Qa antigens, even though they have produced Aa or Qa foals. Other blood factors may also be involved.

Table 13-17
Diagnosis of equine immunodeficiency disorders

| Condition | Lymphocyte counts | Serum concentrations | | B-lymphocytes | T-cell function |
		IgM	IgG		
Combined immunodeficiency (CID)	low	absent	Before 3 mo, low or absent, after 3 mo normal or low depending on adequacy of passive transfer	absent	impaired
Failure of passive transport(FPT)	normal	normal	Below normal until 3 mo of age	normal	normal
Agammaglobulinemia	normal	absent	Low or absent after catabolism of antibodies received in colostrum	absent	normal
Selective IgM deficiency	normal	low or absent	Normal after 3 mo, levels prior to 3 mo depend on adequacy of passive transfer	normal	normal

Prevention may be achieved in many cases by not mating stallions carrying *Aa* or *Qa* genes to mares that lack the *Aa* or *Qa* genes. Because the frequency of *Aa* is high in many populations, it may be difficult to find a suitable stallion. Many foals with blood factors potentially incompatible with their mothers are foaled with no problem. Therefore, an alternative would be to mate to a potentially incompatible stallion but to test the mare's serum for antibodies about three weeks before foaling to determine whether the foal may be in danger of neonatal isoerythrolysis (Trommershausen-Smith, 1980).

When isoerythrolysis is probable, the foal should be kept from nursing its mother for 36 hours and either nursed by a foster mare or bottle fed as shown in Figure 13-20. Transfusion therapy is also used, after which the foal cannot be allowed to nurse until its mother's colostrum is gone, or for 24 to 36 hours when antibodies of the colostrum are no longer absorbed intact by the foal. Otherwise, the condition is generally lethal. The foal should be kept close to the mother during this period to avoid being rejected when her milk becomes safe. The foal should also be fed colostrum from another mare within 12 hours so that protective antibodies in the safe colostrum can be absorbed through the intestine.

Hemophilia (Factor VIII deficiency)

A rare blood-related lethal is the sex-linked recessive factor for bleeding disease (hemophilia) that was discussed in Section 12.2. As illustrated in Figure 12-14, the disease is eventually lethal and is transmitted from a normal carrier mare to half of her sons. Although not definitely proved to be sex-linked in the horse, the same deficiency has been shown to be sex-linked in man and other animals. All of the few reported cases in the horse have involved colts, which supports the theory of sex-linkage. A mare carrying the gene paired with a normal gene will not be hemophilic but is expected to transmit the gene to one-half of her sons, which will be hemophilic. One-half of her daughters also are expected to carry the gene but will not be hemophilic.

The Lethal White Gene, W

As described in Section 13.1, the gene for dominant white (Figure 13-21) was shown to be lethal when homozygous by Pulos and Hutt (1969). Matings by Dr. Pulos's farm in Alfred, New York, among white stallions and mares that had produced at least one colored foal resulted in 28 white and 15 colored foals, a ratio similar to the expected 2:1 ratio when the homozygote is lethal and not observed. The breeder may never realize the existence of

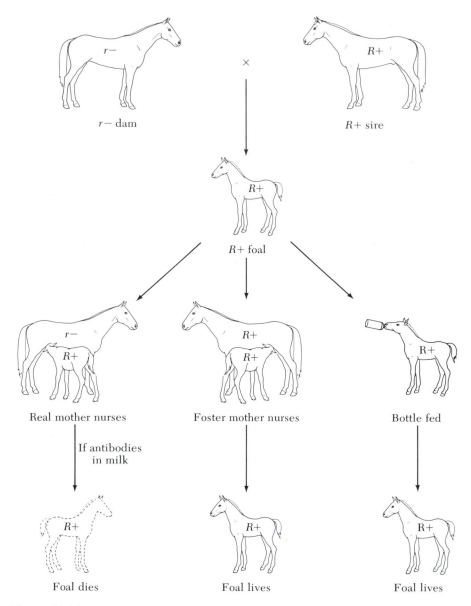

Figure 13-20
Hemolytic disease. An r^- mare with an R^+ foal may produce antibodies against R^+ blood cells. If the newborn foal drinks and absorbs these antibodies from the colostrum, they could destroy its blood cells. R^+ is likely to be either or both blood factors Aa or Qa with r^- being A not a and/or Q not a.

Figure 13-21
The dominant white horse is heterozygous for a gene causing the white color, which, when homozygous, causes early embryonic death. (Photograph courtesy of Ruth Thompson, Naper, Nebraska).

the problem because the *WW* fetus dies early in fetal development and is either resorbed or aborted without a trace. Only an apparently low conception rate is noted. An earlier report by Wriedt (1924) implied the lethality of the gene for white because of the high degree of sterility in matings of white horses in the Royal Fredericksborg Stud in Denmark. The name Fredericksborg lethal was originally given to this phenomenon.

Lethal Dominant Roan, Rn

The gene causing roan appears to be lethal when homozygous as described in Section 13.1 (Figure 13-7). As with lethal dominant white, a breeder of roans may not be aware of the problem because the *RnRn* fetus apparently dies early in fetal development. The homozygote can result only from mating of two roans. Thus, roan by solid matings would not show a reduced fertility rate whereas one-fourth of roan by roan matings would result in the lethal embryo which would lower the apparent conception rate.

White Foal Syndrome

This problem is different from that of dominant white. When two overo horses are mated, the resulting foal will sometimes be white. In addition, other effects are often present, including atresia coli, which is essentially a closing of the large intestine or a narrowing of the lumen of the large intestine. The foal usually dies. The condition seems to be genetically related to the overo pattern although how is not known. The inheritance of the overo pattern also is not known precisely (Section 13.1). Such white foals have been reported from matings of solid-colored parents. Jones and Bogart (1971) have proposed that a recessive allele of the overo series is responsible for the white foal syndrome. The $o^e o^e$ genotype would be affected and could result from matings of overo carriers, $o^e o \times o^e o$, or even from matings of carrier nonspotted horses, $Oo^e \times Oo^e$.

Melanoma

A tumor of the pigment-forming cells (melanocytes) is called a melanoma. Although horses of all colors may be affected, the majority of horses that have melanomas are gray. Thus, the gray gene G apparently predisposes horses to this form of cancer. Older horses are more frequently affected than younger horses. By age 15 most gray horses will have developed melanomas of varying degrees of severity. The growths commonly begin near, and spread from, the anus. Melanomas that spread to the lungs and other organs soon cause death. The average life-span of gray horses is shorter than for other horses, perhaps by two years.

Epitheliogenesis Imperfecta (Missing Patches of Skin)

Either hair or large pieces of skin on the lower limbs or other parts of the body is lacking at birth. Occasionally a hoof or hoof material is also absent. Foals are subject to infection, which usually kills them in a few days. Cases of the condition are rare but have been reported sporadically. Extensive pedigree examination of 33 affected draft-horse foals suggests a recessive gene is the cause (Butz and Meyer, 1957). A similar defect in cattle and pigs had previously been reported to be due to an autosomal recessive gene.

Hereditary Multiple Exostosis

This name is given to numerous bony protrusions that extend from the normal contour of the affected bones. A similar condition occurs in humans and is hereditary. It may affect most long bones as well as the ribs and pelvis of horses. Some horses are barely affected and severe lameness seldom results. Hereditary multiple exostosis appears to be due to an autosomal dominant gene with variable expression in both the horse and man. An affected Thoroughbred stallion produced 11 affected and 6 normal progeny when mated to normal mares (see Figure 13-22). The affected progeny included six males and five females, which rules out sex linkage.

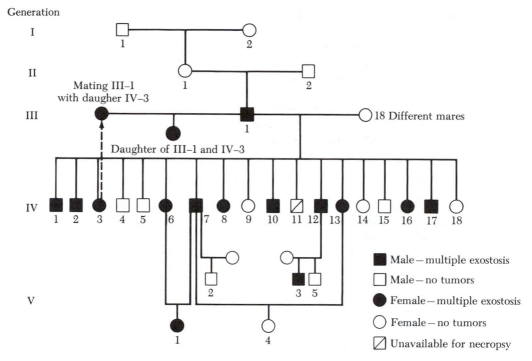

Figure 13-22
Pedigree of progeny and ancestors of stallion (III-1); illustrating the dominant inheritance pattern of multiple exostosis. Adapted from Gardner, E. J., et al. 1975.

Fibular Enlargement

In the condition of fibular enlargement, the ponies are normal except that the hind legs splay outward from the hock joint. The fibula is reduced in modern horses, but in the affected ponies the fibula is a full-length shaft that matches improperly with the tibia. The mechanical pressures that are exerted cause lameness, which does not improve with age. Speed (1958) reported observing this condition in some families of Shetlands and suggested that it may be related to dwarfism. Affected ponies have been foaled by normal parents. Hermans (1970) later studied 50 cases and decided that an autosomal recessive gene was responsible. From 8 matings of affected stallions with affected mares 8 affected foals were born. Affected mares could be used for test matings to identify carrier stallions (Section 12.4).

Aniridia with Cataract

Blindness in a group of Belgian draft horses in Sweden was suggested by Eriksson (1955) to be due to a mutation of a dominant gene in the germ cells of a stallion. Both eyes of affected animals completely lacked the iris. Cataracts developed at about 2 months of age, which caused most of the horses to become blind. Both male and female progeny were affected in equal numbers. Slightly less than half of the stallion's 143 observed offspring were affected. Normal offspring of the stallion used for breeding produced normal offspring, a result that also suggests that a dominant mutation arose and disappeared in one generation.

Relatively little evidence exists to support a genetic basis for the following defects, although for some defects quantitative inheritance may play a partial role in determining susceptibility or resistance to a defect that may be largely controlled by unknown environmental factors.

Wobbles

The symptoms of wobbles, also called equine spinal ataxia, are bilateral muscular incoordination and paddling of the hind feet. It usually occurs in young horses and rarely is reported in older horses. The cause is degeneration of the spinal column due to overgrowth of the articular processes of cervical vertebrae, usually at junction C3-C4 although sometimes at junctions C4-C5 and C5-C6. Severity of the disease depends on the location of the bony lesions along the spinal column. Overgrowth close to the head is most severe. The symptoms often appear suddenly and advance rapidly. In many cases, the horse bends the neck to one side, indicating the presence of neck pain. Whether wobbles is inherited is not clear, but in one study 43 percent of the cases of wobbles in Thoroughbreds were traced to one of the 3 foundation sires. Some mares had more than one foal with wobbles.

Falco et al. (1976), however, paired 67 wobblers and 67 controls and traced the pedigrees for seven generations to show that wobbles is not autosomal dominant, is not sex-linked, is unlikely to be recessive, and is unlikely to be a quantitatively inherited trait. Both the wobbler and control groups had similar blood lines and levels of inbreeding. Nevertheless wobbles is considered a problem in many breeds. The symptoms of the nerve disorder described next resemble superficially the symptoms of wobbles, but whereas wobbles affects the spinal cord, this disease affects the brain directly.

Cerebellar Hypoplasia and Degeneration

The part of the brain called the cerebellum controls coordination of movement. Foals afflicted with cerebellar hypoplasia either have none of the cells responsible for coordination or are deficient in such cells so that normal motor function is prevented. The abnormal function is manifested in incoordination, overreaching, paddling, and head tremor. The tremor is more noticeable when the horse is excited. The disease is often confused with wobbles but occurs at an even younger age (by 4 months), and the horse suffering from this disease has less difficulty in backing and walking when blindfolded than the horse suffering from wobbles. Most reported cases have involved Arabians. Sponseller (1967) has reported cases in which several mares have produced more than one affected foal. Inbreeding had occurred, and the names of two stallions appeared frequently in the pedigrees of 19 affected foals.

Atresia Coli (Severed Large Intestine)

This disease appears to result from more than one mechanism because it has been reported in Thoroughbreds as well as in white foals. Surgical correction is usually unsuccessful. A genetic mechanism may be the cause, as the trait appeared in many foals of the Percheron stallion Superb when he was used in Japan in about 1886 (Yamane, 1927). Affected foals appear healthy for 8 to 24 hours, until increasingly severe "colic" results in death 3 to 4 days later as a result of failure to eliminate by-products of digestion.

Anal Atresia (Blocked Anus)

Because the anus is lacking, no products of digestion can be passed. Other problems may be caused by this condition since surgical correction is rarely of any benefit. The condition is rare and sporadic, so it may be due to a dominant gene that arises by mutation or to a recessive gene that is very rare. Nutritional excesses or deficiencies or the presence or absence of drugs at a critical stage of development may also be the cause.

Hydrocephalus (Water Head)

In this condition, cerebrospinal fluid accumulates, causing the head to enlarge (as shown in Figure 13-23) before birth and the nervous system to be

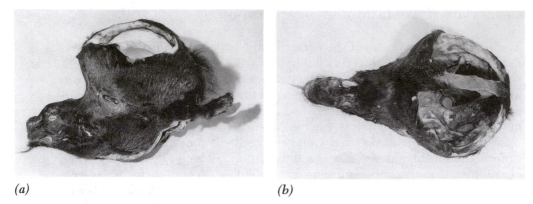

(a)   (b)

Figure 13-23
Hydrocephalus, shown in a section of the head, causes death within 48 hours: (a), side view; (b), top view.

abnormal. The condition is nearly always lethal within 48 hours but fortunately is rare, which indicates that the cause may be a dominant mutation, a rare recessive, or an environmentally caused accident of development.

Shistosoma Reflexum

Fetuses with this abnormal condition, characterized by complete lack of amniotic development, are usually aborted before full term. Inbreeding apparently caused this recessive lethal to become common in a group of horses studied by Weber (1947).

Stiff Forelegs

A Polish Anglo-Arabian stallion sired foals affected with flexed, stiff, or contracted forelegs according to a report by Prawochenski (1936). Seven of the 22 foals sired in 1934 and 1935 were affected. The hind legs and bodies were normal. Since the foal cannot stand to nurse, the condition is lethal. The mares were not closely related to the sire, but all had a common sire in their pedigrees. This may be a recessive characteristic affected by modifier genes or other environmental factors. An affected Appaloosa foal is shown in Figure 13-24.

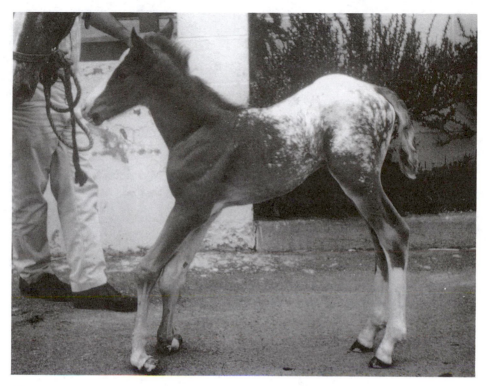

Figure 13-24
Flexed or contracted tendons, which prevent normal extension of the pasterns, is a lethal characteristic in nature because it prevents the foal from nursing.

Hip Dysplasia

When the upper hind leg fits into a flattened, shallow hip joint rather than a normal rounded, cupped joint, hip dysplasia may result. A mincing gait and severe lameness are often the first symptoms. The condition does not appear very frequently, but it has been reported in Standardbreds by Jogi and Norberg (1962) and in Shetlands by Manning (1963). The trait is partially determined by genetics in dogs.

Umbilical Hernia

Since this commonly occurring condition usually persists for only a few months and then disappears, it is not a serious problem (Figure 13-25).

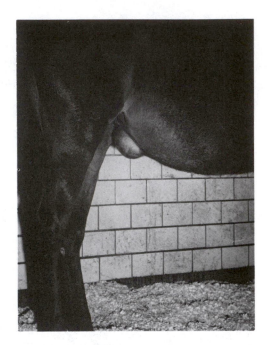

Figure 13-25
Umbilical hernia, as shown by a
2-year-old colt, is usually not a
serious problem. (Photograph by
John E. Lowe, courtesy of Robert H.
Whitlock, Ithaca, New York.)

Surgery is usually successful for those hernias that persist. The inheritance pattern is not known, although Aurich (1959) and Schlaak (1942) have postulated that a recessive gene is the predisposing cause. Whether the chance that it is genetically determined is great enough and the seriousness of the defect is great enough to recommend against mating affected animals must be left to the breeders.

Cryptorchidism

The failure of one or both testes to descend completely into the scrotum is called cryptorchidism. If only one testis fails to descend (unilateral cryptorchidism), the affected stallion is usually fertile. However, if both testes fail to descend (bilateral cryptorchidism), the stallion is sterile. Stanic (1960) presented the unusual data that appear in Table 13-18. These data indicate a surprisingly greater incidence of cryptorchidism on the left side than on the right, particularly in abdominal cryptorchids. The influence of heredity is not clear, but as a precaution, monorchid horses are usually not used for breeding.

Table 13-18
Incidence of bilateral and unilateral cryptorchidism ($n = 417$)

Location	Left side	Right side	Both	Percentage
Abdominal	158/417	77/417	16/417	60.2
Inguinal	75/417	78/417	10/417	39.1
Both	0	0	3	.7
Percentage	55.8	37.2	7.0	

SOURCE: Stanic, M. N. 1960.

Recurrent Uveitis (Equine Periodic Ophthalmia) Moon Blindness

Equine periodic ophthalmia (EPO) is an infection of the eye that was originally observed to recur approximately as often as a new moon. As the disease
progresses, cataract formation and eventual blindness occur. There are many
theories as to the cause of EPO, such as vitamin deficiency or excess, other
nutritional factors, and genetics. There is some evidence that a leptospiral
infection is involved. Another theory is that EPO is the result of an autoim-
mune reaction. Cross (1966) found no evidence for quantitative inheritance.
He did estimate a frequency of 10 to 20 percent in the northeastern United
States. The British, however, reduced the incidence from 20 to 30 percent in
1840 to nearly 7 percent by 1900. Whether or not selection was the cause,
stallions with symptoms of EPO (that is, cataract) were not allowed to breed
as a result of a decree of the Royal Commission for Horse Breeding. This
evidence supports at least a genetic predisposition to the disease, although
some environmental factors may also have changed during such a long
period.

Equine Night Blindness

Equine night blindness (ENB) has been reported predominantly in Appaloosas but has been reported in Quarter Horse and Thoroughbred herds. In
humans, a similar disease, congenital stationary night blindness is known to
be genetically determined. Sight is normal but an affected foal or horse will
bump into objects in dim light. The eye appears to be normal, with the
dysfunction occurring in transmission of nerve impulses. Diagnosis can be
made clinically soon after birth with an electroretinogram. The abnormality
does not become worse with age. Equine night blindness should not be
confused with recurrent uveitis (periodic ophthalmia or moon blindness) nor

with vitamin A deficiency, which is often associated with night blindness in domestic animals. Vitamin A therapy is not helpful in treating equine night blindness.

Sex Reversal

Many infertile mares are found on cytological examination to have an un-paired X chromosome (that is, in the diploid stage they are XO, where the O represents a missing X or a missing Y). The missing X or Y is probably the result of a chance error of segregation in either the mare or stallion. Trommershausen-Smith et al. (1979) in a study of Arabians found no evidence for any pattern of inheritance. This defect is called gonadal dysgenesis—underdevelopment of the ovaries—although the external female genitalia appear normal.

A more unusual case of "mare" infertility is testicular feminization, which appears to result from a "sex-reversal" gene carried on the X chromosome of an animal having an apparently normal XY (male) karyotype, (Kent, 1980). The "sex-reversal" gene overrides the usual effects of genes carried on the Y chromosome so that development of the male reproductive tract ceases early in fetal stage and instead a female reproductive tract develops. Genes on the Y chromosome continue to cause testosterone to be produced but the $X^R Y$ animal does not possess receptors for testosterone and, thus, the testosterone has no effect. An animal lacking testosterone receptors does not show response to injection of testosterone. A normal female would show an effect from testosterone injection. Thus a testosterone receptor test together with karyotyping can be used to identify a "sex-reversal" mare. Apparently a carrier mare ($X^R X$) also does not have testosterone receptors. Testosterone is necessary for the differentiation of the gonad into a testis. Lack of testosterone or its receptor would feminize the embryo even if a Y chromosome was present.

Mares with gonadal dysgenesis also have been found that are XY. Lieberman (1982) and Kent (1980) have proposed that a recessive gene carried on the X chromosome (sex-linked) interferes in some way with the function of genes on the Y chromosome that normally results in development of male genitalia. At least one XY "female" has produced a filly foal from mating to a normal XY stallion (Sharp et al., 1980).

Other Characters

Some other abnormalities not discussed here are mentioned by Jones and Bogart (1971), Stormont (1958), and Lerner (1944).

The relative brevity of the foregoing discussion and the repeated lack of solid facts concerning inheritance patterns emphasize the fact that not much is known about the inheritance of most lethals and abnormalities in horses. Fortunately, few such abnormalities are major problems. Some that are problems can be controlled by a mixture of genetic common sense (color related problems) and careful management (CID, NI). Others tend to eliminate themselves (hemophilia, fibular enlargement, aniridia with cataract).

REFERENCES AND FURTHER READINGS

Coat Color

Adalsteinsson, S. 1974. Inheritance of the palomino color in Icelandic horses. *J. Heredity* 65:15.

Adalsteinsson, S. 1974. Color inheritance in farm animals and its application in selection. *Proc. 1st World Cong. on Genetics Applied to Livestock Production.* Madrid. pp. 29.

Adalsteinsson, S. 1978. A new interpretation of the inheritance of the horse colors dun and Isabella in a Russian stud during the period 1854–1894. *J. Heredity* 69:426.

Adalsteinsson, S. 1978. Inheritance of yellow and blue dun in the Icelandic Toelter horse. *J. Heredity* 69:146.

Anderson, W. S. 1913. The inheritance of coat color in horses. *American Nat.* 47:615.

Anderson, W. S. 1914. The inheritance of coat colors in horses. *Kentucky Agr. Expt. Sta. Bull. 180.*

Anderson, W. S. 1926. The inheritance of coat colour in domestic livestock. *Am. Vet. Med Asso.* 23:338.

Andersson, L., and K. Sandberg. 1982. A linkage group composed of three coat color genes and three serum protein loci in horses. *J. Heredity* 73:91.

Babcock, E. B., and R. E. Clausen. 1918. *Genetics in Relation to Agriculture.* New York: McGraw-Hill.

Berge, S. 1963. Hestefargenes genetikk. *Tidsskr. Det Norske Landbruk.* 70:359.

Blakeslee, L. H., R. S. Hudson, and H. R. Hunt. 1943. Curly coat in horses. *J. Heredity* 34:115.

Blanchard, N. 1902. Inheritance of coat color in Thoroughbred horses. *Biometrika* 1:361.

Blunn, C. T., and C. E. Howell. 1936. The inheritance of white facial markings in Arabian horses. *J. Heredity* 27:293.

Briquet, R., Jr. 1957. So-called "albino horses." *Boletim Indus. Anim.* 16:243.

Briquet, R., Jr. 1959a. Investigations on the relationship between markings on face and limbs in horses. *Rev. Remonta Vet.* 19.

Briquet, R., Jr. 1959b. *Genetica da Pelagem do Cavalo.* Rio de Janeiro: Instituto de Zootechnia.

Bunsow, R. 1911. Inheritance in race horses: Coat color. *Mendel. Jour.* 2:74.

Butaye, R. 1974. Inheritance of coat-colour in Belgian horses. *Vlamms Dier. Tijdshrift* 43:464.

Castle, W. E. 1940. The genetics of coat color in horses. *J. Heredity* 31:127.

Castle, W. E. 1946. Genetics of the Palomino horse. *J. Heredity* 37:35.

Castle, W. E. 1948. The ABC of color inheritance in horses. *Genetics* 33:22.

Castle, W. E. 1951. Dominant and recessive black in mammals. *J. Heredity* 42:48.

Castle, W. E. 1951. Genetics of the color varieties of horses. *J. Heredity* 42:297.

Castle, W. E. 1953. Note on the silver dapple mutation of Shetland ponies. *J. Heredity* 44:224.

Castle, W. E. 1954. Coat color inheritance in horses and other mammals. *Genetics* 39:35.

Castle, W. E. 1960. Fashion in the color of Shetland ponies and its genetic basis. *J. Heredity* 51:247.

Castle, W. E. 1961. Genetics of the Claybank dun horse. *J. Heredity* 51:121.

Castle, W. E., and F. L. King. 1951. New evidence on the genetics of the Palomino horse. *J. Heredity* 42:61.

Castle, W. E., and W. R. Singleton. 1960. Genetics of the "brown" horse. *J. Heredity* 51:127.

Castle, W. E., and W. R. Singleton. 1961. The Palomino horse. *Genetics* 46:1143.

Castle, W. E., and W. H. Smith. 1953. Silver dapple, a unique color variety among Shetland ponies. *J. Heredity* 44:139.

Charles, D. R. 1938. Studies on spotting patterns. *Genetics* 23:523.

Colombani, B. 1964. A case of total albinism in a Sardinian ass. *Ann. Fac. Med. Vet. Pisa* 16:76.

Comfort, A. 1958. Coat-colour and longevity in Thoroughbred mares. *Nature* 182:1531.

Craig, L., and L. D. Van Vleck. 1985. Evidence for inheritance of the red dun dilution in the horse. *J. Heredity* 76:138.

Crew, F. A. E., and A. D. Buchanan-Smith. 1930. The genetics of the horse. *Bibliographica Genetica* 6:123.

Domanski, A. J., and R. T. Prawochenski. 1948. Dun coat color in horses. *J. Heredity* 39:267.

Dreux, P. 1970. The degree of expression on limited piebaldness in the domestic horse. *Ann. Genet. Sel. Anim.* 2:119.

Dring, L. A., H. F. Hintz, and L. D. Van Vleck. 1981. Coat color and gestation length in Thoroughbred mares. *J. Heredity* 72:65.

Estes, B. W. 1948. Lack of correlation between coat color and temperament. *J. Heredity* 39:84.

Geurts, R. 1977. *Hair colour in the horse.* J. A. Allen and Co., Ltd., London.

Gibbon, Helen. 1941. Leopard spotting and color alteration in that recently established breed of horses, the Colorado Ranger. *J. Colorado-Wyoming Acad. Sci.* 3:48.

Gilbey, Sir Walter. 1912. *Horses Breeding to Color.* London: Vinton and Co.

Gregory, W. K. 1926. The horse in the tiger's skin. *Bull. Zool. Soc.* 29:111.

Gremmel, F. 1939. Coat colors in horses. *J. Heredity* 30:437.

Hadwen, S. 1931. The melanomata of gray and white horses. *Canad. Med. Asso. J.* 21:519.

Harper, C. H. 1905. Studies in the inheritance of color in Percheron horses. *Biol. Bull.* 9:265.

Hatley, George. 1962. Crosses that will kill your color. *Appaloosa News* (February).

Heizer, E. E. 1931. Color inheritance in horses. *Proc. 24th Congress of American Society of Animal Production.* Chicago:184.

Hintz, H. F., and L. D. Van Vleck. 1979. Lethal dominant roan in horses. *J. Heredity* 70:145.

Huitema, H. 1904. Archaic pattern in the horse and its relation to colour genes. *Z. Saugetierk.* 29:42.

Hurst, C. C. 1906. Inheritance of coat colour in horses. *Proc. Royal Soc.* 77(B):388.

Jones, W. E., and R. Bogart. 1971. *Genetics of the Horse.* East Lansing, Michigan: Caballus.

Keeler, C. F. 1947. Coat color and physique and temperament. *J. Heredity* 38:271.

Klemola, V. 1933. The "pied" and "splashed white" patterns in horses and ponies. *J. Heredity* 24:65.

Lehmann-Mathildenhoh, E. von. 1941. Beitrag zur Vererbungweissgeborener Pferde. *Zeit. Tierz. Zuchtungsbiol.* 49:191.

Lusis, J. A. 1942. Striping patterns in domestic horses. *Genetica* 23:31.

McCann, L. P. 1916. Sorrel color in horses. *J. Heredity* 7:370.

Miller, R. W. *Appaloosa Coat Color Inheritance.* Moscow, Idaho: Montana State University, Bozeman and Appaloosa Horse Club, Inc.

Odriozola, M. 1940. Where are the Thoroughbred "Palominos"? *J. Heredity* 31:128.

Odriozola, M. 1948. Agouti color in horses: Change of dominance in equine hybrids. *Proc. 8th Int. Cong. on Genetics:* 635.

Odriozola, M. 1951. *A los Colores del Caballo.* Madrid: National Syndicate of Livestock.

Odriozola, M. 1952. The eumelanin horse: Black or brown? *J. Heredity* 43:76.

Pearson, K. 1901. Mathematical contributions to the theory of evolution. Vol. 8. On the inheritance of coat colour in horses. *Philosophical Transactions of the Royal Society* 195:79.

Pocock, R. I. 1903. The coloration of Quaggas. *Nature* 68:356.

Pocock, R. I. 1909. On the colours of horses, zebras, and Tapirs. *Ann. Mag. Nat. Hist.* 4:404.

Pulos, W. L., and F. B. Hutt. 1969. Lethal dominant white in horses. *J. Heredity* 60:59.

Richardson, T. C. 1924. The "pinto" burro. *J. Heredity* 15:73.

Ridgeway, W. 1919. The colour of race horses. *Nature* 104:334.

Salisbury, G. W. 1941. The inheritance of equine coat color. *J. Heredity* 32:235.

Salisbury, G. W., and J. W. Britton. 1941. The inheritance of equine coat color. 2. The dilutes with special reference to the Palomino. *J. Heredity* 32:255.

Searle, A. G. 1968. *Comparative Genetics of Coat Colour in Mammals.* London: Logo Press, Ltd.

Singleton, W. R., and Q. C. Bond. 1966. An allele necessary for dilute coat color in horses. *J. Heredity* 57:75.

Singleton, W. R., and J. N. Dent. 1964. Coat color in small horses of the Philippines. *J. Heredity* 55:220.

Smith, A. T. 1972. Inheritance of chin spot markings in horses. *J. Heredity* 63:100.

Sponenberg, D. P. 1982. The inheritance of leopard spotting in the Noriker horse. *J. Heredity* 73:357.

Sponenberg, D. P., and B. V. Beaver. 1983. *Horse Color*. College Station: Texas A&M University Press.

Sponenberg, D. P., H. T. Harper, and A. L. Harper. 1984. Direct evidence for linkage of roan and extension loci in Belgian horses. *J. Heredity* 75:413.

Stirling, H. B. 1925. Colour heredity in horses. *Scot. Jour. Agric.* 8:32.

Sturtevant, A. H. 1910. On the inheritance of coat color in American Harness horses. *Biol. Bull.* 19:204.

Sturtevant, A. H. 1912. A critical examination of recent studies of color inheritance in horses. *J. Genetics* 2:41.

Trommershausen-Smith, A. 1977. Lethal white foals in matings of overo spotted horses. *Theriogenology* 8:303.

Trommershausen-Smith, A. 1978. Linkage of tobiano coat spotting and albumin markers in a pony family. *J. Heredity* 69:214.

Tubb, P. 1933. Genetiske undersokelser over hestefarver. *Nord. Vet.* 6:28.

Van Vleck, L. D., and M. Davitt. 1977. Confirmation of a gene for dominant dilution of horse color. *J. Heredity* 68:280.

Wentworth, E. N. 1913. Color inheritance in the horse. *Zeit. Induktive Abst. Vererbungslehre* 11:10.

Wilson, J. 1910. The inheritance of coat color in horses. *Proceedings of Royal Dublin Society* 12:331.

Wilson, J. 1912. The inheritance of the dun coat in horses. *Procedings of Royal Dublin Society* 13:184.

Wriedt, C. 1918. Albinisme i hester. (Albinism in horses). *Tiddsskr. f.d. norske Landbruk* 25:396.

Wright, S. 1917. Color inheritance in mammals. Vol. 7. Horse. *J. Heredity* 8:561.

Blood Factors

Andersson, L. 1983. Studies on genetic linkage in domestic animals with special reference to the horse. Ph.D. Thesis. Swedish University of Agricultural Science, Uppsala.

Andersson, L., R. K. Juneja, and K. Sandberg. 1983. Genetic linkage between the loci for phosphohexose isomerase (PHI) and a serum protein (Xk) in horses. *Anim. Blood Groups and Biochem. Genetics* 14:45.

Andersson, L., K. Sandberg, S. Adalsteinsson, and E. Gunnarsson. 1983. Linkage of the equine serum esterase (Es) and mitochrondrial glutamate oxaloacetate transminase (GOT_M) loci: a horse-mouse homology. *J. Heredity* 74:361.

Antczak, D. F. 1984a. Histocompatibility Antigens of the Horse. New York: Dorothy Russell Havemeyer Foundation.

Antczak, D. F., 1984b. Lymphocyte alloantigens of the horse. III. ELY-2.1: a lymphocyte alloantigen not coded by the MHC. *Anim. Blood Groups and Biochem. Genetics* 15:103.

Bailey, E. 1980. Identification and genetics of horse lymphocyte alloantigens. *Immunogenetics* 11:499.

Bailey, E., 1983. Population studies on the ELA system in American Standardbred and Thoroughbred mares. *Anim. Blood Groups and Biochem. Genetics* 14:201.

Bailey, E. 1983. Linkage disequilibrium between the ELA and A blood group systems in Standardbred horses. *Anim. Blood Groups and Biochem. Genetics* 14:37.

Bailey, E., and P. J. Henney. 1984. Comparison of ELY-2.1 with blood group and ELY-1 markers in the horse. *Anim. Blood Groups and Biochem. Genetics* 15:117.

Bailey, E., C. Stormont, Y. Suzuki, and A. Trommershausen-Smith. 1979. Linkage of loci controlling alloantigens on red blood cells and lymphocytes in the horse. *Science* 204:1317.

Bengtsson, S., B. Gahne, and J. Rendel. 1968. Genetic studies on transferrins, albumins, prealbumins, and esterases in Swedish horses. *Acta. Agric. Scand.* 18:60.

Bengtsson, S., and K. Sandberg. 1972. Phosphoglucomutase polymorphism in Swedish horses. *Anim. Blood Groups and Biochem. Genetics* 3:115.

Braend, M. 1964. Serum types of Norwegian horses. *Nord. Vet. Med.* 16:363.

Braend, M. 1967. Genetic variation of horse hemoglobin. *Hereditas* 58:385.

Braend, M. 1970. Genetics of horse acidic prealbumins. *Genetics* 65:495.

Braend, M., and G. Efremov. 1965. Hemoglobins, haptoglobins, and albumins or horses. *Proc. 9th European Conf. on Anim. Blood Groups and Biochem. Polymorphism.* Prague, 1964:253.

Braend, M., and C. Stormont. 1964. Studies on hemoglobin and transferrin types of horses. *Nord. Vet. Med.* 16:31.

Bull, R. W. (ed.) 1983. Joint report of the first international workshop on lymphocyte alloantigens of the horse held 24-29 October 1981. *Anim. Blood Groups and Biochem. Genetics* 14:119.

Clegg, J. B. 1970. Horse hemoglobin polymorphism: evidence for two linked non-allelic α-chain genes. *Proc. Roy. Soc. Lond. B.* 176:235.

Cradock-Watson, J. E. 1967. Immunological similarity of horse, donkey, and mule haemoglobins. *Nature* 215:630.

Deys, B. F. 1972. Demonstration of X-linkage of G6PD, HGPRT, and PGK in the horse by means of mule-mouse cell hybridisation. Ph.D. thesis, University of Leiden.

Franks, D. 1962. Differences in red-cell antigen strength in the horse due to gene interaction. *Nature* 195:580.

Franks, D. 1962. Horse blood groups and hemolytic disease of the newborn foal. *Ann. N.Y. Acad. Sci.* 97:235.

Gahne, B. 1966. Studies of the inheritance of electrophoretic forms of transferrins, albumins, prealbumins, and plasma esterases of horses. *Genetics* 53:681.

Gahne, B. S. Bengtsson, and K. Sandberg, 1970. Genetic control of cholinesterase activity in horse serum. *Anim. Blood Groups and Biochem. Genetics* 1:207.

Juneja, P. K., B. Gahne, and K. Sandberg. 1978. Genetic polymorphism of the vitamin D binding protein and another post albumin protein in horse semen. *Anim. Blood Groups and Biochem. Genetics* 9:29.

Kaminski, M. 1978. The null allele in the horse esterase (Es) system detected by enzyme assay and rocket immunoelectrophoresis in heterozygous animals. *Anim. Blood Groups and Biochem. Genetics* 9:197.

Kelly, E. P., C. Stormont, and Y. Suzuki. 1971. Catalase polymorphism in the red cells of horses. *Anim. Blood Groups and Biochem. Genetics* 2:135.

Kingsbury, E. T., and S. N. Gaunt. 1977. Heterogeneity in whey proteins of mare's milk. *J. Dairy Sci.* 60:274.

Lazary, S., A. L. de Weck, S. Bullen, R. Straub, and H. Gerber. 1980. Equine leukocyte antigen system. I. Serological studies. *Transplantation* 30:203.

Lazary, S., S. Bullen, J. Muller, G. Kovacs, I. Bodo, P. Hockenjos, and A. L. de Weck. 1980. Equine leukocyte antigen system. II. Serological and mixed lymphocyte reactivity studies in families. *Transplantation* 30:210.

Lazary, S., H. Gerber, A. L. de Weck, and P. Arnold. 1982. Equine leukocyte antigen system. III. Non-MHC linked alloantigen system in horses. *J. Immunogenetics* 9:327.

Loen, J. 1939. Fargenedarvingen hos Vestlandshesten (Fjordhesten). Stambok over Vestlandshesten 11:1.

Mathai, C. K., S. Ohno, and E. Beutler. 1966. Sex-linkage of the glucose-6-phosphate dehydrogenase gene in *Equidae. Nature* 210:115.

Mottironi, V. D., L. E. Perryman, B. Pollara, M. R. Mickey, R. Swift, and P. McGrath. 1981. Major histocompatibility locus in the Arabian Horse. *Transplantation* 31:290.

Niece, R. L., and D. W. Kracht. 1967. Genetics of transferrins in burros (*Equus asinus*). *Genetics* 57:837.

Noda, H. 1975. Studies on the relationship between hemolytic icterus of newborn foals and blood groups, and serological diagnosis. *Jap. J. Vet. Res.* 23:103.

Osterhoff, D. R. 1967. Haemoglobin, transferrin, and albumin types in equidae (horses, mules, donkeys, and zebras). *Proc. 10th European Conf. on Anim. Blood Groups and Biochem. Polymorphism,* Paris, 1966:345.

Osterhoff, D. R., D. O. Schmid, and I. S. Ward-Cox. 1970. Blood group and serum type studies in Basuto ponies. *Proc. 11th European Conf. on Anim. Blood Groups and Biochem. Polymorphism,* Warsaw, 1968:453.

Osterhoff, D. R., and I. S. Ward-Cox. 1972. Quantitative studies on horse hemoglobins. *Proc. 12th European Conf. on Anim. Blood Groups and Biochem. Polymorphism,* Budapest 1970:541.

Putt, W., and R. A. Fisher. 1989. An investigation of seven enzymes as possible genetic markers in horse leucocytes. *Anim. Blood Groups and Biochem. Genetics* 10:191.

Sandberg, K. 1968. Genetic polymorphism in carbonic anhydrase from horse erythrocytes. *Hereditas* 60:411.

Sandberg, K. 1970. Blood group factors and erythrocytic protein polymorphism in Swedish horses. *Proc. 11th European Conf. on Anim. Blood Groups and Biochem. Polymorphism,* Warsaw, 1968:447.

Sandberg, K. 1972. A third allele in the horse albumin system. *Anim. Blood Groups and Biochem. Genetics* 3:207.

Sandberg, K. 1973. Phosphohexose isomerase polymorphism in horse erythrocytes. *Anim. Blood Groups and Biochem. Genetics* 4:79.

Sandberg, K. 1973. The D blood group system of the horse. *Anim. Blood Groups and Biochem. Genetics* 4:193.

Sandberg, K. 1974. Linkage between the K blood group locus and the 6-PGD locus in horses. *Anim. Blood Groups and Biochem. Genetics* 5:137.

Sandberg, K. 1974. Blood typing of horses: Current status and application to identification problems. *1st World Cong. on Genetics Applied to Livestock Production* 1:253.

Sandberg, K. 1979. Studies on blood groups and genetic protein polymorphisms of the horse. Ph.D. Thesis, Swedish University of Agricultural Sciences.

Sandberg, K., and L. Andersson. 1984. Genetic linkage in the horse. I. Linkage relationships among 15 blood marker loci. *Hereditas* 100:199.

Sandberg, K., and S. Bengtsson. 1972. Polymorphism of hemoglobin and 6-phosphogluconate dehydrogenase in horse erythrocytes. *Proc. of 12th European Conf. on Anim. Blood Groups and Biochem. Polymorphism*, Budapest. 1970:527.

Sandberg, K., and R. K. Juneja. 1978. Close linkage between the albumin and Gc loci in the horse. *Anim. Blood Groups and Biochem. Genetics* 9:169.

Schmid, D. O. 1966. Further progress in serogenetics in horses. *Proc. 10th European Conf. on Anim. Blood Groups and Biochem. Polymorphism.* Paris, 1966:339.

Scott, A. M. 1972. Improved separation of polymorphic esterases in horses. *Proc. 12th European Conf. on Anim. Blood Groups and Biochem. Polymorphism*, Budapest 1970:551.

Stormont, C. 1972. Genetic markers in the blood of horses. *Proc. Horse Identification Seminar, December 8–9.* Washington State University, Pullman, p. 76.

Stormont, C. 1975. Neonatal isoerythrolysis in domestic animals: A comparative review. In *Advances in Veterinary Science and Comparative Medicine* (C. A. Brandly and C. E. Corneliu eds.) New York: Academic Press. 19:23.

Stormont, C., and Y. Suzuki. 1963. Genetic control of albumin phenotypes in horses. *Proc. Soc. Exper. Biol. Med.* 114:673.

Stormont, C., and Y. Suzuki. 1964. Genetic systems of blood groups in horses. *Genetics* 50:915.

Stormont, C., and Y. Suzuki, 1965. Paternity tests in horses. *Cornell Vet.* 55:365.

Stormont, C., Y. Suzuki, and J. Rendel. 1965. Application of blood typing and protein tests in horses. *Proc. 9th European Conf. on Anim. Blood Groups and Biochem. Polymorphism*, Prague, 1964:221.

Stormont, C., Y. Suzuki, and E. A. Rhode. 1964. Serology of horse blood groups. *Cornell Vet.* 54:439.

Suzuki, Y., and C. Stormont. 1972. Genetic control of an *in vitro* autolytic factor in horse red cells. *Proc. of 12th European Conf. on Anim. Blood Groups and Biochem. Polymorphism*, Budapest 1970:525.

Trommershausen-Smith, A., and Y. Suzuki. 1978. Identity of Xk and Pa systems in equine serum. *Anim. Blood Groups and Biochem. Genetics* 9:127.

Trommershausen-Smith, A., Y. Suzuki, and C. Stormont. 1976. Use of blood typing to confirm principles of coat color genetics in horses. *J. Heredity* 67:6.

Vandeplassche, M., and L. Podliachouk. 1970. Chimerism in horses. *Proc. 11th European Conf. on Anim. Blood Groups and Biochem. Polymorphism.* Warsaw, 1968:459.

Weitkamp, L. R., S. A. Guttormsen, and P. Costello-Leary. 1982. Equine gene mapping: A sex difference in recombination frequency for linkage group II. *Anim. Blood Groups and Biochem. Genetics* 13:305.

Weitkamp, L. R., and Allen, P. Z. 1979. Evolutionary conservation of equine Gc alleles and of mammalian Gc/albumin linkage. *Genetics* 92:1347.

Weitkamp, L. R., S. A. Guttormsen, and P. Costello-Leary. 1982. Equine gene mapping: Close linkage between the loci for soluble malic enzyme and Xk (Pa). *Anim. Blood Groups and Biochem. Genetics* 13:279.

Lethals and Defects

Archer, R. K. 1961. True haemophilia (haemophilia A) in a Thoroughbred foal. *Vet. Rec.* 73:338.

Archer, R. K., and B. V. Allen. 1972. True haemophilia in horses. *Vet. Rec.* 91:655.

Aurich, R. 1959. A contribution to the inheritance of umbilical hernia in the horse. *Berl. Munch. Tierartzl. Wschr.* 72:420.

Bain, A. M. 1969. Foetal losses during pregnancy in the Thoroughbred mare: A record of 2,562 pregnancies. *New Zealand Vet. J.* 17:155.

Baird, J. D., and C. D. MacKenzie. 1974. Cerebellar hypoplasia and degeneration in part-Arab horses. *Aust. Vet. J.* 50:25.

Banks, K. L., T. C. McGuire, and T. R. Jerrels. 1976. Absence of B lymphocytes in a horse with primary agammaglobulinemia. *Clin. Immun. Immunopath.* 5:282.

Basrur, P. K., H. Kanagawa, and L. Podliachouk. 1969. An equine intersex with unilateral gonadal agenesis. *Can. J. Comp. Med.* 33:297.

Basrur, P. K., H. Kanagawa, and L. Podliachouk. 1970. Further studies on the cell populations of an intersex horse. *Can. J. Comp. Med.* 33:294.

Beech, J. 1976. Cervical cord compression and wobbles in horses. *Proc. A. A. E. P.* 22:78.

Behrens, E. 1979. Polydactylism in a foal. *J. A. V. M. A.* 174:324.

Bekschner, H. G. 1969. *Horses' Diseases.* Sydney: Angus and Robertson.

Bielanski, W. 1946. The inheritance of shortening of the lower jaw (brachygnathia inferior) in the horse. *Przegl. hodowl.* 14:24.

Bjorck, G., K. E. Everz, H. J. Hansen, and B. Henrickson. 1973. Congenital cerebellar ataxia in the Gotland pony breed. *Zbl. Vet. Med. A.* 20:341.

Blakeslee, L. H., and R. S. Hudson. 1942. Twinning in horses. *J. Animal Sci.* 1:118.

Blue, M. G., A. N. Bruere, and H. F. Dewes. 1978. The significance of the XO syndrome in infertility of the mare. *New Zealand Vet. J.* 26:137.

Bornstein, S. 1967. The genetic sex of two intersexual horses and some notes on the karyotypes of normal horses. *Acta. Vet. Scand.* 8:291.

Bouters, R., M. Vandeplassche, and A. DeMoor. 1972. An intersex (male pseudohermaphrodite) horse with 64,XX/65,XXY mosaicism. *Equine Vet. J.* 4:150.

Britton, J. W. 1945. An equine hermaphrodite. *Cornell Vet.* 35:373.

Britton, J. W. 1962. Birth defects in foals. *Thoroughbred* 34:288.

Bruere, A. N., M. G. Blue, P. M. Jaine, K. S. Walker, L. M. Henderson, and H. M. Chapman. 1978. Preliminary observations on the occurrence of the equine XO syndrome. *New Zealand Vet. J.* 26:145.

Bruner, D. W., E. R. Doll, F. E. Hull, and A. S. Kinkaid. 1950. Further studies on hemolytic icterus in foals. *Am. J. Vet. Res.* 11:22.

Bruner, D. W., F. E. Hull, and E. R. Doll. 1948. The relation of blood factors to icterus in foals. *Am. J. Vet. Res.* 9:237.

Bruner, D. W., F. E. Hull, P. R. Edwards, and E. R. Doll. 1948. Icteric foals. *J. A. V. M. A.* 122:440.

Buckingham, J. 1936. Hermaphrodite horses. *Vet Record* 48:218.

Butz, H., and H. Meyer. 1957. Epitheliogenesis imperfecta neonatorium equi (Incomplete skin formation in the foal). *Dtsch. tierarztl. Wschr.* 64:555.

Chandley, A. C., J. Fletcher, P. D. Rossdale, C. K. Peace, S. W. Ricketts, R. J. McEnery, J. P. Thorne, R. V. Short, and W. R. Allen. 1975. Chromosome abnormalities as a cause of infertility in mares. *J. Repro. Fertil.* (Suppl. 23):377.

Chandley, A. C., R. V. Short, and W. R. Allen. 1975. Cytogenetic studies to three equine hybrids. *J. Reprod. Fert.* 23:365.

Comfort, A. 1952. Coat colour and longevity in Thoroughbred mares. *Nature* 182:1531.

Cronin, N. T. I. 1956. Hemolytic disease of newborn foals. *Vet. Rec.* 67:474.

Cross, R. S. N. 1966. Equine periodic ophthalmia. *Vet. Rec.* 78:8.

Dimock, W. W. 1950. "Wobbles"—an hereditary disease in horses. *J. Heredity* 41:319.

Doll, E. R. 1953. Evidence of production of anti-isoantibodies by foals with hemolytic icterus. *Cornell Vet.* 43:44.

Doll, E. R., and F. E. Hull. 1951. Observations on hemolytic icterus of newborn foals. *Cornell Vet.* 41:14.

Donahue, M. 1935. Navicular disease in horses. *Vet. Med.* 30:244.

Dungsworth, D. I., and M. E. Fowler. 1966. Cerebellar hypoplasia and degeneration in a foal. *Cornell Vet.* 56:17.

Dunn, H. O., J. T. Vaughan, and K. McEntee. 1974. Bilaterally cryptorchid stallion with female karyotype. *Cornell Vet.* 64:265.

Eaton, O. H. 1937. A summary of lethal characters in animals and man. *J. Heredity* 28:320.

Eldridge, F., and W. F. Blazak. 1976. Horse, ass, and mule chromosomes. *J. Heredity* 67:361.

Eriksson, K. 1955. Hereditary aniridia with secondary cataract in horses. *Nord. Vet.* 7:773.

Eriksson, K., and H. Sandstedt. 1938. Hereditary malformation of the iris and ciliary body with secondary cataract in the horse. *Svensk Vettidskr.* 43:11.

Falco, J., K. Whitwell, and A. C. Palmer. 1976. An investigation into the genetics of "wobbler disease" in Thoroughbred horses in Britain. Equine Vet. J. 8:165.

Finnocchio, E. J. 1970. Congenital patellar ectopia in a foal. *J. A. V. M. A.* 156:222.

Fischer, H., and K. Helbig. 1951. A contribution to the question of the inheritance of patella dislocation in the horse. *Tierzucht.* 5:105.

Flechsig, J. 1950. Hereditary cryptorchism in a depot stallion. *Tierzucht.* 4:208.

Franks, D. 1962. Horse blood groups and hemolytic disease of the newborn foal. *Ann. N.Y. Acad. Sci.* 97:235.

Fraser, W. 1966. Two dissimilar types of cerebellar disorders in the horse. *Vet. Rec.* 78:608.

Gardner, E. J., J. L. Shupe, N. C. Leone, and A. E. Olson. 1975. Hereditary multiple exostosis. *J. Heredity* 66:318.

Gilman, J. P. W. 1956. Congenital hydrocephalus in domestic animals. *Cornell Vet.* 46:482.

Gluhovski, N., H. Bistriceanu, A. Sucui, and M. Bratu. 1970. A case of intersexuality in the horse with type 2A + XXXY chromosome formula. *Brit. Vet. J.* 126:522.

Gonzalez, B. M., and V. Villegas. 1928. "Bighead" of horses: A heritable disease. *J. Heredity* 19:159.

Hadorn, E. 1961. *Developmental Genetics and Lethal Factors.* New York: Wiley.

Hadwen, S. 1931. The melanomata of gray and white horses. *Can. Med. Assoc. J.* 25:519.

Hamori, D. 1940. Genetical notes. Congenital patella sublaxation in horses. *Alatorv. Lapok* 63:141.

Hamori, D. 1940. Inheritance of the tendency to hernia in horses. *Allatorv. Lapok* 63:136.

Hamori, D. 1941. Genetical notes. Myopia. *Allatorv. Lapok* 64:101.

Hamori, D. 1941. Parrot mouth and hog mouth as inherited deformities. *Allatorv. Lapok* 64:57.

Hermans, W. A. 1970. A hereditary anomaly in Shetland ponies. *Netherland J. Vet. Sci.* 3:55.

Hintz, H. F., and L. D. Van Vleck. 1979. Lethal dominant roan in horses. *J. Heredity.* 70:145.

Hitenkov, G. G. 1941. Stringhalt in horses and its inheritance. *Vestn. Seljskohoz. Nauki Zivotn.* 2:64.

Hosoda, T. 1950. On the heritability of susceptibility to wind-sucking in horses. *Jap. J. Zootech. Sci.* 21:25.

Hughes, J. P., K. Benirschke, P. C. Kennedy, and A. Trommershausen-Smith. 1975. Gonadal Dysgenesis in the mare. *J. Reprod. Fertil.* (Suppl. 23):385.

Hughes, J. P., and A. Trommershausen-Smith. 1977. Infertility in the horse associated with chromosomal abnormalities. *Aust. Vet. J.* 53:253.

Huston, R., G. Saperstein, and H. W. Leipold. 1977. Congenital defects in foals. *J. Equine Med. Surg.* 1:146.

Hutchins, D. R., E. E. Lepherd, and I. G. Crook. 1967. A case of equine haemophilia. *Aust. Vet. J.* 43:83.

Jeffcott, L. B. 1975. The transfer of passive immunity to the foal and its relation to immune status after birth. *J. Reprod. Fert.* (Suppl. 23):727.

Jogi, O., and I. Norberg. 1962. Malformation of the hip joint in a Standardbred horse. *Vet. Record* 74:421.

Jones, T. C., and F. D. Maurer. 1942. Heredity in periodic ophthalmia. *J. A. V. M. A.* 101:248.

Jones, W. E. 1979. The overo white foal syndrome. *J. Equine Med. Surg.* 3:54.

Kaleff, B. 1935. Inheritance of flat-hoof in the horse. *Z. Zucht.* 33:153.

Kent, M. 1981. The X-Y sex reversal syndrome. *Horse Illustrated.* 5:30.

Kieffer, N. M., N. Judge, and S. Burns. 1971. Some cytogenetic aspects of an Equus caballus intersex. *Mammalian Chromosome News Letter* 12:18.

Kieffer, N. M., S. J. Burns, and N. G. Judge. 1976. Male pseudohermaphroditism of the testicular feminizing type in a horse. *Equine Vet. J.* 8:38.

Koch, P. 1957. The heritability of chronic pulmonary emphysema in the horse. *Dtsch. tierarztl. Wschr.* 64:485.

Koch, P., and H. Fischer. 1951. Oldenburg foal ataxia as a hereditary disease. *Tierarztl. Umsch.* 6:158.

Koch, W. 1936. Some hereditary diseases in the horse and their practical significance. *Munch. tierarztl. Wschr.* 87:181.

Lambert, W. V., S. R. Speelman, and E. P. Osborn. 1939. Differences in incidence of encephalomyelitis in horses. *J. Heredity* 30:349.

Lauprecht, E. 1935. Inheritance of twinning tendency in the horse. *Zuchtungskunde* 10:433.

Leiberman, B. 1982. Sex reversal and the mutant mare. *Equus* 53:56.

Leipold, H. W. 1971. Adactylia and Polydactylia in a Welsh Foal. Vet. Med. 66:924.

Leipold, H. W., G. W. Brandt, M. Guffy, and B. Blauch. 1974. Congenital atlanto-occipital fusion in a foal. *Vet. Med. Small Anim. Clin.* 69:1312.

Lerner, D. J., and M. D. McCracken. 1978. Hyerelastosis cutis in 2 horses. *J. Equine Med. Surg.* 2:350.

Lerner, I. M. 1944. Lethal and sublethal characters in farm animals. *J. Heredity* 35:219.

Lose, M. P. 1978. A supernumerary leg in a Thoroughbred filly foal. *Vet. Med.* 73:1071.

Mason, T. A. 1981. A high incidence of congenital angular limb deformities in a group of foals. *Vet. Rec.* 109:93.

McChesney, A. E., J. J. England, J. L. Adcock, L. L. Stackhouse, and T. L. Chow. 1970. Adenoviral infection in suckling Arabian foals. *Path. Vet.* 7:547.

McChesney, A. E., J. J. England, and L. J. Rich. 1973. Adenoviral infections in foals. *J. A. V. M. A.* 162:545.

McFeely, R. A., W. C. D. Hare, and J. D. Biggers. 1967. Chromosome studies in 14 cases of intersex in domestic animals. *Cytogenetics* 6:242.

McGuire, T. C., and M. J. Poppie. 1973. Hypogammaglobulinemia and thymic hypoplasia in horses: A primary combined immunodeficiency disorder. *Infect. Immu.* 8:272.

McGuire, T. C., and M. J. Poppie. 1973. Primary hypogammaglobulinemia and thymic hypoplasia in horses. *Fed. Proc.* 32:821.

McGuire, T. C., M. J. Poppie, and K. L. Banks. 1974. Combined (B- and T-lymphocyte) immunodeficiency: A fatal genetic disease in Arabian foals. *J. A. V. M. A.* 164:70.

Mackay-Smith, M. P. 1963. Discussion of pathogenesis and pathology of equine osteoarthritis. *J. A. V. M. A.* 141:1248.

Mahaffey, L. W. 1968. Abortion in mares. *Vet. Rec.* 82:681.

Mayhew, I. G., A. deLahunta, and R. H. Whitlock. 1977. Spinal cord disease in the horse. *Cornell Vet.* 68(Suppl. 6):11.

Mayhew, I. G., A. G. Watson, and J. A. Heissan. 1978. Congenital occipitoatlantoaxial malformations in the horse. *Equine Vet. J.* 10:103.

Manning, J. P. 1963. Equine hip dysplasia-osteoarthritis. *Mod. Vet. Prac.* 44:44.

Mauderer, H. 1938. Hereditary defects in the horse. *Dtsch. Tierartzl. Wschr.* 46:469.

Mauderer, H. 1942. Abrachia and torticollis: Lethal factors in horse breeding. *Zeit. tierarztl. Auzht.* 51:215.

Meacham, T., and C. Hutton. 1968. Reproductive efficiency in 14 horse farms. *J. Anim. Sci.* 27:434.

Miller, J. E. 1917. Horned horses. *J. Heredity* 8:303.

Montali, R. J., M. Bush, and R. M. Sauer. 1974. Spinal ataxia in zebras. 1974. *Vet. Path.* 11:68.

Morgan, J. P., W. D. Carlson, and O. R. Adams. 1962. Hereditary multiple exostosis in the horse. *J. A. V. M. A.* 140:1320.

Noda, H. 1975. Studies on the relationship between hemolytic icterus of newborn foals and blood groups and the serological diagnosis. *Jap. J. Vet. Res.* 23:103.

Palmer, A. C., W. F. Blakemore, W. R. Cook, H. Platt, and K. E. Whitwell. 1973. Cerebellar hypoplasia and degeneration in the young Arab horse: Clinical and neuropathological features. *Vet. Record* 93:62.

Perryman, L. E., and R. L. Torbeck. 1980. Combined immunodeficiency of Arabian horses: Confirmation of autosomal recessive mode of inheritance. *J. A. V. M. A.* 176:1250.

Plank, G. M. van der. 1936. Pathology and inheritance. *Neue Forsch. Tierz. Abstammungsl.* 233:237.

Poppie, M. J., and T. C. McGuire. 1977. Combined immunodeficiency in foals of Arabian breeding: Evaluation of mode of inheritance of estimation of prevalence of affected foals and carrier mares and stallions. *J. A. V. M. A.* 170:31.

Prawochenski, R. T. 1936. A case of lethal genes in the horse. *Nature* 137:869.

Prawochenski, R. T. 1941. *Proc. 7th International Genetics Congress.* Edinburgh: 241.

Pulos, W., and F. B. Hutt. 1969. Lethal dominant white in horses. *J. Heredity* 60:59.

Roberts, S. J. 1956. *Veterinary Obstetrics and Genital Disease.* Ithaca, New York: Author's publications.

Rooney, J. R. 1963. Equine incoordination. *Cornell Vet.* 53:411.

Rooney, J. R. 1966. Contracted foals. *Cornell Vet.* 56:172.

Rooney, J. R., and M. E. Prickett. 1966. Foreleg splints in horses. *Cornell Vet.* 56:259.

Rooney, J. R., and M. E. Prickett. 1967. Congenital lordosis of the horse. *Cornell Vet.* 57:417.

Rooney, J. R., C. W. Raker, and K. J. Harmany. 1971. Congenital lateral luxation of the patella in the horse. *Cornell Vet.* 61:670.

Rossdale, P. D. 1968. Abnormal perinatal behavior in the Thoroughbred horse. *Brit. Vet. J.* 124:540.

Runciman, B. 1940. Roaring and whistling in Thoroughbred horses. *Vet. Record.* 53:37.

Sanger, V. L., R. E. Mairs, and A. L. Trapp. 1964. Hemophilia in a foal. *J. A. V. M. A.* 142:259.

Scekin, V. A. 1973. The inheritance of stringhalt (cock gait) in the horse. *Konevodstru* 2:20.

Schlaak, F. 1942. Investigations on the inheritance of umbilical hernias in a horse breeding region. *Z. Tierz. Zuchtbiol.* 52:198.

Schlotthauer, C. F., and P. E. Zollman. 1956. The occurrence of so-called "white heifer" disease in a white Shetland Pony mare. *J. A. V. M. A.* 129:309.

Schneider, J. E., and H. W. Leipold. 1978. Recessive lethal white in two foals. *J. Equine Med. Surg.* 2:479.

Schulte, M. J. 1979. Positive H-Y antigen testing on a case of X-Y gonadal absence syndrome. *Clinical genetics* 16:438.

Scott, A. M., and L. B. Jeffcott. 1978. Haemolytic disease of the newborn foal. *Vet. Rec.* 103:71.

Severson, B. D. 1917. Cloven hoof in Percherons. *J. Heredity* 8:466.

Severson, B. D. 1918. Extra toes in horse and steer. *J. Heredity* 9:39.

Sharp, A. J., S. S. Wachtel, and K. Benirschke. 1980. H-Y antigen in a fertile XY female horse. *J. Repro. Fert.* 58:1557.

Shupe, J. L., A. E. Olson, and R. P. Sharma. 1970. Multiple exostosis in horses. *Mod. Vet. Pract.* 51:34.

Shupe, J. L., N. C. Leone, A. E. Olson, and E. J. Gardner. 1979. Hereditary multiple exostosis: clinicopathologic features of a comparative study in horses and man. *Am. J. Vet. Res.* 40:751.

Smith, G. A. 1968. A case of convulsive syndrome in a newborn hunter foal. *Vet. Rec.* 83:588.

Speed, J. G. 1958. A cause of malformation of the limbs of Shetland Ponies with a note on its phylogenic significance. *Brit. Vet. J.* 114:18.

Splitter, G. A., L. E. Perryman, N. S. Magnuson, and T. C. McGuire. 1980. Combined immunodeficiency disease of horses: a review. *Dev. Compar. Immunol.* 4:21.

Sponseller, M. L. 1967. Equine cerebellar hypoplasia and degeneration. *Proc. 13th Ann. Meeting Am. Assoc. Eq. Pract.* New Orleans.

Spurrell, F. A., L. V. Baudin, and W. J. L. Felts. 1965. Radiography of the forelimb of the horse. Proc. 11th Ann. Meeting A. A. E. P. Miami Beach.

Stanic, M. N. 1960. *Mod. Vet. Pract.* 41:30.

Stormont, C. 1958. Genetics and disease. In *Advances in Veterinary Science*. C. A. Brandly and E. L. Jungheer, eds. New York: Academic Press 4:137.

Stormont, C., and Y. Suzuki. 1964. Genetic systems of blood groups in horses. *Genetics* 50:915.

Theile, H. 1958. Polydactyly in a foal. *Monatshefte Vet. Med.* 13:342.

Thompson, D. B., M. J. Studdert, R. G. Beilharz, and I. R. Littlejohns. 1975. Inheritance of lethal immunodeficiency disease of Arabian foals. *Aust. Vet. J.* 51:109.

Trommershausen-Smith, A. 1977. Lethal white foals in mating of overo spotted horses. *Theriogenology* 8:303.

Trommershausen-Smith, A., J. P. Hughes, and D. P. Neely. 1979. Cytogenetic and clinic findings in mares with gonadal dysgenesis. *J. Repro. Fertil.* (Suppl. 27):271.

Trommershausen-Smith, A., A. Stormont, and Y. Suzuki. 1975. Alloantibodies: Their role in equine neonatal isoerythrolysis. In H. Kitchen and J. D. Krehbel, eds. *Proc. 1st Int. Symp. Equine Hematology*. Golden, Colorado: American Association of Equine Practitioners.

Tuff, P. 1945. Inheritance of inguinal hernia in domestic animals. *Norsk. Veterinaertidsskrift* 57:332.

Tuff, P. 1948. The inheritance of a number of defects in the joints, bones, and ligaments of the foot of the horse. *Norsk. Veterinaertidsskrift* 60:385.

Vandeplassche, M., L. Podliachouk, and R. Beaud. 1970. Some aspects of twin gestation in the mare. *Can. J. Comp. Med.* 34:218.

Van Der May, G. J. W., E. F. Kleyn, and C. C. Van De Watering. 1967. Investigation of hereditary predisposition to navicular disease. *Tijdschr. Diergenessk.* 92:1261.

Walker, K. S., and A. H. Bruere. 1978. XO condition in mares. *New Zealand Vet. J.* 27:18.

Weber, W. 1947. Congenital cataract, a recessive mutation in the horse. *Schweiz. Arch. Tierheilk.* 89:397.

Weber, W. 1947. Schistosoma reflexum in the horse, with a contribution on its origin. *Schweiz. Arch. Tierheilk.* 89:244.

Weischer, F. 1949. Clarifying the hereditary and environmental relationships in equine mallendars. *Tierarztl. Umsch.* 4:318.

Weisner, E. 1955. The importance of inherited eye defects in horse breeding. *Tierzucht.* 4:310.

Wheat, J. D., and P. C. Kennedy. 1953. Cerebellar hypoplasia and its sequilae in a horse. *J. A. V. M. A.* 131:241.

White, D. J., and D. A. Farebrother. 1969. A case of intersexuality in the horse. *Vet. Rec.* 85:203.

Whitwell, K. E. 1978. Combined immunodeficiency in Arabian foals. *Vet. Rec.* 103:568.

Wille, H. 1945. Is inguinal hernia hereditary in the horse? *Norsk. Veterinaertidsskrift* 57:332.

Wriedt, C. 1924. Vererbungsfaktoren bei weissen Pferden im Gestut *Fredriksborg. Zeit. Tierz. Zuchtungsbiol.* 1:231.

Wriedt, C. 1930. *Heredity in Livestock.* London: Macmillan.

Wussow, W., and W. Hartwig. 1955. Genetic investigations on mallenders in cold-blood horses. *Tierzucht.* 9:195.

Yamane, J. 1927. Atresia coli in the horse. *Zeit. Induktive Abst. Vererbungslehre* 46:188.

CHAPTER 14

RELATIONSHIPS AND INBREEDING

Horses are often said to have so much of the blood of one horse and so much of the blood of another. Naturally this is not literally true, but these estimates may actually apply to the fraction of genes that come from some common ancestor.

For example, the foal in Figure 14-1 received a sample half of the genes of its Appaloosa sire. The relationship between a parent and its offspring is said to be 50 percent. An animal (noninbred) is related to itself by 100 percent. This halving of relationships and genes in common occurs with each generation. The relationship to each parent is 50 percent, to each grandparent is 25 percent, to each great-grandparent is 12.5 percent, and so on. Thus, after only a few generations, any ancestor is likely to be the source of only a small fraction of the genes of its descendants. These rules hold when all the ancestors are unrelated. When some are related, inbreeding results, and the problem of determining relationships is more complex.

Relationships between individuals not in direct line of descent but with a common ancestor can also be determined essentially by a halving for each intervening animal in the pedigree. For example, colts by the same stallion but different mares (paternal half-sibs, Figure 14-2) have an average relationship of 25 percent. Maternal half-sibs are also 25 percent related to each other. Fraternal twins have the same sire and dam as do full brothers or sisters (Figure 14-3).They are related by 50 percent to each other (25 percent through the sire and 25 percent through the dam), as are full sibs born at different times. Identical twins are genetically alike, so they are 100 percent related to each other. The frequency of all twinning (fraternal and identical) in horses, however, is low, so not many of the even less frequent identical twins are encountered.

Figure 14-1
D-R Joker's Playboy, an Appaloosa stallion, and one of his "leopard" foals. (Photograph courtesy of Linda Roberts, South Lyon, Michigan.)

Some samples of common relationships are given in Figure 14-4. Pedigrees are indicated by arrows to increase the reader's understanding of the nomenclature of relationships.

14.1   Importance of Relationships

There are two important reasons to calculate additive relationships between relatives. First, the *additive relationship* is a measure of the fraction of like genes shared by two animals and thus shows how reliable one of the relative's records will be in predicting the genetic value of the other. Second, the *inbreeding coefficient* of an animal, which, if high, is usually undesirable, is calculated as half the additive relationship between the parents. For example, if the non-inbred son in Figure 14-5 is mated to his dam, what will

Figure 14-2
A group of paternal half-sib foals. (Photograph courtesy of Ruth Thompson,
American Albino Association, Crabtree, Oregon.)

be the inbreeding of the resulting foals? The dam and son additive relationship is 50 percent. Since the relationship of the parents is halved, the foals will be 25 percent inbred.

Linebreeding is closely related to inbreeding. Actually, it is a form of inbreeding aimed at trying to maintain a close relationship to a particular animal. Thus, lines usually trace to or are named after the animal that is used as a base. For example, the King Ranch breeding program for Quarter Horses was based on linebreeding to what they considered the ideal Quarter Horse, Old Sorrel. An example of matings to Wimpy, a double grandson of Old Sorrel, is shown in Figure 14-6. A 1945 study of King Ranch Quarter Horses showed an average relationship of 40 percent to Old Sorrel.

Thus, knowledge of relationships can be helpful in selecting animals on the basis of relatives' records or in arranging matings to avoid high levels of inbreeding.

Figure 14-3
A Belgian mare and her full brother-sister progeny. (Photograph courtesy of John Briggs, Cornell University, Ithaca, New York.)

14.2   Computing Relationships

Unfortunately, there is no very easy way to compute the relationships and inbreeding of horses in a herd, but anyone can do the computations by following a few simple rules. The procedure appears to be quite complicated, but working two or three examples will show that many relationships can be calculated more simply and more quickly by using these rules than by any other method.

The next few paragraphs describe a procedure for calculating the relationships among all the animals in a herd. The procedure takes quite a lot of time and a large sheet (or sheets) of paper but is relatively easy. The method

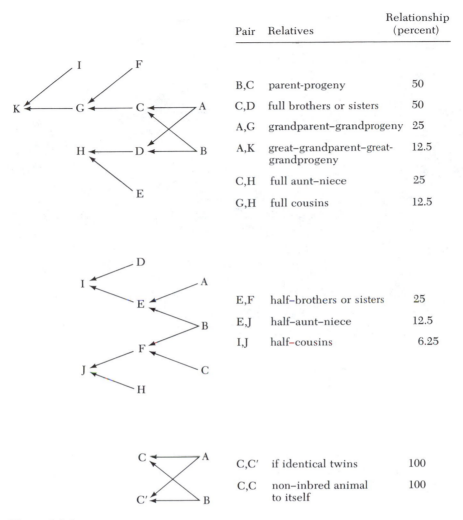

Pair	Relatives	Relationship (percent)
B,C	parent–progeny	50
C,D	full brothers or sisters	50
A,G	grandparent–grandprogeny	25
A,K	great–grandparent–great-grandprogeny	12.5
C,H	full aunt–niece	25
G,H	full cousins	12.5
E,F	half–brothers or sisters	25
E,J	half–aunt–niece	12.5
I,J	half–cousins	6.25
C,C'	if identical twins	100
C,C	non–inbred animal to itself	100

Figure 14-4
Examples of common relationships.

is based on the fact that if two animals are related, then one or both of the parents of the younger of the two must also be related to the older animal of the pair. In fact, if C and D are the two animals and A and B are the parents of D, the additive relationship between animals C and D is one-half the relationship between A and C plus one-half the relationship between B and C. [$a_{CD} = (1/2)a_{AC} + (1/2)a_{BC}$, where a stands for the additive relationship and the subscripts refer to the animals that are related.] The basis for this method is illustrated in Figure 14-7. Similarly, the coefficient of inbreeding of an

Figure 14-5
Leeward Bodick, a perlino Gotland colt, and his buckskin dam, Honung.
(Photograph courtesy of John R. Price, Shady Trail Ranch, Bonner Springs, Kansas.)

animal is calculated as one-half the additive relationship between its parents, as diagrammed in Figure 14-8.

Tabular Method

The procedure is called the tabular method because it requires construction of a table that, when finished, will give the additive relationship of any animal to any other in the herd. For example, in Table 14-1, A and B are parents of D, while C has as one parent animal A. The other parent of C is not related to A or B so it is ignored. The table gives the additive relationships among all pairs of the 4 animals. To find the additive relationship between C and D, go to row C and then across to column D. The additive relationship is 1/4 or 25 percent.

The steps for computing additive relationships and inbreeding are:

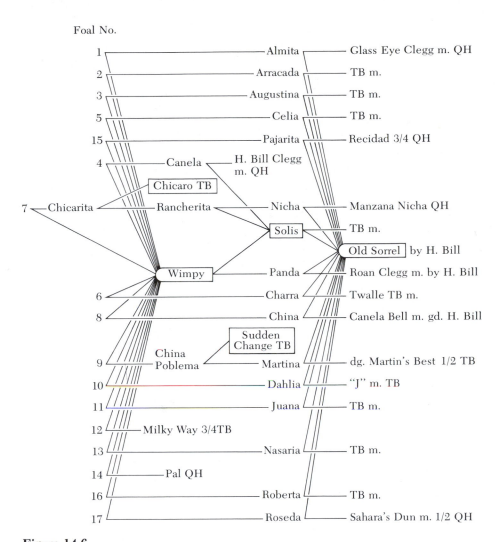

Figure 14-6
Matings to Wimpy, a Quarter Horse stallion. Wimpy, a double grandson of Old Sorrel, was mated to mares whose average relationship to him was equivalent to their being his half-sisters. Wimpy's foals derived nearly half (44.9 percent) of their genes from Old Sorrel. They derived 67 percent of their genes from the Quarter Horse breed and the remaining 33 percent from the Thoroughbred. All sires are boxed in. All horizontal lines trace female lines of descent; all diagonal lines trace male lines of descent. The foals are numbered consecutively in an alphabetical listing of their dams. For example, foal number one is out of Almita by Wimpy; Almita is out of the Glass Eye Clegg mare by Old Sorrel. (Adapted from Rhoad and Kleberg. 1946.)

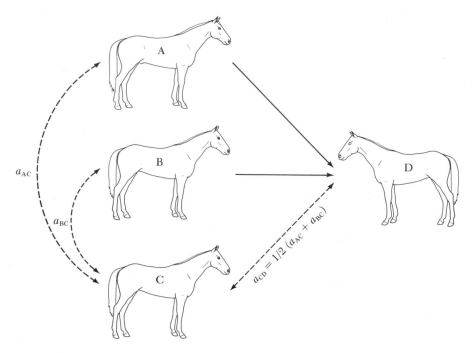

Figure 14-7
Basis for tabular method of computing relationships is that the relationship between C and D is one-half the relationship between C and the sire of D plus one-half the relationship between C and the dam of D.

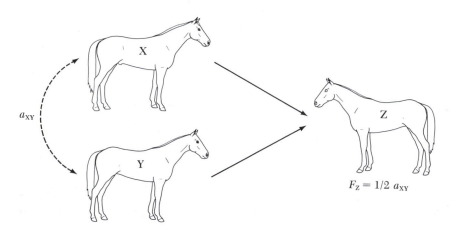

Figure 14-8
The inbreeding coefficient for an animal (F_z for animal Z) is one-half the relationship between the parents (a_{xy} for X and Y).

Table 14-1
Relationship table

			Animals (columns)			
		A	B	A-? C	A-B D
	A	1	0	1/2	1/2
	B	0	1	0	1/2
Animals (rows)	C	1/2	0	1	1/4
	D	1/2	1/2	1/4	1

1. First, determine which animals are to be included. Put them in order by date of birth, oldest first. This step is very important.

2. Write the names or numbers of the animals in order of birth across the top of the table (the columns) and down the left side of the table (the rows) as shown in Table 14-1.

3. Write above the numbers of the animals the names or numbers of their parents, if known.

4. Put a 1 in each of the diagonal cells of the table, such as row 1, column 1; row 2, column 2. This is the animal's basic additive relationship of 100 percent to itself unless the animal is inbred. If the inbreeding coefficient of any of the first or base animals is known, add that to the diagonal for that animal. All other inbreeding coefficients will be computed as in step 6.

5. Compute entries for each off-diagonal cell of row 1 according to the rule of 1/2 the entry for the first parent in this row plus 1/2 the entry for the second parent in the row. When the first row is finished, write the same values down the first column. If this step is not correctly followed, difficulty will almost always result.

6. Go to the next row and begin at the diagonal, which now has a 1 in it. Add to that 1, one-half of the relationship between the animal's parents, which can be found from an earlier entry in the table — or perhaps the parents are known or assumed to be unrelated. This is the inbreeding coefficient described in rule 4. Often the inbreeding coefficient is zero. Continue across the row as before, computing the off-diagonal entries according to rule 5. Put the values for this row down the corresponding column.

7. Continue in this manner until the table is complete, always remembering to do a row at a time and to put the same values down the corresponding column before going to the next row.

In summary, the first basic step is to add to the 1 in the diagonal the inbreeding coefficient, which is one-half the relationship between the animal's parents. This value is found at the intersection of the row and column of the parents, as shown in Figure 14-8. The second basic step is to compute the off-diagonal relationships. In Figure 14-7, the relationship is one-half the sum of the two relationships that appear to the left in the same row.

Because these rules may seem complicated, the following example is given to clarify the procedure.

Example

Suppose that a stallion, sire A, is mated to his own daughter, C, as in Figure 14-9. Note that the only individuals in an animal's pedigree that must be recorded are its parents. The animals, in order by age, are A, B, C, and D, and should be written as described in step 1. Write the parents, if known, above the letters as shown in Table 14-2. The parents of A and B can be ignored unless they are known to be related to each other. Write 1's in all 4 diagonal cells [where the (A,A), (B,B), (C,C), and (D,D) rows and columns meet].

Now begin on row 1, that is, the row for A. B is assumed not to be related to A so enter zero under the column labeled B. If the relationship of A and B is known and it is not zero, enter that instead.

The entry for row A, column C is determined according to rule 5 (1/2 the entry for row A, column A plus 1/2 the entry for row A, column B) since A and B are the parents of C. Thus, $(1/2)(1) + (1/2)(0) = 1/2$.

The entry for row A, column D will be 1/2 the (A,A) entry plus 1/2 the (A,C) entry, which is $(1/2)(1) + (1/2)(1/2) = 3/4$. The first row is now finished.

The last part of rule 5 is to write the values in row A down column A. Column A will be 1, 0, 1/2, and 3/4. The additive relationships of animal A to the other 3 animals have now been computed.

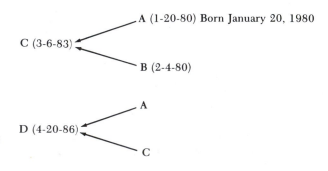

Figure 14-9
Pedigree for a stallion, A, mated to his daughter, C.

Table 14-2
Relationship table for the sample pedigree in Figure 14-9 (The relationships are given in fractions; to obtain percentages, multiply by 100.)

Birthdate:	1-20-80	2-4-80	3-6-83	4-20-86
Parents:	—	—	A-B	A-C
Animal	A	B	C	D
A	1	0	1/2	3/4
B	0	1	1/2	1/4
C	1/2	1/2	$1 + (1/2)(0) = 1$	3/4
D	3/4	1/4	3/4	$1 + (1/2)(1/2) = 1 + 1/4$

The second row, row B, is done similarly, beginning at row B, column B. The parents of B were not related, so B is not inbred. Therefore, do not add anything to the 1 already in the (B,B) cell.

The entry for row B, column C is $(1/2)(0) + (1/2)(1) = 1/2$. The value for row B, column D is $(1/2)(0) + (1/2)(1/2) = 1/4$. Now write the values for row B down column B: 0, 1, 1/2, and 1/4. The 0 and 1 were already there.

Next, row C. The row C, column C entry is the 1 already there plus 1/2 the additive relationship between the parents of C, which are A and B. Look at the entry for row A, column B, and find zero for the additive relationship between A and B. The entry is then $1 + (1/2)(0) = 1$.

The row C, column D entry is $(1/2)(1/2) + (1/2)(1) = 3/4$. Write this value also into row D, column C, according to step 5.

The last entry is row D, column D. Add to the 1 already there, 1/2 the additive relationship between the parents, A and C. A and C are found to be related by 1/2 from the row A, column C entry. The row D, column D entry is then $1 + (1/2)(1/2) = 1 + 1/4$. Thus, D is 1/4 or 25 percent inbred.

Note from this special kind of mating that animal D is related to A, her father, by 75 percent and also by 75 percent to her mother, C, but is related to her grandmother only by 25 percent. The only way that two animals can be related by more than 50 percent is if one or both the animals is inbred. This was true with D and C and also with D and A. D is inbred.

For a herd of any size, the amount of paper required will be large and the amount of time to complete the table will be great. The labor involved, however, is simple and straightforward. Before starting a large herd, it would be a good idea to practice on a small segment of the herd and to work through the examples given here, or to take a particular horse family and find the relationships among the members of the family. Many breeders might be surprised by the results.

Reading the Table

To review, the relationship of any animal to any other animal is obtained by finding the row of the first animal and then going over to the column of the second animal. The value at the intersection of the row and column is the relationship between the first animal and the second animal. The entries in the diagonal (for example, row A and column A) give the relationship of an animal to itself. If the value is greater than 1, the excess over 1 is the inbreeding coefficient for that animal. For example, the value for animal D is found in row D, column D to be $1 + 1/4$. The excess over 1 is $1/4$, so the inbreeding coefficient of D is $1/4$ or 25 percent.

Since an inbred animal will have an additive relationship with itself (diagonal in the table) of greater than 1, the question may arise as to how relationships can be greater than 1. Actually, the additive relationship can be as small as zero and as large as 2. The additive relationship is twice the probability of identical genes occurring in the 2 animals. The maximum of such a probability for identical twins is 1; thus, doubling gives a maximum additive relationship of 2. The additive relationship is, however, the one that gives the fraction of gene effects alike and, thus, is the one used in devising weighting factors for records of relatives in genetic evaluation. The additive relationship between the parents of an animal is also used in computing the inbreeding coefficient for an animal. Even if the additive relationship seems somewhat illogical, it is the most useful measure of relationships. The *coefficient of relationship* is often used to compute a relationship that can have a minimum value of zero and a maximum value of 1. For non-inbred animals, the coefficient of relationship is the same as the additive relationship. For inbred animals (say, A and B), the coefficient of relationship can be computed as $r_{AB} = (a_{AB})/(\sqrt{a_{AA}a_{BB}})$. For example, suppose that we want to find the coefficient of relationship between animals C and D in Table 14-2: $a_{CD} = 3/4$, $a_{CC} = 1$, and $a_{DD} = 1\frac{1}{4}$. Thus, $r_{CD} = (3/4)/[\sqrt{(1)(1\frac{1}{4})}] = .67$ rather than $a_{CD} = .75$. The coefficient of relationship between non-inbred animals A and C (Table 14-2) is $r_{AC} = (1/2)/[\sqrt{(1)(1)}] = .5$, which is the same as a_{AC}.

A Real Example

A more realistic example would be to compute the relationship between one of the foundation sires of the Tennessee Walking Horse breed, Allan F-1, and his great-great-grandson, Hound Allen. The pedigree is shown in Figure 14-10. The horses have been labeled A, B, . . . , J for the purpose of computing the relationships in Table 14-3. The completed table shows all relationships among the 10 animals and their inbreeding coefficients. J. Lane Fletcher computed relationships and inbreeding coefficients for a sample of the Tennessee Walking Horse breed and found average inbreeding to be

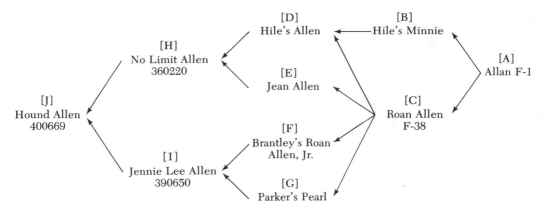

Figure 14-10
Pedigree of Hound Allen, a Tennessee Walking Horse. This pedigree shows the extent of inbreeding to Allan F-1 and to his prominent sons, among whom the most important is Roan Allen. The coefficient of inbreeding of Hound Allen is 14.06 percent (12.5 percent from Roan Allen and 1.56 percent from Allan F-1). (Adapted from Fletcher, 1946.)

low: 1.24 percent for foundation animals and 3.62 percent for those born in 1940.

14.3   Inbreeding In Practice

Studies in the 1980s show low inbreeding levels in Quarterhorses (1 to 3 percent and Morgans (5 to 10 percent). An in-depth study of Thoroughbreds from Weatherby's General Studbook resulted in an average inbreeding coefficient of about 1 percent for calculations based on only five generations and only 12.5 percent when pedigrees were traced to the foundation animals of the late 1600s and early 1700s — an average of less than 1/20 of 1percent per year. A study of Standardbreds showed a similar pattern — most recent animals were inbred but the level was low even when pedigrees were traced as far as possible:

Decade	1900– 1909	1910– 1919	1920– 1929	1930– 1939	1940– 1949	1950– 1959	1960– 1969	1970– 1979	1980
Average inbreeding	2%	2%	3%	4%	5%	6%	6%	8%	9%

Tracing pedigrees only five generations for animals born in 1980 resulted in an inbreeding estimate of only 4 percent. A difference this large is expected

Table 14-3

Relationships in pedigree of Hound Allen (J) show a high relationship of Allan F-1 (A) to his great-great-grandson, Hound Allen (J), as a result of linebreeding, and an even higher relationship of Roan Allen F-38 (C) to his great-grandson (J)

	A	A-? B	A-? C	B-C D	C-? E	C-? F	C-? G	D-E H	F-G I	H-I J
A	1	1/2	1/2	1/2	1/4	1/4	1/4	3/8	1/4	5/16
B	1/2	1	1/4	5/8	1/8	1/8	1/8	3/8	1/8	1/4
C	1/2	1/4	1	5/8	1/2	1/2	1/2	9/16	1/2	17/32
D	1/2	5/8	5/8	1 + 1/8	5/16	5/16	5/16	23/32	5/16	33/64
E	1/4	1/8	1/2	5/16	1	1/4	1/4	21/32	1/4	29/64
F	1/4	1/8	1/2	5/16	1/4	1	1/4	9/32	5/8	29/64
G	1/4	1/8	1/2	5/16	1/4	1/4	1	9/32	5/8	29/64
H	3/8	3/8	9/16	23/32	21/32	9/32	9/32	1 + 5/32	9/32	23/32
I	1/4	1/8	1/2	5/16	1/4	5/8	5/8	9/32	1 + 1/8	45/64
J	5/16	1/4	17/32	33/64	29/64	29/64	29/64	23/32	45/64	1 + 9/64

in most breeds because of the limited number of foundation animals. When inbreeding level accumulates at such low rates a slight amount of selection can easily overcome any deleterious effects of inbreeding. The low levels of inbreeding found for most breeds indicate that breeders consciously avoid inbreeding by not mating close relatives.

The Thoroughbred study showed that 80 percent of the genes in the 1960–1964 population traced to 31 foundation animals and 55 percent to only 12 foundation animals. Table 14-4 also shows the surprising result that a fourth foundation stallion, the Curwen Bay Barb, should be added to three stallions—the Godolphin Arabian, the Darley Arabian, and the Byerly Turk—usually considered the foundation sires of the Thoroughbred.

Cothran et al. (1984) found little relationship between inbreeding level and reproductive performance in Standardbreds at low levels (7 to 10 percent) of inbreeding. Similar results were obtained by Mahon and Cunningham (1980) in their study of Thoroughbreds. High levels of inbreeding are rare and, thus, the effects of high levels of inbreeding on reproduction are unknown, but, based on studies in many classes of livestock, its effects must be assumed to be negative.

Inbreeding appears to have little value in horse-breeding programs because of its many detrimental effects, especially the expected increase in mortality with high inbreeding levels and the increase in fertility problems. However, few systematic studies have been made of the effects of inbreeding. Although the King Ranch followed a linebreeding program in attempt-

Table 14-4
Foundation ancestors of the Thoroughbred

Ancestor	Fraction of genes in 1960–64 population
Godolphin Arabian	.146
Darley Arabian	.075
Curwen Bay Barb	.056
Byerley Turk	.048
Bethell's Arabian	.033
White Darcy Turk	.033
Old Bald Peg (mare)	.031
St. Victor Barb	.031
Lister Turk	.025
Leedes Arabian	.025
Lord Fairfax's Morocco Barb	.025
Shield's Galloway (mare)	.025
Top 12	.553
Top 31	.800

SOURCE: Adapted from Mahon, G. A. T., and E. P. Cunningham. 1980. p. 72.

ing to fix the characteristics of Old Sorrel in the Quarter Horse breed, they also avoided inbreeding as much as possible. The study by Fletcher (1945) showed a higher inbreeding (4.9 percent) for King Ranch horses than for a sample of other Quarter Horses (1.7 percent). The average relationship among the King Ranch Quarter Horses, however, was 20.1 percent, so that random mating of them would have resulted in an average inbreeding coefficient of 10 percent.

Some breeders have stressed crosses between varying sire or dam lines. The amount of inbreeding in the lines, however, has usually been low. Lasley (1972) has suggested that development of inbred lines for crossing may be advantageous. There is no evidence to support or refute such a view.

14.4   Practice Problem in Computing Inbreeding

Practice Problem

Following is an example of systematic inbreeding where a sire, **A**, is mated back to his daughters in each succeeding generation.

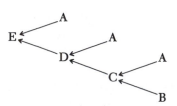

Compute the relationships among the 5 animals as well as the inbreeding coefficients of **D** and **E**. Assume that parents of **A** and **B** are unrelated.

Parents		—	—	A–B	A–C	A–D
Animal		A	B	C	D	E
	A	1	0			
	B	0	1			
	C					
	D					
	E					

Solution to Practice Problem

Parents		—	—	A–B	A–C	A–D
Animal		A	B	C	D	E
	A	1	0	1/2	3/4	7/8
	B	0	1	1/2	1/4	1/8
	C	1/2	1/2	1	3/4	5/8
	D	3/4	1/4	3/4	1 + 1/4	1
	E	7/8	1/8	5/8	1	1 + 3/8

Computations:

(row, column)

$(A,C) = (1/2)(1) + (1/2)(0) = 1/2$
$(A,D) = (1/2)(1) + (1/2)(1/2) = 3/4.$
$(A,E) = (1/2)(1) + (1/2)(3/4) = 7/8$

$(B,C) = (1/2)(0) + (1/2)(1) = 1/2$
$(B,D) = (1/2)(0) + (1/2)(1/2) = 1/4$
$(B,E) = (1/2)(0) + (1/2)(1/4) = 1/8$

$(C,C) = 1 + (1/2)(A,B) = 1 + (1/2)(0) = 1$
$(C,D) = (1/2)(1/2) + (1/2)(1) = 3/4$
$(C,E) = (1/2)(1/2) + (1/2)(3/4) = 5/8$

$(D,D) = 1 + (1/2)(A,C) = 1 + (1/2)(1/2) = 1 + 1/4$
$(D,E) = (1/2)(3/4) + (1/2)(1 + 1/4) = 1$

$(E,E) = 1 + (1/2)(A,D) = 1 + (1/2)(3/4) = 1 + 3/8$

(row, column)

$(C,A) = 1/2$
$(D,A) = 3/4$
$(E,A) = 7/8$

$(C,B) = 1/2$
$(D,B) = 1/4$
$(E,B) = 1/8$

$(D,C) = 3/4$
$(E,C) = 5/8$

$(E,D) = 1$

REFERENCES AND FURTHER READINGS

Cothran, E. G., J. W. MacCluer, L. R. Weitkamp, D. W. Pfenning, and A. J. Boyce. 1984. Inbreeding and reproductive performance in Standardbred horses. *J. Heredity* 75:220.

Cruden, Dorothy. 1949. The computation of inbreeding coefficients. *J. Heredity* 40:248.

Dwyer, D. J., D. F. Antczak, and L. D. Van Vleck. 1982. Inbreeding coefficients in a sample of Standardbred foals. *Abstracts, Am. Soc. Anim. Sci. and Can. Soc. Anim. Sci. Ann. Meeting.* Guelph, Ontario:144.

Emik, L. O., and C. E. Terrill. 1949. Systematic procedures for calculating inbreeding coefficients. *J. Heredity* 40:51.

Fletcher, J. Lane. 1945. A genetic analysis of the American Quarter Horse. *J. Heredity* 36:346.

Fletcher, J. Lane. 1946. A study of the first fifty years of Tennessee Walking Horse breeding. *J. Heredity* 37:369.

Fomin, A. B. 1966. Heterosis obtained by the pure breeding of Orlov Trotters. *Genetika, Mosk.* 11:131.

Gazder, P. J. 1954. The genetic history of the Arabian horse in the United States. *J. Heredity* 45:95.

Lantman, K., and D. J. Balch. 1978. Inbreeding: Key to uniformity at the UVM Morgan Horse Farm. *Morgan Horse Magazine*, Jan–Feb:131.

Lasley, J. F. 1972. *Genetics of Livestock Improvement.* Englewood Cliffs, N.J.: Prentice-Hall.

MacCluer, J. W., A. J. Boyce, B. Dyke, L. R. Weitkamp, D. W. Pfenning, and C. J. Parsons. 1983. Inbreeding and pedigree structure in Standardbred horses. *J. Heredity* 74:394.

Mahon, G. A. T., and E. P. Cunningham. 1980. Inbreeding and infertility in Thoroughbred mares. *Farm and Food Research* 11:72.

Murata, S., and Y. Watanabe. 1975. Genetic analysis of the Thoroughbred population in Japan. IV. Inbreeding and relationship. *Res. Bull. Faculty of Agric. Hokkaido Univ.* 7:20.

Rhoad, A. O., and R. J. Kleberg, Jr. 1946. The development of a superior family in the modern Quarter Horse. *J. Heredity* 37:226.

Steele, D. G. 1944. A genetic analysis of the recent Thoroughbreds, Standardbreds, and American Saddle Horses. *Kentucky Agr. Exp. Sta. Bull.* 462.

Tunnell, J. A., Sanders, J. O., J. D. Williams, and G. D. Potter. 1983. Pedigree analysis of four decades of Quarter Horse breeding. *J. Animal Sci.* 57:585.

Weitkamp, L. R., S. Guttormsen, J. McKnight, N. Wert, J. Witmer, R. Pierson, E. Speece, and J. Egloff. 1980. Equine marker genes: Transferrin mating types and fertility in Standardbreds. *Anim. Blood Groups and Biochem. Genetics* 11(Suppl. 1):45.

Weitkamp, L. R., J. W. MacCluer, S. Guttormsen, J. McKnight, N. Wert, J. Witmer, P. Boyce, and J. Egloff. 1982. Genetics of Standardbred stallion reproductive performance. *J. Reprod. Fert.* (Suppl.) 32:135.

Wright, Sewall. 1922. Coefficients of inbreeding and relationships. *Am. Naturalist* 56:330.

Wright, Sewall. 1923. Mendelian analysis of the pure breeds of livestock. *J. Heredity* 14:339.

Zaher, A. 1948. *A genetic history of the Arabian horse in America.* Ph.D. thesis. Michigan State University, East Lansing.

CHAPTER 15

PRINCIPLES OF SELECTION FOR QUANTITATIVE TRAITS

Most traits of the horse are influenced by many genes, each with a relatively small effect; that is, the traits are not influenced primarily by genes at a single locus. Such traits are called quantitative traits. Many such traits can also be measured quantitatively, as, for example, the speed of a Standardbred for one mile, the speed of the Thoroughbred for a mile and a half or other distances (Figure 15-1), and the speed of a Quarter Horse for a quarter-mile. Measurement is more difficult for certain other traits, such as jumping ability, temperament, or cow sense. Nevertheless, if a measurement scale can be devised, possible measurements will be continuous over a wide range. Each measurement can be thought of as the sum of the genetic value of the horse and the influence of environmental factors. A shorthand model is $P_i = G_i + E_i$, where P_i is the measurement for a trait of a horse named i; G_i is the combined effect of all the genes of horse i on the trait; and E_i is the combined effect of all environmental factors on the trait.

Stating the problem of selection for such a trait is not difficult. Horses are to be selected that have the greatest possible average G. The complication, of course, is that only P is measured. The environmental effects, which can be plus or minus, mask the genetic effects. This makes prediction (estimation) of G from a measurement, P, less than perfect. Improving accuracy of prediction by using records on relatives will be discussed in Section 15.5. In the following section key factors that determine progress in improving quantitative traits by selection will be discussed.

Figure 15-1
The imported Thoroughbred °Nasrullah, at Claiborne, Paris, Kentucky, was the leading sire of stakes and purse winners for many years. (Photograph by J. C. Skeets Meadors, courtesy of Keeneland Library, Lexington, Kentucky.)

15.1   The Key Factors in Selection

Genetic variation is of prime importance. The term "genetic variation" describes the magnitude of differences in genetic values of horses. Obviously, if genetic values are all the same, selection cannot improve the average genetic value. Some must be better and others worse than average so that selection has a chance to improve the average genetic value. The usual measure of variation is the genetic standard deviation, which will be discussed later in this section.

Accuracy of prediction is determined by the method of prediction, by heritability, and by the number and kind of records on a horse and its relatives. Increasing the number of records and relatives will increase accuracy if the best method of predicting genetic value is used.

The method that most accurately predicts which animals are superior is called the selection index. The selection index maximizes accuracy of prediction (the correlation between predicted genetic value and true genetic value), maximizes genetic progress by selection, minimizes errors of predicting genetic value, and maximizes the probability of correctly ranking horses with respect to their genetic value.

The amount of progress to be expected from selection also depends on the *intensity of selection.* Even if genetic value were predicted perfectly, there would be no genetic improvement unless inferior animals were culled. The average genetic value of selected horses increases as more inferior horses are culled. For example, selecting the best 10 percent for breeding instead of the best 20 percent halves the fraction selected and increases the intensity of selection.

These three factors — genetic variation, accuracy of prediction, and selection intensity — determine genetic progress per generation. A fourth factor, *generation interval*, is important in determining rate of genetic improvement per year. Generation interval is the average time between birth of animals and birth of their replacements. Genetic progress per generation divided by the generation interval measured in years gives genetic progress per year.

The equation that summarizes the relationship of these factors to expected genetic progress is:

Genetic Progress per Year =
$$\frac{\text{Accuracy} \times \text{Intensity Factor} \times \text{Genetic Standard Deviation}}{\text{Generation Interval}}$$

An obvious way to maximize genetic progress is to make the numerator parts as large as possible and the generation interval as small as possible. That, indeed, is what to do, but often a change in one factor may cause an undesirable change in another factor. An understanding of the four ingredients of yearly genetic progress will be valuable to the person who must decide how to balance these key factors to increase genetic progress to the optimum for a particular breeding program.

Accuracy

Accuracy of predicting the true genetic value of a mare or stallion can often be increased by using records on relatives. This correlation between true and predicted genetic value can range from 0 percent when genetic value is guessed to 100 percent when genetic value is known exactly, and may even be negative if the predicted genetic value is determined by incorrect procedures. Genetic value is difficult to predict exactly, especially for mares, but for stallions prediction can be nearly perfect if there are enough progeny.

Accuracy depends also on the fraction of differences due to genetic effects and the fraction due to environmental effects. The fraction due to genetic differences is called *heritability*. The symbol σ_G^2 (read as genetic variance) is used for the usual measure of genetic variation. The square root of σ_G^2, σ_G, is the genetic standard deviation that appeared in the genetic progress equation. The symbol for total variance is σ_P^2 (read as phenotypic or total, variance). Total or phenotypic variance is the sum of the genetic variance and the environmental variance: $\sigma_G^2 + \sigma_E^2$. Thus, heritability $= \sigma_G^2/(\sigma_G^2 + \sigma_E^2)$.

As can be seen, if none of the differences are due to genetic differences, heritability is zero, and, of course, genetic progress is impossible. If all differences are genetic, heritability is 100 percent. Then, an animal's record is a perfect measure of its genetic value (accuracy is 100 percent from an animal's record), and genetic progress will be as rapid as the genetic standard deviation, selection intensity, and generation interval will allow.

Accuracy of prediction is greater for traits with high heritability than for traits with low heritability. Figure 15-2 shows the effect of heritability on accuracy of prediction from an animal's own record for traits with different heritability values. As heritability increases, accuracy becomes nearly perfect, even when based solely on an animal's own record. Most quantitative traits, however, have heritabilities in the range of .1 to .5. Within that range, accuracy certainly is not perfect, ranging approximately from 32 percent to 71 percent.

Figure 15-3 shows accuracy of predicting genetic value from using different numbers of half-sib progeny for traits with different heritabilities.

Figure 15-2
Accuracy of predicting genetic value depends on heritability of the trait. The graph shows the accuracy that results when selecting on the basis of a single record of the animal for traits with different heritability.

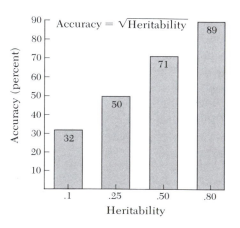

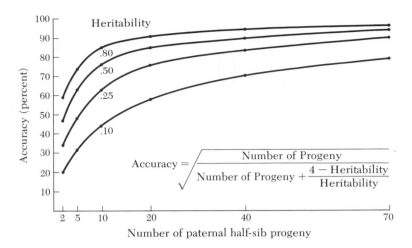

Figure 15-3
Accuracy of predicting genetic value from half-sib progeny depends on heritability of the trait and the number of progeny.

Prediction of a stallion's genetic value from his progeny can be nearly 100 percent accurate if a sufficiently large group of his progeny is used in the prediction, no matter what the heritability of the trait, if the heritability is greater than zero. Accuracy is quite high, however, when only the first few progeny are used in cases of traits with high heritabilities. Three Bars, shown in Figure 15-4, is an example of a stallion that was proved to be great through a progeny proof.

Figures 15-2 and 15-3 can be used to compare the accuracy of prediction from a horse's own records and from progeny records. Generally, for traits with high heritabilities, progeny records have little advantage over an animal's own record. The result is different for traits with low heritabilities. For example, with a heritability of .1, accuracy of prediction is only 32 percent when based on own records alone, which is equivalent to the accuracy of prediction based on four or five progeny records. The use of 30 progeny records will give an accuracy of prediction more than twice as great, 66 percent.

Estimates of heritabilities of quantitative traits of horses are based on very little research. Some representative values are given in Table 15-1.

For traits with higher heritabilities, accuracy of prediction increases very little when relatives' records are used, in comparison with the accuracy obtained when a record of the animal is used. Records on closer relatives, of course, add more to accuracy than do those of relatives two or more generations removed. As shown in Table 15-2, adding one parent record increases accuracy as much as adding all four grandparents, even if heritability is low.

Figure 15-4
Three Bars, a
Thoroughbred, was proved
by his progeny to be an
outstanding sire of racing
Quarter Horses.
(Photograph courtesy of
American Quarter Horse
Association and owner
Sidney H. Vail, Douglas,
Arizona.)

Table 15-1
Likely ranges for heritability for some traits

Trait	Heritability
Thoroughbred races	
Time	.20 – .35
Handicap-performance rate	.30 – .50
Trotting and pacing races	
Time	.20 – .35
Earnings	.20 – .35
Jumping, dressage, eventing	.05 – .25
Conformation	.30 – .50
Weight and height	.50 – .90
Cannon bone circumference	.40 – .80
Legs (score)	.10 – .20
Gallop (score)	.20 – .40
Temperament (score)	.20 – .45
Gestation length	.25 – .40
Fertility	.00 – .20

Table 15-2
Accuracy of predicting genetic value from own and ancestor records (percent)

Records used	Heritability		
	.1	.25	.50
Own	32%	50%	71%
Own + 1 parent or progeny	35	53	73
Only 1 parent or progeny	16	25	35
Own + 2 parents	38	57	76
Only 2 parents	23	35	50
Own + 1 grandparent	32	51	71
Only 1 grandparent	8	12	18
Own + 4 grandparents	35	53	73
Only 4 grandparents	16	25	35

For traits with low heritabilities, adding progeny records when an animal's record is known can increase accuracy substantially, as shown in Figure 15-5.

Intensity of Selection

The factor for intensity of selection is not quite proportional to the fraction culled. The intensity factor increases at an increasing rate as the fraction selected declines. Intensity factors for various fractions of selection are given in Table 15-3. These factors are a relative measure of how much the selected group will exceed the average of the population from which selected animals are taken. For the intensity factors to apply, selection must be based entirely on predicted genetic value. That is, selection of a random 10 out of the top 50 of 100 horses is not 10 percent selection but is equivalent to selection of the best 50 percent—the same as selecting the best 50 of 100—because selection is random out of the best 50.

Intensity of selection can have an important effect on genetic progress, particularly in selection of stallions, because fewer stallions than mares are needed for breeding. Bret Hanover, pictured in Figure 15-6, has had a large influence on the Standardbred breed. His individual performances were outstanding as were those of his progeny. In addition, the accuracy of evaluating stallions may be greater than that for mares because a record is available or because progeny records are available. Mares without records must be evaluated from records of their paternal sibs or their sire. This

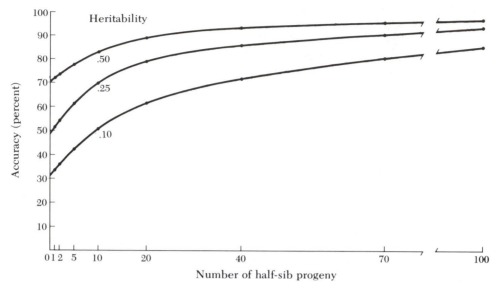

Figure 15-5
Accuracy of prediction that results from using a single record of an animal and the records of several half-sib progeny.

Table 15-3
Selection intensity factor

Select top percentage	Selection intensity factor	Comments
100	.00	No culling
90	.20	
85	.27	
75	.42	Usual level for selection of mares
70	.50	
60	.64	
50	.80	
40	.97	
30	1.16	Range for selecting dams of stud stallions
20	1.40	
10	1.75	
5	2.06	
4	2.15	
3	2.27	Possible range for selecting stallions
2	2.42	
1	2.67	

Figure 15-6
Bret Hanover, a champion pacer, shown (with trainer-driver Frank Ervin) winning the final heat for the 1964 Goshen Cup. He won 62 of 68 starts, was undefeated in 24 starts as a 2-year-old, and was the first 3-time winner of "Harness Horse of the Year" award (1964, 1965, and 1966). (Photograph courtesy of *The Horseman and Fair World*, Lexington, Kentucky.)

one-generation gap in records reduces accuracy considerably, as shown in Table 15-2.

Genetic Standard Deviation

Genetic variation must exist if selection is to be effective. Lack of genetic differences among animals will result in no genetic progress, whether prediction is accurate or selection is intense. Some traits have little genetic variation and others express considerable genetic variation. Variation in traits of horses has been measured neither very often nor for many traits. Quite likely, there is little genetic variation in fertility, but there is a great deal in performance traits such as speed and agility and in size traits such as height at the withers and weight at maturity.

The genetic standard deviation is related to heritability and total or phenotypic variation. Since heritability is defined as the ratio of genetic variance to total variance, the genetic variance is the product of heritability and total variance: $\sigma_G^2 =$ heritability $\times$ phenotypic variance.

The genetic standard deviation is the square root of the genetic variance. Since heritability must have a value between 0 and 1 depending on the trait, the genetic variance is less than or no greater than the phenotypic variance.

Variance is a standard way of describing variation. For example, measurements for many traits have a distribution that is very similar to the normal or bell-shaped distribution. The phenotypic variance describes how close or how far the individual values are likely to be from the overall average. The best time for 1 mile for a group of Standardbreds is plotted in Figure 15-7, which shows the distribution of times. The frequency of horses in each time class was plotted as a vertical bar, and if the midpoints of the tops of the bars were connected, a curve similar to the normal distribution would result.

How spread out or how closely bunched the records are determines the phenotypic variance. In fact, the square root of the variance is called the *standard deviation*, which is the standard measure for determining what size interval above or below the average will contain a specified fraction of all measurements. Tables such as Table 15-4 are available that give the fraction of measurements expected within a certain number of standard deviations from the average.

For example, 34 percent of the measurements are expected to be between the average and the average plus one standard deviation. Similarly, 34 percent are expected to be between the average and the average minus one

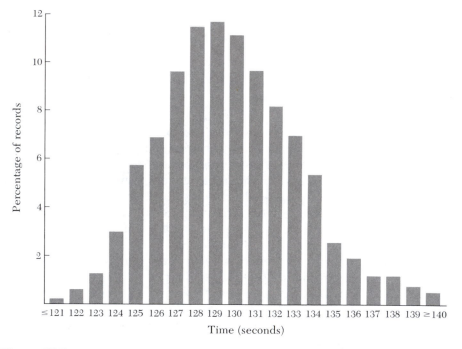

Figure 15-7
An approximation of the normal distribution from the best times (to the nearest second) at trot or pace for the mile on half-mile tracks for 2498 Standardbred dams of progeny with standard times. The standard deviation was 3.45 seconds, and the average was 129.7 seconds. (Data courtesy of Richard Hintz, Cornell University, Ithaca, New York.)

Table 15-4
Fraction of measurements expected to
be between the average and the
average plus a stated number of
standard deviations

Standard deviation (n)	Fraction[a]
.1	.04
.2	.08
.3	.12
.4	.16
.5	.19
.6	.23
.7	.26
.8	.29
.9	.32
1.0	.34
1.5	.43
2.05	.48
2.33	.49

[a]Since the normal curve is symmetrical, this is also the fraction between the average and the average minus that number of standard deviations.

standard deviation. Thus, 68 per cent are in the range of one standard deviation on either side of the average.

Suppose, for example, that the distribution of Standardbred times for one mile is really normal and that the standard deviation of speed for the mile (not just the best times) for a random sample of Standardbreds is 10 seconds and the average is 140 seconds. Then, 68 percent of the times would be expected to be between 140 ± 10 (that is, 130 and 150) seconds. Similarly, 96 percent would be between 140 ± 2.05 (10) (that is, 119.5 and 160.5) seconds. The actual distribution for Standardbred times is not quite normal since the lower limit is approximately 115 seconds. Yet this approximation works quite well.

The phenotypic variance and, hence, the phenotypic standard deviation can be computed directly from records. The procedure is not too complicated and can be found in any elementary statistics text.

The genetic values cannot be observed directly, so the genetic variance can be computed only indirectly from the correlation between relatives according to quantitative genetic theory. These techniques are also not difficult and can be found in most animal breeding textbooks.

Generation Interval

A general statement of the influence of generation interval on genetic progress is that the shorter the generation interval, the more genetic progress per year. One reason for relatively slow progress in breeding horses is the long generation interval compared with that of other farm animals. Generally, low fertility, a long gestation period, and lack of financial incentive may be some of the reasons. The generation interval, however, commonly becomes shorter when a breed is expanding rapidly. Studies made in the early 1940s of the Tennessee Walking Horse, the Quarter Horse, the Standardbred, and the Thoroughbred showed a range in generation intervals from approximately 9 years for the Quarter Horse to 12 years for the Thoroughbred. The interval for sires was longer than that for dams. Since then, the Quarter Horse breed has expanded rapidly, so their average generation interval may have become even less.

When males and females have different generation intervals, different accuracies of prediction, and different selection intensities, overall genetic progress per year can be written as:

Genetic Progress per Year =

$$\frac{\text{Selected Male Superiority} + \text{Selected Female Superiority}}{\text{Generation Interval for Males} + \text{Generation Interval for Females}}$$

where

Selected Male Superiority =
Accuracy for Males × Intensity Factor for Males × Genetic Standard Deviation

and

Selected Female Superiority =
Accuracy for Females × Intensity Factor for Females × Genetic Standard Deviation

As mentioned earlier in this section, changes in some of the key factors for genetic progress may affect other factors. For example, decreasing generation interval will probably decrease accuracy, because fewer records would be available to use in selection. Often, accuracy needs to be balanced against intensity of selection, particularly if stallions are being progeny-tested on a fixed number of mares. The genetic standard deviation is a fixed part of the equation; that is, it is a constant value for a particular trait in a specified population and thus is not affected by changes in the other key factors.

15.2   Traits to Measure and Use in Selection

Records are essential in any effective selection program. Performance records, of course, are obviously of interest, but other information is also needed to adjust performance records to a common base. For example, age influences racing speed, and post position affects the chance of winning, particularly for trotters and pacers.

The primary trait for racehorses is speed for a desired distance (Figure 15-8). Conformation and temperament are important to the extent that they influence racing ability, but racing ability is a composite measure of all such factors. There are fast horses of varying conformation and temperament, which implies that no one ideal standard for conformation and temperament is best. Racing performance will be discussed in Section 15.3.

Some factors that may influence performance are age, nutrition, skill of the trainer, starting position, condition of the track, handicap weight, sex,

Figure 15-8
Man O' War, winner of 21 of 22 races (second by ½ length) from 1919 to 1920, was the holder in 1920 of records for 1, 1⅛, 1⅜, 1½, and 1⅝ miles. Shown with groom, Will Harbut. (Photograph by J. C. Skeets Meadors, courtesy of Keeneland Library, Lexington, Kentucky.)

and, of course, level of competition. Adjustments for these factors generally have not been developed, but such adjustments are needed to apply accurately the basic principles of selection. Biases may result from the use of unadjusted records and may slow the rate of progress.

Jumpers, cutting horses, and polo ponies, for example, all can be measured in the same manner as racehorses for the performance that is required of them. Horses shown in harness and riding classes are judged on action and style, which are difficult to score on a quantitative basis. Winnings or some type of score card measure may be used as traits for selection. Figure 15-9 shows Tropical Gale in typical parade horse action. The Hackney Pony, Fernwick's Brown Jet, is shown in a Cob-Tail harness class in Figure 15-10.

Figure 15-9
Tropical Gale, a Pinto being ridden by Michelle MacFarlane, shows typical action in a parade-horse class. (Photograph by Jack Holvoet, Fort Madison, Iowa, courtesy of Ellen S. Davis, San Diego, California.)

Figure 15-10
An international champion Cob-Tail Hackney Pony, Fernwick's Brown Jet, driven by Mrs. David LaSalle, showing the action and animation selected for by Hackney breeders. (Photograph courtesy of David LaSalle, Harness Makers, North Scituate, Rhode Island.)

Horses in halter classes constitute another category (Figure 15-11). Their conformation and temperament traits are many and varied. Winnings may be a reasonable measure of composite ability to place well. The general rule of thumb is that if selection is for n traits simultaneously, progress for any one is only $1/\sqrt{n}$ as much as if selection were for that trait alone. Of course, if all n traits are important, then selection for only one trait is less preferable than simultaneous selection for all of them. The important point is to place the proper emphasis on each trait — more on valuable traits and less on fine points that have little value.

Horses used for pleasure should have traits that actually make riding them a pleasure — even temperament, smoothness of gait, and endurance. Freedom from abnormalities is also essential.

Draft horses need to have pulling power and endurance. If used commercially, they also need a pleasant disposition. A team of Percherons is shown in action in Figure 15-12.

The coat color and pattern of the coat color of the color breeds must be recorded. Horses of color breeds used for racing or pleasure should have the same traits as racers or pleasure horses. Breeders of the color breeds are thus handicapped somewhat because they must select for color patterns, which may be quite difficult, in addition to selecting for other traits of economic value.

Some breeders have attempted to breed for the small or miniature horses. Size is easy to select for because of high heritability and because of the wide range of sizes of various breeds, as illustrated in Figure 15-13.

Figure 15-11
The Quarter Horse stallion Poco Bueno appears in the pedigree of many breed champions. (Photograph courtesy of United States Department of Agriculture.)

15.3   Racing Performance

Many studies of factors affecting racing performance have been made (see Hintz, 1980; Langlois, 1980; and Tolley, et al., 1984 for reviews). Most have been with national Thoroughbred populations and with national trotter breeds. Racing performance is moderately heritable, although there is little agreement as to the best measure of racing performance.

Thoroughbred measures of performance generally fall in three categories—*earnings, handicaps,* and *time*—with several variations of these. Estes (1948, 1960) introduced the average earnings index (AEI),

Figure 15-12
A hitch of Percheron horses pulling a load of hay at the Meyring Livestock Company Ranch, Walden, Colorado. Draft horses have been selected for their ability to start and pull heavy loads. (Photograph courtesy of Oliver Meyring, Walden, Colorado.)

which is essentially the yearly earnings of a horse or average progeny as a proportion of the total amount of purses available in that year per horse. Modifications include the standard starts index (SSI) (Clark, 1977) which separates colts and fillies as well as 2-year-olds and older horses. Because many horses have little or no earnings, the distribution of earnings is extremely skewed. To make the distribution more normal and less skewed, logarithmic or square root transformations of earnings are often made.

Handicaps are of two kinds: weight carried during the race and lengths (length of an average horse). Handicap weights are assigned subjectively so that theoretically all horses would finish a race in a dead heat. Handicap weights for individual races are not useful for research but the principle can be extended to a national population [e.g., the time form ratings used in analyses by Field and Cunningham (1976) and More O'Ferrall and Cun-

(a)

(b)

Figure 15-13
Size and weight are traits that can be increased or decreased by selection. The wide differences in size among breeds are illustrated by (a) Drake Farms Leonet, an imported Percheron stallion weighing more than one ton; and (b) Little Bummer, an American Miniature Horse stallion that weighs 190 pounds, contrasted with Elai, a yearling Arabian filly. (Photograph (a) courtesy of Glenn Burns, Meadville, Pennsylvania; photograph (b) by Franks Photo House, courtesy of Jim and Loretta Murphy, Hilltop Farm, Huron, South Dakota.)

ningham (1974)]. Handicaps measured in lengths are ratings of how many lengths a horse will finish behind the winner. Again, measurements on a single race are not useful, but when the quality of horses in the race, the track conditions, and all races for a country in a year are considered, the principle can be extended to a national basis. Gillespie (1971) included these factors in his calculation to arrive at *performance rates*. Performance rate is an estimate of how many lengths in front, or behind, the horse would finish compared to an average horse in an average race for a particular year.

Times for races are difficult to work with for Thoroughbreds because of different race lengths and age classes. Nevertheless, average times for more usual race lengths or best times are sometimes used.

Trotters and pacers are usually evaluated on earnings or times. Researchers in different countries have made different recommendations as to whether to use some function of earnings or of time. For the Netherlands, Minkema has recommended earnings in order to include nonstarters in the evaluations. He has found similar heritabilities for time and earning traits. Ojala, on the other hand, has recommended yearly or lifetime best times for evaluating racing performance of Finnish trotters. (Best time is currently maintained in the Finnish records system, whereas average time is more difficult to obtain from their records.) He found correlations greater than .90 between best and average times as well as a repeatability of about .70 between times at finish within the same year. More importantly he found heritability of .30 for best time while heritabilities for earnings and rank at finish were smaller and more variable. Heritabilities for best time and average time were similar (Ojala and Van Vleck, 1981).

Horses are not entered randomly to races. Usually a horse is entered in a class of race consistent with recent performance. This procedure helps to ensure competitive races. Ferguson and Harvey (1982) and Tolley et al. (1983) have reasoned that the class of race affects the performance of every horse in the race. A fast pace may increase the speed of naturally slower racemates and a slow pace may decrease the speed of all entrants in the race. The class of race is difficult to adjust for directly, but adjusting times of all racemates for the pace of the race by regression on time of the winner effectively accounted for the class of race. Thus, they have recommended adjusting all times for the time of the race winner.

15.4   Special Problems with Selection

Many characteristics of the horse can be changed by selection. For example, racing performance, size and growth, and gestation length can be measured, heritabilities are moderate to high, and phenotypic variation is sufficient for selection to be effective. Nevertheless in some cases intense selection may not be advisable. For example, gestation length is moderately heritable (.25 to .40) and exhibits considerable variation. Thus, selection for longer (or shorter) gestation length would be expected to be very effective. During the process of evolution a gestation length of about 11 months with a standard deviation of 7 to 10 days has become a characteristic of the horse. Should gestation length selected by nature be increased or decreased by man-made selection? Probably not, especially under natural conditions where the reproductive cycle of the horse has evolved to match the seasonality of its environment. Even with close management, selection for longer gestation length may not be advisable. Longer gestation periods would result in disruption of the yearly foaling pattern. Shorter gestation periods, possibly with smaller, more immature foals, may result in less viable foals that are more likely to die shortly after birth. There is no evidence that longer gestation length is associated with better racing performance or other desirable characteristics.

Similarly, growth rate, and body weight, wither height, and body measurements such as cannon bone circumference at different ages are moderately to highly heritable (Hintz et al., 1978). Differences in breeds attest to the effectiveness of selection for changing weight and size traits. In many cases selection for size or weight may not be desirable — for example, heavy draft horses cannot compete with race horses for speed. Usually an intermediate optimum and balance of parts is desirable for any particular function. Extremes, except for speed in race horses, are not often what is wanted.

Heritabilities for Thoroughbred or Standardbred racing times are sufficiently high to expect substantial genetic improvement for speed if selection is actually practiced. Cunningham (1975) reports intense selection in English and Irish Thoroughbreds — effectively the top 2 percent of stallions and 44 percent of mares. Langlois (1980) calculated similar intensities of selection for the French population (5 percent for stallions and 50 percent for mares). Even less intense selection would be expected to result in major improvement in racing times. Cunningham (1975) also reported that the average winning times by decade at three classic races for 3-year-olds (the Derby, the Oaks, and the St. Leger) where records have been kept for a century and a half have not decreased for the last 60 years. In the United States very little decrease in record times in Thoroughbred or Standardbred races occurred for nearly the same length of time. Do these results imply that selection has been ineffective? Not necessarily. For a trait where a decrease is wanted there is a natural limit. As the limit is approached, it becomes more difficult to make large advances toward the limit; improvements come in smaller and smaller increments. Although world and track record times may have changed little, what has happened is that because of selection a larger fraction of horses today than a few years ago have times superior to a standard time. For example, in 1970 a larger fraction of Standardbred horses racing had a time less than two minutes for the mile than did in 1950 or 1960 (R. L. Dobec, 1974, personal communication).

15.5   Application of Relatives' Records to Selection

Records of relatives can be used for selection, although their importance varies according to the heritability of the trait, the closeness of relationship of the relatives to the animal being evaluated, and which other relatives' records are also used.

In selection of mares, relatives' records, in general, have value in the following order: (1) own record, (2) progeny average, (3) average of progeny of animal's sire, (4) parent's or parents' records, and (5) average of progeny of a grandsire.

In the evaluation of a stallion, his own record or his progeny average is most valuable; his relatives' records rank in importance in the same order as those for mares. Usually, no more than the best two of these kinds of records need to be used in an evaluation. Additional records will add only a little to the accuracy of the prediction.

The general selection index procedure for finding the best weights for records on any number of relatives is given in Box 15-1. Some specific weights and accuracy values are given in Table 15-5. A few combinations will now be discussed.

Evaluation Using Own Record Only :

When only a record on the animal is available, genetic value can best be predicted as $\hat{G} = h^2(\text{record} - \overline{X})$, where $\hat{G}$ is the estimate of genetic value as a difference from the population average, $\overline{X}$ is the average of the population with which the horse is being compared, and h^2 is the symbol for heritability. The prediction, including the population average, is $\overline{X} + \hat{G}$. The accuracy of the prediction is $\sqrt{h^2}$, as given in Table 15-5.

For some traits, an animal may have several records. (For example, several times may be recorded for a certain length of race.) Then, repeatability is needed in addition to heritability. The symbol used is r, and it represents the correlation between records of the same animal. Repeatability is as large as or larger than heritability, and like heritability, is a fixed value for a particular trait in a specified population.

Box 15-1
General Selection Index Procedure for Use of Records of Relatives

The selection index weights for records of relatives are obtained by minimizing the average squared error of prediction of genetic value; that is, the average of $(\hat{G} - G)^2$, where G is the true genetic value and $\hat{G}$ is the prediction of genetic value, or $\hat{G} = b_1(X_1 - \overline{X}) + \ldots + b_n(X_n - \overline{X})$. The b's are the best weighting factors for the average of records (X_i) on relatives $1, 2, \ldots, n$. For example, X_1 may be the average of records on the animal itself; X_2 may be a record of the dam; X_3 may be the average of p_3 paternal half-sibs of the animal; and so on.

The general equations to solve to find the b's to predict genetic value for some symbolic animal α can be simplified to the form:

$$d_1b_1 + a_{12}b_2 + a_{13}b_3 + \ldots + a_{1n}b_n = a_{1\alpha}$$
$$a_{21}b_1 + d_2b_2 + a_{23}b_3 + \ldots + a_{2n}b_n = a_{2\alpha}$$
$$a_{31}b_1 + a_{32}b_2 + d_3b_3 + \ldots + a_{3n}b_n = a_{3\alpha}$$

$$\cdot \qquad \cdot \qquad \cdot \qquad \qquad \cdot \qquad \cdot$$

$$\cdot \qquad \cdot \qquad \cdot \qquad \qquad \cdot \qquad \cdot$$

$$\cdot \qquad \cdot \qquad \cdot \qquad \qquad \cdot \qquad \cdot$$

$$a_{n1}b_1 + a_{n2}b_2 + a_{n3}b_3 + \ldots + d_nb_n = a_{n\alpha}$$

There will be one equation for each kind of record (X_i) available for use in the index. If there is only one, let it be X_1, and the equation to find b_1 is $d_1b_1 = a_{1\alpha}$. If there are two, X_1 and X_2, the equations are:

$$d_1b_1 + a_{12}b_2 = a_{1\alpha}$$
$$a_{21}b_1 + d_2b_2 = a_{2\alpha}$$

The prediction of genetic value for an animal with n records is

$$\hat{G} = \{nh^2/[1 + (n - 1)r]\}\{\text{average of } n \text{ records} - \overline{X}\}$$

The accuracy of prediction is $\sqrt{nh^2/[1 + (n - 1)r]}$, as given in Table 15-5.

As an example, assume that heritability, h^2, for birth weight of foals as a trait of the mare is .20 and repeatability is .40. The genetic value for foal

When more X's are available, the pattern continues.

The a's are the additive relationships among the animals: a_{12} between relative type 1 and relative type 2, $a_{1\alpha}$ between relative type 1 and the animal to be evaluated, α.

The d's are the diagonal coefficients:

$$d_i = [\{[1 + (n_i - 1)r]/n_ih^2\} + \{[p_i - 1]a_{ii'}\}]/p_i$$

This procedure assumes that X_i is the average of exactly n_i records on each of p_i animals in the group, all related to each other by the additive relationship, $a_{ii'}$, and all related to the animal being evaluated, α, by $a_{i\alpha}$. As in Table 15-5, h^2 is heritability and r is repeatability of the trait.

Note that when each animal in the group has only one record, $n_i = 1$ and $d_i = [(1/h^2) + (p_i - 1)a_{ii'}]/p_i$, and that when there is only one animal in the group with n_i records (for example, the dam's records), then $p_i = 1$ and $d_i = [1 + (n_i - 1)r]/n_ih^2$. The accuracy (correlation between G and $\hat{G}$) of the index, $r_{G\hat{G}}$, can be found by using the equation $r_{G\hat{G}} = \sqrt{b_1a_{1\alpha} + b_2a_{2\alpha} + \ldots + b_na_{n\alpha}}$.

If a future record of animal α, rather than the genetic value of the animal, is to be predicted, the equations to find the b's are the same except that if the animal already has a record, or an average of several records (for example, X_1), the corresponding part of the right-hand side of the equation changes from $a_{1\alpha} = 1$ to $a_{1\alpha} = r/h^2$. If the animal has no previous records, the equations are the same as those for predicting genetic value and the index is also the same.

The accuracy, however, changes because instead of predicting G_α, the index is trying to predict $P_\alpha = G_\alpha + E_\alpha$, and the E_α is random and essentially nonpredictable. In this case, the accuracy is $r_{P\hat{P}} = \sqrt{h^2(b_1a_{1\alpha} + b_2a_{2a} + \ldots + b_na_{n\alpha})} = r_{G\hat{G}}\sqrt{h^2}$. Specific examples using these general equations to find the b's and accuracy values for predicting G are given in the text.

weight is to be predicted for three mares. (Assume that the foal weights have already been adjusted for age of the mare.) Mare A has three foals that at birth weighed 70, 60, and 80 pounds. Mare B has one foal that weighed 45 pounds. Mare C has two foals that weighed 65 and 75 pounds. Assume that the farm average for foal birth weight is 60 pounds. Then,

Table 15-5

Weights and accuracy values for predicting additive genetic value from records of various relatives (h^2 = heritability; r = repeatability.)

Records		Weights	Accuracy
Individual	(1)	h^2	$\sqrt{h^2}$
	(n)	$nh^2/[1 + (n-1)r]$	$\sqrt{nh^2/[1 + (n-1)r]}$
Dam or sire or progeny	(1)	$h^2/2$	$\sqrt{h^2}/2$
	(n)	$nh^2/2[1 + (n-1)r]$	$\sqrt{nh^2/[1 + (n-1)r]}/2$
Sire and dam	(1)	$h^2/2;\; h^2/2$	$.71\,\sqrt{h^2}$
	(n)	$nh^2/2[1 + (n-1)r];$ $nh^2/2[1 + (n-1)r]$	$.71\,\sqrt{nh^2/[1 + (n-1)r]}$
One grandparent		$h^2/4$	$\sqrt{h^2}/4$
Four grandparents		All $h^2/4$	$\sqrt{h^2}/2$
One great-grandparent		$h^2/8$	$\sqrt{h^2}/8$
Eight great-grandparents		All $h^2/8$	$.35\,\sqrt{h^2}$
Individual and one parent or progeny		$[h^2 - (h^2/2)^2]/[1 - (h^2/2)^2];$ $[h^2(1 - h^2)/2]/[1 - (h^2/2^2)]$	$\sqrt{(5h^2 - 2h^4)/(4 - h^4)}$
Individual and both parents		$h^2(h^2 - 2)/(h^4 - 2);$ $h^2(h^2 - 1)/(h^4 - 2)\ \ldots$	$\sqrt{h^2(2h^2 - 3)/(h^4 - 2)}$

Relationship	Formula(s)	
Individual and one grandparent or grandprogeny	$h^2(h^2-16)/(h^4-16)$; $4h^2(h^2-1)/(h^4-16)$	$\sqrt{h^2(2h^2-17)/(h^4-16)}$
Individual and four grandparents	$h^2(h^2-4)/(h^4-4)$; $h^2(h^2-1)/(h^4-4)$	$\sqrt{h^2(2h^2-5)/(h^4-4)}$
Parent and progeny	$2h^2/(4+h^2)$; $2h^2/(4+h^2)$	$\sqrt{2h^2/(4+h^2)}$
Progeny (p half-sibs)	$2ph^2/[4+(p-1)h^2]$	$\sqrt{ph^2/[4+(p-1)h^2]}$

Let $A = [1+(n-1)r]/n$, $D = \{1+[(p-1)h^2/4]\}/p$, and $C = AD - (h^4/16)$

Relationship	Formula(s)	
Individual (n) and paternal half-sibs (p)	$[h^2D - (h^2/4)^2]/C$; $h^2(A-h^2)/4C$	$\sqrt{b_1+(b_2/4)}$
Individual (n) and his paternal half-sib progeny (p)	$[h^2D - (h^2/2)]/[C-(3h^4/16)]$; $h^2(A-h^2)/2[C-(3h^4/16)]$	$\sqrt{b_1+(b_2/2)}$
Dam (n) and paternal half-sibs (p)	$nh^2/2[1+(n-1)r]$; $ph^2/[4+(p-1)h^2]$	$\sqrt{(b_1/2)+(b_1/4)}$
Dam (1), sire (1), and progeny (1)	$[h^2-(h^4/16)]/[2-(h^4/64)]$; $[h^2-(h^4/16)]/[2-(h^4/64)]$; $[h^2-(h^4/8)]/[2-(h^4/64)]$	$\sqrt{(b_1+b_2+b_3)/2}$
Paternal half-sibs (m), dam (n), and paternal half-sibs (p)	$mh^2/[4+(m-1)h^2]$; $h^2[D-(h^2/16)]/2C$; $h^2(A-h^2)/8C$	$\sqrt{(b_1/4)+(b_2/2)+(b_3/8)}$

$$\hat{G}_A = \{(3)(.20)/[1 + (3 - 1)(.40)]\}\{[(70 + 60 + 80)/3] - 60\}$$

$$= .333(70 - 60) = 3.33 \text{ pounds}$$

$$\hat{G}_B = (.20)(45 - 60) = -3.00 \text{ pounds}$$

$$\hat{G}_C = \{(2)(.20)/[1 + (2 - 1)(.40)]\} \{[(65 + 75)/2] - 60\}$$

$$= .2875(70 - 60) = 2.88 \text{ pounds}$$

Note that these are not predictions of the genetic values for birth weight of the mares but of their genetic values for birth weight of their foals as a difference from the farm average. Adding the farm average gives the predicted foal weights, $\overline{X} + \hat{G}$: 63.33 for A, 57.00 for B, and 62.88 for C.

This procedure allows comparison of animals with different numbers of records.

Evaluation from Progeny Average Alone

Prediction of a stallion's genetic value from his progeny average is simple if each of the paternal half-sib progeny has only one record:

$$\hat{G} = [2p/\{p + [(4 - h^2)/h^2]\}] [\text{average of his } p \text{ progeny} - \overline{X}]$$

Even when the progeny have more than one record, the above prediction is a good approximation. The accuracy of prediction is $\sqrt{p/\{p + [(4 - h^2)/h^2]\}}$. The weight and accuracy in Table 15-5 are algebraically the same but are written differently.

If the average of future progeny of a stallion is being predicted from his present progeny, the weighting factor is half that for predicting the stallion's genetic value: predicted future progeny = $[p/\{p + [(4 - h^2)/h^2]\}]$ [average of his p progeny $- \overline{X}$].

For example, assume that racing speed for the quarter-mile has a heritability of .50 and that the appropriate average is 25 seconds. Two stallions are to be evaluated for racing speed from age-adjusted records of their paternal half-sib progeny. Stallion A has 25 progeny that averaged 24.50 seconds. Stallion B has 5 progeny that averaged 25.75 seconds. Then,

$$\hat{G}_A = [(2)(25)/\{25 + [(4 - .50)/.50]\}] [24.5 - 25]$$

$$= 1.5625(-.5) = -.78 \text{ seconds}$$

$$\hat{G}_B = [(2)(5)/(5 + 7)] [25.75 - 25]$$

$$= .8333(+.75) = +.62 \text{ seconds}$$

The prediction of averages of future progeny would be

$$A: .78125(-.5) = -.39 \text{ seconds}$$

$$B: .4167(+.75) = +.31 \text{ seconds}$$

After adding the population average of 25 seconds, the progeny predictions are 24.61 and 25.31 seconds, respectively.

Evaluation from the Sire's and Dam's Predicted Genetic Values

The equation for prediction of the genetic value of the progeny from the sire's and dam's predicted genetic values is $\hat{G}_{progeny} = (\hat{G}_{sire} + \hat{G}_{dam})/2$. The accuracy is

$$\sqrt{(\text{Accuracy for sire})^2 + (\text{Accuracy for dam})^2}/2$$

when the sire and dam are not related.

Other Combinations

Table 15-5 gives weights for predicting genetic value from most possible combinations of relatives. The procedure is to find the weights for each relative or group of relatives, (b_1 for relative 1, b_2 for relative 2, and so on) and to use the equation, $\hat{G}_{animal} = b_1(\text{average for first relative} - \overline{X}) + b_2(\text{average for second relative} - \overline{X}) + \ldots + b_n(\text{average for } n^{th}$ relative $- \overline{X})$. To avoid using the predictions as differences (some positive and some negative), the population average, $\overline{X}$, can be added to $\hat{G}$.

15.6   Selection for Several Traits

The preceding section gave methods of predicting genetic value for a single trait by using records for that trait on the animal and its relatives. These methods ignore other traits; in fact, they assume that other traits have no economic value. Yet, many breeders must select to improve more than one trait because each of several traits may contribute to the economic value of the animal.

The problem of predicting total economic value by considering all economically important traits is complicated for several reasons. The genetic value for each trait is unknown and can only be predicted. Economic values are difficult to determine for many traits (for example, temperament and cow sense). The correlations between genetic values and between environ-

Box 15-2
Computation of an Index of Economic Value

Overall economic value can be defined as $G = v_1 G_1 + v_2 G_2 + \ldots + v_n G_n$, where G_i $(i = 1, \ldots, n)$ are the genetic values for the n economically important traits and v_i is the net economic value for a one unit increase in trait i. The procedure described here is based on minimizing the average of squared errors of prediction by the same method that was used when records of relatives were used to predict the genetic value for a single trait.

First, estimate the genetic value of the animal for each trait that is considered economically important. The availability of one record on each of n traits will be assumed. All records should be adjusted for fixed factors and should be expressed as differences from the appropriate herd or population average if necessary.

For trait 1, the equations used to find the weights (b's) for the records are:

$$V_{11}b_{11} + V_{12}b_{21} + V_{13}b_{31} + \ldots + V_{1n}b_{n1} = h_1^2 V_{11}$$

$$V_{21}b_{11} + V_{22}b_{21} + V_{23}b_{31} + \ldots + V_{2n}b_{n1} = r_{g_{12}} \sqrt{h_1^2 V_{11}(h_2^2 V_{22})}$$

$$V_{31}b_{11} + V_{32}b_{21} + V_{33}b_{31} + \ldots + V_{3n}b_{n1} = r_{g_{13}} \sqrt{h_1^2 V_{11}(h_3^2 V_{33})}$$

$$. \qquad . \qquad . \qquad\qquad . \qquad .$$
$$. \qquad . \qquad . \qquad\qquad . \qquad .$$
$$. \qquad . \qquad . \qquad\qquad . \qquad .$$

$$V_{n1}b_{11} + V_{n2}b_{21} + V_{n3}b_{31} + \ldots + V_{nn}b_{n1} = r_{g_{1n}} \sqrt{h_1^2 V_{11}(h_n^2 V_{nn})}$$

where V_{ii} is the phenotypic (total) variance of trait i; V_{ij} is the phenotypic covariance between traits i and j, and $V_{ij} = r_{p_{ij}} \sqrt{V_{ii}V_{jj}}$ where $r_{p_{ij}}$ is the phenotypic correlation between traits i and j; b_{11} is the weight for the record on trait 1 in predicting the genetic value for trait 1, b_{21} is the weight for the record on trait 2 in predicting trait 1, $\ldots$, and b_{n1} is the weight for the record on trait n in predicting trait 1; h_i^2 is the heritability for trait i; and $r_{g_{ij}}$ is the genetic correlation between traits i and j.

The second subscript of the weights refers to the trait being predicted; the first subscript refers to the trait being used. The index for genetic value for trait 1 is thus $I_1 = b_{11}X_1 + b_{21}X_2 + b_{31}X_3 + \ldots + b_{n1}X_n$, where the X's are records on the n traits. This will also be the index for trait 1 when all traits are used if selection is for trait 1 alone.

The equations are the same for trait 2 except that b_{i2} $(i = 1, \ldots, n)$ are substituted for the b_{i1}'s, and the right sides of the equations become

$$r_{g_{21}} \sqrt{h_2^2 V_{22}(h_1^2 V_{11})}$$

$$h_2^2 V_{22}$$

$$r_{g_{23}} \sqrt{h_2^2 V_{22}(h_3^2 V_{33})}$$

.

.

.

$$r_{g_{2n}} \sqrt{h_2^2 V_{22}(h_n^2 V_{nn})}$$

The index for trait 2 is $I_2 = b_{12}X_1 + b_{22}X_2 + b_{32}X_3 + \ldots + b_{n2}X_n$.

Use similar substitutions for the other traits—new b's and new right sides of the equations—so that the general term on the right side of the equation is $r_{g_{ki}} \sqrt{h_k^2 V_{kk}(h_i^2 V_{ii})}$ for equation i when selecting for trait k. Note that when $k = i$ (selecting for trait k and for equation k), $r_{g_{kk}} = 1$, and the right side of the equation becomes $h_k^2 V_{kk}$.

After the indexes for each trait have been computed ($I_1, I_2, \ldots, I_n$), they can be combined by weighting by the appropriate economic value per unit for each trait. If v_1 is the economic value for trait 1, v_2 is the value for trait 2, $\ldots$, and v_n is the value for trait n, then the overall index for total economic value is $I = v_1 I_1 + v_2 I_2 + \ldots + v_n I_n$.

By substitution, the overall index can be written as $I = \beta_1 X_1 + \beta_2 X_2 + \ldots + \beta_n X_{Xn}$, where

$$\beta_1 = v_1 b_{11} + v_2 b_{12} + v_3 b_{13} + \ldots + v_n b_{1n}$$

$$\beta_2 = v_1 b_{21} + v_2 b_{22} + v_3 b_{23} + \ldots + v_n b_{2n}$$

$$\beta_3 = v_1 b_{31} + v_2 b_{32} + v_3 b_{33} + \ldots + v_n b_{3n}$$

.   .   .   .   .

.   .   .   .   .

.   .   .   .   .

$$\beta_n = v_1 b_{n1} + v_2 b_{n2} + v_3 b_{n3} + \ldots + v_n b_{nn}$$

The correlated genetic response for the genetic value of trait c, G_c, can be predicted from the following expression: change in G_c = selection intensity factor $\times$ [covariance (G_c, I)]/standard deviation of I.

The formula for variance of I is $\beta_1^2 V_{11} + \beta_2^2 V_{22} + \ldots + \beta_n^2 V_{nn} + 2\beta_1\beta_2 V_{12} + 2\beta_1\beta_3 V_{13} + \ldots + 2\beta_1\beta_n V_{1n} + 2\beta_2\beta_3 V_{23} + \ldots + 2\beta_{n-1}\beta_n V_{n-1,n}$, which is the sum of the products of the overall index weights squared and the corresponding variances plus the sum of twice the product of all covariances and the corresponding weights. The square root of the variance of the index gives the standard deviation of the index. The expected superiority in overall economic value by selection is the standard deviation of I times the selection intensity factor.

The formula for the covariance between the genetic value for trait c and the index I is $\beta_1 r_{g_{1c}} \sqrt{h_1^2 V_{11}(h_c^2 V_{cc})} + \beta_2 r_{g_{2c}} \sqrt{h_2^2 V_{22}(h_c^2 V_{cc})} + \ldots + \beta_n r_{g_{nc}} \sqrt{h_n^2 V_{nn}(h_c^2 V_{cc})}$.

mental effects for the economically important traits are also necessary for accurate prediction of total economic value but are difficult to estimate. The problem is especially difficult in horse breeding because so little research has been devoted to estimating these relationships—the correlations between genetic values for different traits and the correlations between phenotypic values for different traits.

Method of Selection for Total Economic Value

The method of selection for more than one trait is similar to that for selection for a single trait. The result is an index of economic value made up of records on traits of the animal and its relatives. The weight for each of the records depends on the relationships among the relatives, economic values of the traits, genetic and environmental correlations among the traits, and heritabilities of the traits. A full explanation of this procedure is beyond the scope of this book and also depends on correlations that are probably not accurately estimated. Box 15-2 does, however, illustrate the procedure for those who have the patience and interest to investigate selection for overall economic value further.

Some research has suggested that when genetic and environmental correlations are likely to be poorly estimated, a simpler procedure may be appropriate. The genetic value of each trait is predicted from records for that trait as described in the preceding section. Those predicted genetic values are then weighted by their economic values, which in many cases will have to be "educated guesses" based on the good sense of the breeder. For example, suppose that a one-second decrease in time to run a quarter-mile is worth $400 and that cow sense is scored on a point scale of 1 to 20 with a one-point increase worth $50. Suppose also that the genetic values for speed and cow sense for two horses have been predicted as shown in Table 15-6. These predictions are differences from the population averages.

Table 15-6
Example of selecting for overall genetic value using two traits

Horse	Predicted genetic value	
	Speed (seconds)	Cow sense (points)
A	−.5	−1
B	1.0	+2
Economic value of one-unit increase in the trait	−$400	$50

Economic value for a longer time is negative because a lower time is desirable. The predicted overall economic values are, for A: $(-\$400)(-.5) + (\$50)(-1) = \$150$, and, for B: $(-\$400)(1.0) + (\$50)(2) = -\$300$. There is a difference of \$450 in predicted genetic value of the two horses after the value of two economically important traits is considered.

This example illustrates two points. First, the procedure predicts the difference between two horses or their differences from an average horse. Second, economic values are difficult to assign. Whether the economic values used in this example make sense is debatable.

This method of predicting overall genetic value can, in general, be summarized as $I = v_1\hat{G}_1 + v_2\hat{G}_2 + \ldots + v_n\hat{G}_n$, where I is the overall index of economic value, $\hat{G}_1$ is the prediction of genetic value for trait 1, $\hat{G}_2$ is the prediction of genetic value for trait 2, $\ldots$, $\hat{G}_n$ is the prediction of economic value for trait n, and the v's are corresponding economic values.

Relative Progress

The usual rule of thumb is that if selection is jointly for n traits, the progress for one of them is $1/\sqrt{n}$ of the progress from selection for that trait alone. This approximation is exact when the traits have equal economic values per

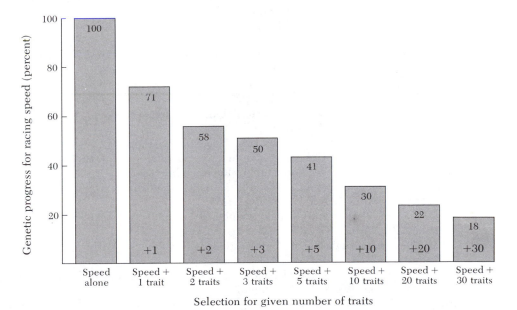

Figure 15-14
Relative progress in breeding for racing speed if selection emphasis is equal for several traits.

standard deviation, equal heritabilities, and zero genetic and phenotypic correlations. Such assumptions are not likely to be completely true, but the rule does provide a rough guideline. The rule is illustrated in Figure 15-14, where racing speed is the trait for which progress is being compared.

REFERENCES

Arnason, T. 1983. *Genetic studies on conformation and performance of Icelandic Toelter horses.* Report 59. Department of Animal Breeding and Genetics, Swedish University of Agricultural Science, Uppsala.

Artz, W. 1961. A contribution on the evaluation of performance tests in Thoroughbred breeding with special reference to the racing performance of individual stallion progeny groups. *Giessen. SchrReihe Tierz. Haustiergenet.* 2:1 (*Anim. Breed. Abstr.* 31:313).

Bade, B., P. Glodek, and H. Schormann. 1975. The development of selection criteria for the breeding of riding horses. I. Genetic parameters of performance-testing criteria for young stallions tested at a station. *Züchtungskunde* 47:67.

Bade, B., P. Glodek, and H. Schormann. 1975. The development of selection criteria for the breeding of riding horses. II. Genetic parameters of progeny-tested criteria for stallions tested in the field. *Züchtungskunde* 47:154.

Baumohl, A., and J. A. Estes. 1960. Racing class and sire success. *The Blood-Horse* 80:48.

Bormann, P. 1966. A comparison between handicap weight and timing as measures of selection in Thoroughbred breeding. *Züchtungskunde* 38:301.

Bruns, E. 1980. Results on a more objective measurement of performance traits of stallions in performance testing at station. *Livest. Prod. Sci.*, 7:607.

Cape, L., and L. D. Van Vleck. 1981. Heritability of training scores for Standardbred yearlings. *J. Heredity* 72:437.

Clark, S. 1977. Standard starts index. *The Blood-Horse* 103:1612.

Cunningham, E. P. 1976. Genetic studies in horse populations. In *Proc. Int. Symp. on Genetics and Horse Breeding.* Dublin: Royal Dublin Soc. p. 2.

Dring, L. A., H. F. Hintz, and L. D. Van Vleck. 1981. Coat color and gestation length in Thoroughbred mares. *J. Heredity* 72:65.

Dusek, J. 1965. A contribution to the study of heritability of some properties in horses. *Ziv. Vyroba* 10:449. (*Anim. Breed. Abst.* 33:532)

Dusek, J. 1970. Heritability of conformation and gait in the horse. *Z. Tierzucht. Züchtbiol.* 87:14.

Dusek, J. 1970. Repeatability of pace length and speed in horses. *Zivoc. Vyroba* 15:755. (*Anim. Breed. Abstr.* 39:643).

Dusek, J. 1971. Inheritance of pace length and speed in commercial horse breeds. *Bayer. landw. Jb.* 48:43.

Dusek, J. 1971. Some biological factors and performance in the study of heredity in horse breeding. *Sci. Agric. Bohemoslov* 3:199.

Dusek, J. 1975. The effect of some biological and performance factors on heritability in horse breeding. *Bayer. landw. Jb.* 52:223.

Dusek, J. 1978. The objectivization of selection criteria for estimation of genetic parameters in breeding of the English full-blooded horse. *Sci. Agric. Bohemoslovaca* 10:137. (*Anim. Breed. Abstr.* 49:518).

Dusek, J. 1979. Estimation of the heritability of speed in Hungarian and Italian Trotters. *Bodenkultur* 30:79. (*Anim. Breed. Abstr.* 47:577).

Dusek, J., H. J. Ehrlein W. von Engelhardt, and H. Hornicke. 1970. Correlations between pace length, pace frequency and speed in horses. *Z. Tierzucht. Züchtungsbiol.* 87:177.

Estes, B. W. 1952. A study of the relationship between temperament of Thoroughbred broodmares and performance of offspring. *J. Genetic Psyc.* 81:273.

Estes, J. A., and A. Baumohl. 1960. Racing class and sire success. *The Blood-Horse.* 80:48.

Fedorski, J. 1977. The heritability of racing performance in Thoroughbred horses in Poland. *Prace i Materialy Zootechniczne* 14:121. (*Anim. Breed. Abstr.* 46:549).

Foye, D. B., H. C. Dickey, and C. J. Sniffen. 1972. Heritability of racing performance and a selection index for breeding potential in the Thoroughbred horse. *J. Anim. Sci.* 35:1141.

Field, J. K., and E. P. Cunningham. 1976. A further study of the inheritance of racing performance in Thoroughbred horses. *J. Hered.* 67:247.

Galizzi-Vecchiotti, A. G., and G. Pazzaglia. 1976. Heritability of racing performance in Thoroughbreds studied in Italian flat races. *Atti della Societa Italiana delle Scienze Veterinarie* 30:490. (*Anim. Breed. Abstr.* 46:497).

Galton, F. 1898. An examination into the registered speeds of American trotting horses, with remarks on their value as hereditary data. *Proc. Royal Society, London* 62:310.

Gillespie, R. H. 1971. A new way to evaluate race horses—performance rates. *The Thoroughbred Record* 193:961.

Gopka, B. M. 1971. Heritability of speed in Orlov trotters. *Genetika i Selektsiya Ukraine.* (*Anim. Breed. Abstr.* 40:222).

Gopka, B. M., V. M. Klok, and A. V. Derev'yanchuk. Color and pregnancy duration in trotter mares. *Nankovi Pratt USGA.* 41:169. (*Anim. Breed. Abstr.* 41:594).

Green, D. A. 1969. A study of growth rate in Thoroughbred foals. *Brit. Vet. J.* 125:539.

Hamori, D. 1963. The inheritance of performance in Thoroughbreds. *Acta. Agron.* 12:19.

Hamori, D., and G. Halasz. 1959. The effect of selection on the development of speed in horses. *Z. Tierz. Züchtungsbiol.* 73:47 (*Anim. Breed. Abstr.* 27:394).

Hartwig, W., and U. Reichardt. 1958. The heritability of fertility in horse breeding. *Züchtungskunde* 30:205.

Hazel, L. N. 1943. The genetic basis for constructing selection indexes. *Genetics* 28:476.

Hecker, W. 1975. The inheritance of speed. Állattenyésztés 24:117. (*Anim. Breed. Abstr.* 43:323).

Hellman, T. 1978. Heredity of performance traits in the Standardbred trotter in Finland. *Hevosurheilu, Jalostuskuvasto* 1:106.

Hildebrand, M. 1959. Motions of the running cheetah and horse. *J. Mammals* 40:481.

Hintz, R. L. 1980. Genetics of performance in the horse. *J. Anim. Sci.* 51:582.

Hintz, H. F., R. L. Hintz, D. H. Lein, and L. D. Van Vleck. 1979. Length of gestation periods in Thoroughbred mares. *J. Equine Med. Surgery.* 3:289.

Hintz, R. L., H. F. Hintz, and L. D. Van Vleck. 1978. Estimation of heritabilities for weight, height and front cannon bone circumference of Thoroughbreds. *J. Anim. Sci.* 47:1243.

Hintz, R. L., and L. D. Van Vleck. 1978. Factors influencing racing performance of the Standardbred pacer. *J. Anim. Sci.* 46:60.

Hollingsworth, K. 1967. Conclusion: The greater a colt's earnings on the track, the greater his chance of success as a stallion. *The Blood-Horse* (November 4):3267.

Iorov, I., and M. Kis'ov. 1976. Heritability of some body measurements and speed in Thoroughbred horses. *Genetika i Selektsiya* 9:480. (*Anim. Breed. Abstr.* 46:72).

Kalmykov, A. N. 1973. Heritability of economically important traits in the Orlov trotter. *Genetica* 9:50.

Kalmykov, A. N. 1975. Heritability of quantitative traits in trotters. *Uchenye Zapiski Kasanskogo Gosudartsvennogo Veterinarnogo Instuta* 121:23. (*Anim. Breed. Abstr.* 46:363).

Katona, O. 1979. Genetical-statistical analysis of traits in the German Trotter. *Livestock Prod. Sci.* 6:407.

Katona, O., and K. Osterkorn. 1977. Genetic-statistical analysis of racing time in the German trotter population. *Züchtungskunde* 49:185.

Kieffer, N. M. 1973. Inheritance of racing ability in the Thoroughbred. *Thoroughbred Rec.* 198:50.

Kownacki, M., M. Fabiani, and K. J. Jaszczak. 1971. Genetical parameters of some traits of Thoroughbred horses. *Genetica Pol.* 12:431.

Langlois, B. 1975. Statistical and genetic analysis of winnings of horses in French equestrian competitions. *Livestock Prod. Sci.* 2:191.

Langlois, B. 1975. Statistical and genetic analysis of earnings of 3-year-old Thoroughbreds in French flat races from 1971 to 1973. *Ann. Genet. Sel. Anim.* 7:387.

Langlois, B. 1976. French studies on inheritance of jumping and three day event ability. In *Proc. Int. Symp. on Genetics and Horse-Breeding* Dublin: Royal Dublin Soc. p. 33.

Langlois, B. 1980. Heritability of racing ability in Thoroughbreds — a review. *Livest. Prod. Sci.* 7:591.

Langlois, B., J. Froidevaux, L. Lamarch, C. Legault, P. Legault, L. Tassencourt, and M. Theret. 1978. Analysis of relationships between morphology and gallop, trot and jumping abilities in horses. *Ann. Genet. Sel. Anim.* 10:443.

Langlois, B., D. Minkema, and E. Bruns. 1983. Genetic problems in horse breeding. *Livestock Prod. Sci.* 10:69.

Lasley, J. F. 1976. *Genetic Principles in Horse Breeding.* Cordovan Corp., Houston, Texas.

Laughlin, H. H. 1934. Racing capacity in the Thoroughbred horse. *Sci. Monthly* 38:210.

Legault, C. 1976. Can we select systematically for jumping ability? In *Proc. Int. Symp. on Genetics and Horse-Breeding.* Dublin: Royal Dublin Soc. p. 71.

Linner, M. T., and K. Osterkorn. 1974. Breeding evaluation of the racing performances of trotters on 1963 and 1964 in the German Federal Republic. *Züchtungskunde* 46:168.

Minkema, D. 1975. Studies on the genetics of trotting performance in Dutch Trotters. I. The heritability of trotting performance. *Ann. Génét. Sél. Anim.* 7:99.

Minkema, D. 1975. Studies on the genetics of trotting performance in Dutch Trotters. II. A method for the breeding value estimation of trotter stallions. *Ann. Génét. Sél. Anim.* 8:511.

Minkema, D. 1975. Studies on the genetics of trotting performance in Dutch Trotters. In *Proc. Int. Symp. on Genetics and Horse-Breeding.* Dublin: Royal Dublin Soc. p. 24.

Minkema, D. 1981. Studies on the genetics of trotting performance in Dutch Trotters. III. Estimation of genetic change in speed. *Ann. Génét. Sél. Anim.* 13:245.

More O'Ferral, G. J., and E. P. Cunningham. 1974. Heritability of racing performance in Thoroughbred horses. *Livest. Prod. Sci.* 1:87.

Neisser, E. 1976. Evaluation of several criteria to measure performance potential in the Thoroughbred. II. *Internat'l Wissenschaftliches Symp.*, Leipzig. (*Anim. Breed. Abstr.* 47:5291).

Ocsag, I. 1963. Which methods are suitable for evaluating the inheritance of speed potential in stud stallions of racing breeds? *Acta. Agron.* [Budapest] 12:181.

Ocsag, I., and I. Toth. 1959. The heritability of speed in horses. *Agrartud. egy. Mezogazdas Tud. Karanak Kozl.*, p. 61. (*Anim. Breed. Abst.* 30:320).

Ojala, M. J., and L. D. Van Vleck. 1981. Measures of racetrack performance with regard to breeding evaluation of trotters. *J. Animal Sci.* 53:611.

Pern, E. M. 1970. The heritability of speed in Thoroughbred horses. *Genetika Mosk.* 6:110. (*Anim. Breed. Abstr.* 38:380).

Pern, E. M. 1971. Variability and heritability of speed in Thoroughbred horses. *Nausch. Trudy vses. nauchoaoissled.* Inst. Konev. 25:98. (*Anim. Breed. Abstr.* 40:643).

Philipsson, J. 1975. Estimates of heritability for performance of Swedish riding horses. *EAAP*, Warsaw, Poland, June 23–27.

Philipsson, J. 1976. Studies on population structure in Swedish horses. In *Proc. Int. Symp. of Genetics and Horse-Breeding.* Dublin: Royal Dublin Soc. p. 59.

Pirri, J., Jr., and D. G. Steele. 1951. Heritability of racing capacity in Thoroughbreds. *J. Anim. Sci.* 10:1029.

Pirri, J., and D. G. Steele. 1952. The heritability of racing capacity. *The Blood-Horse.* 63:976.

Pirchner, F. 1969. Population Genetics in Animal Breeding. New York: Freeman.

Porter, R. A., 1971. Performance Rates. *Thoroughbred Rec.* 193:960.

Rodero Franganillo, A., and R. Pozo Lora. 1960. Heritability and repeatability of gestation length in Spanish and Arabian mares. *Arch. Zootec.*, 9:132.

Rollins, W. C., and C. E. Howell. 1951. Genetic sources of variation in the gestation length of the horse. *J. Animal. Sci.* 10:797.

Rønningen, K. 1975. Genetic and environmental factors for traits in the North-Swedish trotter. *Z. Tierz. Züchtungsbiol.* 92:164.

Sato, K., M. Miyaki, K. Sugiyama, and T. Yoshikawa. 1973. An analytical study on the duration of gestation in horses. *Jap. J. Zoo. Tech. Sci.* 44:375.

Schermerhorn, E. C., L. D. Van Vleck, and T. R. Rounsaville. 1980. Heritabilities of reproductive performance and gestation length in a sample of Standardbred horses. *J. Animal Sci.* 52:128.

Schwark, H. J., and E. Neisser. 1971. Breeding of English Thoroughbred horses in the G.D.R. 2. Results of estimates of heritability and breeding value. *Arch. Tierz.* 14:69.

Sloan, F. 1949. Color and temperament in horses. *J. Heredity* 40:12.

Stevens, R. W. C. 1969. Artificial insemination in horses: Genetic potential and control. *Can. Vet. J.* 10:203.

Strom, H., and J. Philipsson. 1978. Relative importance of performance tests and progeny tests in horse breeding. *Livestock Prod. Sci.* 5:303.

Tolley, E. A., D. R. Notter, and T. J. Marlowe. 1983. Heritability and repeatability of speed for two- and three-year-old Standardbred racehorses. *J. Anim. Sci.* 56:1294.

Varo, M. 1965. On the relationship between characters for selection in horses. *Ann. Agr. Fenn.* 4:38.

Varo, M. 1965. Some coefficients of heritability in horses. *Ann. Agr. Fenn.* 4:223.

Walton, A., and J. Hammond. 1938. The maternal effects on growth and conformation in Shire Horse-Shetland Pony crosses. *Proc. Royal Soc.* 125(B):311.

Watanabe, Y. 1969. Timing as a measure of selection in Thoroughbred breeding. *Jap. J. Zootech. Sci.* 40:271.

Watanabe, Y. 1974. Performance rates of Thoroughbreds as a criterion of racing ability. *Jap. J. Zootech. Sci.* 45:408.

Yorov, I., and M. Kissyov. 1976. Heritability of some body measurements and speed in Thoroughbred horses. *Genetika i Selektsiya* 9:480.

6

HEALTH

CHAPTER 16

DISEASES

"Give my roan horse a drench," says he.
Henry IV, Part I, Act 2, Scene 4

A good health program is one of the most important components of horse management. To maintain an effective health care program, a horse owner must

1. Be able to identify problems.
2. Keep accurate records.
3. Have knowledge of equine diseases and recent research results.
4. Organize a vaccination and parasite control program.
5. Avoid conditions or situations that make the animal a high risk for disease or injury.
6. Be prepared to handle emergencies.
7. Have advice and service from a veterinarian and a farrier.

16.1   Identification of Problems

A physical examination of the horse and an accurate history are necessary if the horse owner is to decide whether the horse is sick, and if so, whether to call the veterinarian, treat the horse himself, or simply assume that the horse will recover without treatment.

The general impression and appearance of the animal are important indicators of its state of health. Does the animal appear to be active? How

593

does he respond to normal stimuli such as noise? Is he alert or depressed? Excessive excitability may also indicate a problem. The posture should be noted. For example, continual shifting of weight from limb to limb may indicate serious foot lameness. "Sawhorse position" or straddling of legs is characteristic of severe abdominal pain or tetanus.

What is the body condition of the animal? Has there been a recent significant loss of weight? Does the hair coat shine? Did the winter coat shed late or fail to shed?

Behavior may also indicate the state of health. An animal that remains away from its group is often ill. What are its eating habits? Inappetence indicates that something is wrong. Routine examination of the teeth is important. Is there any pain, swelling, or heat (i.e., inflammation)? Chewing may be slow or one-sided when teeth are affected. Swallowing may be painful with strangles. Does the animal defecate and urinate normally? Is the animal's respiration rate normal, or does it have trouble breathing?

After the general impression is noted, a more detailed examination should be made. For example, are the eyelids swollen? Is there any discharge from the eyes or nose?

These criteria are often subjective, and a certain amount of experience is required to interpret them properly. However, there are some objective measurements that even the novice can readily use.

Body temperature is usually taken with a rectal thermometer. The normal temperature for a horse is 100.5°F (99° to 101°F). Low readings may result if the temperature is taken immediately after defecation or if the thermometer is stuck into a fecal ball. Temperature may be low in the morning and peak in the afternoon. High environmental temperature and humidity may increase the body temperature as much as 2°F. Working may increase the temperature up to 3°F, and up to 2 hours may be required to return it to normal. A body temperature above 101°F in a resting horse should be considered abnormal.

The pulse can be taken from the facial artery. This artery is felt in a shallow groove on the lower border of the jaw beneath the last cheek tooth. You can practice feeling the pulse by using your own facial artery. Pulse rate is dependent on the heart, and normally the rate is 30 to 40 beats per minute. The rate may be much greater in the newborn foal (see Table 3-1). Tachycardia (rapid heart rate) often may occur when the animal is in pain or suffering from toxemia (toxic substances in the blood) or septicemia (virulent microorganisms in the blood). The most common cause of tachycardia is exercise or excitement. Bradycardia (slow heart rate) is unusual. The respiration rate can be determined by observing the rise and fall of the flank or rib cage. The normal rate is 8 to 16 respirations per minute. Pain or fever will increase the rate, as will exercise, poor ventilation in buildings, hot weather, or excitement.

16.2   Accurate Records

Records should include the age of animals, as well as a health history including vaccination and worming schedule. Weights of the animals should be taken periodically, either by scales or by heart-girth measurement tapes. Dates of trimming and shoeing should be recorded, and the breeding history of mares should also be recorded (see Section 11.5).

16.3   Infectious Diseases

More information on infectious diseases can be found in *Equine Medicine and Surgery* (Mansman et al., 1982) and *Current Therapy in Equine Medicine* (Robinson, 1987).

Foalhood or Neonatal Septicemias

Brewer and Koterba (1987) emphasized the importance of prevention of septicemia when they reported that even with intensive supportive care the survival rate can be as low as 26 percent.

The foal is very susceptible to infection. Septicemia, frequently localized in the joints, is often caused by *Escherichia coli, Actinobacillus equuli, Salmonella abortivoequina, Streptococcus pyogenes equi,* or *Salmonella typhimurium. Clostridium perfringens* may cause enteritis (inflammation of the intestine). Bacteria may enter the foal's blood system via the umbilical cord and thus the navel should be disinfected at birth.

The umbilical cord should not be broken too soon after birth because if it is, the foal will be deprived of substantial quantities of blood. The cord will usually break without any help from the attendant.

Early recognition and appropriate treatment are the best defenses. Conditions that are often associated with septicemia include placentitis, fever in the mare, dystocia, induced parturition, premature foals, delayed suck reflex, unsanitary foaling conditions, and poor ventilation. Brewer and Koterba (1987) also state that failure to acquire sufficient colostral antibodies is probably the leading contributory cause of neonatal infection. Several quick and inexpensive tests to determine if a foal has received colostral antibodies are now available.

The early signs of septicemia are decreased appetite, generalized weakness, weakened suck reflex, and mild dehydration. Fever may be present but normal temperatures are just as likely. Foals at risk (i.e., those exposed to any of the conditions listed above) should be watched closely.

Strangles

Strangles is an acute contagious disease caused by infection with *Streptococcus equi*. The upper respiratory tract is inflamed and the adjacent lymph nodes, especially those between the jaws close to the skin, become enlarged and abscessed. The swelling of the nodes may take 3 to 4 days to develop. If not effectively treated, the nodes may rupture after about 10 days and discharge a thick, cream-colored pus. When the infection is severe, many other nodes may also abscess. Local abscesses can even occur on the body surface or in internal organs. This is referred to as bastard strangles. In the first stage of the disease, the horse usually goes off feed, has a high temperature (103° to 105°F), and emits a watery nasal discharge that rapidly becomes purulent (thick and cream-colored). The larynx and pharynx become inflamed and swallowing may be difficult. Body temperature may decrease to normal after 2 to 3 days, but increases again when abscesses develop in the lymph nodes.

Strangles usually affects horses of 1 to 5 years of age, although the condition may develop in horses of any age. Outbreaks are most likely to occur after horses are moved, particularly during inclement weather.

The disease is spread when the nasal discharge of the infected animals contaminates pasture, feed troughs, or water troughs. Infected horses may spread the disease for at least 4 weeks, and the organisms may survive in the environment for a month or more after the infected animals are removed.

Thus, infected animals should be isolated as soon as possible. The stall, feed troughs, water buckets, brushes, and so on, should be disinfected. If animals can be treated before an abscess is formed, penicillin is effective. Inadequate treatment may result in recurrence of the infection. Horses that have recovered from natural infection are usually immune, but the condition may repeat itself after approximately 6 months.

Preventive measures include isolating new animals for 4 weeks, but carriers can exist for 6 weeks after an outbreak. These horses look perfectly normal but they are "typhoid Marys" spreading *Streptococcus equi* at all times from normal-looking nasal secretions. Without repeated nasal cultures it is impossible to identify these animals. Isolation is an alternative but it is often impractical or too expensive. A vaccine made from killed bacteria is also available for animals 12 weeks of age or older. The vaccine is given two or three times at 8-week intervals. Annual booster injections are usually recommended. Vaccination does not prevent outbreaks of the disease but may decrease its severity. The effectiveness of the vaccine is controversial, and it is not uniformly accepted by the veterinary profession.

Another vaccine made from the "M" protein portion of the bacteria is also marketed. Its ability to protect is also suspect.

Pneumonia

The term "pneumonia" refers to any inflammatory disease of the lungs. There are several types of pneumonia, which can be caused by a bacteria or a virus or a combination of the two. There is often a predisposing factor, such as chilling or previous infection with influenza or strangles. Affected horses usually have a temperature of 102° to 105°F, have difficulty breathing, have a nasal discharge, go off feed, and suffer from chest pains and lung congestion. Prompt treatment with antibiotics is usually effective if the infection is of bacterial origin.

Inhalation pneumonia is caused by improper oral administration of liquids. For example, a stomach tube might be wrongly placed in the lungs rather than in the stomach, and materials such as worm medicine or mineral oil may be poured into the lungs—a fatal mistake. Therefore, stomach tubes should be inserted only by veterinarians or upon the instruction of a veterinarian.

Arabian foals or part-Arabian foals may develop adenovial pneumonia. The condition is acute or progressively chronic and fatal in an animal with an inherited combined immunodeficiency (CID) McGuire et al., 1974; Thompson et al., 1975). Foals with this immune deficiency have a small or hypoplastic thymus, a very low number of lymphocytes in the blood, and low levels of gamma globulin in the blood. Foals that do not receive protective antibodies in the colostrum and therefore lack gamma globulin may also suffer from adenoviral pneumonia (McChesncy ct al., 1974). These latter animals will respond to treatment. CID is inherited as an autosomal recessive trait (Section 13-3).

Influenza

Influenza in horses is caused by myxoviruses and has many of the characteristics of influenza in man. The horse develops a high temperature (101° to 106°F), has a depressed appetite, and emits a watery nasal discharge. Dry, hard coughing is often a sign. The condition is usually not fatal, but diseased horses that are worked, shipped, or exposed to bad weather may develop fatal complications. Young foals are particularly susceptible and may develop pneumonia and die.

Some strains of human influenza virus have produced influenza in horses, but the most common viruses in the United States are myxovirus influenza A/Equi 1 and myxovirus influenza A/Equi 2. The incidence of the disease varies greatly from year to year. It has been estimated that in 1963, 50 to 90 percent of the horses in some areas of the United States suffered from the disease (Blood and Henderson, 1972). Influenza is spread by droplet inhalation, and the aerosol form of the virus can survive for from 24 to 36 hours. Fortunately, the virus is usually transmitted only short distances, and isolation of new animals prevents spreading of the condition. An effective vaccine made from killed virus is now available. It contains both types of equine influenza viruses.

Equine Viral Rhinopneumonitis (EVR)

This disease is caused by one of several herpes viruses — type 1, type 2, or type 3. Type 1 is the major cause of rhinopneuminitis in the United States. Secondary bacterial infection is often a complicating factor. A fever of up to 106°F may last 2 to 5 days, and may be accompanied by a clear nasal discharge and coughing. Coughing may persist up to 3 weeks. Abortions may occur up to 4 months after the respiratory signs are noted, although in some cases the latter are so mild that they may escape observation. Abortions occur most frequently during the last third of the gestation period. If one abortion occurs, a high percentage of the mares on the farm may also have abortions (that is, there may be an abortion storm). There can also be perinatal mortality (death soon after birth). Sporadic outbreaks of paralysis have also been reported.

Although there is no treatment, antibiotics are often given to control secondary bacterial infections. Recovered animals are immune for several months but may be reinfected.

A killed vaccine and a modified live virus vaccine are available to prevent EVR. The killed vaccine (Pneumobost K®) is the only one labeled as being useful to prevent viral abortion. It appears that either vaccine is capable of controlling abortion storms but as with most vaccines neither is 100 percent effective. The vaccination interval is controversial among veterinarians.

Equine Viral Arteritis (EVA)

This is an acute upper-respiratory tract infection caused by a specific herpes virus that may also causes abortion. Although the disease has been identified for many years, it received little attention until a limited outbreak occurred in Kentucky in Thoroughbreds during the summer of 1984. Viral arteritis is similar to influenza and rhinopneumonitis in that there is a fever (102° to 106°F) and a nasal discharge, but is more serious to the mature horse than either, because horses of all age groups are susceptible.

The disease also can produce edema of the limbs, an increased respiratory rate, and occasionally a skin rash. Some strains of the virus appear to cause only subclinical infection (Powell, 1986).

Studies suggest that the virus is prevalent among Standardbred and Saddlebred horses in the United States, whereas only about 2 percent of Thoroughbred horses are positive.

Powell (1987) reported that majority of infected animals recover without therapy but antibiotics may reduce the possibility of secondary infection.

Under certain conditions mortality may be as high as 33 percent and the abortion rate is approximately 50 percent. Laboratory examinations using complement-fixation or serum neutralization tests may be necessary to diagnose the disease. The disease spreads rapidly. It can be transmitted by inhalation of droplets from infected horses or sexually. A live modified EVA vaccine has been developed commercially and was licensed for general distribution during 1985.

Equine Encephalomyelitis

This disease has received great public attention because it can be transmitted to humans. The disease at present is restricted to the American continents. Several different viruses cause equine encephalomyelitis and are usually designated by geographical terms; Venezuelan (VEE), Eastern (EEE), and Western (WEE) are the three most common types. The disease is transmitted primarily by mosquitoes (especially *Culex* spp.) from infected horses to susceptible horses or to man. Other insects, such as ticks, bloodsucking bugs, mites, and lice, may also occasionally transmit the disease. The Venezuelan type can also be spread by aerosol when animals are in close contact. The viruses have been isolated from many species of animals, such as horses, mules, donkeys, man, monkeys, snakes, and frogs, in natural conditions and calves, dogs, mice, and guinea pigs under experimental conditions. Wild birds, however, are the principal reservoir of the virus.

The Western strain usually has the lowest mortality rate: 20 to 30 percent of the infected animals die. The Eastern and Venezuelan strains have mortality rates of 70 to 90 percent. Young horses are most susceptible. Immunity can last for approximately 2 years after infection, but recovered animals often have suffered permanent brain damage that makes them useless.

The incubation period may last from 1 to 3 weeks. One of the earliest signs is a fever that may reach 106°F. The fever usually lasts for no more then 24 to 28 hours and may not be detected. Nervous signs that appear during the period of peak fever include hypersensitivity to sound, a transitory period of excitement, and restlessness. Shortly thereafter, the signs associated with brain lesions — such as drowsiness, drooping ears, abnormal gait, and circling—appear. The horse may stand with head held low and food hanging from the lips.

The next stage is paralysis. The horse loses the ability to raise its head, the lower lip drops, and the tongue may hang out. The horse also has difficulty walking. Defecation and urination are difficult. The final stage—complete paralysis and death—may occur 2 to 4 days after the first signs appear.

Encephalitis can be prevented by vaccination. A combination of killed Eastern and Western viruses given in two doses provides protection for about one year. A modified live virus product from the Venezuelan strain gives protection for several years. A killed Venezuelan virus vaccine has only recently become available.

Equine Infectious Anemia (EIA) (Swamp Fever)

This viral disease has received considerable attention in recent years. Incidences of it have been reported in many parts of the world and in all of the contiguous United States. Equine species are the only known natural hosts.

There is no treatment to eliminate EIA virus from an infected horse and horses surviving clinical EIA can be a source of infection for horses and may have relapses (Warner and Morris, 1987).

The acute form of the disease is characterized by the sudden onset of a high fever of 104° to 108°F, severe depression, depressed appetite, and loss of weight. The animal usually becomes weak and loses coordination. Jaundice and edema of the ventral abdomen, sheath, scrotum, and limbs may develop. The spleen becomes enlarged, as can sometimes be verified by rectal examination. There is rapid destruction of red blood cells. The anemia is characterized by low hematocrit and low hemoglobin levels, low red-blood-cell count, and a high sedimentation rate. In severe cases the mortality rate is high.

The subacute form is reported to be the most common form. The signs are similar to the acute form but are not severe and death seldom occurs.

Horses with a chronic form of the disease usually appear to be unthrifty and lack stamina. They may have acute or subacute attacks. Anemia may develop periodically. Such animals are usually not satisfactory for work or breeding. Some horses may appear to recover after 7 to 20 days but may have another attack weeks, months, or even years later, usually during periods of stress. These animals are dangerous to the welfare of other horses because they are carriers of the virus but may not be recognized as such.

EIA is transmitted in the blood from infected horses. Bloodsucking flies are probably the most common vectors but mosquitoes, hypodermic needles, surgical instruments, dental floats, and bridle bits may also transmit the virus.

No vaccine is available but a reliable diagnostic method is. The Coggins test, named after Dr. Leroy Coggins of Cornell University, is an agar gel-im-muno-diffusion test that is now the official USDA-approved test. Only 10 cc of the animal's blood is needed to conduct the test and the results can be available within 48 hours. Previously, it was necessary to take blood from a suspected animal and inject it into a horse or pony that had been in isolation. Thirty to sixty days were then required to detect evidence of infection in the test animal. Regulations and testing intervals vary from state to state.

Tetanus (Lockjaw)

Tetanus is caused by *Clostridium tetani*. The bacteria can be found in the feces of horses and in soil contaminated by horse feces throughout the world. The bacteria, which enter the body through a wound or the naval cord, are anaerobic, and thus deep puncture wounds are more likely to result in tetanus than are surface lacerations. A neurotoxin is produced, which reaches the central nervous system by traveling up peripheral nerve trunks rather than by passing through the blood stream. The exact mode of action of the toxin is not known, but it brings about overreaction to reflex and motor stimuli, resulting in spasmodic or constant muscular rigidity. Thus the animal has great difficulty walking and may develop a "sawhorse" posture. The prolapse of the third eyelid is a characteristic sign of tetanus. Spasms of the masseter muscles can occur early in the disease, hence the term "lockjaw." Death usually occurs by asphyxiation because of rigidity of the muscles of respiration.

Mortality is almost 100 percent in untreated animals and 75 to 80 percent in treated animals. Treatment includes relaxation of muscles by frequent injections of muscle-relaxing drugs. Tranquilizers have also been used. Slinging and stomach tubing or intravenous feeding are necessary in some cases because the animals cannot eat.

The best approach is prevention. Active immunity is best achieved by injection of tetanus toxoid with annual boosters. Any horse not immunized that has suffered a puncture wound or deep laceration should be given tetanus antitoxin.

Scratches

Scratches is an eczema that affects the fetlock and heel areas. Repeated exposure to sweat, mud, and filth predisposes the development of the condition. There may be complications caused by secondary bacterial or fungal infection. The affected area should be thoroughly cleaned, long hair clipped, and skin debris removed. Application of ointments and topical antibiotics is helpful.

Thrush

Thrush is a degenerative condition of the frog. A black discharge, the offensive odor of which identifies the condition, is emitted from the frog. In severe cases most of the frog may be eroded and lameness may occur. The frog is normally resistant to bacterial infection, but continued exposure to bacteria (such as standing in manure) overcomes the resistance. The foot should be thoroughly cleaned with soap and water, the rotting frog trimmed away, and an antiseptic applied.

Potomac Horse Fever

Potomac horse fever was so named because it was first recognized (in 1979) in the Potomac Valley. The first signs are usually mild depression and decrease in appetite. Body temperature may be elevated and gut sounds (borborygmi) are decreased or absent. Within 24 to 48 hours, a profuse watery diarrhea may develop. Palmer (1989) pointed out, however, that horses may respond in different ways. Some horses may just become depressed and develop a fever whereas others develop severe problems of the digestive tract including abdominal distention and severe colic. Some of the horses have developed laminitis. Potomac horse fever is caused by a *Rickettsia Ehrlichia* hence some people prefer the name acute equine ehrlichial enterocolitis which is more descriptive than Potomac horse fever. Clinical diagnosis can be determined by isolation of *Ehrlichia* in tissue culture.

The disease has now been reported in several other areas such as Pennsylvania, New York, Idaho, Illinois, West Virginia, Indiana, Ohio, Kentucky, and Minnesota (Perry et al., 1986).

The carrier of the rickettsia has not been identified but blood-sucking arthropods such as the American dog tick are considered as likely suspects (Palmer, 1987).

Palmer (1987) reported that the fatality rate of untreated horses with Potomac horse fever ranged from 17 to 36 percent but that prompt veterinary care — such as correcting the fluid, electrolyte, and acid-base imbalance and treatment with certain antibiotics — can greatly increase survival rate. A commercial vaccine is now available. Complete prevention however may not be realized until more information concerning the carrier of the disease, and the method by which the disease is spread, is available.

Rabies

Rabies is caused by a rhabdovirus. It can be transmitted to horses by a bite from carnivores such as skunks, raccoons, foxes, or dogs, or by bats.

West (1985) reported that the signs of rabies in horses can differ greatly. Sometimes there are no signs; the horse is simply found dead. Facial paralysis, lameness, anorexia, hindleg paralysis, grinding of teeth, colic, faulty vision, and viciousness have all been reported in affected horses prior to death.

Fortunately the incidence of rabies in horses is quite low. The Center for Disease Control in Atlanta reported that in the United States in 1984 there were 2081 reported cases in skunks, 1038 in raccoons, 139 in foxes, 159 in cattle, 135 in dogs and 59 in other domestic animals. The data illustrate the value of rabies shots for dogs. In 1946, there were 8000 cases of rabies reported in dogs.

Botulism

Botulism is caused by toxin produced by the gram-positive bacterium *Clostridium*.

Although the disease can be found in young and mature horses, it is probably of greatest concern in foals in the United States. The disease in foals is called "shaker foal syndrome" because of marked muscle tremors. The foals have difficulty swallowing and muscular weakness is present throughout the body. Death may result within 12 to 48 hours because of respiratory failure.

In adult horses, the disease usually appears three to seven days after the ingestion of the toxin. Ricketts et al. (1984) reported that affected horses have muscular paralysis of the limbs, jaw and tongue, and pharynx. Food may accumulate in the mouth and the tongue may become flaccid. Saliva may drool from the mouth. Because of the muscular weakness, the animals may move with a shuffling stilted gait, drag their toes along the ground, or stand with head and neck hanging down. Ricketts et al. (1984) reported 13 cases of botulism in horses fed big-bale silage with inadequate fermentation. Kelley et al. (1984) reported several cases of botulism in horses fed contaminated oat chaff. Animal carcasses or soil can be the source of the toxin in forages.

Vaccination with *Cl. botulinum* toxoid can protect against botulism. Johnston and Whitlock (1986) reported that maximum protection of foals involves vaccinating the mare three times before foaling. A commercial vaccine is now available.

Contagious Equine Metritis (CEM)

Contagious equine metritis is a venereal disease, recently reviewed by Brewer (1983). It is characterized by pus or fluid coming from the uterus. The vagina may be inflamed. Infection causes infertility or early abortion.

Carriers of either sex show no clinical signs and thus allow the infection to persist—that is, the organism can remain undetected for long periods in the clitoral fossa of the mare or the urethral fossa of the stallion. No clinical signs are observed in the stallions and only clinical signs related to the reproduction tract are seen in mares.

CEM is not present in the United States. The disease was first identified in England and France and has since spread to many other countries in Europe and other parts of the world. Two outbreaks occurred in the United States, but they were eradicated by strict control measures—including hygiene, culturing, and quarantine. The disease can be eliminated by use of an antiseptic on the reproduction tract plus surgical removal of the clitoral sinus in the mare. Repeated cultures are used to determine if the animal is free of the disease.

By following a strict protocol that includes preventive treatment and culturing of mares plus breeding a stallion to two test mares, animals from CEM-infected countries are now allowed to enter the United States if they pass the entire quarantine protocol. The quarantine measures are supervised by the USDA and the individual state animal disease control divisions.

16.4   Vaccination Program

This program can be started at 3 months of age or at any age thereafter. Some veterinarians may prefer to start earlier.

All Animals

Tetanus (Lockjaw)   Initially requires two doses 4 to 8 weeks apart, with yearly booster dose thereafter. If vaccination status is unknown, tetanus antitoxin can be administered.

Selected Animals

Equine Influenza   Vaccine is recommended for weanlings, yearlings, and horses that are shown, raced, or, in general, brought in direct contact with groups of strange horses. Initially requires two doses 4 to 12 weeks apart with one to four booster injections annually, depending on exposure to outbreaks.

Eastern and Western Equine Encephalomyelitis (EEE, WEE)   Follow your veterinarian's advice. Vaccination is usually carried out each year early in the spring before the mosquito season begins.

Venezuelan Equine Encephalomyelitis (VEE)   Requires one dose only, good for 3 years or longer. Venezuelan encephalitis vaccine is no longer recommended except in those states bordering Mexico. If an outbreak occurs again in the United States, widespread vaccination will be utilized. Protection is afforded within 14 days or less after initial vaccination. A combined Eastern, Western, and Venezuelan equine encephalitis vaccine is also available.

Equine Viral Rhinopneumonitis (Viral Abortion)   One form of protection is an intranasal "planned infection program." It should be carried out only on those farms presently on the program or where specifically recommended by the local veterinarian. The owner must understand that there is an attendant abortion risk whenever this "planned infection program" is initiated for the first time on a given farm. An injectable vaccine is also available. Two doses are required, 4 to 8 weeks apart with a yearly booster. The vaccine can be used on any horse at any age over 3 months.

Strangles   Two types of killed vaccine are available. Much controversy exists over the value versus the danger of strangles vaccine. Consult your veterinarian.

Anthrax   A number of vaccines are available. They are for use only under federal control in an epidemic area.

16.5   Metabolic and Miscellaneous Disease

Colic

Colic is a general term indicating abdominal pain. The pain may cause the horse to become restless, paw, kick at its belly, or get up and down frequently. The horse may roll, lie on its back, or sit like a dog. The saw-horse posture is often observed. Geldings may exhibit the penis without urinating. The pain may be intermittent but, in severe cases, is constant and unrelenting. The pulse rate is increased but temperature may be normal or elevated from 101° to 103°F as a result of physical activity.

There are many types and many causes of colic. Digestive colic may be caused by overfeeding, sudden change in type of feed, moldy feed, or feed that was not properly chewed because of poor teeth or bolting. Heavy work after a large meal may also produce digestive colic. Spasmodic colic is caused by severe contraction of the intestines. The condition is so named because there are intervals between the spasms of pain. The noise caused by the severe contractions can sometimes be heard while standing near the flank of the horse.

Intestinal obstruction or blocking of the bowel, although a less frequent cause of colic, is a severe condition that causes extreme pain and is a threat to the life of the horse.

Intestinal obstruction is usually caused by blockage with food but can be caused by eating foreign objects. A technical term for eating unnatural material is *allotriophagy*. Large objects may directly block the intestine, causing a very acute, extremely painful, and usually fatal illness. For example, S. M. Getty and coworkers at Michigan State University reported such a case of intestinal obstruction (Getty et al., 1976). A 5-month-old filly was presented to the university clinic because her feed intake was greatly reduced and she was losing weight, was lethargic, and suffered from recurring attacks of colic. She died shortly after admission. At autopsy, large masses of cord from a rubber fence were found in the stomach, small intestine, and large intestine.

Small objects might become the center of balls of ingesta called fecaliths. The fecaliths "grow" by accumulating material on their surfaces. For example, recently several fecaliths each approximately 2 inches in diameter were surgically removed from the intestine of a horse with chronic colic. A piece of wood the size of one-third of a toothpick was found in the center of each of the fecaliths. (The horse was a habitual wood chewer.) Fecaliths composed primarily of minerals called enteroliths (Lloyd et al., 1987) may also be found.

Enteroliths were quite common in the late 1880s but few were reported the first half of the twentieth century. However, since the 1970s the incidence seems to be greatly increasing in certain areas such as southern California. Enteroliths are usually composed of magnesium ammonium phosphate. Factors leading to enterolith formation need further study but the presence of a nidus such as a hard stone or metallic object and adequate concentrations of ammonia, magnesium, and phosphorus could be critical. Surgical removal of enteroliths is highly successful unless there has been severe damage to the intestinal wall.

Horses that are fed on the ground may develop sand colic because they ingest so much sand that the intestine becomes obstructed. This condition is common in the Southwestern United States, Texas, and Florida. Gravel or small stones are also eaten by horses and may cause intestinal obstruction.

Parasites are the most common cause of colic Migrating strongyle larvae damage the blood vessels by producing aneurisms, that is, by weakening the walls. Hence the blood supply to the intestines is decreased and the cells become anoxic, resulting in decreased motility and pain. Roundworms in large numbers can cause impaction or obstruction of the intestine and tapeworms have been associated with intussusception (telescoping of the intestine). A good parasite control program is essential for the prevention of colic.

Twisted intestines also result in obstruction and colic. The causes of the twisting are not clearly defined.

Horses have twists of the large colon more frequently than do ponies. Some breeds appear to be more subject to twists. Anatomy and chance must play a prominent role but the precise circumstances are complex and intertwined. The end result is always the same — severe colic and death if surgery is not performed to correct the condition. Time is critical. Unfortunately, many cases are beyond help by the time surgery can be performed.

A horse with colic should be observed closely and if rolling should be walked to prevent injury during rolling. If after 15 to 20 minutes of walking, the horse acts as if the pain is unrelenting or increasing and makes repeated attempts to lie down or breaks out in a cold sweat, a veterinarian should be called. The treatment for colic varies with the type of colic. For example, recovery may be spontaneous. Analgesics (pain-relieving drugs) may be adequate in some cases, but in other cases, such as a horse with a twisted intestine or foreign material obstruction, surgery may be necessary.

Heaves (Chronic Obstructive Pulmonary Disease)

The disease develops gradually and the clinical signs are most commonly seen in horses 5 years old or older. At first the animal has a cough or wheeze that may be more pronounced after exercise. As the disease progresses, the coughing becomes more frequent and a nasal discharge may be present. With chronic heaves, a "heave line" may develop because of the extra effort needed to exhale. The heave line is a prominent ridge of abdominal muscle that is seen as a nearly straight line from high in the flank diagonally forward and down. It appears with each expiration following normal relaxation of the rib cage because the abdominal muscles give an extra squeeze to force the intestines against the diaphragm, which in turn forces air out of the affected lungs.

Horses with clinical heaves are of little use for performing work because of the reduced capacity for oxygen uptake. Horses with mild heaves, however, can be used as pleasure horses if they are not pushed too hard and, with proper care can lead a long and useful life.

The causes of heaves are not well defined. Some reports indicate that it is an allergic response that is perhaps due to molds. Dusty feeds greatly aggravate the condition. Heaves is seldom seen in horses pastured the year round.

No treatment is known that will result in complete recovery. The condition is alleviated by resting the animal, providing fresh air, and eliminating dust. Pasturing the horse is often the best solution. Wetting the hay and grain or using a complete pelleted ration may be helpful in reducing dust. The bedding should not be dusty. Shavings might be substituted for straw. The animal should be protected against extreme environments such as high humidity or high temperature. Antihistamine drugs may be of value in some situations. But, to repeat, the most important remedies are rest, elimination of dust, and provision of fresh air.

Roaring (Laryngeal Hemiplegia)

Roaring is caused by damage to the recurrent laryngeal nerve and results in lack of muscular control of the vocal cords (usually the left cord but sometimes both). The affected vocal cord vibrates with inspiration, causing a roaring sound. On expiration the vocal cord and sac associated with it are passively pushed aside by the exiting air. The severity of the condition varies but the symptoms are usually most noticeable after exercise. Roaring greatly impairs the use of working horses, such as racehorses, cow ponies, hunters, and jumpers. If the horse is used only for light pleasure riding, roaring might not impair its serviceability, but the roaring sound may be annoying to the rider. Roaring can usually be corrected by surgery. The etiology of the damage to the nerve remains obscure, although the left nerve is almost always damaged, probably because it travels around the arch of the aorta and along the deep face of that vessel, where it receives pressure from the strong aortic pulsations.

16.6   First Aid

Two questions that plague horseowners are: (1) when to call the veterinarian, and (2) what to do until the veterinarian arrives. Several guidelines can be given but the keys to answering both questions are keeping calm, using common sense, and being prepared. Preparation includes being informed, having adequate supplies on hand, and developing a program with your veterinarian.

When should the veterinarian be called? As mentioned earlier, the owner should know the proper procedures to measure body temperature, respiration rate and heart rate. *Call when temperature is elevated.* Elevated body temperatures (103°F or above) may indicate infection but they can also be caused by stress or pain. Also remember that a normal temperature does not prove that the horse is healthy. Under many conditions seriously ill horses can have normal temperatures.

Whenever there is a serious wound the veterinarian should be called. The type of wound determines the action that should be taken. Wounds are classified as follows:

1. Abrasion—multiple superficial scratches that do not penetrate the full thickness of the skin (e.g., rope burn)

2. Incision—clean wound caused by a very sharp object (e.g., glass or metal gate)

3. Laceration—a wound that penetrates the full thickness of the skin and is caused by a less-sharp object (e.g., barbed wire), resulting in both cutting and tearing of skin.

4. Puncture—a wound caused by a more or less pointed object (e.g., nail)

5. Avulsion—a wound characterized by tearing of skin to cause a loose flap

When a horse has been wounded, the primary duties that should be accomplished until the veterinarian arrives are: (1) stop the hemorrhage, and (2) keep the wound clean.

Hemorrhage Control   It is very unusual for a horse to bleed enough after a wound to go into shock from blood loss. The most dangerous locations are wounds over the lower part of the neck (jugular vein, carotid artery) and over the sides of the pastern (digital arteries). If a wound has squirting vessels of large size (e.g., half the diameter of a pencil) or is bleeding copiously from damage to a number of smaller vessels, hemorrhage should be brought under control by applying a pressure bandage. The wound should be covered with clean material such as a Telfa pad, clean cloth, or disposable diaper. Cotton or material with loose fibers should be avoided because these have a marked tendency to stick in the wound, making later cleansing difficult. Following the application of clean material to the wound, hemorrhage is controlled by wrapping the wound snugly with an Ace bandage, or derby wrap if the wound is on the leg. Hemorrhage from wounds in other locations is controlled by firmly pressing the dressing with the hand. The dressing may become soaked with blood and still be controlling severe hemorrhage. Unless the firmly applied dressing drips copious amounts of blood, leave it in place. If you think too much blood is dripping add another layer of dressing.

Most wounds of the leg with the exception of those with squirting, pulsating bleeding vessels associated with them can be "hosed" with cold water if it is available. This does three things: (1) it soothes the site and helps decrease pain; (2) the cold helps constrict vessels and control bleeding; and (3) the high volume of water and gentle action of the water pressure help cleanse the wound and allow you to get a complete look at it without getting your hands into it.

Wounds of the legs should be kept clean by dressing the wound as described above and applying a moderately snug wrap. Do not put anything into the wound unless you have previously checked the material with your veterinarian. Materials that should NOT be used include dyes such as Coppertox or Blue-coat, pine-tar-based medications, and hydrogen peroxide, because these are all damaging to fragile tissues in the wound. The only thing ever to put into a fresh wound is bland antibiotic in powder or ointment form. A rule of thumb to follow in deciding whether or not to place a given material in a wound is to place nothing in a wound that you would not place into your own eye! If a veterinarian has been called to examine and treat a wound, do not put anything into it.

If a horse's tetanus vaccination status is up to date, veterinary attention is probably unnecessary for most abrasions and for superficial wounds of the upper legs, trunk, and neck. These wounds can be treated by clipping the area around the wound with a fine clipper blade followed by gentle washing of the wound with a nonirritating soap. The wound is then dressed daily with bland antibiotic powder or ointment. In the summer, insect repellents should be placed around the wound but never in the wound. Wounds of the legs below the knee or hock, if small (e.g., less than 1½ inches long), can usually be treated in the manner described above.

Veterinary attention is necessary for large wounds of the upper legs, neck and trunk and for wounds of the lower legs exceeding 1½ inches in length and cut through the full thickness of skin. If these wounds are fresh, they are usually sutured up. If they are not fresh, they are ordinarily treated by thorough cleansing and by debridement (surgical removal of damaged tissue). Wounds of the limb are usually kept under a firm bandage during the healing process.

Veterinary treatment is necessary for any wound located near a joint, tendon, or coronary band. Additionally, any wound that may contain foreign material such as wood and any puncture wounds should be examined by a veterinarian.

It is mandatory that tetanus vaccination be brought up to date in any horse that has incurred a wound, regardless of the size of the wound.

Whenever there is a fracture, a veterinarian should be called. Fractures in horses usually involve the bones of the legs and can occur anywhere from the coffin bone to the bones of the shoulder or hip. They are characterized by an acute onset of severe (i.e., non-weightbearing) lameness. Angulation of the leg may be evident in some cases.

A horse that has fractured its leg should not be moved until the leg is splinted. If the site of the fracture can be identified by swelling, angulation, or crepitation (grating on movement), that area should be immobilized as well as the joint above and the joint below the fracture. Fractures of the long or short pastern bones can be successfully splinted by applying a very heavy pressure bandage consisting of several quilt or cotton wraps and several Ace or derby bandages. For fractures of the cannon bone, radius, or tibia, a heavy splint should be placed. Satisfactory wraps include a very heavy cotton and Ace bandage wrap (Robert Jones bandage), a pillow splint, a heavy wrap splinted with broomsticks, or a pipe splint. Splinting above the hock or above the knee is difficult at best. If possible, just keep the horse quiet until the veterinarian arrives.

If the horse is down, try to keep him down, until the leg is splinted or the veterinarian arrives, by sitting on his neck just behind the head and holding the nose in the air. DO NOT try this if the horse is violent from excitement and/or pain, as the risk of injury to handlers may be great. What about colic? As mentioned earlier, colic is a general term used to describe pain in the abdomen. Many conditions can cause colic in horses but common ones are spasm of the intestine, blockage of the intestine by feedstuffs or foreign material, or twisting of the intestine. The horse should be carefully observed, and if the problem persists longer than 15 to 20 minutes, or if the horse shows signs of severe pain, a veterinarian should be called.

Colic   Signs of colic vary considerably, depending on what is causing the problem. They may include restlessness, kicking at the belly, looking around at the flanks, sweating, dropping to the ground and rolling, or unmanageable violence.

Do not allow the horse to eat or drink water. Many animals showing mild abdominal pain will recover without treatment.

The major duty of horsepeople in attendance is to prevent the horse from injuring itself. If a horse is uncomfortable but is not trying to roll, just observe him until the veterinarian arrives. If the horse is unmanageable or trying to roll on the ground, he should be walked. DO NOT put yourself in a position where you may be injured by a violent horse.

Tying-up   Tying-up disease is usually seen in very fit horses that are on a regular daily exercise schedule but have missed a few days of work and have remained on full feed. Signs of tying-up generally appear at from 15 to 30 minutes into exercise, but may not be seen until exercise is over. The horse usually is reluctant to move and may seem to be stiff all over. Sweating, painful breathing, and muscle tremors may also be seen. Some horses have tense, painful muscles over the back and hips. Urine may be brownish.

This condition requires immediate veterinary attention. Do not move the horse any further or attempt to ride it home. Leave the horse where it is, blanket it and wait for the veterinarian.

Choke   Choke is partial or complete blockage of the esophagus, causing inability to swallow. Choke is usually due to blockage of the esophagus by feedstuffs (hay or grain) or other material (apples, wood chips). Choke may be associated with greedy eating, bad teeth, or lack of water. Signs of choke include arching and stretching the neck, anxiety, and the appearance of saliva or saliva mixed with feed at the nostrils.

Do not allow the horse to eat or to drink water. Observe the horse closely. If the horse has not returned to normal in 1 or 2 hours, veterinary attention should be sought.

The veterinarian should also be called when the mare has difficulty foaling. Problems with foaling were discussed earlier (Section 11.5).

Of course, veterinary attention might be needed whenever a horse has excessive discharge from the nostrils or has a persistent cough, is lame, or when there is something "not quite right" with the horse.

What medical supplies should be kept on hand?

A nitrofurazone salve can be used for treating burns and wounds. An antibiotic eye ointment can be helpful. Fly repellents and petroleum jelly for mild abrasions and to protect the skin below draining wounds should also be on hand.

The medicine cabinet should have bandaging supplies such as a box of 3-inch-square gauze pads; a nonstick pad; a roll of conforming (stretch) gauze; several 4-inch-wide elastic bandages, adhesive elastic, or knitted bandages; a 2-inch-wide roll of adhesive tape; a 1-pound roll of cotton or sanitary pads (for larger wounds or pressure bandages); padding (sheet cotton, quilted pads, or disposable diapers); a roll of black electrician's tape (to hold ice packs and poultices on the leg); cotton swabs; bandage scissors; and a sharp knife.

Injectable medications such as antibiotics and analgesics, if kept on hand, should be used only in consultation with your veterinarian. Refrigeration may be required for some medications and most have expiration dates, after which they should be discarded.

REFERENCES

Belschner, H. G. 1969. *Horse Diseases*. London: Angus and Robertson.

Blood, D. C., and J. A. Henderson. 1972. *Veterinary Medicine*. London: Bailliure, Tindall and Cassel.

Brewer, B. D., and A. M. Koterba. 1987. Neonatal septicemia. In *Equine Medicine Surgery*. N. E. Robinson, ed. Philadelphia: W. B. Saunders.

Brewer, R. A. 1983. Contagious equine metritis: A review/summary. *Vet. Bull.* 53:881.

Bryans, J. T., and H. Gerber, eds. 1973. Equine Infectious Diseases. In *Proc. 3rd International Conf. Equine Infectious Diseases*, Basel, Switzerland. New York: Karger.

Catcott, E. J., and J. F. Smithcors. 1972. *Equine Medicine and Surgery*. Wheaton, Illinois: Am. Vet. Publ.

Getty, S. M., D. J. Ellis, J. D. Krehbiel, and D. L. Whitenak. 1976. Rubberized fencing as a gastrointestinal obstruction in a young horse. *Vet. Med. Small Animal Clin.* 71:221.

Hayes, J. ed. 1984. *Yearbook of Agriculture. Animal Health—Livestock and Pets.* U.S.D.A.

Johnston, J., and R. H. Whitlock. 1987. Botulism. In *Current Therapy in Equine Medicine*. 2nd ed. N. E. Robinson, ed. Philadelphia: W. B. Saunders.

Kelly, A. P., R. T. Jones, J. C. Gillick, and L. D. Sims. 1984. Outbreak of botulism in horses. *Equine Vet. J.* 16:519.

Lloyd, K., H. F. Hintz, J. D. Wheat, and H. F. Schryver. 1987. Enteroliths in horses. *Cornell Vet.* 77:172–186.

Mansmann, R. A., E. S. McAllister, and P. W. Pratt, ed. 1982. *Equine Medicine and Surgery*. Santa Barbara, California: American Vet. Publ.

McChesney, A. E., J. J. England, C. E. Whiteman, and J. L. Adcock. 1974. Experimental transmission of equine adenovirus in Arabian and non-Arabian foals. *Amer. J. Vet. Res.* 35:1015–1023.

McGuire, T. C., M. J. Poppie, and K. L. Banks. 1974. Combined (B- and T-lymphocyte) immunodeficiency: A fatal genetic disease in Arabian foals. *J.A.V.M.A.* 164:70.

Palmer, J. E. 1987. Potomac horse fever. In *Current Therapy in Equine Medicine* 2nd ed. N. E. Robinson, ed. Philadelphia: W. B. Saunders.

Perry, B. D., J. E. Palmer, H. F. Troutt, J. B. Birch, D. Morris, M. Ehrich, and Y. Rikihisa. 1986. A case-control study of Potomic horse fever. *Prev. Vet. Med.* 4:69

Powell, D. G. 1987. Viral respiratory disease, In *Current Therapy in Equine Medicine*, 2nd ed., N. E. Robinson, Philadelphia: W. B. Saunders.

Robinson, N. E., (ed) 1987. *Current Therapy in Equine Medicine*, 2nd ed., Philadelphia: W. B. Saunders.

Ricketts, S. W., et al. 1984. Thirteen cases of botulism in horses fed by bale silage. *Equine Vet. J.* 16:515.

Rossdale, P. D., and S. M. S. Wreford. 1974. *The Horse's Health from A to Z*. London: David and Charles.

Stear, R. L. 1976. Diagnosis and control of equine rhinopneumonitis. *Proc. 1st National Horsemen's Seminar*. Virginia Horse Council, Fredericksburg, Virginia.

Thompson, D. B., M. J. Studdert, R. G. Beilharz, and I. R. Littlejohns. 1975. Inheritance of a lethal immunodeficiency disease of Arabian foals. *Austr. Vet. J.* 51:109.

Timoney, P. J., and W. H. McCollun. 1986. The epidemiology of equine viral arteritis. *Proc. Ann. Conv. Am. Assoc. Equine Pract.* 31:545–51.

Traub-Dargatz, J. L., S. L. Ralston, J. K. Collins, D. G. Bennett, and P. J. Timoney. 1985. Equine viral arteritis. *Comp. Cont. Educ. Vet.* 7:S490.

Warner, A., and D. Morris. 1987. Hemolytic anemia, In *Current Therapy in Equine Medicine*, 2nd ed., p. 295. N. E. Robinson, ed., Philadelphia: W. B. Saunders.

West, G. P. 1985. Equine rabies. *Eq. Vet. J.* 17:280.

Table 16-1
Summary of infectious diseases

Disease	Cause	Primary characteristics	Prevention	Treatment
Influenza	Myxovirus	Fever of 101° – 105°F; watery nasal discharge	Vaccination; isolation of new animals	Good nursing care
Strangles	*Streptococcus equi*	Enlarged lymph nodes; nasal discharge—watery at first, then purulent; fever of 103° – 105°F	Isolation of new animals; vaccinate all animals with killed vaccine (controversial)	Antibiotics
Pneumonia	Virus; bacteria	Fever of 102° – 105°F; chest pains; lung congestion; difficulty in breathing	Good management; avoid chills and stress; provide proper ventilation	Antibiotics when cause is bacterial

Disease	Cause	Symptoms	Prevention/Control	Treatment
Rhinopneumonitis	Herpes virus	Fever of 102°–106°F; coughing; clear nasal discharge, abortion primarily in last third of gestation	Modified live virus vaccine	No treatment
Viral arteritis	Herpes virus	Fever of 102°–106°F; abortion	Vaccine available. Some states have specific regulations	Good nursing care
Equine encephalomyelitis	Virus	Brain lesions; drowsiness; fever; lower lip drops; difficulty in walking	Vaccination	Good nursing care
Equine infectious anemia	Virus transmitted in blood of infected horses	Fever of 104°–108°F; weakness; jaundice; edema of ventral abdomen	Use of Coggins test to detect carriers	No treatment
Tetanus	Clostridium tetani	Muscular rigidity; prolapse of third eyelid	Tetanus toxoid or antitoxin	(Mortality rate is high)
Anthrax	Bacillus anthracis	High fever; edema about throat, lower neck, and chest, lasts approximately 48–96 hours	Vaccines—use only in an epidemic area. Regulated procedure	Antibiotics, anti-anthrax serum

Table 16-2
Summary of metabolic and miscellaneous diseases of the horse

Condition	Cause	Primary characteristics	Prevention	Treatment
Colic	Parasites; overfeeding; moldy feed; twisted intestines	Abdominal pain; kicking at belly; rolling; sawhorse posture	Parasite control; proper feeding methods	Walking; laxatives; analgesics; surgery in some cases
Heaves	Not well defined—perhaps allergic response	Coughing; heave line	Prevent and treat respiratory infections; control stable dust	Control dust; provide fresh air; give antihistamine drugs
Roaring	Damage to recurrent laryngeal nerve	Noise during inspiration	Prevent and treat respiratory infections	Surgery
Recurrent uveitis	Immunological response, particularly to infection with leptospirosis	Eye is periodically cloudy; after several attacks, permanent blindness may result	Recommendations cannot be made for adequate prevention because of insufficient knowledge concerning the condition	Steroid therapy may be helpful
Azoturia (Tying-up disease, myositis)	Unknown, but condition occurs when animal is exercised after being full fed during period of inactivity	Stiff gait; reluctance to move; dark-colored urine	Reduce feed intake when animal is not working	Injections of thiamin, vitamin E, and selenium are frequently used, further studies needed to determine their value
Founder (Laminitis)	Overeating; retained placenta	Lameness; heat in hoof; "seedy toe"; drop of coffin bone	Maintain a sound feeding program; do not keep horses too fat	Acute: obtain immediate veterinary attention; chronic: perform corrective trimming and shoeing

CHAPTER 17

PARASITES OF THE HORSE

17.1 Parasites and Parasitism

A *parasite* is a small organism that lives on or in and at the expense of a larger organism called the host. Parasites that attack the skin and body openings of the host are called external parasites or *ectoparasites*, and those that live in the internal organs, body cavities, and tissues are called internal parasites or *endoparasites*.

Parasitic adaptation is a product of evolution, and parasite species have evolved special anatomical and physiological modifications that render them capable of exploiting the energy source embodied in their host species. These special modifications have no alternative applications, so most parasites rely entirely upon their hosts and perish without them. For example, *Gasterophilus* species, the stomach botflies, have evolved intricate mechanisms for gaining access to the stomach of a horse and for resisting digestion there for up to 9 months, but they are completely incapable of making a living in any other way. The adult botflies do not even have mouthparts and so cannot eat anything; they must mate and lay their eggs using energy that they stored when they were parasitic larvae (stomach bots). Most of the parasites of horses are equally committed to the parasitic way of life, although many have so-called *free-living stages* that represent a period of development outside the host. This period of development invariably culminates in an *infective stage*, a stage that is anatomically and physiologically adapted to gaining access to and taking up residence in (that is, infecting) the host.

Certain parasites of horses also utilize other species of hosts as sources of energy. Examples of these include the "horseflies," mosquitoes, certain ticks, faceflies, and one species of stomach worm called *Trichostrongylus*

617

axei. Other parasites are entirely faithful to the horse, and the most important among these are the strongyle and ascarid intestinal worms and the stomach bots. This chapter is devoted mainly to the biology of this latter group.

17.2   The Nematodes or Roundworms

Nematodes have slender, unsegmented, cylindrical bodies that taper toward each end (Figure 17-1). With a single exception (*Strongyloides westeri*), both male and female sexes are represented in all species of nematodes that parasitize the horse. Male nematodes are smaller than the females of their species. The posterior ends of the males of many species expand to form a *copulatory bursa* (Figure 17-2) for grasping the female and are equipped with somewhat rigid rod-shaped *spicules* for dilating the vulva. The female reproductive tract consists of two ovaries, oviducts, and uteri connected to a single vagina and vulva (Figure 17-1). The uteri of female worms contain enormous numbers of eggs that, in most cases, leave the host with the feces to undergo a period of development in the outside world or in the body of an *intermediate host* before becoming infective. A tiny worm, or first-stage larva as it is called, develops within the egg (Figure 17-3). Depending on the species of worm, this larva may hatch out of the egg, as do the strongyles, or it may remain within the egg to undergo further development and await ingestion by a horse, as do the ascarids (Figure 17-4). All nematodes, however, undergo a series of four molts characterized by a *lethargus* or resting stage, *metamorphosis* or restructuring, *ecdysis* or casting of the old larval skin, and, finally, emergence as a new worm adapted anatomically and physiologically to overcome the obstacles that are to be encountered and to achieve the goals that must be achieved if that individual is to play a role in the perpetuation of its species. For example, in the first and second stages of the development of *Strongylus vulgaris*, these nematodes feed on the bacteria that abound in the feces that constitute their environment. In the third, infective stage, they no longer feed on bacteria but, consuming stored energy, move in water films on the surface of vegetation until they are ingested by a horse, whereupon they burrow into the mucous membrane of the intestinal tract, curl up, and molt to the fourth stage. In the fourth stage, *Strongylus vulgaris* larvae have great powers of penetration and rapidly bore through tissue, enter the small arterioles in the wall of the gut, and, crawling along the lining of the arteries, migrate at random for several months. Finally, they return to the wall of the large intestine, molt to the fifth or adult stage, and enter the lumen of the large intestine to mate and repro-

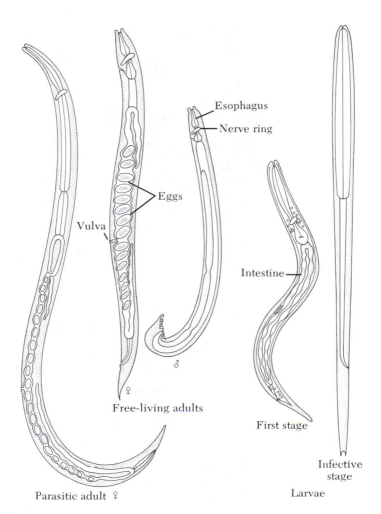

Figure 17-1
General structure of male, female, and larval stages of a nematode worm, *Strongyloides. Strongyloides* is unique among nematode parasites because it has both parasitic and free-living generations. *Strongyloides westeri*, a species that parasitizes the horse, is capable of causing illness in nursing foals. (Adapted from Craig and Faust, 1945.)

duce. Thus, each stage in the life history of this nematode is peculiarly and uniquely adapted to overcome the particular obstacles that confront that stage, and a major "overhaul" or "retooling" is accomplished by each metamorphosis. The life cycle of *Strongylus vulgaris* is summarized diagrammatically in Figure 17-5.

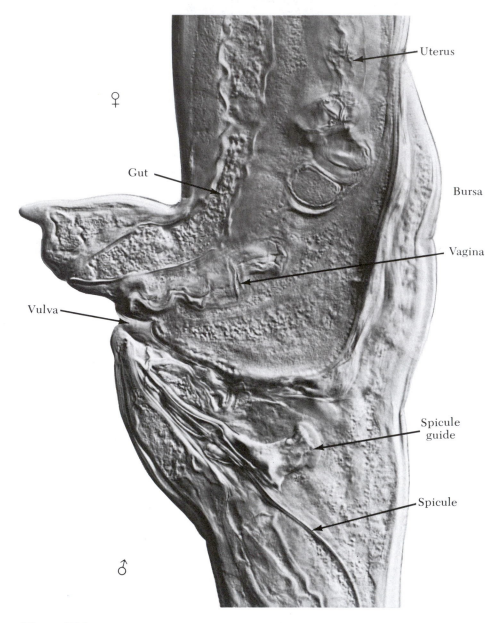

Figure 17-2
Male and female nematodes copulating. The bursa is a special male organ used for
clasping the female. Only the dorsal ray of the bursa is visible in the optical plane
of this photomicrograph, but there are also lateral rays that are applied to each
side of the female. The spicules of the male worm have not yet entered the vagina
of the female. The worms shown here are *Cylicocerus catinatus*, a species of small
strongyle that parasitizes the large intestine of the horse.

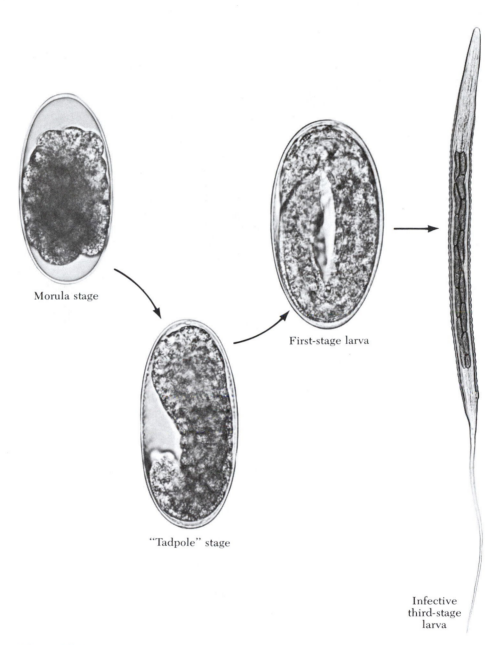

Morula stage

"Tadpole" stage

First-stage larva

Infective
third-stage
larva

Figure 17-3
Development of a strongyle from egg to infective larva.

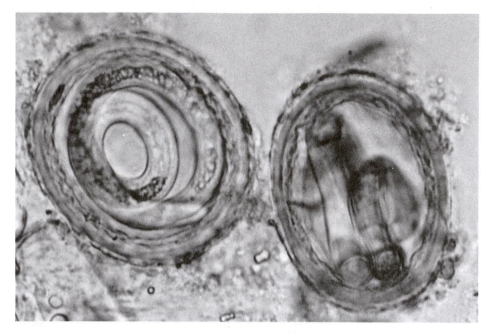

Figure 17-4
Two infective eggs of the ascarid *Parascaris equorum*. The infective second-stage larva does not hatch until the egg is ingested by a horse. Thus, in the infective stage, *P. equorum* is less mobile than the strongyle larva, but, encased in its sticky and protective shell, it may remain alive and stuck to a surface for years.

Strongyles

The horse is host to the approximately 54 species that belong to the family Strongylidae. The most important of these are the "big three," *Strongylus vulgaris, S. edentatus,* and *S. equinus,* which, along with several species of *Triodontophorus* and a few rare species, constitute the large strongyles. The rest of the species, collectively called small strongyles, are usually dismissed as having little or no pathological significance, although invasion of the intestinal mucous membrane by large numbers of them in their immature stages can cause severe diarrhea.

As adult worms, all of the "big three" are bloodsucking parasites of the large intestine (Figure 17-6) and are capable of causing various degrees of anemia, depending upon the number of worms present and the constitutional vigor and state of nutrition of the horse. As immature or larval worms, however, each of these three species undergoes a characteristic, prolonged, and destructive migration through the organs and tissues of its host that may

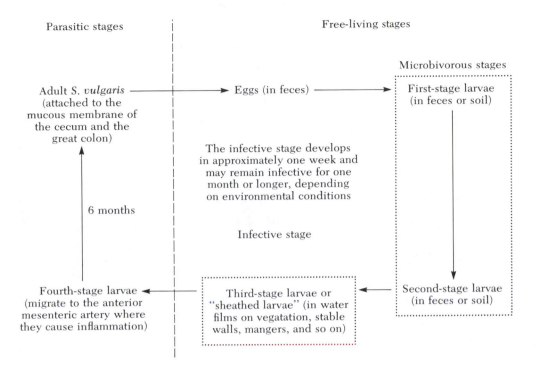

Parasitic stages | Free-living stages

Microbivorous stages

Adult *S. vulgaris* —————→ Eggs (in feces) —————→ First-stage larvae
(attached to the (in feces or soil)
mucous membrane of
the cecum and the
great colon)

The infective stage develops
in approximately one week and
may remain infective for one
month or longer, depending
on environmental conditions

6 months

Infective stage

Fourth-stage larvae ◄———— Third-stage larvae or ◄———— Second-stage larvae
(migrate to the anterior "sheathed larvae" (in water (in feces or soil)
mesenteric artery where films on vegatation, stable
they cause inflammation) walls, mangers, and so on)

Figure 17-5
Life cycle of *Strongylus vulgaris*. (From Georgi, 1974.)

prove even more harmful to the health of the horse than the blood-sucking
activities of the adult worms.

Strongylus vulgaris When ingested by a horse, the infective third-stage
larvae of *Strongylus vulgaris* cast off their sheaths in the stomach and small
intestine and enter the wall of the cecum and ventral colon. Here the larvae
curl up under the mucous membrane, become encapsulated by the inflammation of the surrounding tissues, and prepare to molt. After 8 days, the
molt is completed and the fourth-stage larvae resume their migrations, leaving behind only a few host inflammatory cells and their third-stage cuticles.
The fourth-stage larvae now penetrate the walls of small arterioles lying in
the submucosa and wander upstream into progressively larger arteries by
crawling along the intima or innermost layer of the arterial wall. Inflammation and a slight deposit of fibrin trace the course of each larva, and wherever many such paths cross, summation of the individual reactions leads to
gross thrombosis that may even occlude the vessel lumen. Because the
cranial mesenteric artery is the stem trunk connecting the abdominal aorta

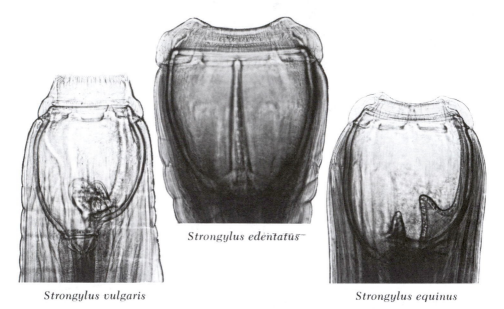

Strongylus edentatus

Strongylus vulgaris   *Strongylus equinus*

Figure 17-6
The "big three" strongyle nematodes of the horse, *Strongylus vulgaris, S. equinus,* and *S. edentatus.* Note the large mouths, by which they attach themselves to the lining of the large intestine.

with all of the branch arteries that supply blood to the intestines, all possible migration routes converge upon this inch-long vessel. In almost every horse, the cranial mesenteric artery bears some mark of the effect of the larvae of *Strongylus vulgaris.* A few larvae excite a superficial inflammatory reaction that results in the formation of a *thrombus* (an intravascular clot). Continued invasion of the inflamed area by larvae causes progressive growth of the thrombus and extension of the inflammatory process into the deeper layers of the arterial wall. Pieces of thrombus called *emboli* (singular, *embolus*) may break away and be carried by the bloodstream until they enter a branch that is too small to accommodate them. Obstruction of various branches of the anterior mesenteric artery by emboli is probably a common occurrence; its effect on the maintenance of arterial circulation to the intestinal tract is discussed in the following paragraphs of this discussion of *Strongylus vulgaris.* Deep inflammation weakens the arterial wall and causes it to yield to the pressure of the blood within, leading to the formation of a saclike dilatation called an aneurysm. Finally, the larvae leave the thrombi, return to the wall of the intestine, molt to the fifth (adult) stage, and enter the lumen of the intestine to mate and reproduce. This final molt may occur as early as 90 days after invasion but may occur much later. Larvae that are enmeshed and detained in thrombi may molt even before returning to the intestinal wall. Thus, there is much variation in the generation time of *S. vulgaris* from egg to egg; the minimum is approximately 6 months. These *S. vulgaris* larval migrations were demonstrated experimentally in 1950 and 1951 by Karl Enigk of the Tierärztliche Hochschule, Hanover. An English translation of Enigk's brilliant work has been published (Georgi, 1973).

Before we further consider the effects of thrombosis and embolism of the arteries that supply the wall of the bowel with blood, we must first define the familiar but frequently misunderstood term "colic."

Colic is acute abdominal pain characterized by restlessness (see Section 8.4). Anything capable of distending or obstructing the lumen of the intestinal tract is capable of causing colic. In 1870, Otto Bollinger hypothesized that occlusion of the intestinal arteries by worm thrombi or emboli could account for the majority of equine colic cases, both fatal and nonfatal. Occlusion of intestinal arteries would, according to Bollinger, always lead to partial or complete paralysis of the musculature of the bowel wall with consequent slowing or stoppage of the flow of intestinal ingesta and gases. The accumulation of gas, aggravated by accelerated fermentation, would then painfully distend the affected portions of the intestine. Unless the circulatory impairment were overcome and intestinal motility restored, the horse would die an excruciatingly painful death. Bollinger interpreted a variety of postmortem lesions as direct manifestations of the thromboembolism of the intestinal arteries and pointed out that many cases of fatal displacements and obstructions of the bowel were probably the result rather than the cause of the colic victim's agonized rolling and thrashing about. Thus, Bollinger believed that virtually all pathological changes observed at a postmortem examination of fatal colic cases had their origins in verminous thrombosis or embolism of the intestinal arteries. However, his views were challenged on two counts.

First, the classical morbid anatomists (pathologists) argued that mere occlusion of an artery, even a large one, is not proof that the blood supply has been completely cut off. They insisted that evidence of infarction or massive death of the affected region of the bowel wall must be present to justify a diagnosis of thrombo-embolic colic. Second, experimental efforts to induce such changes by surgical ligation of major arterial branches also failed to produce fatal infarction of the bowel wall. The reason that occlusion of intestinal arteries either by verminous thrombo-embolism or by surgical ligation infrequently leads to complete cessation of blood flow is that there exists in the intestinal arterial tree of the horse a truly amazing network of interconnections called *anastomoses*. These anastomoses afford rapid development of *collateral circulation* by means of which blood is "detoured" around the obstruction and nutrition of the deprived area is thereby restored. Figures 17-7 and 17-8 show several kinds of anastomoses. The larger the caliber of the anastomotic connection, the more rapidly will collateral

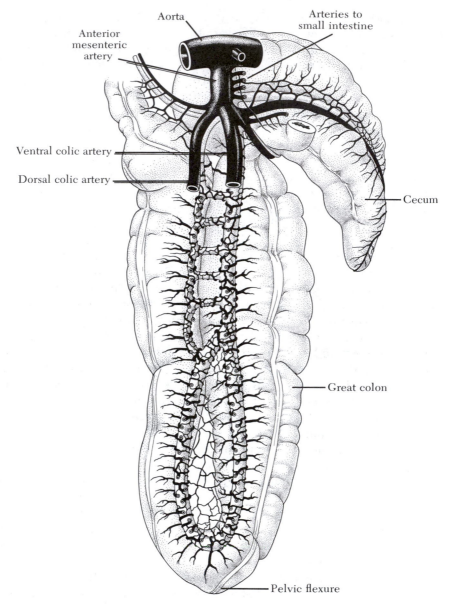

Aorta

Arteries to
small intestine

Anterior
mesenteric
artery

Ventral colic artery

Dorsal colic artery

Cecum

Great colon

Pelvic flexure

Figure 17-7
The large intestine of the horse and its principal arteries. The cranial mesenteric
artery, frequently the site of verminous thrombosis and aneurysm formation,
supplies virtually all of the arterial blood to the small intestine, great colon, and
cecum. Examples of anastomoses depicted here include (1) the fusion of dorsal
and ventral colic vessels at the pelvic flexure, and (2) the cross-anastomoses
connecting the dorsal and ventral colic arteries along most of their lengths.
(Adapted from Dobberstein and Hartmann, 1932.)

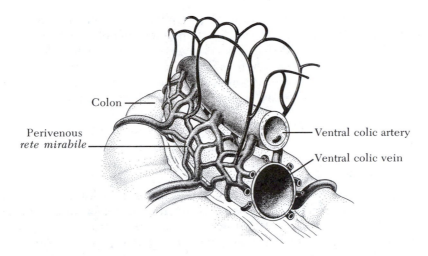

Colon

Perivenous
rete mirabile

Ventral colic artery

Ventral colic vein

Figure 17-8
Arterial network (*rete mirabile*) encasing the ventral colic vein. Dilation of this
arterial network can establish a detour (collateral circulation) around an obstructed
region in the parent ventral colic artery and thus maintain the blood supply to the
wall of the colon. (Adapted from Dobberstein and Hartmann, 1932.)

circulation be established. Thus, obstruction of either the dorsal or the
ventral colic artery will be rapidly overcome via the anastomosis of these two
arteries at the pelvic flexure (Figure 17-7). On the other hand, the establishment of collateral circulation to the area of bowel wall deprived of circulation by double embolism of either the dorsal or the ventral colic artery
would necessitate great dilation of the small-caliber perivenous network
(Figure 17-8) and cross-anastomoses, and consequently would require more
time.

Of all domestic animals, the horse has by far the most highly developed
system of anastomoses in the arterial tree serving the intestine. In an evolutionary context, this may be interpreted as evidence that *Strongylus vulgaris*,
which has no direct counterpart in other domestic animals, has probably
been making a mess of horses' arteries for millions of years. In other words,
the extraordinary network of anastomoses in the arterial supply to the intestines of the horse may logically be interpreted as an evolutionary adaptation
to a selection pressure imposed by this parasite whose larval migrations tend
to cause the occlusion of these particular vessels.

The arteries that supply certain portions of the bowel wall are less well
supplied with anastomoses, and occlusion of these branches is more likely to
lead to severe circulatory disturbances. For example, the medial and lateral
arteries of the cecum do not join at the tip of the cecum, and thus obstruction of either cecal artery is far more serious than obstruction of either the

dorsal or ventral colic artery. Fortunately, verminous embolism of the cecal arteries happens much less frequently than does embolism of the colic arteries. It has been claimed that this is because the cecal artery leaves its parent vessel at a right angle, whereas the colic arteries represent more or less direct continuation of the parent trunk, and that solid emboli are less likely to negotiate sharp turns than the liquid blood transporting them (Figure 17-7).

Although it is probably true that obstruction of intestinal arteries by worm thrombi or emboli leads to infarction and death of the intestinal wall in only a small proportion of cases, it appears equally likely that temporary curtailment of blood flow pending establishment of collateral circulation could very well produce the sequence of morbid events hypothesized by Bollinger and expressed clinically as colic. Further, the fatal intestinal displacements often interpreted at postmortem examination to be the cause of colic symptoms are more likely to be the *result* of abnormalities of intestinal tone and motility brought about by overeating or verminous thrombo-embolism and by the horses' violent efforts to obtain relief.

Karl Enigk (1950, 1951) showed that as few as several hundred infective *S. vulgaris* larvae could cause widespread and fatal thrombosis of the very fine branches (arterioles) of the intestinal arteries of young foals that were reared parasite free. Thrombosis of these small vessels occurred very early in the course of infection, even before a single larva had reached the anterior mesenteric stem artery. In such cases, infarction of the bowel wall is present at postmortem examination, but no thromboses or emboli can be demonstrated by gross dissection.

These are some of the reasons that *Strongylus vulgaris* is the most important parasite of the horse.

Strongylus Edentatus and Strongylus Equinus Adult *Strongylus edentatus* and *S. equinus* worms are twice as large as adult *S. vulgaris*. There is no doubt that the prolonged migrations of these parasites are destructive and capable of producing disease, but their pathogenic importance is overshadowed by that of *S. vulgaris*. The migration routes followed by *Strongylus edentatus* and *S. equinus* were the subject of years of patient research by Rudolph Wetzel and his coworkers (1942) at the University of Berlin. The exsheathed third-stage larvae of *S. edentatus* migrate to the liver, where they become encapsulated and molt to the fourth stage in approximately 2 weeks. After molting, the fourth-stage in approximately 2 weeks. After molting, the fourth-stage larvae wander about aimlessly in the liver tissue for approximately 2 months, during which they continue to grow. Leaving the liver by way of the ligaments that hold this organ in position, the larvae wander for months in the connective tissue layer that lies immediately beneath the peritoneal lining of the abdominal cavity. Eleven months after infection, the mature *S. edentatus* worms may be found attached by means of their enormous suction cuplike mouths to the lining of the cecum and colon.

Third-stage *Strongylus equinus* larvae, like those of *S. vulgaris*, encyst and undergo their third molt in the wall of the large intestine, principally the cecum. Many of these larvae are found in nodules under the serosa or peritoneal covering of the intestinal wall. After molting, they bore through the intestinal wall and enter the right half of the liver, which, in the living horse, lies in contact with this portion of the large intestine. The larvae tunnel about in the liver tissue for 6 to 7 weeks or more before emerging and entering the pancreas and the abdominal cavity in which locations they complete their development into adult male and female worms. Finally, these adult worms penetrate the wall of the intestine and re-enter the lumen of the large intestine to mate and produce eggs.

The wounds inflicted by the migrating "big three" larvae heal but seldom without the formation of scar tissue and a consequent reduction in the functional capacity of this tissue. Because there is no cure for such damage once it has occurred, every effort must be made to prevent it.

The Small Strongyles   About 40 distinct species of small strongyle nematodes (Cyathostominae) parasitize the cecum and colon of horses and as many as 15 or 20 of these often infect the same horse at the same time. Cyathostomin larvae do not migrate beyond the mucous membrane of the large intestine and so their pathogenic effects are considerably less dramatic than those of the "big three." However, they are important in their own right.

From 75 to 100 percent of the strongyle eggs passed in the feces of naturally infected horses are produced by the small strongyles because these greatly outnumber the large strongyles both in numbers of species and in numbers of individuals. Several species of small strongyles have developed resistance to a several of the more popular worming compounds so that, after treatment and the affected compounds have ceased to be useful in their control, eggs of these species continue to be passed in the manure of horses. At this time writing, the large strongyles, fortunately, have not developed resistance to these compounds.

Heavy infections with small strongyles may cause severe and persistent diarrhea. Massive invasions of the bright-red fourth-stage larvae of *Cylicocyclus insigne* riddling the mucosa of the large intestine are particularly impressive in this regard. Mirck (1977) described verminous enteritis in young horses and ponies in which large numbers of these worms were discharged in the feces. This form of *cyathostominosis* occurs in the Netherlands from November to May, is characterized by watery diarrhea associated with severe inflammation of the mucous membrane of the cecum and colon, and often terminates fatally. Most of the worms are immature, and egg counts are therefore misleadingly low. Worming compounds have no influence on the course of the disease or on the number of worms being passed in the feces. Apparently, the larvae encysted in the mucosa are unaffected by currently available anthelmintic medication. There are many more larvae than can be accommodated as adult parasites and, as they mature, many are swept out with the manure.

The Ascarid, Parascaris equorum

Parascaris equorum is a very large (5 to 15 inches long), yellowish-white nematode parasite of the small intestine. Extraordinarily severe infections caused by adult worms may result in perforation of the bowel wall and fatal peritonitis, but such accidents are fortunately rare. More moderate degrees of infection may cause chronic enteritis and result in subnormal growth by interfering with digestion and absorption of nutrients, notably protein. A foal with heavy ascarid burdens is smaller than it would have been, its haircoat is dull, its skin dry and leathery, and its abdomen distended. In short, it is a victim of malnutrition. A major portion of the blame, however, can be ascribed to the ascarid larvae, whose destructive migrations through the liver and lungs are discussed in the following two paragraphs.

The life cycle of *Parascaris equorum* (Figure 17-9) is described as follows. When infective eggs are swallowed, they hatch and liberate infective larvae, which burrow into the wall of the small intestine and are carried to the liver by the portal vein. After migrating through the liver tissue, these larvae enter the hepatic veins and are carried by the posterior vena cava to the lungs, where they break into the alveoli, molt and are coughed up and

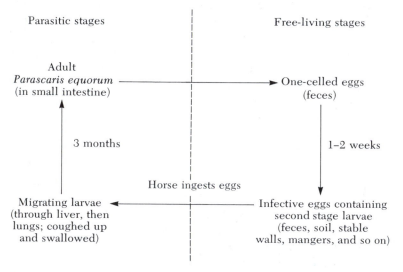

Figure 17-9
Life cycle of *Parascaris equorum*. (From Georgi, 1974.)

swallowed, thus returning to the small intestine where they mature. The first waves of larvae mainly inflict mechanical injury that appears to the unaided eye to be tiny hemorrhages. However, as the host becomes immunized to *Parascaris,* the immunity is expressed, in part at least, as allergic inflammation, which increases the damage inflicted by subsequent waves of migrating larvae. The damage done to the liver and lungs eventually heals, but the chronic reduction in functional capacity suffered during what is normally a period of rapid growth leaves a permanent mark on the yearling, which will never be what it could have been.

As indicated in Figure 17-9, only 1 to 2 weeks are required for the *P. equorum* eggs shed in the feces of an infected horse to develop to the infective stage. The outer, proteinaceous layer of the *P. equorum* eggshell is very sticky and causes the eggs to adhere to stall walls, mangers, water buckets, and the like. For these reasons, stall sanitation for the control of ascarids must, in order to prove effective, include weekly removal of all manure and bedding and thorough cleaning of all surfaces with a high-pressure sprayer of the sort used in car wash stalls or with a steam jenny. Most chemical disinfectants do not kill ascarid eggs, and therefore thorough cleansing with very hot water or live steam (steam under pressure) is necessary. After the surfaces are physically clean, they may be mopped or sprayed with 1 percent sodium hypochlorite (three cups Clorox per gallon of cool water) to strip off the outer protein coat of the ascarid eggs so they can no longer stick to surfaces and can be rinsed away. The preliminary cleaning is absolutely essential because any appreciable amount of residual organic matter will neutralize the sodium hypochlorite and render it ineffective in stripping the ascarid eggs. Notice that nothing has been said about killing the ascarid eggs. The above treatment does not kill ascarid eggs, it just knocks them loose. The removal of manure before its burden of parasite eggs has had time to develop to the infective stage is particularly important when foals are being raised because foals deliberately eat substantial amounts of manure. This phenomenon, termed *coprophagia,* is considered in greater detail in Section 8.5.

A Peculiar Parasite, Strongyloides Westeri

Strongyloides literally means "strongyle-like," but this worm is really not at all like a strongyle. Members of this genus are unique among parasites of domestic animals in that they have both parasitic and free-living generations. The larvae stemming from parasitic females develop from unfertilized eggs. Parasitic males do not exist and parasitic females contain no male gonads. The tiny (less than 1 cm long) parasitic females are thus parthenogenetic, not hermaphroditic, and their progeny are termed homogonic (arising from one gonad) to distinguish them from the heterogonic offspring of the sexual generation. A homogonic *Strongyloides* larva may develop into either a male or a female free-living (nonparasitic) worm or, in some species, directly into an infective larva. The heterogonic larva of the free-living generation develops into an infective larva. These somewhat complex relationships are summarized diagrammatically in Figure 17-10. Because the complete *S. westeri* life cycle can be completed in less than 2 weeks and because there is multiplication in the free-living stage, populations of this species can grow to gigantic proportions in a relatively short time, especially in stables.

Strongyloides westeri, like other members of the genus, develops rapidly in passed feces to the infective filariform stage, which usually enters the host by penetrating its skin or oral mucous membranes. The remarkable fact about *Strongyloides westeri* is that the adult worms are encountered principally in suckling and weanling foals; the dam of an infected foal sheds no *S. westeri* eggs even though she is the source of infection. Infection is transmitted from the dam to the foal by way of the mammary gland (Lyons et al., 1969, 1973), and the foals begin to shed eggs in their feces at 10 days to two weeks after birth. Diarrhea rather frequently afflicts foals between the ninth and thirteenth day of life, thus occurring coincidentally with the first postparturient estrus of the mare. Enigk and colleagues (1974) presented convincing evidence that this so-called "foal heat diarrhea" is caused by *S. westeri* and is not related to any alteration in the chemical composition of the mare's milk. Heavy infections in foals persist for 10 weeks; lighter infections may last two or three times as long. Occasionally, very light infections are observed in yearlings and older horses. These may represent percutaneous infections in hosts that were not exposed as sucklings (Enigk et al., 1974).

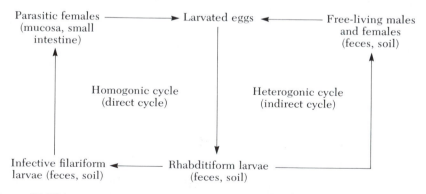

Figure 17-10
Life cycle of *Strongyloides*. The terms *filariform* (threadlike) and *rhabditiform* (rodlike) refer to the shape of the esophagus of the parasitic and free-living generations, respectively.

17.3   Stomach Bots

Bots are the larvae of botflies of the genus *Gasterophilus*. The adult flies have vestigial mouth parts and never feed; they are propelled by stored energy and the instinctive urge to mate and deposit their eggs on the hairs of a horse. The three principal North American species, *Gasterophilus nasalis*, *G. hemorrhoidalis*, and *G. intestinalis*, differ importantly in the area of the skin selected for egg laying and in the tactics adopted by the larvae that hatch from these eggs to gain entry to the stomach of the horse.

Gasterophilus nasalis female flies deposit their yellowish-white eggs (Figure 17-11) on the hairs of the space between the jawbones. First-stage larvae (Figure 17-12) develop within 5 or 6 days and hatch in the absence of any external stimulus. The larvae crawl downward toward the chin, proceed from there directly toward the mouth, and pass between the lips. In the mouth, the larvae burrow into pockets between the molar teeth and develop for a period of several weeks, after which they molt to the second stage, are swallowed, and attach themselves to the lining of the stomach. Following further development and a final molt, the fully developed thirdstage larvae (Figure 17-13) may be found attached by means of their formidable mouth hooks to the mucous membrane of the duodenum. Bots spend approximately 9 months clinging to the insides of a horse waiting for spring and storing the energy needed for the metamorphosis into adult flies, the mating flight, and finally the laying of eggs. When the time comes, the bots let go, pass out with the manure, and burrow into the ground to pupate. How the bots know when the right time has arrived is a mystery. During these months they

Gasterophilus nasalis   *Gasterophilus intestinalis*   *Gasterophilus hemorrhoidalis*

Figure 17-11
Eggs of *Gasterophilus* species. Each species of *Gasterophilus* lays its eggs on the hairs of a particular region of the horse's body. (Adapted from Wells and Knipling, 1938.)

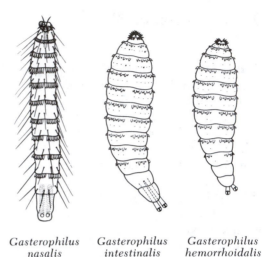

Figure 17-12
First-stage larvae of
Gasterophilus species.
(Adapted from Wells and
Knipling, 1938.)

Gasterophilus   *Gasterophilus*   *Gasterophilus*
nasalis   *intestinalis*   *hemorrhoidalis*

remain in total darkness, immersed in digestive juices with hydrogen ion
concentration as high as 0.1 moles per liter and a temperature of 38°C.
Either some signal of spring must get through to them (perhaps the horse
eats something that gives them a clue), or they have some sort of intrinsic
"fuse" that tells them when it is time to leave. We do know that no adult fly
survives killing frost and that each species of *Gasterophilus* is dependent on
its parasitic larvae to overwinter and perpetuate its kind. Is it not remarkable
that, in spite of the obvious vulnerability of such a life history to annihilation
and in spite of a long list of extraordinarily effective antiparasitic drugs,
almost every horse is host to at least a few bots every winter?

The *Gasterophilus hemorrhoidalis* female deposits her black eggs on the
short hairs that adjoin the lips. After 2 to 4 days of incubation, these eggs

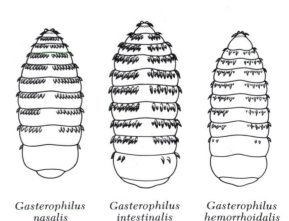

Figure 17-13
Third-stage larvae of botfly
of *Gasterophilus* species.
(Adapted from Wells and
Knipling, 1938.)

Gasterophilus   *Gasterophilus*   *Gasterophilus*
nasalis   *intestinalis*   *hemorrhoidalis*

hatch on contact with moisture and penetrate the epidermis of the lips, then burrow in the mucous membrane of the mouth. Thereafter, the life cycle of *G. hemorrhoidalis* is much like that of *G. nasalis*, just described.

The most common botfly, *Gasterophilus intestinalis*, deposits its eggs on the hairs of the forelimbs and shoulders. After an incubation period of 5 days, these eggs are prepared to hatch rapidly in response to the sudden rise in temperature that occurs when the horse brings its warm lips and breath in contact with them. Then the larvae pop out of their egg cases, enter the horse's mouth, and bore into the upper surface of the tongue near its tip. For the next month or so, these bot larvae bore their way toward the root of the tongue, stopping occasionally to make breathing holes and rest. At last they emerge from the tongue and enter pockets between the molar teeth where the first molt occurs. Second-stage larvae leave these interdental pockets, attach to the root of the tongue for a short while, and finally reach the stomach by being swallowed (Cogley, Anderson, and Cogley, 1982). Except that they prefer the nonglandular portion of the gastric mucous membrane, these *G. intestinalis* bots behave like *G. nasalis* for the balance of their parasitic careers. The life cycle of *Gasterophilus* species is summarized diagrammatically in Figure 17-14.

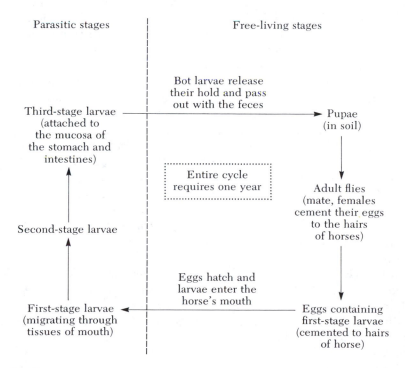

Figure 17-14
Life cycle of *Gasterophilus* species. (From Georgi, 1974.)

17.4   Development of Worm Infections in Foals

Ann F. Russell, a research fellow of the Ministry of Agriculture and Fisheries, Weybridge, England, has written a remarkable account of the development of strongyle and other helminth infections in Thoroughbred foals reared "under the ordinary conditions of stud management." Her report (Russell, 1948), brilliant enough in its own right, is invaluable to us now for the unique picture it presents of the natural sequence of changes in the composition of parasite species that invade the foal during its first year of life. Today, the almost universal application of periodic or continuous anthelmintic medication on stud farms precludes duplication of her work.

Figure 17-15 shows the five kinds of worm eggs existing in the manure of horses that can be distinguished on the basis of their microscopic structure. Unfortunately, there are approximately 54 species of worms that lay indistinguishable "strongyle" eggs. Some of these (the "big three," for example) are extremely destructive; others, the smaller representatives of the Cyathostominae, are relatively harmless, but they all lay "strongyle" eggs. One way of determining which species of strongyles are infecting a living horse is to incubate the manure for 7 to 12 days in order to allow the eggs to hatch and develop to the third or infective stage. An expert can identify these third-stage larvae, and this is how Russell determined the sequential changes in the composition of worm populations in 26 foals from seven different studs. She made observations on foals every week from the age of 4 weeks to at least 6 months of age and in a few cases to more than one year of age. Diagnostic morphometry and computer analysis (Georgi and McCulloch, 1989) provide a new and better means of differentiating strongyle eggs but they were not available to Russell.

In Figure 17-16, the number of eggs per gram of manure is plotted against time for *Strongyloides westeri*, for *Parascaris equorum*, and for the "strongyles" collectively. Note that the *Strongyloides westeri* infection was at a maximum during the early weeks of life, rapidly dropped to a low level, and finally disappeared at approximately 5 months of age. Infection with *Strongyloides* occurs soon after birth by ingestion of larvae shed in the dam's milk and progresses until the foal develops sufficient immunity to reject these parasites.

Parascaris equorum behaved similarly except that evidence of infection did not appear until the twelfth week, and the infection persisted indefinitely at a low level after the main peak had been reached. The 12-week delay in the appearance of *P. equorum* eggs in the feces corresponds exactly to the period during which this parasite is migrating through the liver and lungs, and we may deduce from this that significant infection of the foals was acquired immediately after birth. Anthelmintic medication of the pregnant mare with piperazine compounds, careful bathing of her udder and teats, and thorough cleaning of her stall at foaling time are thus logical measures

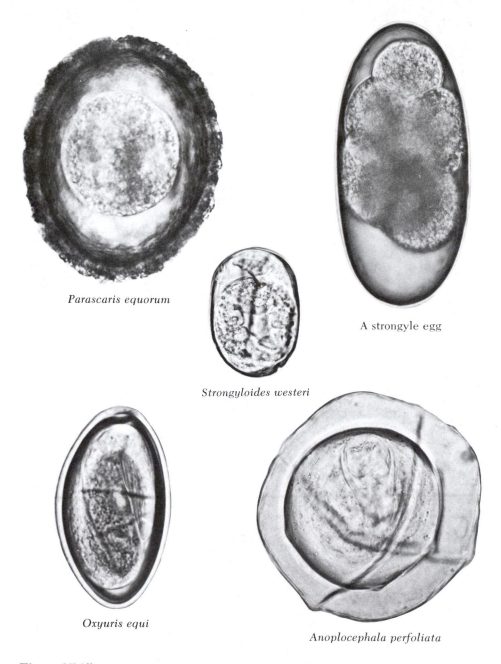

Parascaris equorum

A strongyle egg

Strongyloides westeri

Oxyuris equi

Anoplocephala perfoliata

Figure 17-15
Eggs of worm parasites of the horse. *Parascaris equorum* eggs have a sticky
surface layer that causes them to adhere to stall walls, mangers, and the like.
These eggs are extremely difficult to destroy. Strongyle eggs present a problem.
There are approximately 54 species of strongyle worms. They vary greatly in
pathogenicity, but their eggs cannot be told apart. *Strongyloides westeri* eggs
contain a larva when deposited in the manure. *Oxyuris equi* eggs are not usually
found in the manure; they adhere to the skin of the anus. *Anoplocephala perfoliata*
is a tapeworm; all of the others shown in this figure are nematodes.

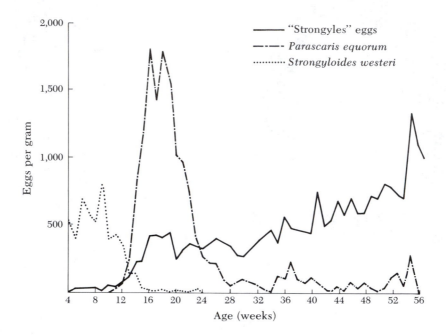

Figure 17-16
Average number of eggs of *Parascaris equorum*, "strongyles," and *Strongyloides westeri* counted per gram of manure. Data obtained from weekly observations of 26 foals. (Adapted from Russell, 1948, p. 112.)

for the prevention of significant infection of foals with *Parascaris equorum*. The persistence of infection at a low level after the peak has passed and the extreme resistance of the infective egg to the rigors of the external environment make *Parascaris equorum* a difficult parasite to control.

The third and most important curve shown in Figure 17-16 represents a gradual increase in the number of strongyle eggs per gram of manure during the first year of life. In order to make sense out of this curve, we must be aware of the species of strongyles being considered. This was determined by culture of these eggs and identification of the resulting larvae. The findings are summarized in Figure 17-17. It is immediately obvious from Figure 17-17 that the eggs of small strongyles (Cyathostominae) always predominate in the feces. This is what might be expected in view of the 6- to 11-month period required for the "big three" to develop to the egg-laying stage. It is interesting, therefore, that small numbers of strongyle eggs are already being shed at 4 to 12 weeks of age and that some of these are eggs of *Strongylus vulgaris* and *S. edentatus*. Russell observed this phenomenon in every one of her 26 foals and interpreted it as a confirmation of Wetzel's

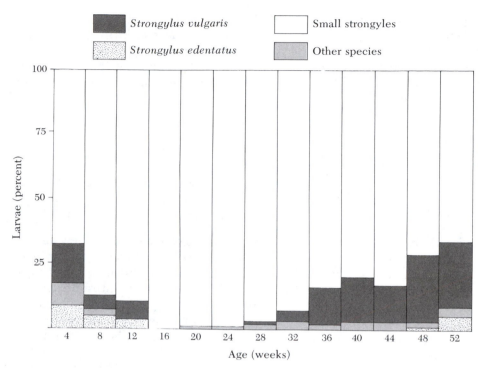

Figure 17-17
Percentage of larvae of different species of strongyles in fecal cultures. Data obtained from monthly observations of 26 foals. (Adapted from Russell, 1948, p. 112.)

(1942) suggestion that the appearance of strongyle eggs in the manure of foals less than 6 weeks old results from a foal's proclivity for nibbling at its dam's manure (coprophagia). Russell even calculated the amount of manure ingested by a particular foal: "For example, the egg count of foal 11 at five weeks was 100 e.p.g., while that of its dam for the same week was 1,200 e.p.g. If the foal was passing per day 2 lb of feces with a total of approximately 98,800 eggs it must have eaten about 2½ oz. of the dam's feces in order to produce the count of 100 e.p.g." This ingestion of feces (coprophagia) by foals may be related to the normal "seeding" of the cecum and great colon with the proper species of microorganisms essential for digestion of forage, but it is also a clear opportunity for parasites to invade virgin territory. Only eggs that are undeveloped when ingested will pass through the foal's digestive tract undamaged; infective eggs and larvae from similar sources will have found their goal in life—a susceptible host.

As Figures 17-16 and 17-17 show, strongyle eggs are passed in ever increasing numbers, and *Strongylus vulgaris* and *S. edentatus* make their appearances on schedule at 6 and 11 months, respectively. This information clearly indicates that worm infection of foals starts at birth and proceeds without interruption. Because young foals are much more susceptible to the pathogenic effects of these parasites than older horses, it follows that the greatest efforts should be directed toward preventing excessive exposure, especially during the first months of life.

17.5   Other Parasites

Pinworms

The horse is host to two species of pinworm, so called because their tails are slender and terminate in a point. the small pinworm, *Probstmayria vivipara*, measures less than 3 mm in length. Although it is capable of completing all stages of its life cycle within the confines of its host's large intestine and thus developing enormous infections, this pinworm is apparently quite harmless. The large white pinworm, *Oxyuris equi*, is 50 times as large as *P. vivipara* but of only slightly greater importance to the health of the horse. It has been reported that severe infections of third- and fourth-stage larvae may produce significant inflammation of the cecum and colon manifested by vague signs of abdominal discomfort. However, the most common complaint attributable to *Oxyuris* is *pruritus ani*, or itching of the anus caused by the adhesive egg masses that are deposited on the skin of the anus and surrounding area by the female worm. In its efforts to relieve the itching the horse will habitually rub its tail against posts, mangers, and the like until the tail head becomes disheveled, bare of hair, and even scarified or lacerated. Disfigurement and discomfort are thus the principal undesirable effects of *Oxyuris* infection.

Adult male and female *Oxyuris equi* live in the large intestine. When ready to deposit eggs, the female does not simply discharge them into the intestinal contents of the host in the same manner as do the strongyles, ascarids, and other intestinal nematodes. Instead, the gravid female *Oxyuris* migrates down the intestine and out through the anal opening to cement her egg masses to the skin of the anus and its immediate surroundings. These egg masses consist of a tenacious, yellowish-gray fluid containing 8000 to 60,000 eggs. The eggs develop to the infective stage in the course of 4 or 5 days, during which the cementing fluid dries, cracks, and detaches from the skin in flakes containing large numbers of infective eggs. These flakes are still sticky enough to adhere to mangers, pails, walls, and the like, thus contaminating the environment of the horse and increasing the likelihood of reinfection or the spread of infection to other horses (see Figure 17-18). It should be

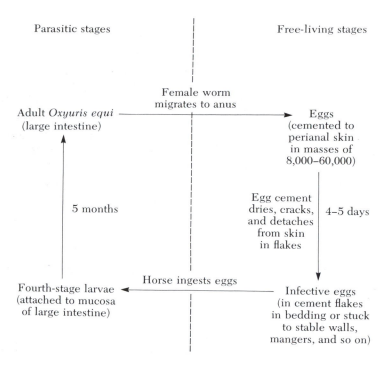

Parasitic stages | Free-living stages

Female worm
migrates to anus

Adult *Oxyuris equi* ————————→ Eggs
(large intestine)   (cemented to
                                     perianal skin
                                     in masses of
                                     8,000–60,000)

5 months   Egg cement   4–5 days
                  dries, cracks,
                  and detaches
                  from skin
                  in flakes

Horse ingests eggs

Fourth-stage larvae ←———————— Infective eggs
(attached to mucosa   (in cement flakes
of large intestine)   in bedding or stuck
                               to stable walls,
                               mangers, and so on)

Figure 17-18
Life cycle of *Oxyuris equi*, the large pinworm of the horse. (From Georgi, 1974.)

obvious from these considerations that the same cloth or sponge should never be applied to the area under a horse's tail and then to the muzzle and nostrils. Any effort to eliminate or control pinworms should certainly include regular cleansing of the anal area with warm water and mild soap to remove the egg masses before they have an opportunity to develop and scale off. However, paper towels or disposable cloths are to be preferred for this purpose because any nonexpendable object, such as a sponge or terrycloth, will inevitably become heavily contaminated with eggs and thus become a source of infection.

Another cause of *pruritus ani* to be differentiated from that caused by pinworms is infestation with the tail mange mite *Chorioptes equi* or the biting louse *Damalinia equi* (see the following sections on lice and mites).

Neither pinworm of the horse has anything to do with pinworm infection in children. The human pinworm, *Enterobius vermicularis*, is a distinct species that does not occur in any domestic animal, including the dog, although many people are misinformed on this point.

Tapeworms

Horses are hosts to three species of tapeworm: *Anoplocephala magna, A. perfoliata*, and *Paranoplocephala mamillana*. Infection is acquired at pasture where the horse accidentally ingests free-living mites that are infected with tapeworm larvae. Ordinarily, tapeworm infections have no measurable deleterious effects on horses and tapeworms are infrequently considered in the planning of worming programs. Exceptionally, however, severe infections of *A. perfoliata*, which have a remarkable tendency to group together in clusters near the ileo-cecal junction, may cause chronic diarrhea, and have sometimes been accused of precipitating intersusception, a telescoping of the intestine that has fatal consequences.

Face Flies

The face fly *Musca autumnalis* is a close relative of the common housefly *M. domestica*. Face flies feed on secretions around the eyes, nostrils, and mouth of horses and on the blood that continues to ooze from wounds made by horse flies; they annoy horses to distraction on warm, sunny days. Face-fly maggots only develop in cow manure and so horses pastured with or near cattle are the ones subject to discomfort from this pest. Control may be attempted by application of stirofos, coumaphos, or pyrethrins to the entire horse, and elimination or insecticidal treatment of breeding sites (i.e., cow manure) when feasible. Face flies do not pursue their victims indoors, so stabling horses during hours of peak fly activity often proves to be the best solution.

House Flies

Control of house flies is based on elimination of breeding sites (manure and decaying organic matter) and regular application of insecticides. Dichlorvos, naled, stirofos, fenthion, Ravap, and permethrin may be sprayed on horses. Pyrethrin sprays may be applied to the entire horse two to four times a week. Residual sprays (dimethoate) applied to the walls of animal quarters every two weeks throughout the summer are quite effective in reducing fly populations. Baits may be set out wherever flies congregate. Vapona resin strips may be hung from the ceiling of closed barns.

Lice

Two species of lice infest the horse, the bloodsucking louse, *Haematopinus asini* (Figure 17-19), and the biting louse, *Damalinia equi* (Figure 17-20). These lice spend their entire life from egg to adult clinging to the hairs of a

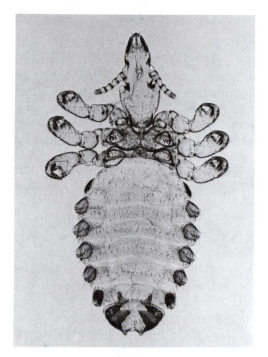

Figure 17-19
The bloodsucking louse,
Haematopinus asini. (From Georgi,
1974.)

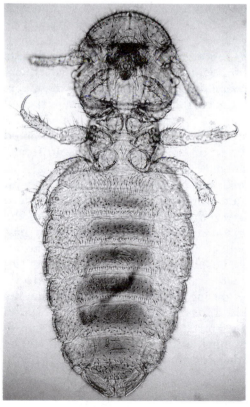

Figure 17-20
The biting louse, *Damalinia equi.*
(From Georgi, 1974.)

horse. Without a hair to grasp, a louse is relatively helpless. It passes from host to host most efficiently when the hosts' haircoats are in direct contact with one another. The practical consequence of this is that louse infestation is usually transmitted by direct contact between horses, although the blanket of a lousy horse could conceivably transmit the infestation if it were placed on another horse within a matter of hours. Lice of other domestic animals, such as chickens and cattle, do not remain long on a horse, nor is there any hazard of an exchange of lice between humans and horses in either direction.

The life cycle of a louse is simple. The rather large eggs, called *nits*, develop in the abdomen of the female. When the female lays an egg, a sticky secretion is passed that glues the egg to the shaft of a hair. In a few days a tiny nymph, closely resembling its parents in appearance, emerges through a lid or *operculum* at the distal end of the egg. As the nymph grows in size, its rigid skin becomes too small. The louse inside then becomes quiescent (undergoes a lethargus), reorganizes its tissues, and develops a new, more pliable skin. When ready, the nymph splits open its old skin and emerges, now capable of growing and expanding for a while until its new skin hardens and the whole process of molting has to be repeated. Because no remarkable change occurs from one nymphal stage or *instar* to the next, the sequence of molts culminating in an adult male or female louse is termed gradual or incomplete metamorphosis. A female louse lives for approximately a month and a half and may lay 20 to 30 eggs at intervals of 1½ days. This contrasts with the extraordinary alteration in form (complete metamorphosis) experienced by the pupating larva of *Gasterophilus* that culminates in the emergence of a winged, six-legged fly from the barrel-shaped puparium or tanned last larval skin.

Biting lice cause rather severe itching that may result in self-inflicted abrasions and lacerations when the horse attempts to scratch itself. Heavy burdens of bloodsucking lice cause anemia.

Prevention of lousiness is accomplished by avoidance of contact with infested horses. Sunshine and fresh air are inimical to lice, and therefore severe louse infestation is rarely observed in pastured horses during the warmer months. Louse populations sometimes increase rapidly on pastured horses late in the fall. Treatment is available in the form of various powders and sprays containing organophosphorus compounds (Vapona, Ciodrin, and so on). Powders are satisfactory if only a few horses are to be treated and are to be preferred for use during cold winter weather when the thorough saturation of the coat required for effective application of sprays would surely result in chilling of the horse, perhaps with serious consequences. If the horse can be clipped, much of the louse population will be removed with the hair and the lice that remain will be more accessible to treatment.

Mange Mites

Mange mites are distant cousins of spiders. Barely visible to a person with excellent eyesight, these tiny 8-legged nuisances live on or in the epidermis and cause mange, a form of dermatitis characterized by the formation of crusts or scabs, and by intense itching. As is true of louse infestation, mange is spread mainly by direct contact with infested individuals. Diagnosis and treatment are tasks for the veterinarian.

Habronemiasis

Habronema species are of little consequence as nematode parasites of the stomach. One species, *Habronema megastoma* (often called *Draschia megastoma*), stimulates the formation by the host of fibrous nodules that project from the wall into the lumen of the stomach. These nodules are riddled with intercommunicating galleries filled with pus and with the rather small (approximately 1-cm) worms. Although impressive when described, these nodules and their occupants usually do not affect the health of the horse in any observable way. Larvae of *Habronema* species, on the other hand, are incriminated in the production of persistent, disfiguring, and intensely itching skin lesions called swamp cancer, summer sores, and a variety of other colloquial names. Houseflies and stableflies serve as intermediate hosts for *Habronema* species. When these infected insects visit fresh wounds or areas of skin that are subjected to somewhat continuous moisture, *Habronema* larvae enter the tissues and cause an inflammatory reaction characterized by severe itching and the production of "proud flesh," or "exuberant" granulation tissue, as it is sometimes called. When infected flies feed in the corner of a horse's eyes, *Habronema* larvae may enter the conjunctival membranes and engender a chronic conjunctivitis.

Cutaneous and conjunctival habronemiasis are chronic, persistent and very difficult to cure, especially during summer and early fall. Professional veterinary treatment for the patient and extreme patience on the part of the owner are essential.

17.6   Anthelmintic Medication

An important point with regard to worming is frequently overlooked or simply misunderstood: Worm remedies as routinely administered are most effective against the adult parasites living in the lumen or attached to the mucous membrane lining the intestinal tract and much less effective against migrating larvae. This means that the main damage inflicted on the horse by strongyles and ascarids has already been done, and instead of treating a disease we are, in a sense, merely exacting retribution. What benefit, then, can the drug bestow beyond relieving the horse on the comparatively less serious physiological burden imposed by the lumen-dwelling adult parasites? The anthelmintic drug helps to reduce the contamination of the environment with eggs by reducing the numbers of adult parasites (not even the best anthelmintics remove them all) and by depressing the reproductive functions of the worms that happen to survive. Thus, the principal objective of worming horses is really to prevent environmental contamination and thereby reduce the degree of reinfection and the transmission of infection to young, highly susceptible horses. In the cavalry and on rich estates of old, this contamination was dealt with by an unremitting sweeping and raking of horse droppings into bushel baskets by squads of troopers or grooms. The bottle of anthelmintic drug thus represents a relatively inexpensive substitute for a laborer with broom and bushel basket. With the true objective of anthelmintic medication thus defined, two strategies for the control of parasites suggest themselves.

Routine Worming of All Horses

The administration of full therapeutic doses of anthelmintic drugs at specified intervals usually suppresses strongyle egg production sufficiently to achieve satisfactory control. Empirical observation indicates that strongyle eggs tend to reappear in a horse's feces approximately 6 to 8 weeks after administration of an effective anthelmintic drug. The adult worms that survived the attack have regained their reproductive powers, and larvae that were migrating through the tissues at the time, thus protected from the effects of the drug, have arrived in the lumen of the bowel and are now starting to lay eggs. Therefore, if a full dose of an effective anthelmintic drug is administered at 2-month intervals (six times a year) to *all* horses on a premises, the rate of contamination of the environment will be greatly reduced and serious infection of susceptible foals will, it is hoped, be forestalled. The advantage of this strategy is that laboratory examination of fecal specimens is not required to determine the extent of infection. The disadvantages of this strategy are that (1) *all* horses on the premises must be treated no less than six times a year, thus multiplying the expense and hazard of toxic reaction inherent in anthelmintic medication, and that (2) persistent application of antiparasitic chemicals inevitably leads to the selection of strains of parasites possessing resistance to those chemicals and thus ultimately ceases to be effective.

Phenothiazine, thiabendazole, cambendazole, mebendazole, fenbenda-zole, oxfendazole, and febantel are no longer as effective against small strongyles as they were when first introduced. Drudge and colleagues (1979) identified five species of small strongyles that exhibited cross resist-ance to cambendazole, fenbendazole, mebendazole, oxfendazole, and thia-bendazole. Resistant populations may be controlled with oxibendazole, dichlorvos, pyrantel pamoate, or ivermectin, or by administering a benzimi-dazole with piperazine. The benzimidazole anthelmintics continue to retain excellent potency against the large strongyles and other nematode parasites of horses.

Based upon epidemiological evidence, the most essential of all strategic treatments is that administered in spring, around foaling time. It is during this period that the adult worm population is greatly augmented through maturation of arrested and migrating larvae. The fecundity of these worms is also increased, and larger numbers of larvae reach the infective stage, thus posing a threat to young, susceptible horses. Elimination of these egg-laying worms in spring thus renders the pastures safer for grazing horses (Duncan, 1974). Herd et al. (1985) demonstrated the importance of spring and sum-mer anthelmintic treatments for the control of strongyle parasites in adult horses and suggested that this strategy should be a key part of any program designed to keep pasture infectivity at low levels.

Because resistance is the inevitable result of frequent, regular anthel-mintic medication, a better course might be to worm only those horses with significant fecal egg counts. The objective, after all, is to reduce the rate of contamination of the environment with eggs. This approach is discussed in greater detail below.

Quantitative Fecal Examination and Individual Treatment

Recognizing that the objective of any worming program is to reduce the contamination of the environment with eggs, a reasonable strategy would be to determine what the output of eggs from each horse actually is and to administer an anthelmintic drug only to those individuals with more than some specified number of eggs per gram of feces. The number of eggs tolerated depends on the stocking rate. For example, 500 eggs per gram would perhaps be acceptable when a few horses occupy a relatively large pasture, but the limit should be lowered to 100 eggs per gram when many horses are confined to a small paddock because of the consequent concen-tration and accumulation of manure. The limit, in fact, need not be an arbitrary constant but might be adjusted periodically in response to trends observed in the egg count data. The advantage of this plan is that it mini-mizes the expense and hazards of treatment; the disadvantage is that it requires expert analysis of the feces of all horses on the premises as well as maintenance of accurate records and interpretation of trends in the data. The choice of strategies depends on the relative costs of anthelmintic medication on the one hand and of expert fecal analysis on the other.

A practical qualitative test for the presence of strongyle eggs in horse manure consists of the following. Put a ball of manure in a quart jar, set the lid loosely on top, and place the jar in a dark place at room temperature for approximately one week. Each day examine the inner walls of the jar for the presence of droplets of condensed moisture (evidence that a suitable degree of humidity exists in the culture). If the walls of the jar are dry, add 10 to 20 drops of water directly to the mass of manure, cover the jar, and return it to the dark. After a week to 10 days, examine the walls of the jar with a hand lens in bright daylight. Unless the horse that provided the culture medium had very few strongyles, infective larvae will be easily visible as they writhe and thrash about in the droplets of condensed moisture. Cultures from heavily infected horses contain so many larvae that the walls of the jar will appear to be frosted. A subjective impression of the degree of infection can thus be gained by comparing cultures from a number of different horses. If a compound microscope is available, transfer some larvae to a microslide with an eyedropper and kill them by gentle warming. Hold a lighted match below the slide, view the cessation of motion and extension of the larvae from above, and remove the match as soon as motion ceases, otherwise, the larvae will be cooked. Now add a coverslip and examine the specimen with the lower powers of the microscope. With experience, a stable manager can use this simple culture procedure, requiring only quart jars, eyedroppers, a hand lens, and water, as a practical means of monitoring strongyle infection in horses.

Some Thoughts about Management   In 1965, Karl Enigk, of the Tierärztliche Hochschule in Hanover, asserted without qualification that the practice of keeping horses in box stalls rather than tie stalls is to blame for the high degree of worm infection observed in riding horses. For this reason, claimed Enigk, those kinds of worm infection that are almost exclusively acquired in the stable (for example, ascarids, pinworms, and *Strongyloides*) are particularly common. He pointed out that strongyles (especially the pathogenic large strongyles) are the most important parasites of the horse, ascribed their great prevalence to a combination of box-stall stabling and exposure at pasture, and prescribed tie stalls, zero grazing, and exclusive feeding of hay for control of these parasites. Strongyle larvae do not survive for longer than 8 months in well-cured hay, and the halter and shank assure that all manure will be deposited several feet from the manger and water bucket. Because the pasture, especially on stud farms, is the greatest source of strongyle infection, zero grazing is a logical measure in strongyle control. These measures advocated by Enigk might appear extreme to the modern pleasure-horse owner, but they correspond exactly to the routine husbandry practices employed for draft and cavalry animals a few decades ago. In the days before phenothiazine, there was no really effective drug for removing strongyles, yet civilization depended on horsepower for transportation and agriculture, and great horses performed great feats. Then, any system of horse husbandry that did not, either intentionally or unintentionally, keep strongyles in check could not compete with those that did in producing healthy, useful horses. The same is true today except that in general, anthelmintic chemicals are substituted for rigorous sanitation and for husbandry devices (such as tie stalls) that, by their nature, interfere with the strongyle life cycle.

In arid regions, scattering of droppings on pasture with a tractor and harrow every few days should theoretically reduce strongyle larva populations by breaking up the manure and causing it to dry out before the larvae have achieved the dessication-resistant third stage. However, in the more humid regions that prevail over much of North America, the interior even of scattered manure will remain sufficiently moist long enough for completion of this development. The infective larvae will be capable of withstanding dessication for several weeks if they are large strongyles and for up to 8 months if they are small strongyles. It follows from this that hay harvested from meadows in which horses have been allowed to graze or meadows that have recently been fertilized with horse manure should be aged for at least 8 months before being fed to horses. Alternate grazing of pastures by horses and either cattle or sheep is also of theoretical value in parasite control because, except for *Trichostrongylus axei*, horses and ruminants share no parasites. Therefore, grazing cattle remove larvae from the pasture that are infective for the horse (and vice versa) while time and environmental stresses wear away at the larvae that remain. Of course, advantage can be taken of this principle only when both horses and ruminants are kept, when the terrain and fence construction are suitable for both, and when the pasture actually recovers sufficiently after the first grazing to be of further use that season. (There are many pastures that do not.)

As J. H. Drudge and E. T. Lyons (1966) of the Department of Veterinary Science of the University of Kentucky have pointed out, "An effective parasite control program must be regarded as a long-term undertaking before desired results are achieved and maintained." The extreme hardiness of infective ascarid eggs and the prolonged tissue migrations of the "big three" strongyles introduce an element of inertia that can be overcome by perseverance in the use of effective measures. The beneficial effect of a dose of thiabendazole on a wormy colt is certainly prompt and dramatic, but it is just as certainly temporary.

A knowledge of the life cycles of parasites and of the nature and habits of their infective stages provides a sound basis for parasite control. Antiparasitic chemicals form an integral part of any practicable program, but their application to the exclusion of appropriate measures of sanitation and management leads to the emergence of resistant parasites and to ultimate failure. As a general rule, a parasite control program, to be effective, must be tailor-made to the farm or stable in question — that is, properly suited to its objectives, landscape, physical resources, horses, and the people managing it.

REFERENCES

Bollinger, O. 1870. *Sitzungsb. k.-Bayer. Akad. Wissensch. München* 1:539.

Cogley, T. P., Anderson, J. R., and L. J. Cogley. 1982. Migration of *Gasterophilus intestinalis* larvae (Diptera: Gasterophilidae) in the equine oral cavity. *Int. J. Parasitol.* *12*(5):473–480.

Craig, E. F., and E. C. Faust. 1945. *Clinical Parasitology.* Philadelphia: Lea & Febiger.

Dobberstein, J., and H. Hartmann. 1932. Über die Anastomosenbildung im Bereich der Blind- und Grimmdarmarterien des Pferdes und ihre Bedeutung für die Entstehung der embolischen Kolik. *Berliner tierärztliche Wochenschrift* 48:399.

Drudge, J. H., and E. T. Lyons, 1966. Control of internal parasites of the horse. *J.A.V.M.A.* 148(4):378–383

Drudge, J. H., Lyons, E. T., and S. Tolliver. 1979. Benzimidazole resistance of equine strongyles — critical tests of six compounds against population B. *Am. J. Vet. Res.* 40:590–594.

Duncan, J. L. 1974. Field studies on the epidemiology of mixed strongyle infection in the horse. *Vet. Rec.* 94:337–345.

Enigk, K. 1950. Zur Entwicklung von *Strongylus vulgaris* [Nematodes] im *Wirtstier*. *Z. Tropenmed. Parasitol.* 2(2):287–306.

Enigk, K. 1951. Weitere Untersuchungen zur Biologie von *Strongylus vulgaris* (Nematodes) im *Wirtstier*. *Z. Tropenmed. Parasitol.* 2:523–535.

Enigk, K. 1965. Behandlung und Vorbeuge des Parasitenbefalles der Pferde. *Deutsch. tierarztl. Wochenschrift* 72:493.

Enigk, K., Dey-Hazra, A., and J. Batke, 1974. Zur Klinischen Bedeutung und Behandlung des galaktogen erworbenen *Strongyloides* Befalls der Fohlen. *Dtsch. Tierärztl. Wochschr.* 81:605–607.

Georgi, J. R. 1973. The Kikuchi-Enigk model of *Strongylus vulgaris* migrations in the horse. *Cornell Vet.* 63 (2):220–263. [English translation of Enigk's articles listed above].

Georgi, J. R. 1974. *Parasitology for Veterinarians.* 2nd ed. Philadelphia: W. B. Saunders.

Georgi, J. R., and McCulloch, C. E. 1989. Diagnostic morphometry: Identification of helminth eggs by discriminant analysis of morphometric data. *Proc. Helminthol. Soc. Wash.*, 56:44–57.

Herd, R. P., Willardson, K. L., and A. A. Gabel. 1985. Epidemiological approach to the control of horse strongyles. *Eq. Vet. J.* 17:202–207.

Lyons, E. T., J. H. Drudge, and S. Tolliver. On the life cycle of *Strongyloides westeri* in the equine. *Jour. of Parasit.* 59 780–787.

Lyons, E. T., J. H. Drudge, and S. C. Tolliver. 1969. Parasites from the mare's milk. *Blood Horse,* 95:2270–2271.

Mirck, M. H. 1977. Cyathostominose: een vorm van ernstige strongylidose. *Tijdschr. Diergeneesk.* 102(15):932–934.

Russell, Ann F. 1948. The development of helminthiasis in Thoroughbred foals. *J. Comp. Path.* 58:107–127.

Wells, R. W., and E. F. Knipling, 1938. A report of some recent studies of *Gasterophilus* occurring in horses in the United States. *Iowa State College Journal of Science* 12:201.

Wetzel, R. 1940. Zur Entwicklung des grossen Palisadenwormes (*Strongylus equinus*) im Pferd. *Arch. wissensch. u. prakt. Tierh.* 76:81.

Wetzel, R., and W. Kersten. 1956. Die Leberphase der Entwicklung von *Strongylus edentatus. Wien. tierärztl. Monatsschrift.* 11:664.

7

MANAGEMENT

CHAPTER 18

BEHAVIORAL PRINCIPLES OF TRAINING AND MANAGEMENT

At least two people always point to a champion performance horse with pride. One is the breeder, who would suggest that he selectively bred this animal to be a superior performance horse. The other is the trainer, who maintains that his influence was what made the horse a champion. Both are partially correct. Behavior in a performing horse is the result of both hereditary and environmental influences. Nature versus nurture is an old controversy that has been studied with regard to several species by proponents of both views, and effects of both genetics and environment have been observed to affect performance.

18.1 Heritability of Performance

Psychologists have been more concerned with environmental effects on behavior than with genetics because of the human application, which lends itself more to environmental manipulation than to genetic control. However, the animal breeder has the opportunity through selective breeding to breed for superior genetic effects on behavior. Therefore, horse breeders are interested in improving performance in their horses through selection for desirable behavioral characteristics. The problem lies in the ability to separate genetic from environmental effects in the performance of a given horse.

This leads to difficulty in recognizing animals with superior genotypes for a given type of performance. If a real difference in learning ability in horses does exist, one of the prime goals of horse breeding should be selection for this learning ability. This is because horses are unique among domestic livestock in that the value of horses can be greatly increased with training. Obviously, learning ability is one of the most significant factors in the final performance achieved by any horse.

There is considerable evidence that performance behavior is heritable in several species. Some of the classic experiments demonstrated a difference in the maze-running ability of selected strains of rats. Strains of "bright" and "dull" rats have been developed experimentally, which strongly suggests that behavior has a heritable component. Other types of behavior that have been shown to be partially controlled by genetic influence include (1) level of activity of rats in a rotating cage, (2) emotionality in rats, (3) maze-running ability in mice, (4) sex drive in guinea pigs, (5) nest-building instinct in rabbits, (6) aggression in rabbits, (7) behavior during conditioning in dogs, and (8) avoidance learning in pigs. There have even been reports of hybrid vigor for some of these behavioral characteristics.

Even though little research information is available on the heritability of performance in horses, data obtained with other species strongly suggest that performance in horses is likely to be heritable to some degree. Heritability of racing ability in Thoroughbreds is quite high, but in racehorses, psychological training is not as important as physical training, physiology, conformation, and so on. On the other hand, heritability of performance in cutting horses has been reported to be quite low. This is to be expected since environmental input (ability of the trainer) may vary widely and subsequent performance is likely to be reflective of the trainer and not heredity. Superior trainers can make horses with average ability reach a high level of performance. Other trainers have difficulty getting any horse to reach its potential. All cutting-horse trainers believe that there is a heritable component in cutting horses. For example, trainers of cutting horses like to ride sons and daughters of the Quarter Horse stallion Doc Bar because his sons and daughters have done well in the cutting arena.

Every time a mare is bred to a stallion with superior performance records, it is assumed that the foal will inherit some of the characteristics of the stallion. The notion is accepted that part of the stallion's performance is due to inheritance and that this can be transmitted to the foal. Unfortunately, it is difficult at best to separate that portion of the stallion's performance and behavior which is due to genetics from that which is due to the trainer. Most of the criteria available to horsepeople to measure performance are biased because the effects of the environment or trainer cannot be held constant between horses. A stallion whose offspring are trained by the best trainers will show an advantage over another stallion whose offspring have average trainers. Choice of trainer is often determined by the financial or other position of the stallion owner. If a given stallion's offspring appear to be superior performance horses, there is always a clamor among trainers to get one of these horses. This only biases further the use of performance records as selection criteria because the best trainers usually have the best performing horses.

Figure 18-1
Colonel Alois Podhajsky of the Vienna Spanish Riding School mounted on one of the school's Lippizaner stallions. The Lippizaners are among the most highly trained horses in the world, but their amenability to training has been influenced to some extent by breeding. (Photograph courtesy of The Bettmann Archive.)

There is a real need to refine methods of measuring learning ability in horses and develop indices that are unbiased, objective, inexpensive, easy to use, and useable at an early age. Meanwhile, the breeder must make the best possible use of existing performance information when selecting for learning ability in horses. The breeder must try hard to ascertain the relative contribution of genetics and environment to each horse's performance. Limited research with horses and many studies of other species indicate that there is considerable genetic variance in behavioral characteristics of horses and that the potential for improving learning ability in horses through selection may be substantial.

18.2   Early Experiences

An animal's ability to learn, to solve problems, or to survive the effects of severe stress has been shown to be greatly influenced by the environment in which that animal was raised. Since these effects precede training, they affect trainability in much the same manner as does heredity.

The effects of early experience have been shown to influence behavior in many animals. Early stimulation may occur before birth, after birth, before weaning, and after weaning. Continuous stimulation, such as radical changes in environment (enriched, deprived, or isolated), and discontinuous stimulation, such as varying periods of daily handling, have both been found to affect behavior. Rats that were raised in an environment where discrete circles and triangles were painted on the walls were superior to rats that were exposed to no designs when both groups were later tested on circle-triangle discrimination tests. Similar tests with diffuse stimuli have shown that such stimuli increase emotionality. Emotionality is also affected by early handling. When subjected to stress, rats that were handled daily were less emotional than a control group of rats that received no handling. Chimps that were raised in darkness and thus suffered perceptual deprivation had problems later in perceiving stimulus movement, form, and direction.

In the period from before weaning through part of the yearling year, the horseman has ample opportunity to influence a horse's learning ability by varying its environment (Figure 18-2). However, the early experience of the horse has not been studied to any extent. Such variables as critical time to stimulate, most effective method of stimulation, and how to manipulate this stimulation effectively to affect the adult performance have not been analyzed. There has been a general observation that the horses of today are more docile, more trainable, and more tractable than were horses 30 years ago. Some of this change may be due to heredity, but it can easily be attributed to increased handling of horses at earlier ages. Horses that have

Figure 18-2
Desirable early experiences enhance learning and trainability in later stages of development.

been trained only as adults are wilder and much less dependable than those with which human contacts have been established early in life. Docility enhances learning in that the horse is less easily distracted. A social relationship between horse and handler can be established much more easily at an early age. Most authorities on the breaking and training of horses recommend that the handling and preparation for riding begin as early as possible. Halter breaking can start before weaning, followed by longeing, saddling, driving, and other activities limited only by the horse's physical capabilities.

18.3   Environmental Influences: Training

There is an abundance of research information on the effects of the environment on laboratory animals that is directly applicable to horse training. Horse training deals with the modification of behavior, and almost all of the principles of learning that apply to other species also apply to horses. When training begins, the horse has a certain operant level of performance. The trainer must determine this level and then progressively modify the behavior in such a manner as to achieve a desired level of performance. At this point the genetic potential of the horse has already been determined, and the hope is that the horse will be capable of reaching the desired level of performance. What the trainer has to manipulate is the environment. He divides this environment into stimuli and reinforcement and then uses them to obtain the response he desires. Stimuli, responses, and reinforcement and their relationships to each other are the basis of the psychology of horse training. The aspiring trainer must develop a clear understanding of each.

Responses

The term "response" is used to refer to specific types of behavior. Responses are the acts or movements of the horse, and the goal of training is to teach the horse to make the desired response. On occasion, the term "response" is used to refer to major maneuvers such as stopping or jumping. However, for the most part, responses should be thought of as smaller segments of these major maneuvers, which are composed of many responses chained together. The trainer must learn to identify the smaller units of the maneuvers (Figure 18-3). For example, the initial reaction of a horse to bit pressure is a response that, properly reinforced, will eventually be part of a stop, which is a major maneuver (Figure 18-4).

Figure 18-3
Learning to cross the front legs is the first step in the process of learning to do the spin or roll-back.

Figure 18-4
Learning to flex at the poll when pressure is applied on the bit is a first step in learning to stop or back correctly.

Stimuli

Stimuli can be divided into conditioned and unconditioned categories, depending upon their natural effect on the horse. If a stimulus can naturally cause a response with no prior practice, it is said to be unconditioned. Few of these stimuli are used in horse training. A stimulus that has been learned through practice is called a conditioned stimulus. These stimuli are frequently used in horse training and are known to horsemen as cues (Figure 18-5). Remember, *most cues are learned*; very few natural stimuli will cause the response sought in the trained horse.

Basic Cues   Since few of the cues used in training are natural and cues for the most part must be learned, where does the trainer start? He starts with those basic cues that are closest to being natural. These are cues that practically show the horse what to do. They are very obvious and readily learned. For example, direct rein pressure response is obvious because the horse is shown the direction in which it is supposed to go, and it can learn this quickly. If these obvious cues were all that the horse ever acquired, it would not be considered a very highly trained horse.

Presenting New Cues   The trainer will use these basic cues to advance the horse to new cues. This advancement is accomplished by pairing the new cue

Figure 18-5
Taking a lead as the result of a leg cue must be learned by the horse. Any cue can be used if the trainer is specific and consistent in teaching the cue to the horse.

with the basic one in the daily training routine. The rate at which the horse learns is affected by the manner in which these two cues are paired.

The new cue should always be presented first, followed by the old cue, which the horse already knows. For best results, there should be only a very slight time lapse between the two cues, but they must not be presented exactly at the same time. The old cue should be paired closely enough with the new one that the two slightly overlap. In teaching a horse the neck rein, the basic cue of the direct rein is introduced first. Once the horse has learned to respond to this cue, the bearing rein is introduced. The bearing rein means nothing to the horse in the beginning, but if the trainer follows its presentation with the direct rein, the horse soon learns to make the desired response to the bearing rein. Assuming that from early training the horse knows that the cue "whoa" means stop, the command "whoa" can enhance learning to stop in response to bit pressure—pull on the bit (new cue) and, while pressure is still intact, say "whoa" (old cue). The horse should learn to stop from the bit pressure faster and more efficiently if this method is used. The trainer continually uses this method to teach the horse less obvious cues, but the method is successful only if the horse has learned the old cue well. The old cue must reliably cause the response before it can be used to encourage the horse to acquire a new cue. In the preceding example, if the bit is getting the correct response, then leg pressures may be introduced to teach stopping from leg cues.

Teaching a horse to make a response to a combination of cues using different sensory modalities is also a common practice. This method has the advantage of presenting a stronger total stimulus to the horse and helps to ensure that the horse gets the message.

Specificity of Cues   The horse can receive stimuli through its sensory systems in many ways, but man's most effective method of communication with the horse seems to be through the horse's senses of touch and hearing. Voice commands and pressures are common stimuli (cues) used in horse training. Pressures are used on the horse's nose by the hackamore, in the mouth by the bit, on the neck by the reins, and on various parts of the body by the rider's feet and legs. Unpublished results of a study at Texas A & M University indicated that a horse makes greater use of hearing than touch or sight. Therefore, the use of voice commands should be utilized whenever possible.

Indiscriminate presentation of cues only confuses the horse. Cues must be specific so that the horse can identify them and separate them from other things that are happening to it at the same time that the cues are presented. It takes ability and skill to deliver specific identifiable cues to the horse. This helps to explain why some horsemen are better trainers than others. Good horsemanship is a must. A rider cannot use poor, inconsistent technique with a horse and yet apply discriminate and specific cues that the horse is expected to identify.

It is important to remember that horses must be taught to respond to cues. The trainer should start with the most basic cues, present new cues to be learned just before the old cues are learned, and be very specific in presenting a cue. A trainer can use any cue he wants for a given response if he is specific and consistent in presenting that cue. The horse learns and responds from truth, not confusion.

Reinforcement

The principle of reinforcement means that certain events are capable of strengthening responses to certain stimuli. It is doubtful that any learning can be accomplished without reinforcement. Reinforcement can be divided into the broad categories of primary and secondary reinforcement. Primary reinforcers have natural reinforcing properties. Feed is a primary reinforcer in that the horse naturally appreciates feed, and it can be used to strengthen certain behavior. Only a few primary reinforcers are used directly in training.

Secondary reinforcers are learned by the horse and are acquired over a longer period. General acts of kindness to the horse can acquire secondary reinforcing properties. Learning that the training period will end if the horse performs well is secondary reinforcement for the horse. There are many more examples, but they are all distinguished from primary reinforcements in that they are *learned*. A horse trainer who can condition his horse for secondary reinforcement has another tool with which to work.

Positive Reinforcement   Reinforcers, whether primary or secondary, can be further divided into two general types; positive and negative reinforcers. Positive reinforcement has often been referred to as reward training. The mechanics of using positive reinforcement are not always readily apparent in horse training. What rewards can be bestowed upon the horse during the training process other than a pat on the neck or a spoken word, and how rewarding are they? Any rewarding effects that these positive reinforcements have are of an acquired or learned nature (Figure 18-6). An untrained horse does not appreciate them until they are associated with primary reinforcers such as food and water. Therefore, it is hard for trainers to implement positive reinforcement that is contingent upon desired responses.

Many experienced horsepeople report having obtained results from reward training. It is nearly always secondary or acquired reinforcement and sometimes is so vague that it is difficult to describe. The presence of the

Figure 18-6
A horse must learn to appreciate secondary positive reinforcement. It is not effective unless the horse learns to appreciate it.

trainer or the sight of a working ring or a riding trail may acquire reinforcing properties for the horse. Positive reinforcement can be a valuable asset for the horseperson if it can be implemented. There is too much evidence of successful results obtained from reward training in other species for the horse trainer to ignore its use.

Negative Reinforcement   These are aversive stimuli that the horse will work to avoid or get rid of if given a choice. There are at least three different methods of training by using negative reinforcement or aversive stimuli:

1. Punishment. In this type of negative reinforcement, the horse makes a response in the absence of a cue and then is punished immediately upon making the response. The aim of this method is not the acquisition of a new response but rather the weakening or elimination of a response that is already in the horse's repertoire. Traditionally, horsepeople use this method to break bad habits or to correct vices. The intensity of punishment is critical. It has been shown that light punishment does not break a habit but merely suppresses it. This is why horsepeople do not recommend that horses be pecked gently for correction. If the horse must be corrected, it should be punished sharply and quickly. There is much variation among individual horses concerning punishment. A light tap with a bat may mean practically nothing to one horse and be fairly severe punishment for another. To be most effective, punishment must cause the horse to select a desirable alternate response that leads to reward.

2. Escape. In this type of negative reinforcement, the aversive stimulus is applied with little or no cue and independent of what the horse is doing. The execution of a specific response by the horse is necessary for the termination of the stimulus. When a spur is pushed into a horse's ribs to get it to move, the horse makes an escape response that will result in the spur being removed. When a direct rein is applied to the right, the horse turns to the right to escape the right rein pull.

3. Avoidance. In this method the horse is first given a cue to which to respond. If the response is correct, there is no punishment. If the response is incorrect or latent, then aversive stimuli or negative reinforcement is applied (Figure 18-7). This method constitutes the bulk of the negative reinforcement used in horse training. Sharp and precisely trained horses performing with almost imperceptible cues are actually making avoidance responses. The threat of aversive stimuli is what keeps them alert. A well-trained horse will require very little negative reinforcement because it has learned to make the correct response on cue. For example, the response to neck reining is actually an avoidance response. The horse moves as a result of the bearing rein pressure because the threat of the direct rein is always there. Backing when light bit pressure is applied is also avoidance

Figure 18-7
Negative reinforcement applied at the correct time and with the proper intensity is very effective in increasing both the effort that a horse will put into a performance and the speed of the performance.

conditioning because the threat of hard pressure on the mouth is always present. This is an example of a stimulus that is a cue at one intensity and negative reinforcement at an increased intensity. Many stimuli fit this description, and are used by horsepeople quite frequently. A horse moves when slight heel pressure is applied because failure to do so in the past has been followed by a hard kick.

Contingent Reinforcement—Alternative Response   For any reinforcement to have maximum effectiveness, it must be contingent upon the response; that is, it must be given immediately following the response. Contingency between response and reinforcement is limited in reward training. However, in negative reinforcement there are many more possibilities, as was shown in the previously discussed examples. Contingent punishment enables a horse to know what response is being punished, whereas noncontingent punishment causes the horse to have a general fear, and all its behavior will be abnormal.

The availability of alternative responses is also important when negative reinforcement is used as a training tool. This means that when a horse is punished for making a wrong response, the trainer should make sure that the response desired is available to the horse. If alternative responses are not available, it is difficult to change behavior permanently with any kind of aversive stimuli. When alternative responses are available and the punishment is contingent on the behavior being reinforced, negative reinforcement is more effective than any other type of training in causing one response to be abandoned and another adopted.

The effects of contingent reinforcement and alternative response are very evident in the training of cutting horses. A cutting horse must be trained to move with a cow and to make hard stops and turns on its own with only the cow as its cue. When the cutting horse makes a bad turn, the horse should be corrected immediately. This is usually done with the kick of a spur or a sharp pull on the bit. The more contingent this punishment is on the bad turn, the more quickly the horse learns to make the correct maneuver. However, it should be noted that the alternative response (correct turn) is available. The cutting horse may also exhibit abnormal behavior as a result of overly severe and noncontingent punishment. Punishment of this kind will cause the horse to watch the rider much more than the cow and the horse will thus be more likely to make mistakes. Many cutting-horse trainers are relying more on reward training and less on severe punishment. Also, they have learned that any punishment used must be contingent and that alternative responses must always be available.

A good trainer is, by necessity, also a good rider because he must be skillful enough to administer contingent negative reinforcement. Realization of the significance of contingent negative reinforcement, combined with the skill to administer it, is one of the key differences between good trainers and mediocre ones. Many novice horsepeople do not realize that punishment must be contingent. For example, a barrel-racing horse will frequently be punished after a run is over because it failed to turn the first barrel correctly. The horse does not know why the punishment is being administered, and certainly at this point it cannot make the alternative response—turning the barrel correctly. Punishment of this kind will only make the horse dread the entire ordeal. The correct way to administer the punishment would be to work the horse several times at home, beginning slowly and gradually increasing the speed, and administer punishment when the horse first makes the wrong move around the barrel. It would be more effective to administer no punishment at all than for the punishment to be noncontingent. A wise trainer will let a mistake pass but will be alert for its future occurrences and will then try to administer negative reinforcement contingent upon the mistake.

Punishment must be administered with care so that it will not become a stimulus for bad behavior. If the trainer inflicts punishment on a horse that causes him to become so unmanageable that the punishment must cease, the unmanageable behavior is then reinforced by cessation of the punishment. Every time this particular punishment is initiated the unmanageable condition will reappear. No punishment should be given until all consequences have been considered and there is assurance that control of the horse can be maintained and the desired response is obtained.

Schedules of Reinforcement

Schedules refer to when and how often a response is reinforced. These schedules of reinforcement have a direct effect both on horses being trained and on previously trained horses that are being ridden. Early training, when all behavior is being taught for the first time, is called the acquisition phase. During acquisition, most desirable responses are reinforced. This is referred to as continuous reinforcement and results in faster learning by the horse. During this phase, the trainer continually either rewards or punishes the horse for every response it makes to cues given it. If cues are presented without any reinforcement, the learned responses will gradually diminish to the pretraining level. Eventually the horse will make no correct responses. This is referred to as extinction.

Continuous reinforcement during acquisition and no reinforcement resulting in extinction represent the two extremes in reinforcement schedules. Most older trained horses are reinforced between these two extremes, on some intermittent schedule, to keep them performing correctly. Intermittent schedules of reinforcement mean that the horse is reinforced not after each response, but at irregular intervals. This change from continuous to intermittent reinforcement usually occurs gradually as the horse progresses in its training. As the horse progresses, the trainer applies less reinforcement, which means he is not reinforcing every response and is moving toward an intermittent schedule.

It is important that a horse be trained with intermittent reinforcement because this will result in resistance to extinction. A horse trained on an intermittent schedule will perform longer with no reinforcement than a horse trained completely on continuous reinforcement. This is what is referred to as a "finished" or fully trained horse. The finished horse is frequently presented as a horse that requires no reinforcement. The implication is that this horse will perform indefinitely in accordance with its cues and that all that is necessary is to present the cues to it. This is not true. If there is no reinforcement, the horse will cease to perform as it was trained regardless of the cues given it. However, this horse can be ridden longer, still responding to cues by a novice rider who lacks reinforcing ability, than a horse trained on continuous reinforcement. Eventually, the novice rider must learn the reinforcement for the horse or a more experienced rider will have to ride it again. Horsemen refer to this situation when they talk about getting a horse "tuned up."

A horse that requires continuous reinforcement is not a very well-trained horse. In fact, it is still in the acquisition phase. Horsepeople should try to advance the horse to intermittent reinforcement by applying no more reinforcement than is necessary after the initial acquisition phase. This horse should appear to perform repeatedly with only the presentation of cues. The amount of reinforcement necessary should be so minimal that it goes unnoticed by the average horseman.

The principle of extinction is also used to a good advantage in horse training. Horsepeople recognize this principle and ignore certain acts of the horse, knowing that the horse will cease doing them. If these acts are not reinforced, they will extinguish themselves. A similar problem arises when a horse is first being reinforced for responses. Very seldom are responses so accurate that there are not some undesirable actions being made at the same time. In some situations, the horse is not sure which response was reinforced. Several trials of contingent reinforcement may be required for this undesirable behavior to extinguish itself as the desirable behavior becomes stronger. The trainer should be aware of what is happening to the horse and be tolerant during these stages. It is important to be sure that some undesired response or behavior is not unknowingly reinforced.

Shaping

A term used by behaviorists that is analogous to horse training is shaping. In the initial stages, behavior is shaped by reinforcing each successive approximation of the desired response. Trainers use different techniques, but each will try to get a horse first to do something similar to the desired response; they will then progressively reinforce only the more desired behavior. The importance of recognizing small responses is paramount at this point. The trainer must recognize and be able to apply reinforcement to these small responses as he progresses to advanced performance. That first step taken backward, properly reinforced, will eventually lead to a good backing horse.

A rollback is a difficult maneuver to teach a horse. It includes a series of movements that must be learned one at a time. The trainer usually starts by teaching a stop. Then he teaches another maneuver designed to get the horse moving laterally, especially in the front. Then the horse is taught some short pivots. Finally, the horse combines all these movements and does a rollback. Throughout, the trainer applies reinforcement to any behavior that approximates that which is desired; however, he must expect more from the horse during each series of trials before reinforcing him or else no progress will be made. As learning progresses, only desired responses will be reinforced. Other undesired responses will drop out because of lack of reinforcement.

Effects of Effort on Learning

The more effort that is required of a horse to make a particular response, the harder it will be for the horse to learn the response. The time and number of trials (amount of practice) required to learn a response will increase as the amount of effort required increases. Less time is usually required to train a pleasure horse for a horse-show performance than a reining or jumping horse. The amount of effort required to perform the tasks is the main reason for this time difference. Extinction of the responses is also faster when the response is difficult. To horsepeople this means that a reining horse will come untrained more easily than a pleasure horse.

Many of the tasks for which horses are trained require great effort. Horses that have athletic ability should learn faster than those with little or no athletic ability because the former require less effort to perform the tasks. The trainer should be aware of the effort required to do a task and realize that some horses with little athletic ability will learn slowly and may never learn to perform the task well. Apparently, behavior that must be learned is sometimes learned less raedily by the horse than by other animals. Because the porpoise has been trained to perform many feats that are quite remarkable, that animal appears to be very intelligent. However, much of the behavior required for these tasks is easily learned by this animal.

Many of the maneuvers and tasks that horses are asked to perform are far removed from their natural behavior in the wild state. We do not see untrained horses in pastures doing sliding stops, performing difficult rollbacks, jumping high obstacles, or backing. One real natural behavioral characteristic of the horse is ease of locomotion or, more specifically, an instinctive tendency to run. Consequently, it is not hard to teach a horse to run. It may be difficult to condition it to run well, but any novice can get a horse to run. That this is a natural instinct should be borne in mind when training timed-event horses. Teaching a horse to run will not be difficult, but it will be difficult to teach hard turns, flying lead changes, and stops. For this reason, the training of timed-event horses should concentrate on the hard-to-teach maneuvers.

Massing of Trials

Massing of trials, or prolonged practice, has been proven to be an inefficient method of training. Students who cram for exams are inefficient, as is the football kicker who kicks 12 hours one day and then does not practice again for a week. Fatigue is related to inefficient learning and is thought to have much the same effect as effort. Horses that have been overworked to the point of fatigue will learn, but an an inefficient rate.

During the very early stages of training, it may be necessary to bring about light fatigue in order to gain control of the horse. After this initial period, the trainer will enforce learning by using shorter training periods. This is not to say that the horse cannot be ridden for longer periods, but rather that the time devoted to practice for certain maneuvers should be shortened. For example, the horse may be ridden for one hour but practice rollbacks or jumps for only a few minutes. There might be one or more short practice sessions during each ride, broken by periods of relaxation or simple riding.

Inhibition

One intervening variable between a stimulus and a response, often seen with repeated practice, is inhibition. Such a variable is inferred from observed behavior and is used in a theoretical sense only, with no physiological reference. Suppose that animals were allowed 100 free-choice trials down a T maze. We would probably record 50 to the left and 50 to the right. If we close the right side and run left only for 100 additional trials, then open up both sides again, we would expect all choices to be left because of the extra practice in this direction. What we would probably get is all right choices because of the theoretical proposition of inhibition. Eventually, we would again get an equally distributed choice of one-half each way. The influences of inhibition can be seen in horse training. What a horse has learned on one side he must also learn on the other. Consequently, training must be directed at both sides or in both directions. Repeated trials to one side only may actually inhibit the horse from going that direction. To balance the effects of this phenomenon and still train both sides, the trainer should periodically work both sides of the horse and should concentrate on the problem side.

18.4   Intelligence

There are many different kinds of intelligence, and there is no one general measure of intelligence. Not all animals excel in all tests of intelligence, which indicates that some have different abilities for different kinds of activities.

Intelligence includes the ability to reason and some have suggested that horses do not reason. However, some research results indicate an ability of horses to reason. In problem-solving intelligence tests, the horse usually places low, but so does the dog. Rats, cats, monkeys, and birds commonly score higher in these tests. The donkey is thought to possess much more reasoning power than the horse, but some horsepeople would be hard to convince of that assessment.

This seeming lack of high reasoning ability does not necessarily condemn the horse to a state of helplessness, because it has other talents. The horse learns quite readily. Therefore, the horse can best be evaluated on problems that require learning and skills of locomotion. New tasks are not difficult to learn if the cues affect the stronger senses (tactile, sight, hearing). The great performance horses are examples of the high degree to which these senses can be developed. As early as the nineteenth century, famous cases were recorded in the history of animal psychology that were proof of the horse's ability to respond to stimuli that were almost too weak to be perceived.

The successful horse trainer must be aware of limitations in the horses ability to reason and must concentrate training on the positive assets of the horse. The horse's greatest asset is its ability to learn and discriminate between the slightest of cues. No matter how intelligent a horse may appear when working, the trainer should remember that somewhere behind the performance are cues that the horse has learned. In general, the horse does not logically decide what is desired, but responds to stimuli or cues.

18.5   Training Summary

Not all of the principles of horse training have been presented here; however, those presented are encountered daily in training routines. On occasion, the trainer may apply all of the principles that have been mentioned almost simultaneously. Horse trainers may not recognize some of these principles by the terms used here. However, if the trainer is successful, he is using most of these principles.

Most of this information has been presented to support the following concept. A cue is presented to signal a desired response, and this association is strengthened by reinforcement. The following guidelines are based on this concept and are suggested for training horses.

1. Learn to recognize and appreciate small responses or segments of the final performance during the learning phase.
2. Learn how to start with basic cues and advance to new ones.
3. Be specific and consistent in the presentation of all cues.
4. Be contingent with all reinforcement.
5. Provide alternative responses when using punishment.

6. Learn to shape behavior by reinforcing approximations of the desired response.

7. Advance to intermittent schedules of reinforcement.

8. Do not try to train with negative reinforcement only; it has too many pitfalls and can cause too many undesirable effects.

9. Learn how to use positive reinforcement; try to develop secondary positive reinforcement.

10. Do not expect the horse to have reasoning powers that may be common to humans.

18.6   Behavior and Management

Because of the close social relationship that exists between man and his horse, and because horse behavior is usually modified to some extent during training, an understanding of horse behavior is essential to effective management. There is also a "generation gap" in the transmission of knowledge about horses, many people who are currently part of the horse industry have only a limited notion of the difference between normal and abnormal horse behavior. Management of horses in close confinement (as opposed to open ranges) also necessitates an awareness of how horses react to each other.

Categories of Behavior

Although the research that has been conducted on horse behavior is limited, several reports of research on the behavior of other species of animals have made possible the organization of information into concepts that are applicable to horses. Social behavior in animals has been considered by several researchers, and some have classified this behavior into 9 categories. These categories will be convenient for the following discussion of horse behavior.

Contactual Behavior   This type of behavior is generally considered the result of seeking affection or protection, although horses sometimes exhibit contactual behavior for other reasons. Contactual behavior is also seen in many other animals. Horses will huddle together during inclement weather or at times of suspected danger. Although contactual behavior is not as highly developed in the horse as in some other species, horses do exhibit variations of this type of behavior.

Ingestive Behavior   Ingestive behavior is the taking of food or water into the digestive tract. Ingestion of solid food by the horse begins at a very young age. Foals will start nibbling grass and will begin to eat grain within the first week of life if given the opportunity. Such behavior as chewing the bark of trees and tail chewing is not considered normal ingestive behavior and is generally a result of deprivation of roughage or nutrients, boredom, lack of exercise, or a combination of these. Horses vary tremendously in their rate of feed consumption. Some horses eat very quickly even to the extent that feed is not thoroughly masticated, whereas others eat very slowly.

The anatomy and physiology of the digestive system are such that horses will, if given a chance, take in small amounts of food at one time and eat at frequent intervals. Grazing horses prefer to cover a large area, taking small bites of grass at one time and seldom taking more than one or two bites before taking a step. Consequently, horses will usually bite the top of a bunch of grass and leave the bottom portion if the pasture is not contaminated with large amounts of feces. In pastures where stocking rate is high and in pastures that are not routinely harrowed, horses will graze portions of the pasture very close and will leave defecation areas to grow tall.

Horses are very apt to overeat, particularly if given access to feed after periods of low feed intake. Consequently, horses are very susceptible to digestive disorders caused by overeating.

Eliminative Behavior   In general horses will interrupt any activity to urinate or defecate. Unless forced they will not urinate or defecate while walking, eating, or drinking.

Horses are likely to establish an elimination area in the pasture or paddocks and will not graze that area until there is no other source of feed (Figure 18-8). Horses will generally walk some distance to the elimination area to urinate or defecate.

Many horses will defecate in the same place when being moved. For example, some horses will routinely defecate when going through a particular gate between pastures or paddocks. This may be a result of an attempt at territorial marking by the horse.

Some horses will readily urinate in a stall or trailer whereas others will urinate immediately after being let out of a trailer or stall. Almost all horses will defecate when approaching a trailer or immediately after they are inside it. This is probably because of the nervous action resulting from anticipation.

Elimination patterns often differ between the sexes. Characteristically, stallions will smell the elimination area and walk up on it or back down on it before defecating or urinating. As a result, the elimination area of stallions usually does not become appreciably larger in diameter. Mares and geldings will usually smell the spot and urinate or defecate without walking upon the spot. Therefore, the elimination area of mares and geldings usually becomes larger.

Adult horses will generally reject feces of their own kind, but foals will repeatedly eat feces of mares, especially when kept in dry lots. Coprophagy in adult horses is not believed to be a normal phenomenon.

Figure 18-8
Horses will establish an elimination area in a paddock or
pasture. In this area the grass grows tall, and horses will not
readily graze it.

Sexual Behavior   Sexual behavior includes all acts associated with the ultimate fertilization process, including courtship and copulation. Sexual behavior is not confined to mares and stallions. Geldings will definitely show signs
of sexual behavior, particularly if castrated after puberty, because sexual
behavior is controlled by both hormonal and neural influences. Many geldings will tease mares in estrus and will even mount. Also, many geldings are
very possessive of mares that are pastured with them and will fight for a
particular mare. Horsepeople should not assume that geldings will not cause
trouble around mares. Usually they will not. However, putting a gelding in
the presence of mares in estrus or vice versa can cause serious behavioral
problems and upset management of horses.

Discussion of normal sexual behavior in mares and stallions is omitted
here since it is discussed in the chapters on reproductive physiology.

Epimeletic Behavior   The giving of care and attention is common between
mare and foal, but other behavioral acts of this nature are also observed
among horses. Such behavior is not common in mature stallions but is commonly seen in mares, geldings, and young horses. For example, horses will
routinely stand in the shade head-to-tail and mutually fight flies for each
other. Licking another animal is not very common in horses, but many horses

will simultaneously scratch each other on the neck and over the withers and back.

Et-Epimeletic Behavior   Both young and adult horses signal for care and attention by calling or movement. This behavior is commonly seen when horses are separated from each other. Foals will call very excitedly for their dams when accidently separated or when weaned. Even adult horses that are accustomed to being together will call repeatedly for their companion.

Strong pair-bonding relationships exist between horses, and breaking these relationships can complicate management. Horses will often go through or over fences that they normally would not bother with to get together. Some horses, when separated at a show, will become excited and call for their companion to such an extent that the action interferes with their performance.

As a result of conditioning, many horses will call for the handler at feeding time. It is common for horses to call on sight of the manager when feeding time is near. Horses being supplementally fed in pastures are likely to go to the feeding place at or near feeding time.

Allelomimetic Behavior (Mimicry)   This type of behavior is also referred to as contagious or infectious behavior. Horses are not thought to be likely to copy the behavior of another horse to the extent that other animal species copy each other. There are, however, some examples of this kind of behavior in horses. Horses in a group are likely to run simply because another horse is running. This may be mimicry, or it may be the result of fright. Some horses seem to run and play because other horses run and play. Having a horse that is hard to catch in a group will likely result in other horses in that group becoming hard to catch. Horses stalled adjacent to another horse that chews wood have a tendency to copy this behavior. Suckling foals will chew on a feed trough after being exposed to this behavior in their dams.

Investigative Behavior   Sensory inspection of the environment is very highly developed in horses and generally involves movement and one or all of the horse's senses—sight, touch, smell, hearing, and sometimes taste. Horses are very curious, particularly of new surroundings or objects, and will use all necessary senses to investigate them (Figure 18-9). Horses do not see detail very efficiently and therefore are seldom satisfied to investigate something new by sight alone. They cannot resist smelling, listening, touching, and sometimes tasting a new object. This investigation apparently must be completed before a horse will accept something new without sustained apprehension. Even strange horses are investigated. During the investigative processes, horses are very excitable and are likely to overreact to sudden movement or sounds. Horses have frequently injured themselves as a result of running into a fence or other object after being excited by a new object.

Figure 18-9
Horses are very curious and investigative of new objects. During the investigative process, horses are very excitable and may run into fences or other objects as a result of a sudden movement or sound.

Horses put into a new paddock or pasture next to strange horses may run into the fence in their attempt to investigate these new horses. It is not uncommon for a horse to fear being caught or trapped as a result of investigating a strange object.

Agonistic Behavior   This type of behavior includes all actions that are a result of or associated with conflict or fighting and includes aggression, submission, and attempts to escape. Agonistic behavior in horses is very pronounced and highly variable. It causes many problems for horsepeople, but when carefully observed it can be managed. Awareness of agonistic behavior in horses and implementation of management practices to minimize its adverse effects are essential to successful horse management.

Dominance   Agonistic behavior in horses is the result of highly developed dominance hierarchies. Dominance hierarchies or "pecking orders" are seen in many species of animals and are definitely a significant part of the normal behavior of horses (Figure 18-10). Dominance in horses is established through aggression, but for the purpose of preventing severe fighting. Once the dominance hierarchy is established and so long as all horses in a group stay together, only the threat of aggression is needed to maintain the dominance hierarchy (Figure 18-11). Seldom will the dominant horse have to do more than pin its ears, bare its teeth, or make a sudden move toward the other horse to maintain the dominant position. The submissive horse will seldom challenge the dominant horse for position. Horses learn to live together without serious conflict as a result of their recognition of the dominance hierarchy. The dominance hierarchy is thought to be linear — from the most dominant down to the most submissive animal in the group. However, there are instances in which an animal that is low in the order may

Figure 18-10
In every group of horses
there is a "boss" horse.

appear dominant over some animal at the top of the order. Most researchers believe this to be an example of a dominance hierarchy that has not been completely explained.

In establishing dominance, animals may have very traumatic experiences, or they may fight little or not at all. For example, stallions in the wild state often fought to the death of one or until crippling injury resulted. Putting two "boss" mares together may result in severe fighting. Many groups of horses can resolve the dominance problem without serious consequence *if* the submissive animal is allowed plenty of room for retreat and is not forced to stand and fight. Therefore, when grouping strange horses it is

Figure 18-11
When the dominant horse
makes a threat, submissive
horses will retreat if possible.

very important that they have plenty of room and are watched closely. Many considerations affect the degree of aggression in horses and may influence the expression of dominance. Hormone levels in males obviously affect the degree of aggression, because stallions are normally more difficult to manage than geldings. However, blood estrogen levels in mares do not appear to cause changes in dominance. Even in deep estrus, the timid mare is not likely to challenge the dominant mare for position, feed, or presence near the stallion. Some authorities believe that dominant mares are more aggressive when in estrus, but there is no clear evidence to support or refute this belief. Some geldings that are normally very tolerant when in a group of other geldings become very dominant and aggressive in the presence of a mare. Other variables such as size, maturity, home territory, old age, and injury affect dominance to some degree and may mask the real dominance hierarchy. Some horses that appear dominant in a stall may actually be submissive when outside and vice versa.

The presence of the manager will often cause a false expression of the dominance hierarchy. Many horses that are actually in the lower end of the order will appear aggressive and dominant in the presence of the manager. On the other hand, some horses that are very dominant in the herd will not express their dominance in the presence of the manager. It is important to realize that dominance between horses must be established by the horses themselves. There is little the manager can do to change or reduce dominance in his absence. Therefore, the manager should group horses according to compatibility and closely observe their behavior. Horses should be grouped to allow orderly establishment of the dominance hierarchy and to minimize fighting and injury.

Management Suggestions

The following management practices are suggested to help the perceptive manager, who is aware of social and other behavior in horses, keep any adverse effects of this behavior to a minimum.

1. Use fences that prevent injury in horses that huddle across the fence from each other.
2. Feed in individual feeders whenever possible. Space feeders well apart and provide at least one feeder for each horse in paddock or pasture.
3. Keep feed rooms locked, and use entrances outside the paddock or stall area.
4. Feed balanced rations according to requirements and include either hay or pasture.

5. Make provision for horses that eat very slowly.

6. Discourage rapid eating.

7. Routinely clean stalls and paddocks, and compost manure before spreading.

8. Mow and harrow pastures to scatter feces and keep grass young and growing.

9. Use rotational grazing whenever possible. Graze heavily for short periods and then change pastures.

10. Provide a mechanism for teasing timid mares.

11. Keep foals in a safe place and in sight of the mare during breeding.

12. Be careful when separating horses that are accustomed to being together.

13. Never turn horses into a new area late in the afternoon or at night. Observe closely when putting a horse in a new area.

14. Never attempt to catch a horse until it can be done. Never let a group of horses start to run when trying to catch them. Stop and pen them.

15. Use only treated wood or metal when building stalls, feeders, fences, and so on, that horses may chew on. Prevention is easier than cure.

16. Keep pastures, pens, and other facilities free of foreign objects. Avoid exposing horses to new objects that may be dangerous.

17. Be careful when turning horses out to pasture with halters on them.

18. Gates should be well secured and hard to open. Horses will investigate them.

19. Group horses by age and sex whenever possible.

20. Observe closely for aggressive behavior when changing groups.

21. Use box stalls with solid partitions between horses to prevent aggression.

22. Be aware of changes in dominance when horses are changed from one type of horse management to another.

23. Keep fences in good repair and avoid using barbed wire.

REFERENCES

Baer, K. L., C. A. McCall, T. H. Friend, and G. D. Potter. 1979. Comparison of methods used in determing social dominance. *ASAS Abstracts*. p. 145.

Baer, K. L., G. D. Potter, T. H. Friend, and B. V. Beaver. 1983. Observation effects on learning in horses. *Applied Anim. Ethology* 11:123.

Bingham, W. E., and W. J. Griffiths. 1952. The effect of different environments during infancy on adult behavior in the rat. *J. Comp. Physiol. Psychol.* 45:307.

Collins, R. A. 1970. Aggression in mice selectively bred for brain weight. *Behavior Gen.* 2:169.

Denenberg, V. H., P. B. Sawin, G. P. Frommer, and S. Ross. 1958. Genetic, physiological and behavioral background of reproduction in the rabbit. IV. An analysis of maternal behavior at successive parturitions. *Behavior* 13:131.

Fiske, J. C., and G. D. Potter. 1979. Discrimination reserval learning in yearling horses. *J. Anim. Sci.* 49:583.

Fox, M. W. 1968. *Abnormal Behavior in Animals.* Philadelphia: W. B. Saunders.

Hafez, E. S. E. 1969. *The Behavior of Domestic Animals,* 2nd ed. Baltimore: Williams and Wilkins.

Hagerbaumer, J. M., G. D. Potter, and L. M. Schake. 1979. Discrimination learning in young horses. *SASAS Abstracts.* p. 25.

Heron, W. T. 1941. The inheritance of brightness and dullness in maze learning ability in the rat. *J. Genet. Psychol.* 59:41.

Ingram, R. S., C. A. McCall, G. D. Potter, and T. H. Friend. 1979. Comparative learning performances of horses and cattle in the Hebb-Williams closed field maze. *ASAS Abstracts.* p. 147.

James, W. T. 1941. Morphological form and its relation to behavior. In *The Genetic and Endocrine Basis for Differences in Form and Behavior.* C. R. Stockard, ed. Philadelphia: Wisar Institute.

Kieffer, N. M. 1968. Heritability of cutting in horses. *Proc. Horse Short Course.* Texas A & M University, p. 46.

Kieffer, N. M. 1974. Methods of estimating the breeding values of potential sires and dams. *Proc. Horse Prod. Short Course.* Texas A & M University, p. 95.

Klinghammer, E., and M. W. Fox. 1971. Ethology and its place in animal science. *J. Animal Sci.* 32:1278.

Kratzer, D. D. 1971. Learning in farm animals. *J. Anim. Sci.* 32:1268.

McCall, C. A., T. H. Friend, G. Dellmeier, and G. D. Potter. 1981. Comparison of ranking procedures and influences of group size and number of observations on social rank orders. *Proc. Ann. Meeting Am. Behavior So.*

McCall, C. A., G. D. Potter, T. H. Friend, and R. S. Ingram. 1981. Learning abilities in yearling horses using the Hebb-Williams closed field maze. *J. Anim. Sci.* 53:928.

McCall, C. A., G. D. Potter, and J. L. Kreider. 1985. Locomotor, vocal and other behavioral responses to varying methods of weaning foals. *Applied Behavior Sci.* 14:27.

Miller, R. W. 1974. *Horse Behavior and Training.* Bozeman, Montana: Big Sky Books.

Rundquist, E. A. 1933. Inheritance of spontaneous activity in rats. *J. Comp. Psychol.* 16:415.

Sawin, P. B., and D. D. Crary. 1953. Genetic and physiological background of reproduction in the rabbit. II. Some racial differences in the pattern of maternal behavior. *Behavior* 6:128.

Scott, J. P. 1956. The analysis of social organization in animals. *Ecol.* 37:213.

Scott, J. P. 1962. Introduction to animal behavior. In *Behavior of Domestic Animals.* E. S. E. Hafez, ed. Baltimore: Wilkins and Wilkins.

Tolman, E. C. 1924. The inheritance of maze-learning ability in rats. *J. Comp. Psychol.* 4:1

Tyler, P. A., and T. E. McClean. 1970. A quantitative genetic analysis of runway learning in mice. *Behavior Gen.* 1:1.

Vicari, E. M. 1921. Heredity of behavior in mice. Washington, D.C.: Carnegie Inst. p. 132.

Wieckert, D. A. 1971. Social behavior in farm animals. *J. Anim. Sci.* 32:1274.

Willham, R. L., D. F. Cox, and G. E. Karas. 1953. Genetic variations in a measure of avoidance learning in swine. *J. Comp. and Physiol. Psychol.* 56:294.

Yeates, B. F. 1974a. Applying principles of psychology to horse training. *Proc. Horse Prod. Short Course.* Texas A & M University, p. 77.

Yeates, B. F. 1974b. Recognition and use of social order in horse management. *Proc. Horse Prod. Short Course.* Texas A & M University, p. 71.

CHAPTER 19

ANATOMY, PHYSIOLOGY, AND CARE OF THE FEET AND LEGS

To obtain maximum performance and prolong the useful life of a horse, we must understand the anatomy, physiology, and care of the horse's feet and legs. When an injury occurs to a bone, muscle, nerve, tendon, ligament, or blood vessel, the horse owner should be able to understand the seriousness of the injury and its potential effect on later performance.

19.1 Anatomy

Bones

The bones of the limbs are illustrated in Figures 3-1 and 19-1. The foreleg is attached to the horse's body not by bone, but by muscles and ligamentous material (Figure 19-2). This type of attachment serves to absorb concussion transmitted up the foreleg when the foot strikes the ground. It also permits the scapula to move freely over the underlying ribs. Without this freedom of scapula movement, the horse would not be able to run at fast speeds, because it has a relatively rigid spine.

The *scapula* is divided longitudinally by the scapular spine, which can be found by palpation. The *suprascapular nerve* crosses the spine. This arrangement allows the nerve to be easily damaged if the horse's shoulder strikes a

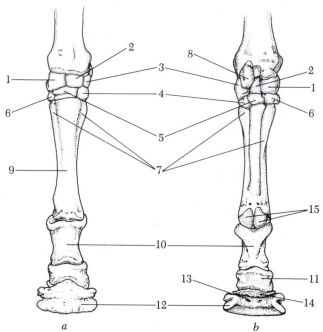

Figure 19-1
Bones of the left forelimb below the knee: (a), anterior view; (b), posterior view. 1, radial carpal; 2, intermediate carpal; 3, ulnar carpal; 4, fourth carpal; 5, third carpal; 6, second carpal; 7, second and fourth metacarpals; 8, accessory carpal; 9, third metacarpal; 10, first phalanx; 11, second phalanx; 12, third phalanx; 13, arterial foramen; 14, distal sesamoid; 15, proximal sesamoid bones.

hard object. Subsequent degeneration of the muscles innervated by the nerve is termed *shoulder sweeney*. The upper end of the scapula is not ossified and is cartilage, the cartilage of prolongation. At the lower end of the scapula, the scapular tuberosity projects out at the point of the shoulder; this is an area where injuries from contusions may occur. Length of the scapula is important for proper movement of the foreleg and advancement of the humerus. If the spines of the thoracic vertebrae that form the withers are long, the scapula must be longer still because it is attached to them. The humerus forms the skeleton of the arm. Relatively short in the horse, it is one of the strongest bones in the horse's body, and has many muscle insertions on it.

The *forearm* of the horse is composed of two bones, the radius and the ulna. The *radius* is the larger bone and articulates with the humerus and carpal bones. The *ulna* is short and closely attached to the radius. It projects above and behind the radius as the *olecranon*, which serves as an attachment for extensor muscles of the elbow. It is unique in that it has no marrow cavity, which the other long bones possess.

The *knee* or *carpus* is composed of seven or eight carpal bones arranged in two rows (Figures 19-1 and 19-6). The first carpal in the distal row may be absent. The accessory carpal bone serves as a pulley block to enable several

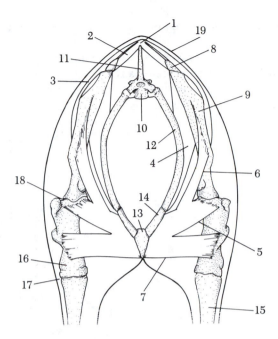

Figure 19-2
Diagrammatic section through the chest at the level of the scapula. The thorax is suspended between the forelimbs entirely by muscular attachments and not by bone. This enables the rib cage to move within its cradle of muscles and, as the thoracic spine is an almost rigid column, the body of the horse moves with it. 1, withers; 2, thoracic part of rhomboideus muscle; 3, trapezius muscle; 4, thoracic part of serratus ventralis muscle; 5, posterior deep pectoral muscle; 6, anterior deep pectoral muscle; 7, superficial pectoral muscle; 8, cartilage of prolongation of the scapula; 9, scapula; 10, body of thoracic vertebra; 11, spinous process of thoracic vertebra; 12, rib; 13, sternum; 14, costal cartilage; 15, radius; 16, humerus; 17, elbow joint; 18, shoulder joint; 19, body surface. (From Smythe and Goody, 1975.)

muscles to work at a mechanical advantage. Overextension of the knee is prevented by several ligaments.

In each foreleg or hind leg, respectively, horses have one large functional *metacarpal* or *metatarsal* bone, the third, which is referred to as the *cannon bone*. The second and fourth metacarpal or metatarsal bones are small and rudimentary and are referred to as *splint bones*. They extend about two-thirds of the way down the lateral and medial borders of the posterior surfaces of the cannon bones, and serve to provide some support to part of the carpal bones. Initially, they are connected to the cannon bones by an interosseous ligament that later becomes ossified as the horse matures. Disturbance of the interosseous ligament prior to ossification can lead to "splints."

The digit of the horse consists of three *phalanges* and three *sesamoid bones* (Figure 19-3). The first phalanx is the longest and is commonly called the *long pastern*. The second phalanx or *short pastern* belongs partly to the leg and partly to the hoof. It contains no medullary cavity, so it is solid throughout. The third phalanx, referred to as the *pedal* or *coffin bone*, is inside the hoof and resembles a small hoof. The bone is very porous and

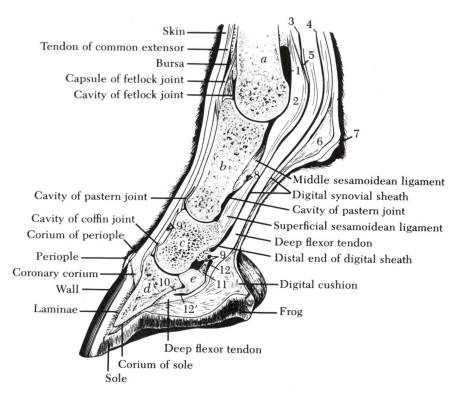

Figure 19-3
Sagittal section of digit and distal part of metacarpus of horse: (*a*), metacarpal bone; (*b*), first phalanx; (*c*), second phalanx; (*d*), third phalanx; (*e*), distal sesamoid bone. 1, volar pouch of capsule of fetlock joint; 2, intersesamoidean ligament; 3,4, proximal end of digital synovial sheath; 5, ring formed by superficial flexor tendon; 6, fibrous tissue underlying ergot; 7, ergot; 8,9, branches of digital vessels; 10, distal ligament of distal sesamoid bone; 11, suspensory ligament of distal sesamoid bone; 12, proximal end of navicular bursa; 12′, distal end of navicular bursa. The superficial flexor tendon (*behind 4*) is not shown. (From Sisson and Grossman, 1953.)

blood vessels pass into and out of the pores. Lateral cartilages are attached to the wings of the third phalanx (Figure 19-3). They rise above the coronary band and extend approximately one-third of the way around each side of the hoof from the heel. The two proximal sesamoids are located on the posterior surface of the fetlock joint. These pyramid-shaped bones serve as a pulley block for the deep flexor tendon so that its leverage is increased. Numerous ligaments are also attached to the proximal sesamoids, so they are an important part of the suspensory apparatus. Because of their function, they are exposed to excessive strain when we subject horses to racing, jumping, cutting, and other strenuous exercises. The *distal sesamoid bone* or *navicular bone* is located at the posterior surface of the *digital interpharangeal* or *coffin joint*. Its function is to increase the articulatory surface of the joint and to serve as a smooth pulley for the deep flexor tendon.

The hind leg is attached to the spine with a bone to bone connection. It has the same number of segments at approximately the same levels as the foreleg.

The first bone of the hind leg is the *os coxae* or *pelvic girdle*. It is made up of two halves that are fused together in the adult at the *pelvic symphysis*. Each half is composed of three bones fused together. The *ilium* attaches to the spine and the *pelvic floor* is composed of the *pubis* at the front and *ischium* at the back of the girdle. All three bones meet at the *acetabulum*, which houses the head of the femur to form the hip joint. One corner of the triangular ilium forms the *tuber coxae* or hip bone.

Because the os coxae is relatively immobile, the *femur* is considered the first moving bone. Situated in the thigh, it is the heaviest and strongest bone in the horse's body. Because of the number of muscles attaching to it, the os coxae has a number of processes, tuberosities, trochanters, fossas, and impressions to increase its surface area. The patella is located at the stifle joint and is discussed in the section on joints.

The *tibia* and *fibula* are the anatomical base of the *gaskin*. The tibia is a long bone with the fibula attached to its back. The lower end of the tibia has two deep articular grooves separated by an articular ridge that fits into the tibial tarsal bone. These ridges and grooves must be set correctly or the movement of the lower end of the hind leg will not be correct. Incorrect set will cause the hind feet to be toe wide (splayed) or toe narrow (pigeon toed). If the set is toe wide, the hind legs will move in an outward and forward direction, resulting in loss of motion and power. Any required movement straight forward will be impaired by an improper set in the joint.

There are six bones in the *tarus* or *hock* of the horse. Occasionally seven bones may be present but the first and second tarsals are usually fused together. The bones are arranged roughly in three rows.

The remainder of the bones in the hind legs are similar to those in the foreleg and have been previously discussed. The hind cannons are usually longer than the front cannons and the splint bones are longer and more developed.

Chestnuts are semi-horny structures located above the knees and on the lower portion of the hocks on the medial side of the legs. The pattern of the chestnuts is distinct for each horse, so they are used as a means of positive identification. The *ergots* are semi-horny structures located on the posterior-ventral surface of the fetlock (Figure 19-3). There is no scientific evidence to suggest that the chestnuts or ergots are reduced vestiges of toes.

Joints

The joints of the limbs of the horse are *diarthroses*—true joints that have a joint cavity and synovial membrane (Figure 19-4). They are formed by articular surfaces, articular cartilages, a joint capsule, and ligaments. Two types of this joint are found in the horse's leg. The *ginglymus* is a hinge joint such as the elbow joint. The *enarthrosis* is a ball-and-socket joint, exemplified by the hip joint.

The shoulder joint between the scapula and humerus is unique in that it is the only leg joint without collateral ligaments. They are unnecessary here because of the powerful muscles surrounding the joint. The joint is a typical ball-and-socket joint. However, adduction and abduction are very restricted, so the chief movements are flexion and extension. At rest, the angle formed

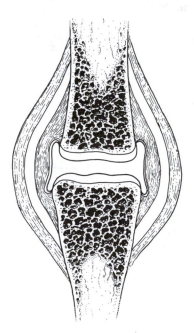

Figure 19-4
Typical diarthrotic articulation.

between the scapula and humerus on the posterior side is 120 degrees to 130 degrees. During flexion the angle is reduced to about 80 degrees, and during extension it is increased to about 145 degrees.

The elbow joint (Figure 19-5) is a hinge joint. An angle of 140 degrees to 150 degrees from the front is normal when the horse is standing. The range of movement of the joint is 55 degrees to 60 degrees. The shape of the articular surfaces and collateral ligaments prevents any lateral movement of the joint. Overextension is prevented by the passage of the anconeal process into the olecranon fossa, strong collateral ligaments, the biceps muscle, and the cranial ligament.

The *carpal* or *knee joint* (Figure 19-6) is a hinge-type joint made up of three subsidiary joints. The *radiocarpal joint* occurs between the radius and proximal row of carpal bones, the *intercarpal joint* occurs between the two rows of carpal bones and the *carpometacarpal joint* occurs between the distal row of carpal bones and the metacarpus. Practically all movement occurs at the radiocarpal and intercarpal joints, the articular surfaces of which are widely separated in front during flexion, but remain in contact behind.

The three phalanges are arranged to form three joints (Figure 19-3). The *metacarpophalangeal joint* is the *fetlock joint*. This joint is subjected to the greatest strain of all joints because it may be required to support the full weight of the horse during certain phases of some gaits. From the front, the articular angle is approximately 140 degrees in the foreleg and approximately 145 degrees in the hind leg. The *proximal interphalangeal joint* or *pastern joint* is the least movable of the three joints. Because of its immobility, concussion exposes the pastern joint to more wear and tear. The *distal interphalangeal joint* or *coffin joint* articulates between the second and third

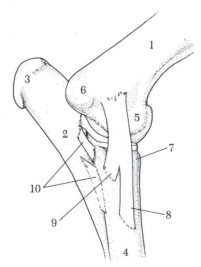

Figure 19-5
Elbow joint of the left side, medial view: 1, humerus; 2, shaft of the ulna; 3, olecranon process; 4, radius; 5, medial condyle; 6, medial epicondyle; 7, Radial tuberosity; 8, long medial collateral ligament; 9, short medial collateral ligament; 10, medial transverse radioulnar ligament. (From Smythe and Goody, 1975.)

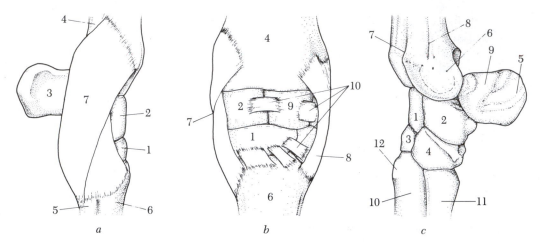

Figure 19-6
Carpus of the left side: (*a*), medial view showing the main ligaments; (*b*), anterior view showing the main ligaments; 1, Third carpal; 2, radial carpal, 3, accessory carpal; 4, radius; 5, small metacarpal bone; 6, large metacarpal bone; 7, medial collateral ligament; 8, lateral collateral ligaments; 9, intermediate carpal; 10, short dorsal carpal ligament. (*c*), lateral view. 1, intermediate carpal; 2, ulnar carpal; 3, third carpal; 4, fourth carpal; 5, accessory carpal; 6, radius; 7, groove for common extensor tendon; 8, groove for lateral extensor tendon; 9, groove for ulnaris lateralis tendon; 10, large metacarpal bone; 11, small metacarpal bone; 12, metacarpal tuberosity. (From Smythe and Goody, 1975.)

phalanges and the navicular bone. The *navicular bone* is subjected to considerable stress and as a result, is commonly involved in unsoundnesses.

The *sacroiliac* articulation is a diarthrosis joint formed between the auricular surfaces of the sacrum and ilium. Its movement is inappreciable in the adult since stability, not mobility, is the main function of the joint. The angle formed by the long axis of the ilium on the horizontal plane varies from 30 degrees to 40 degrees.

The *hip joint* is composed of the head of the femur and the acetabulum. It cannot be palpated, and the major trochanter of the femur is often mistaken for the hip joint. The joint can be moved in all directions. However, movement in some directions is almost nonexistent. The accessory ligament limits the horse's ability to move the limb away from his body and thus "cow-kick." Inward rotation of the leg is also restricted by the accessory ligament. Forward movement is limited, so it is difficult for the horse to move its hind feet further forward than its navel. The angle of the hip joint is 110 degrees to 115 degrees from the front. The femur is set at an angle of 80 degrees to the vertical and the ilium is set about 30 degrees to 35 degrees to the horizontal.

The *stifle joint* (Figure 19-7) is the largest joint in the horse and consists of two articulations. The *femorotibial articulation* is between the tibia and femur. The *femoropatellar* articulation is between the femur and patella. The *patella*, a sesamoid bone, serves to provide mechanical advantage for the quadriceps muscle. During overextension of the stifle joint, the patella can be pulled upwards and forward over the upper trochlear ridge. This locks the stifle in a rigid position and prevents flexion. The hind leg is carried out behind the horse. This condition is referred to as being "stifled."

The *hock joint* (Figure 19-8) is a hinge joint composed of several joints and three planes of articulation. However, almost all movement takes place in the tibial-tarsal articulation between the tibia and the tibial tarsal bone. The hock is a joint where most of the propulsive force is located, so it receives lots of stress and strain. Collateral ligaments prevent complete extension of the hock. Excessive stress and strain during extreme extension or flexion causes the plantar ligament to enlarge, a condition known as a *curb*. The *plantar ligament* is part of the joint capsule that is extremely thick

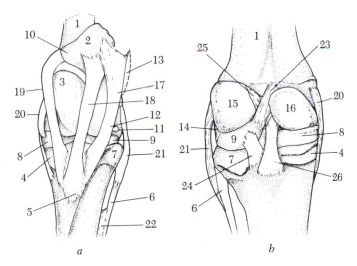

a   *b*

Figure 19-7
Stifle joint of the left side: (*a*), anterior view; (*b*), posterior view. 1, femur; 2, patella; 3, medial ridge of trochlea; 4, medial condyle of tibia; 5, tibial tuberosity; 6, fibula; 7, lateral condyle of tibia; 8, medial meniscus; 9, lateral meniscus; 10, accessory cartilage of patella; 11, stump of extensor longus and peroneus tertius; 12, stump of fascia lata; 13, insertion of biceps femoris muscle; 14, stump of popliteus tendon; 15, lateral condyle of femur; 16, medial condyle of femur; 17, lateral patellar ligament; 18, middle patellar ligament; 19, medial patellar ligament; 20, medial femorotibial ligament; 21, lateral femorotibial ligament; 22, interosseous ligament; 23, femoral ligament of lateral meniscus; 24, posterior ligament of lateral meniscus; 25, anterior cruciate ligament; 26, posterior cruciate ligament. (From Smythe and Goody, 1975.)

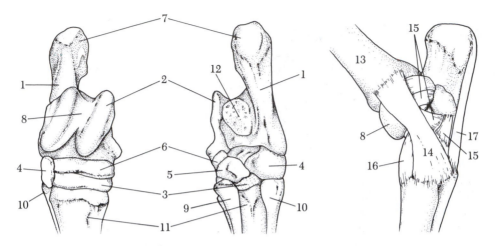

Figure 19-8
Tarsus of the right side: (*a*), anterior view; (*b*), posterior view, (*c*), medial view showing the main ligaments of the hock. 1, fibular tarsal; 2, tibial tarsal; 3, third tarsal; 4, fourth tarsal; 5, first tarsal fused with second tarsal; 6, central tarsal; 7, tuber calcis; 8, trochlea; 9, second metatarsal; 10, fourth metatarsal; 11, large metatarsal; 12, tarsal groove for deep flexor tendon; 13, tibia; 14, long medial collateral ligament; 15, short medial ligaments; 16, dorsal tarsal ligaments; 17, plantar ligament. (From Smythe and Goody, 1975.)

and is cartilaginous in part. It gives rise to the subtarsal *check ligament*, which unites with the deep flexor tendon.

The other joints are similar to those of the foreleg.

Foot

The foot is defined as the hoof wall and all the structures it contains. The external structures of the foot are illustrated in Figure 19-9, and the internal structures are illustrated in Figure 19-3. The forefoot is almost round in shape. The medial wall is usually not so round and tends to be somewhat straight compared with the lateral side. The sole is slightly concave medial to lateral and anterior to posterior. The shape of the hind foot is slightly different from that of the forefoot. The toe of the hind foot is more pointed and the sole is more concave.

Hoof Wall and Laminae   The bulk of the hoof wall is composed of epithelia cells that have been keratinized. The cells are arranged in tubules that extend from the coronary band to the bottom surface of the hoof wall. The relationship of the tubules to the coronary band is illustrated in Figure

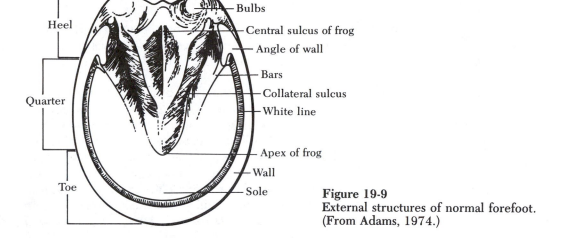

Figure 19-9
External structures of normal forefoot.
(From Adams, 1974.)

19-10. Papillae projecting from the coronary corium form the tubules and are responsible for growth of the hoof wall. Each tubule is filled with cells that absorb and conduct water. *Intertubular horn*, a cementlike substance of loosely packed cells that holds the tubules together, is formed between the papillae. To aid in preventing evaporative water loss from the hoof wall, the

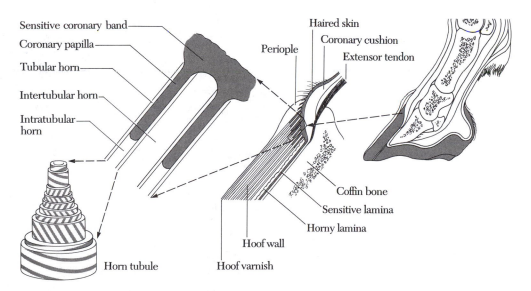

Figure 19-10
Magnified diagrammatic section of papillae of the coronary band and horn tubules of the hoof wall. (From Butler, 1974.)

tubules toward the outer surface are flattened and packed closely together (Figure 19-11).

The cells of the intertubular horn that are located next to the outer surface of the hoof wall contain pigment and are responsible for the color of the hoof wall. The same cells adjacent to the white line do not contain pigment.

The outer surface of the hoof wall is covered by the periople and the stratum tectorium. The *periople* extends below the coronary band for approximately three-quarters of an inch except at the heels, where it covers the bulbs and blends in with the frog. When the hoof wall is very dry, the periople may tend to curl up and is easily recognized. The periople serves to protect the junction of the hoof wall with the coronary band (Figure 19-10). As the hoof wall grows, the underside of the periople flakes off. The material that flakes off forms the *stratum tectorium*, which is a waxy substance. The stratum tectorium extends from the periople to the bottom of the hoof wall

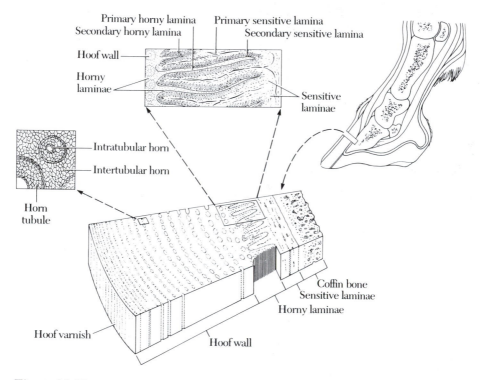

Figure 19-11
Diagrammatic representation of the cross section of the interlocking of the sensitive and horny laminae, showing the relative size of the horn tubules of the hoof wall. (From Butler, 1974.)

and aids in preventing evaporative water loss. Friction from dirt, grass, and so on usually wears off this thin layer of cells, so it is usually not present on the lower part of the hoof wall. Removal of the stratum tectorium and the outer layer of flattened tubules with the rasp during trimming or shoeing of the hoof allows the hoof wall to dry out at a faster rate than normal.

The inner layer of the hoof wall is the insensitive laminar layer. The insensitive laminar layer intermeshes with the sensitive laminae. The sensitive laminae are attached to the top surface of the third phalanx. The *sensitive* and *insensitive laminae* are composed of primary and secondary leaves (Figure 19-11). There are about 600 primary leaves in each foot. From each primary leaf extend about 100 secondary leaves. The primary and secondary sensitive and insensitive laminae intermesh in such a way that the hoof wall is held to the third phalanx. Therefore, the weight of the horse is not exerted on the sole surface, but is borne by the hoof wall. The third phalanx (major bone in the foot) is suspended from the inside of the hoof wall. The area where the laminae intermesh is commonly referred to as the *white line* (Figure 19-9). The white line is actually yellowish in color as a result of the combination of laminar structures and the interlaminar horn produced by papillae at the base of the sensitive laminae. The white line is often confused with the white (unpigmented) area of hoof-wall tubules adjacent to the white line.

The hoof wall is thickest at the toe and becomes thin at the quarters. At the angle of the wall, the wall is reflected forward to form the bars of the foot. The bars serve as a brace structure to prevent overexpansion of the hoof wall. The white line separates the hoof wall from the sole.

Sole   The composition of the sole is similar to that of the hoof wall. The *sensitive sole* or *sole corium* (Figure 19-3) covers the bottom of the third phalanx and is responsible for growth of the sole. The sole is composed of horn tubules, similar to those in the hoof wall, which are secreted by papillae. The sole is easily bruised, particularly when it bears the weight of the horse. To prevent the sole from bearing weight, it is concave and has self-limiting growth. Flakes of the sole slough off after the sole gets about as thick as the hoof wall. The sole varies in thickness from about ⅜ inch adjacent to the hoof wall to about ¼ inch at the heel area.

Frog   The frog is the elastic, wedge-shaped mass that occupies the area between the bars. The bottom surface is marked by a depression called the *central sulcus* or *cleft*. Internally, the frog stay is formed as a result of the wedge shape. The point of the frog toward the toe is referred to as the apex. Papillae projecting from the surface of the frog corium are responsible for the growth of the frog. The frog is an elastic structure because the horn tubules produced by the frog corium are not completely keratinized, moisture content remains at about 50 percent, and there are greasy secretions from fat glands that pass from the digital cushion into the frog. About every six months it will shed itself.

Coriums   There are five coriums of modified vascular tissue that furnish nutrition to the hoof. The *perioplic corium* lies within the perioplic groove in the coronary band and serves the periople. The *coronary corium*, responsible for the growth and nutrition of the bulk of the hoof wall, lies within the coronary band. Because the coronary band is responsible for wall growth, injury to the coronary band is quite serious and usually leads to a defect in wall growth and structure. The *laminar corium* is attached to the top of the third phalanx and bears the sensitive laminae. Consequently, it transports the blood supply and nutrition to the sensitive and insensitive laminae as well as to the white line. The *sole corium*, on the lower surface of the third phalanx, nourishes the sole. The frog is nourished by the frog corium.

Digital Cushion   The back half of the foot contains the digital, or *plantar cushion*. This fibro-elastic, fatty cushion acts as a shock absorber for the foot.

Bones   Three bones are located in the horse's foot and are discussed in the section on bones.

Lateral Cartilage   The lateral cartilages are attached to the wings of the third phalanx (Figure 19-3). They rise above the coronary band and extend approximately one-third of the way around each side of the hoof from the heel.

Muscles

The muscles of the horse's foreleg and hind leg are shown in Figures 19-12 to 19-14. These muscles are skeletal muscles; by their contraction and relaxation, the horse is able to move its limbs in a wide variety of movements, such as flexion, extension, adduction, abduction, protraction, and rotation. Muscles are usually classified by their major function of extension or flexion (Table 19-1). They are also classified by other systems. Muscles that work together to accomplish a movement are called *synergists* and those that work against each other are called *antagonists*. The synergists initiate a movement while antagonists limit the movement.

Each muscle is composed of muscle fibers (Figure 19-15). Each fiber is innervated by a nerve and is composed of contractile units. The contractile units are actin and myosin filaments. When a muscle functions by contracting to produce a movement, several controls must be exercised to get the desired movement. The range of movement is controlled by a *unit system*. A unit consists of several fibers controlled by branches of a common nerve

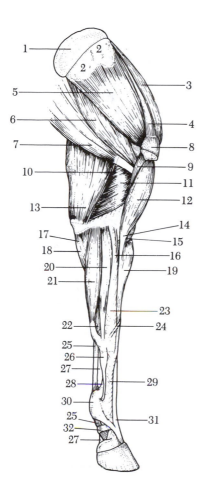

Figure 19-12
Muscles of the pectoral limb, medial view: 1, cartilage of scapula; 2, serrated face; 3, subclavius; 4, supraspinatus; 5, subscapularis; 6, teres major; 7, latissimus dorsi; 8, deep pectoral; 9, coracobrachialis; 10, long head of triceps branchii; 11, medial head of triceps brachii; 12, riceps brachii; 13, tensor faciae antebrachii; 14, lacertus fibrosus; 15, brachialis; 16, medial collateral ligament; 17, ulnar head of deep digital flexor; 18, ulnar head of flexor carpi ulnaris; 19, extensor carpi radialis; 20, flexor carpi radialis; 21, humeral head of flexor carpi ulnaris; 22, accessory ligament of superficial digital flexor muscle; 23, radius; 24, tendon of abductor pollicus longus; 25, tendon of superficial digital flexor; 26, metacarpal II; 27, tendon of deep digital flexor; 28, interosseous (suspensory ligament); 29, metacarpal III; 30, palmar annular ligament; 31, tendon of common digital extensor; 32, proximal digital annular ligament. (From Pasquini, Reddy, and Ratzlaff, 1978.)

fiber. Each muscle has hundreds of these units. Therefore, a certain number of units are simultaneously activated to produce a given degree of movement. *Duration*, or the time required for the movement, is controlled by the sequential contraction of different fibers or units. Because the elapsed time required for movement is usually longer than the time a fiber can contract, units take turns contracting and relaxing. *Power* is controlled by the number of units that contract simultaneously. As more power is required more fibers contract simultaneously. *Speed* of a movement is controlled by the rate at which the muscle fibers contract. The faster the nerves stimulate the fibers, the faster the muscle contracts.

The function of individual muscles will be discussed in the next section.

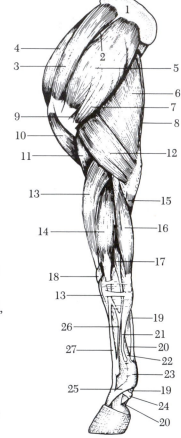

Figure 19-13
Muscles of the pectoral limb, lateral view: 1, cartilage of scapula; 2, tuber spinae; 3, supraspinatus; 4, subclavius; 5, infraspinatus; 6, long head of triceps brachii; 7, deltoideus; 8, tensor fasciae antebrachii; 9, teres minor; 10, biceps brachii; 11, brachialis; 12, lateral head of triceps brachii; 13, extensor carpi radialis; 14, common digital extensor; 15, ulnar head of deep digital flexor; 16, ulnaris lateralis; 17, lateral digital extensor; 18, abductor pollicis longus; 19, tendon of superficial digital flexor; 20, tendon of deep digital flexor; 21, metacarpal IV; 22, interosseous (suspensory ligament); 23, palmar annular ligament; 24, proximal digital annular ligament; 25, extensor branch of interosseous; 26, tendon of lateral digital extensor; 27, tendon of common digital extensor. (From Pasquini, Reddy and Ratzlaff, 1978.)

Nerve Supply

The *brachial plexus* is a large motor nerve and central nerve plexus for the shoulder and foreleg (Figure 19-16). It is formed by the ventral branches of the first and second thoracic and sixth, seventh, and eighth cervical nerves. It is located anterior to the first rib and between two parts of the ventral scalenus muscle. The plexus has eleven branches or main nerves. Six of these innervate the shoulder area: subscapular, suprascapular, anterior pectoral or thoracic, posterior pectorial or external thoracic, long thoracic and thoraco-dorsal muscles. Therefore, they innervate muscles that bind the foreleg to the body or shoulder muscles. The *suprascapular nerve* is significant because its anatomical location makes it readily susceptible to injury. It crosses the

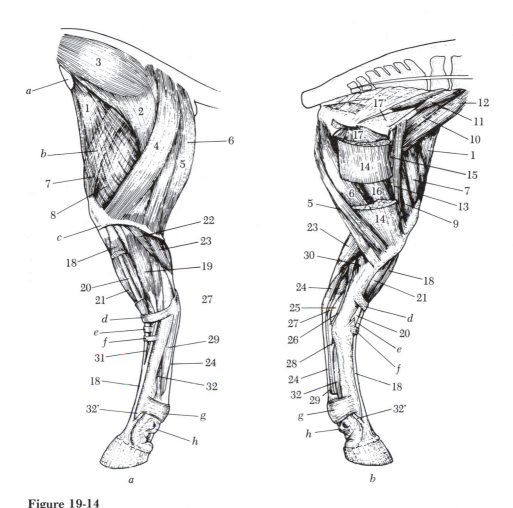

Figure 19-14
Muscles of the pelvic left limb: (*a*), lateral view; (*b*), medial view. 1, tensor fasciae
latae; 2, superficial gluteal; 3, middle gluteal; 4, biceps femoris; 5,
semitendinosus; 6, semimembranosus; 7–9, quadriceps femoris (vastus intermedius
not shown); 7, rectus femoris; 8, vastus lateralis; 9, vastus medialis; 10, psoas
major; 11, psoas minor; 12, iliacus; 13, sartorius; 14, gracilis; 15, pectineus; 16,
adductor; 17, internal obturator; 17′, iliac head; 18, long digital extensor; 19, lat.
digital extensor; 20, peroneus (fibularis) tertius; 21, cranial tibial; 22–23, triceps
surae; 22, soleus; 23, gastrocnemius; 24, superficial digital flexor; 25–27, deep
digital flexor; 25, superficial head (causal tibial); 26, medial head (long digital
flexor); 27, deep head (flexor digit I longus); 28, accessory ligament; 29, deep
digital flexor tendon; 30, popliteus; 31, short digital extensor; 32, interosseous
(suspensory ligament); 32′, extensor br.; *a*, tuber coxae; *b*, fasciae latae; *c*, crural
fascie; *d,e,f*, proximal, medial, and distal extensor retinacula; *g*, plantar annular
ligament; *h*, proximal digital annular ligament. (From Pasquini, Reddy, and
Ratzlaff, 1978.)

Table 19-1
Muscles: Function, innervation, and blood supply

Function[°]	Muscles	Innervation	Blood supply (arteries)
Suspensory muscles of foreleg	1. Pectoral group: Anterior and posterior superficial pectoral and anterior and posterior deep pectoral	Cranial, pectorals, thoraco-dorsal, long thoracic of the brachial plexus, and posterior cervical	Inferior cervical, carotid, vertebral, internal and external thoracic, thoraco-dorsal, intercostals, and deep cervical
	2. Serratus ventralis: Cervical and thoracic portions		
	3. Trapezius: Cervical and thoracic portions		
	4. Rhomboideus: Cervical and thoracic portions		
	5. Latissimus dorsi		
	6. Brachiocephalicus complex		
Shoulder flexors	1. Lateral side: (a) Deltoideus, (b) infraspinatus, and (c) teres minor	a,c) Axillary b) Suprascapular	a,b,c,) Subscapular
	2. Medial side: (a) Subscapularis, (b) teres major, (c) coraco-brachialis, (d) capsularis	a) Subscapular b) Axillary c) Musculo-cutaneous	a,b) Subscapular c) Anterior circumflex
Shoulder extensors	Supraspinatus	Suprascapular	Suprascapular and posterior circumflex
Elbow flexors	1. Biceps brachii 2. Brachialis	Musculo-cutaneous Median	Brachial Brachhial
Elbow extensors	1. Tensor fascia antibrachii 2. Triceps brachii 3. Anconeus	Radial Radial Radial	Subscapular, ulnar, deep brachial, and posterior circumflex of the humerus
Carpus flexors	1. Flexor carpi radialis 2. Flexor carpi ulnaris 3. Ulnaris lateralis	1,2) Ulnar and median 3) Radial	1,2,3) Median and interosseous

700

Carpus extensors	1. Extensor carpi radialis 2. Extensor carpi obliquus	1,2) Radial	1,2) Median and interosseous
Digital flexors	1. Superficial digital flexor 2. Deep digital flexor	1,2) Median and ulnar and radial	1,2) Median
Digital extensors	1. Common digital extensor 2. Lateral digital extensor	1,2) Radial	1,2) Radial and interosseous
Sublumbar—Pelvic flexors	1. Psoas minor 2. Quadratus lumborum	1,2) Thoracic and lumbar segmented	1,2) Ventral branches of thoracic and lumbar segmented, the circumflex iliac and some branches of femoral
Sublumbar—Hip flexors	1. Psoas major 2. Iliacus	1,2) Thoracic and lumbar segmented	1,2) Same as above
Hip lateral flexor	Tensor fascia lata	Anterior gluteal	1) Circumflex iliac
Hip medial flexor	1. Sartorius 2. Pectineus (adductor function) 3. Capsularis coxae	1) Saphenous 2) Obturator 3) ?	1,2) Femor and deep femoral arteries 3) ?
Hip lateral extensors	1. Superficial gluteal 2. Gluteus medius 3. Gluteus profundus 4. Biceps femoris 5. Semitendinosus 6. Semimembranosus 7. Piriformis	1–7) Anterior and posterior gluteal, sciatic, and peroneal	1–7) Gluteals, lumbars, deep femoral and posterior femoral arteries
Hip medial extensors (also adduct limb)	1. Quadratis femoris 2. Adductor (main function is adduction) 3. Obturator internis 4. Obturator externis (adductor function) 5. Gemellus 6. Gracilis (adductor function)	1–6) Sciatic and obturator (obturator in adductor of rear leg)	1–6) Femoral, deep femoral, obturator, and internal pubic
Stifle flexor	Popliteus	Tibial	Popliteal and posterior tibial

continued

701

Table 19-1
Muscles: Function, innervation, and blood supply — (*continued*)

Function[*]	Muscles	Innervation	Blood supply (arteries)
Stifle extensors	Quadriceps femoris consisting of rectus femoris, vastus mediatis, vastus intermedius and vastus lateralis	Femoral	Anterior femoral and femoral
Hock extensors	1. Gastrocnemius 2. Soleus 3. Tibialis posterior	1–3) Tibial	1–3) Popliteal
Hock flexors	1. Tibialis anterior 2. Peroneus tertius	1,2) Peroneal (fibular)	1,2) Anterior tibial
Digital flexors of hind leg	1. Superficial digital flexor 2. Deep digital flexor	1,2) Tibial	1,2) Posterior femoral, posterior tibial, and saphenous
Digital extensors of hind legs	1. Long digital extensor 2. Lateral digital extensor 3. Short digital extensor 4. Medial digital extensor	1–4) Peroneal	1–4) Anterior tibial

[*]All muscles are listed only under their primary functions.

702

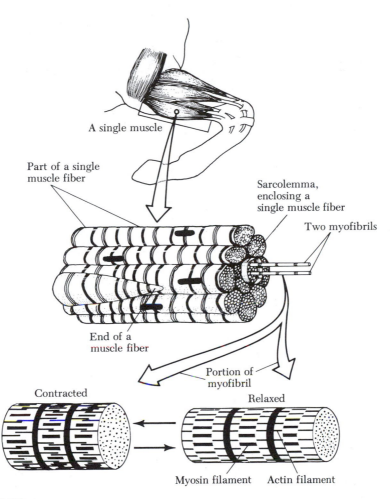

Figure 19-15
Basic anatomy of a muscle. (From *Practical Horseman,* July, 1979.)

spine of the scapula, and contusions to the scapula will cause paralysis of the nerve. This in turn causes disuse atrophy of the supraspinatus and infraspinatus muscles that results in the condition commonly referred to as "sweeney." These muscles play a role in shoulder flexion and extension but shoulder function is not greatly impaired when the horse has a sweeney. Several other muscles perform the same function. The remaining five nerves are distributed to the foreleg distal to the elbow. The *radial nerve* is the main extensor nerve since it innervates the extensor muscles of the elbow, carpal, and phalangeal joints. The *axillary nerve* cooperates with the radial nerve in

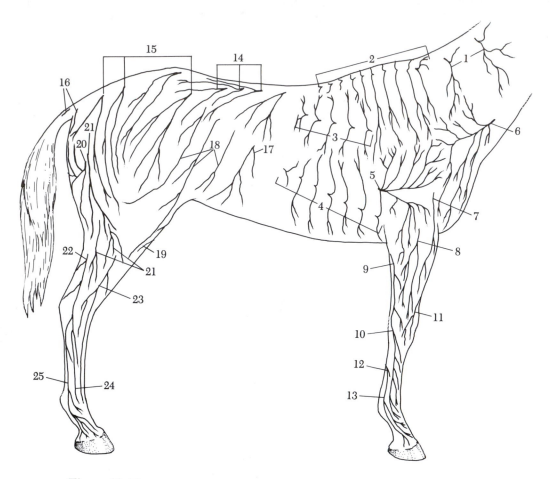

Figure 19-16
Superficial nerves in horse's body: 1, medial dorsal branches of the 7th and the 8th cervical nerves; 2, medial dorsal branches of the first 10 thoracic nerves; 3, lateral dorsal branches of thoracic nerves; 4, superficial banches of ventral thoracic nerves; 5, intercostobrachial nerve, which is the lateral cutaneous branch of the 2nd intercostal nerve; 6, lateral ventral branch of the 6th cervical nerve; 7, cranial cutaneous antebrachial branch of the axillary nerve; 8, lateral cutaneous antebrachial branch of the radial nerve; 9, caudal cutaneous antebrachial branch of the ulnar nerve; 10, dorsal branch of the ulnar nerve; 11, a branch of the medial cutaneous antebrachial from the musculocutaneous nerve; 12, anastomotic branch connecting the medial and the lateral volar metacarpal nerves; 13, lateral volar metacarpal nerve formed by the union above the carpus of the lateral branch of the medial nerve and the volar branch of the ulnar nerve; 14, cutaneous branches of lumbar nerves; 15, cutaneous branches of sacral nerves; 16, cutaneous branches of coccygeal nerves; 17, cutaneous branches of sixteenth and seventeenth thoracic nerves; 18, cutaneous branches of last thoracic and first and second lumbar nerves; 19, end of lateral cutaneous nerve of the thigh; 20, posterior cutaneous nerve of thigh; 21, cutaneous branches of great sciatic nerve; 22, posterior cutaneous nerve of the leg; 23, superficial peroneal nerve; 24, terminal part of deep peroneal nerve; 25, lateral plantar nerve.

that it innervates the flexors of the shoulder. The *ulnar* and *medial nerves* cooperate to activate the flexor muscles. The *musculocutaneous nerve* also plays a role in flexion since it supplies flexors of the elbow.

The digit is supplied by two dorsal and two volar nerves (Figure 19-17). The *dorsal nerves* are represented by dorsal branches of the terminal portions of the ulnar and median nerves. The *volnar nerves* are derived from the median and ulnar nerves.

The *lumbosacral plexus* supplies nerves to the hind leg (Figure 19-17). There are five major nerves of the plexus and three other important ones to serve the motor functions. The *sciatic nerve* is the largest nerve in the horse's body and is a composite of two important nerves in the hind limb, the peroneal and the tibial. The *tibial nerve* is prominent in activating the muscles to thrust the limb backwards by extending the hock and flexing the stifle and digit. It divides into the medial and lateral *plantar nerves*. They descend the cannon bone and form the two plantar proper digital nerves for the digit which innervate the deep structures of the foot. The *peroneal nerve* activates muscles that project the limb forward by flexing the hock and extending the digits. It divides into the superficial and deep branches. The superficial branch of the peroneal nerve innervates the hock, whereas the deep peroneal divides below the hock to form the two *deep dorsal metatarsal nerves*. The deep dorsal metatarsal forms the *dorsal proper digital nerves*,

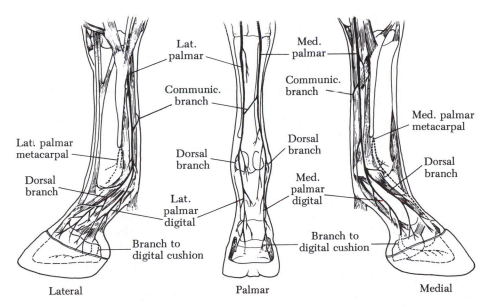

Figure 19-17
Nerves on the right forefoot. (From Sack and Habel, 1982.)

which innervate the sensitive corium. The *anterior gluteal nerve* innervates the gluteal muscles. The *posterior gluteal nerve* innervates the biceps femoris and gluteal muscles. The *obturator nerve* supplies the adductor muscles of the thigh. The fifth major nerve is the *femoral* which gives rise to another important nerve, the *saphenous*.

In addition to motor function, the nerves have a sensory function. In the foreleg, nerves from the brachial plexus supply the skin and superficial structures of the forearm (Figure 19-16). The medial aspect of the forearm is supplied by the median and *musculocutaneous nerve*. The posterior and posteriolateral aspects are supplied by the ulnar whereas the axillary and radial nerves supply the anteriolateral, anterior and anteromedial aspects. Sensory innervation of the knee and below is by the median, ulnar and radial nerves. In the hind leg, the posterior cutaneous nerve of the thigh, posterior cutaneous nerve of the leg and lateral cutaneous nerve of the leg supply the lateral and posterior aspects of the femoral, tibial, hock, and cannon areas (Figure 19-17). The anterolateral aspect of the thigh is supplied by the lateral cutaneous nerve of the thigh. The medial aspect of the tibial area and some of the medial aspects of the hock and cannon areas are supplied by the saphenous nerve.

Blood Supply

The names given to the blood vessels of the limbs vary depending upon the text of reference. Figure 19-18 shows the arrangement of the major arteries and veins of the forelegs. The *axillary artery* is the main artery; it gives off major branches to form the other major arteries. Eventually, after several branches are given off, it becomes the *common digital artery*, which supplies the phalanges and foot. At the level of the fetlock, it divides into the *lateral and medial digital arteries*. Several branches are given off before they unite to form the terminal arch within the semicircular canal of the third phalanx.

In the hind leg, the *femoral artery* is the main arterial source. As it passes down the leg, it branches and divides into several important arteries as shown in Figure 19-19. The major artery below the hock is the *metatarsal artery*. Prior to giving off several branches to supply the phalanges and foot, it reunites with the superficial and deep *plantar metatarsal arteries* to form the *distal plantar arch*. Vessels from this area are the lateral and medial digital arteries. They anastomose in the semilunar canal in the third phalanx.

The veins draining the foreleg are shown in Figure 19-20 and those draining the hind limb are shown in Figure 19-21. Veins in the limbs have valves above the knees and hocks to assist in returning blood to the heart. The massaging effect of contraction and relaxation of skeletal muscles also assists in returning the blood to the heart. One of the reasons that fluid

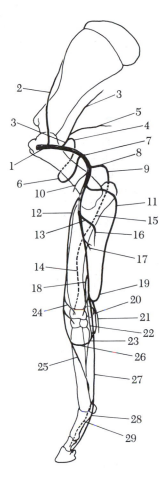

Figure 19-18
Arterial supply to forelimb: 1, axillary; 2, suprascapular; 3, subscapular; 4, caudal circumflex humeral; 5, thoracodorsal; 6, cranial circumflex humeral; 7, brachial; 8, deep brachial; 9, collateral radial; 10, bicipital; 11, collateral ulnar; 12, median cubital; 13, common interosseous; 14, cranial interosseous; 15, recurrent interosseous; 16, caudal interosseous; 17, deep antebrachial; 18, proximal radial; 19, palmar branch of median; 20, deep branch of #19; 21, superficial branch of #19 (continues down as palmar common digital artery III); 22, radial; 23, deep palmar arch; 24, rete carpi dorsale; 25, medial dorsal metacarpal (dorsal metacarpal artery II); 26, anastomotic branch of radial to medial dorsal metacarpal; 27, medial palmar (palmar common digital artery II); 28, medial digital; 29, lateral digital. (From Pasquini, Reddy, and Ratzlaff, 1979.)

accumulates below knees and hocks ("stocking up") when horses stand for long periods is the absence of valves in those areas.

19.2   Function

Muscles

When the horse is standing still, it is able to relax such that there is little fatigue. Except for a few minutes each day when it is in deep sleep, the horse can remain upright. If necessary, it can remain upright for several days

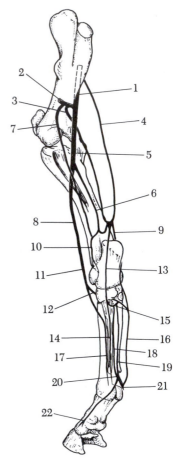

Figure 19-19
Arteries of the left hind leg, caudolateral view: 1, femoral; 2, caudal femoral; 3, descending branch of caudal femoral; 4, saphenous; 5, popliteal; 6, caudal tibial; 7, anastomotic branch of caudal lateral malleolar artery to caudal femoral artery (former recurrent tarsal artery); 8, cranial tibial; 9, medial plantar; 10, caudal lateral malleolar; 11, dorsal pedial; 12, perforating tarsal; 13, lateral plantar; 14, lateral plantar (plantar common digital artery III); 15, deep plantar arch; 16, medial plantar (plantar common digital artery II); 17, dorsal metatarsal artery III; 18, lateral plantar metatarsal (plantar metatarsal artery III); 19, medial plantar metatarsal (plantar metatarsal artery II); 20, perforating branch; 21, medial plantar digital; 22, lateral plantar digital.

before it lies down. It can "rest" in the upright position because of the "stay mechanism" of the forelegs and hind legs. The joints are locked in position by a system of muscles and ligaments (Figures 19-22 and 19-23). Part of the system functions the same in the forelegs and hind legs. The *suspensory ligament* passes down the back of the cannon bone, divides into two branches and inserts onto the proximal sesamoids. Ligaments from the proximal sesamoids go to the first and second phalanges. The main function of this system is to prevent excessive overextension of the fetlock joint. It also ties down the common extensor tendon by passing to the front of the phalanges. This suspensory ligament is the first to tighten as weight is passed to the fetlock. Next, the superficial tendon tightens, and finally the deep flexor tendon. The superficial and deep flexor tendons are the second and third components of the forelegs and hind legs. As they reach their limits of extension, the check

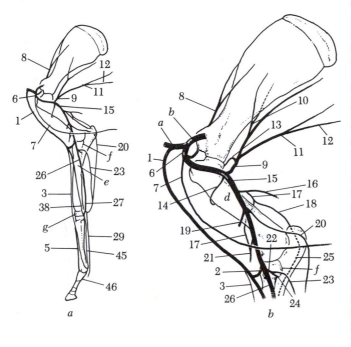

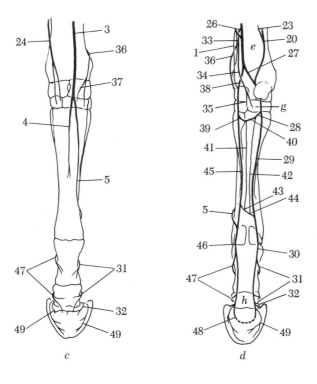

Figure 19-20
Veins in (*a*) right forelimb, medial view; (*b*) right shoulder and arm, medial view; (*c*) right manus, dorsal view; and (*d*) right manus, palmar view. 1–5, superficial veins: 1, cephalic; 2, medial cubital; 3, accessory cephalic; 4, lateral branch of 3; 5, medial branch of 3. 6–51, deep veins: 6, axillary; 7, external thoracic; 8, suprascapular; 9, subscapular; 10, circumflex scapular; 11, thoracodorsal; 12, superficial thoracic (spur vein); 13, caudal circumflex humeral; 14, cranial circumflex humeral; 15, brachial; 16, deep brachial; 17, collateral radial; 18, middle collateral; 19, bicipital; 20, collateral ulnar; 21, transverse cubital; 22, common interosseous; 23, caudal interosseous; 25, cranial interosseous; 26, medial; 27, palmar branch of 26; 28, deep branch of 27; 29, palmar common digital vein III (superficial branch of 27); 30, lateral palmar (proper) digital vein III; 31, dorsal branches of 30; 32, coronary; 33, proximal radial; 34, palmar branch of 33; 35, rete carpi palmare; 36, dorsal branch of 33; 37, rete carpi dorsale; 38, radial; 39, deep branch of 38 (anastomoses with 28 to form 40); 40; proximal deep palmar arch; 41, palmar metacarpal vein II; 42, palmar metacarpal vein III; 43, distal deep palmar arch; 44, superficial palmar arch (anastomotic branch from 43 to 29 & 45); 45, palmar common digital vein II (superficial branch of 38); 46, medial palmar (proper) digital vein III; 47, dorsal branches of 46; 48, terminal arch; 49, venous plexus; *a*, external jugular vein; *b*, subclavian vein; *c*, scapula; *d*, humerus; *e*, radius; *f*, antebrachial interosseus space; *g*, carpus; *h*, digit. (From Pasquini, Reddy, and Ratzlaff, 1979.)

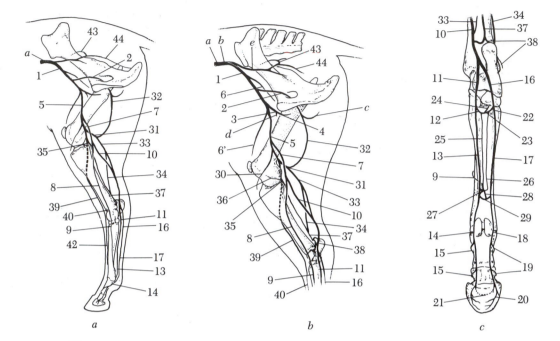

Figure 19-21

Venous drainage of (*a*) the right hindleg, medial view; (*b*) the hip and thigh, medial view; and (*c*) the right pes, plantar view. 1, external iliac; 2, obturator; 3, deep femoral; 4, medial circumflex femoral; 5, femoral; 6, ileacofemoral; 6', lateral circumflex femoral; 7, medial saphenous; 8, cranial branch of 7; 9, dorsal common digital vein II; 10, caudal branch of 7; 11, medial plantar; 12, deep branch of 11; 13, plantar common digital vein II (superficial branch of 11); 14, medial plantar proper digital vein III; 15, dorsal branches of 14; 16, lateral plantar; 17, plantar common digital vein III (superficial branch of 16); 18, lateral plantar proper digital vein III; 19, dorsal branches of 18; 20, venous plexuses; 21, terminal arch; 22, deep branch of 16; 23, proximal deep plantar arch; 24, perforating tarsal; 25, plantar metatarsal vein II; 26, plantar metatarsal vein III; 27, distal deep plantar arch; 28, anastomotic branch from 27 to 29; 29, superficial plantar arch; 30, descending genicular; 31, caudal femoral; 32, ascending branch of 31; 33, descending branch of 31; 34, lateral saphenous; 35, popliteal; 36, genicular; 37, caudal tibial; 38, lateral caudal malleola; 39, cranial tibial; 40, dorsal pedal; 41, dorsal metatarsal vein II; 42, cranial gluteal; 43, caudal gluteal; *a*, caudal vena cava; *b*, common iliac vein (lateral); *c*, middle vein of penis (clitoris); *d*, pudendoepigastric vein; *e*, internal iliac. (Adapted from Pasquini, Reddy and Katzlaff, 1978.)

Figure 19-22

Check and stay apparatuses of fore and hind limbs: 1, cervical part of rhomboid (romboideus cervicis); 2, cervical part of ventral serrate (serratus ventralis cervicis [levator angulae scapulae]); 3, thoracic part of ventral serrate (serratus ventralis thoracis [serratus magnus]); 4, supraspinatus; 5, biceps brachii; 6, long head of triceps (triceps brachii caput longum); 7, lateral head of triceps (triceps brachii

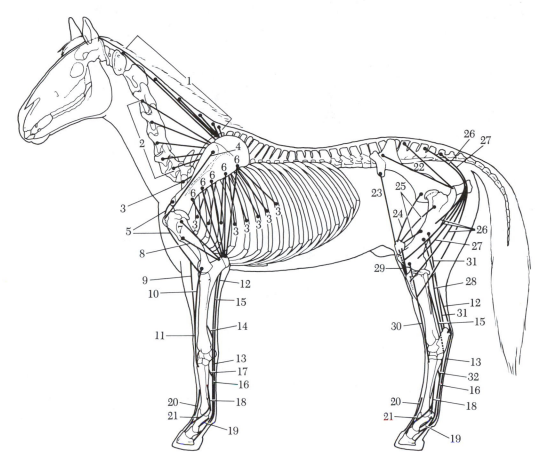

caput laterale); 8, medial head of triceps (triceps brachii caput mediale); 9, lacertus fibrosus; 10, radial carpa; extensor (extensor carpi radialis [extensor metacarpi magnus]); 11, conjoint tendon of radial carpal extensor and lacertus fibrosus; 12, superficial digital flexor (flexor digitorum superficialis [flexor perforatus]); 13, tendon of superficial digital flexor; 14, radial check ligament (superior check ligament [radial or tendinous head of superficial digital flexor muscle]); 15, deep digital flexor (flexor digitorum profundus [flexor perforans]); 16, tendon of deep digital flexor muscle; 17, carpa; check ligament (inferior check ligament [carpal or tendinous head of deep digital flexor muscle]); 18, suspensory ligament (superior sesamoidean ligament derived from the modified interosseous muscle); 19, distal sesamoidean ligament; 20, tendon of common digital extensor (extensor digitorum communis [extensor pedis]); 21, extensor branch of suspensory ligament attaching to common digital extensor tendon; 22, gluteal muscles; 23, tensor muscle of lateral fascia of thigh (tensor fasciae latae); 24, rectus femoris (one of the four parts of the quadriceps femoris); 25, vastus muscles (three parts of the quadriceps femoris); 26, biceps femoris (the most important of the three components of the hamstring group); 27, semitendinosus (one of the three components of the hamstring group); 28, accessory or tarsal tendon from the biceps femoris and semitendinosus of the hamstring group); 29, straight patellar ligaments; 30, peroneus tertius (tendinous or superficial portion of flexor metatarsi); 31, gastrocnemius (part of the triceps surae); 32, tarsal check ligament (tendinous head of deep digital flexor). (From Goody, 1976.)

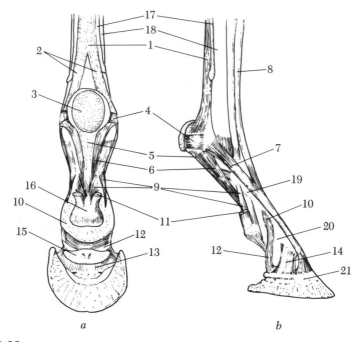

Figure 19-23
Left limb below the knee showing the suspensory apparatus, (a), posterior view; (b), lateral view. 1, suspensory ligament; 2, bifurcation of ligament; 3, intersesamoidean ligament; 4, collateral sesamoidean ligament; 5, straight distal sesamoidean ligament; 6, oblique distal sesamoidan ligament; 7, suspensory ligament branch to common extensor tendon; 8, common digital extensor tendon; 9, volar ligaments of pastern joint; 10, lateral ligaments of pastern joint; 11, cut stump of superficial digital flexor tendon; 12, suspensory ligament of navicular; 13, distal navicular ligament; 14, collateral ligament of coffin joint; 15, navicular bone; 16, fibrous plate; 17, small metacarpal; 18, large metacarpal; 19, first phalanx; 20, second phalanx; 21, third phalanx. (From Smythe and Goody, 1975.)

ligaments make the tendons function as ligaments by cutting off the muscular action above them.

Several muscles are components of the foreleg stay mechanism. The front of the body is suspended from the dorsal end of the scapula and this arrangement tends to flex the shoulder. The *biceps branchii* muscle prevents the shoulder flexion, provided it is prevented from flexing the elbow. This is accomplished by the deep and superficial *flexor muscles* and the *triceps muscle*. Contraction of the *supraspinatus muscle* also serves to extend the shoulder joint. The knee is prevented from bucking forward by the *lacertus fibrosus tendon*.

In the hind leg, the *peroneus tertius* counteracts flexion of the stifle but causes a flexion of the hock. The superficial digital flexor muscle counteracts the peroneus tertius muscle to stabilize the hock. The tendinous part of the *gastrocnemius* also helps stabilize the hock. For this system to function, the stifle must first be locked in an extended position. This is probably accomplished by the biceps femoris and semitendinosus muscles without the aid of the quadraceps femoris muscles.

When a horse moves forward, the muscles in the legs go through a series of precisely timed contractions and relaxations. These movements can be broken down into several general stages.

Stage 1: The initial movement in preparation to move the foreleg forward is a shift of the center of gravity backwards by a slight flexion of the hock. This transfers part of the weight borne by the foreleg to the hindleg.

Stage 2: The shoulder and elbow joints flex. Several muscles—*deltoid, teres minor,* and *teres major*—contract to flex the shoulder. The elbow joint flexes when the biceps and brachialis and the triceps relax. Flexion of the two joints moves the knee forward and upward. As the knee moves, contraction of the flexors at the back of the knee and distal joint flex the knee and distal joint. At the end of this stage of movement, the foreleg is off the ground and the foot is suspended from the withers through the dorsal scapular ligament and the tendinous middle part of the trapezius muscle.

Stage 3: The foreleg is swung forward by the brachiocephalic and thoracic part of the *serratus ventralis* muscles. This action is accomplished by the serratus ventralis muscle pulling the upper end of the scapula down and back while the brachiocephalic pulls the lower end of the humerus forward.

Stage 4: Toward the end of the forward movement, the leg is straightened again by muscles extending several joints: shoulder by supraspinatus muscle, elbow by triceps muscle, carpal and digital joints by carpal and digital extensors. At the last moment, contraction of the common digital extensor muscle causes an overextension of the coffin and pastern joints. This serves to increase the tension in the superficial and deep flexor tendons and their check ligaments.

Stage 5: The leg begins its backward movement. Relaxation of the common extensor muscle and the elastic rebound of the superficial and deep flexor tendons align the coffin and pastern joints in the proper position for contact of the foot with the ground.

Stage 6: The foot strikes the ground with the leg in a rigid position and the body starts to pass over the limb. The foot remains in contact with the ground and does not move. The shoulder and elbow are fixed by simultaneous contraction of the biceps and triceps muscles. This is necessary because forces exerted by the body weight acting through the thoracic part of the serratus ventralis muscle tend to pull the scapula back and down and the humerus vertical.

Stage 7: The propulsive phase begins after the limb has passed the vertical. Moving the body over the foreleg is accomplished by the contraction of several muscles. There muscles rotate the leg as a rigid frame around the muscular attachment of the scapula to the thorax. The humerus is pulled back by the *latissimus dorsi* and the major part of the deep pectoral muscles. The upper end of the scapula is rotated forward by the cervical part of the serratus ventralis, the *rhomboideus* and anterior deep pectoral muscles. The net effect of this stage is rotation of the body forward on the rigid leg (a lever) with the foot serving as the fulcrum. Propulsion occurs as a result of the limb retractor muscles and by the release of kinetic energy stored in several muscles as a result of the displacement of the scapula and humerus. Additional propulsive thrust is gained by the action of the fetlock joint during Stages 6 and 7. When the forefoot strikes the ground the fetlock moves downward. This downward movement is checked by several structures acting in sequence: suspensory ligament, superficial flexor tendon, deep flexor tendon, and lastly by check ligaments or the superficial and deep flexor tendons. Therefore, at the time of contact with the ground, the phalanges are straight with the leg. Then they become overextended as the leg becomes vertical, and recover as the leg passes the vertical. As the fetlock joint straightens out again, forward propulsive force is gained. Contraction of the deep flexor muscle pulls the third phalanx back with considerable force, which also imparts considerable propulsive action.

Stage 8: As the foot leaves the ground and the leg starts its forward swing, the natural elasticity of the suspensory ligament and stored kinetic energy in the displaced flexor tendons move the fetlock forward. The lower end of the humerus is pulled forward by the elastic rebound of the biceps tendon as the triceps relaxes and the tendinous thoracic part of the serratus ventralis as the rhomboideus and cervical part of the serratus ventralis muscles are relaxing.

The hind legs of the horse provide the main propulsive force for forward movement. This is accomplished by a slight upward movement of the forehand and hock and by flexion and extension of the stifle joint. The push involves an action that is similar to that of the foreleg. The hind leg is protracted, brought to the ground, and serves as a rigid frame until it is past the vertical and joint extension occurs. These movements can also be broken down into a series of several general stages:

Stage 1: Initially, the hind limb is moved forward by flexion of the hip joint by the psoas minor, iliacus, superficial gluteal, sartorius and tensor fasciae latae. This carries the femur and stifle forward. Movement of the femur is limited by the heavy thigh muscles and close attachment of the thigh to the horse's body. Simultaneously, the stifle is being directly flexed by the biceps femoris and semitendinosus muscles and indirectly by the tendinous part of the gastrocnemius muscles. As the stifle flexes, the hock flexes as a result of mechanical action of the peroneus tertius. Contraction of the tibialis anterior muscle also flexes the hock. The digit is flexed by pure mechanical action of the superficial flexor.

Stage 2: Toward the end of the forward moving stage and prior to the foot striking the ground, the joints start to extend to lock the hind leg in a rigid position. This is brought about by the action of the quadriceps femoris group of muscles on the stifle, the reciprocal action of the tendinous superficial flexor and tendinous cord of the gastrocnemius on the hock and the relaxation of the superficial flexor tendon of the digit. The common digital extensor muscle causes the pastern and coffin joints to overextend, as occurs in the forefeet.

Stage 3: Retraction (backward movement) begins with contraction of several muscle groups. The femur and tibia are pulled backward by the biceps femoris, semitendinosus, semimembranosus, and part of the gracilis, which all attach to the back of the femur and tibia. This movement is enhanced by the tarsal tendon of the biceps femoris and semitendinosus muscles pulling on the tuber calcis of the hock. The stifle joint is stabilized by these muscles and the insertion of the biceps femoris onto the patella plays a critical role. The reciprocal mechanism stabilizes the hock joint. The hip joint is extended by the pull of the middle gluteals on the head of the femur.

Stage 4: The hind leg is retracted and approaches a vertical position. Kinetic energy is stored as the stifle and hock flex slightly. The femur is moving toward a vertical position as the body moves over the rigid leg.

Stage 5: The leg passes the vertical position. The fetlock goes through the same changes as it does in the foreleg. Conversion of the potential energy back to kinetic energy aids in extension of the hock, stifle and fetlock joints. Contraction of the quadriceps femoris pulls on the patella extending the stifle joint. Muscles of the rump and hamstring groups extend the hip, stifle, and hock joints. Extension of the hock joint is aided by the reciprocal system, the gastrocnemius, deep flexor muscles, the tarsal tendon of the biceps femoris, and the semitendinosus. Active extension of the fetlock occurs as a result of contraction of the deep flexor muscle.

Stage 6: The hind leg becomes fully extended. All the leg joints are prevented from overextending by the tensor fasciae muscle.

Stage 7: The foot clears the ground and the leg starts forward. Semiflexion of the stifle and hock joints occur as a result of elastic rebound of the peroneus tertius and tibialis anterior muscles. The system then starts over again.

When a horse jumps, it must raise the forehand and then develop tremendous power in the hind quarters (Figure 19-24). Movement over the obstacle can be generalized as five stages:

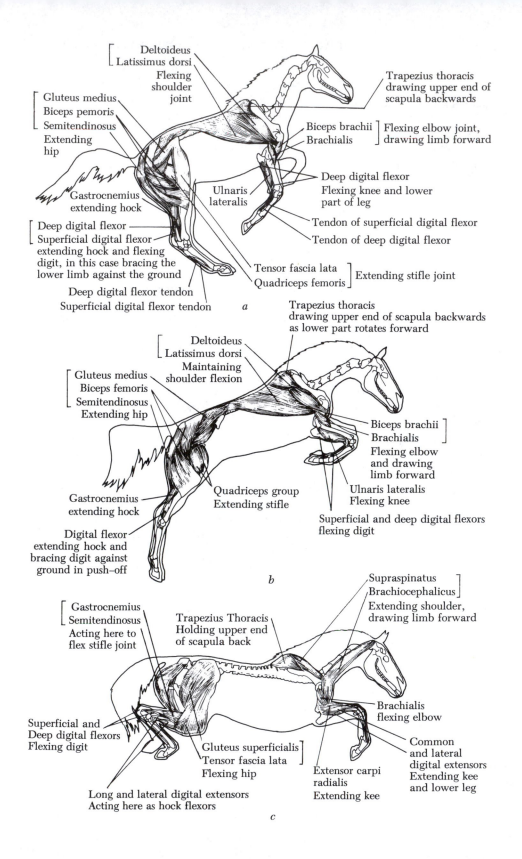

Deltoideus
Latissimus dorsi
Flexing
shoulder
joint

Trapezius thoracis
drawing upper end of
scapula backwards

Gluteus medius
Biceps pemoris
Semitendinosus
Extending
hip

Biceps brachii
Brachialis

Flexing elbow joint,
drawing limb forward

Gastrocnemius
extending hock

Deep digital flexor
Flexing knee and lower
part of leg

Ulnaris
lateralis

Deep digital flexor
Superficial digital flexor
extending hock and flexing
digit, in this case bracing the
lower limb against the ground

Tendon of superficial digital flexor

Tendon of deep digital flexor

Deep digital flexor tendon
Superficial digital flexor tendon

Tensor fascia lata
Quadriceps femoris

Extending stifle joint

a

Trapezius thoracis
drawing upper end of scapula backwards
as lower part rotates forward

Deltoideus
Latissimus dorsi
Maintaining
shoulder flexion

Gluteus medius
Biceps femoris
Semitendinosus
Extending hip

Biceps brachii
Brachialis
Flexing elbow
and drawing
limb forward

Ulnaris lateralis
Flexing knee

Gastrocnemius
extending hock

Quadriceps group
Extending stifle

Superficial and deep digital flexors
flexing digit

Digital flexor
extending hock and
bracing digit against
ground in push–off

b

Supraspinatus
Brachiocephalicus
Extending shoulder,
drawing limb forward

Gastrocnemius
Semitendinosus
Acting here to
flex stifle joint

Trapezius Thoracis
Holding upper end
of scapula back

Brachialis
flexing elbow

Superficial and
Deep digital flexors
Flexing digit

Gluteus superficialis
Tensor fascia lata
Flexing hip

Common
and lateral
digital extensors
Extending kee
and lower leg

Extensor carpi
radialis
Extending kee

Long and lateral digital extensors
Acting here as hock flexors

c

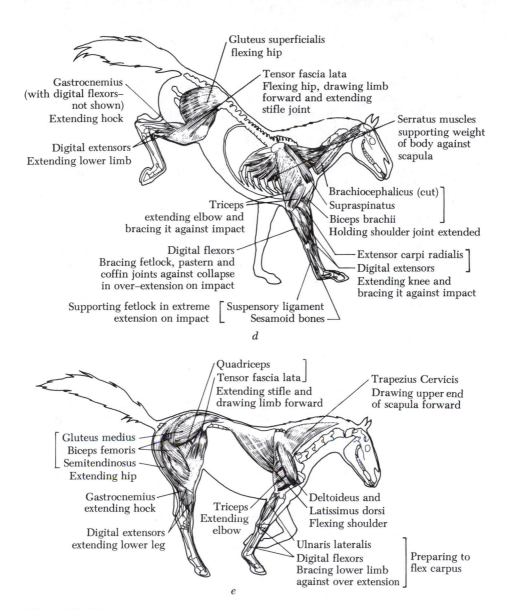

Gluteus superficialis
flexing hip

Tensor fascia lata
Flexing hip, drawing limb
forward and extending
stifle joint

Gastrocnemius
(with digital flexors–
not shown)
Extending hock

Serratus muscles
supporting weight
of body against
scapula

Digital extensors
Extending lower limb

Brachiocephalicus (cut) ⎤
Supraspinatus
Biceps brachii ⎦
Holding shoulder joint extended

Triceps
extending elbow and
bracing it against impact

Digital flexors
Bracing fetlock, pastern and
coffin joints against collapse
in over–extension on impact

Extensor carpi radialis ⎤
Digital extensors ⎦
Extending knee and
bracing it against impact

Supporting fetlock in extreme ⎡ Suspensory ligament
extension on impact ⎣ Sesamoid bones

d

Quadriceps ⎤
Tensor fascia lata ⎦
Extending stifle and
drawing limb forward

Trapezius Cervicis
Drawing upper end
of scapula forward

⎡ Gluteus medius
⎢ Biceps femoris
⎣ Semitendinosus
Extending hip

Gastrocnemius
extending hock

Triceps
Extending
elbow

Deltoideus and
Latissimus dorsi
Flexing shoulder

Digital extensors
extending lower leg

Ulnaris lateralis ⎤
Digital flexors
Bracing lower limb
against over extension ⎦

Preparing to
flex carpus

e

Figure 19-24
Major muscles involved in jumping an obstacle: (*a*), crouched for spring: The
forelegs are being flexed and drawn forward (protracted). The hind limbs have
been drawn forward (protracted). The hind limbs have been drawn under the
body, flexing the back, and they are braced in a fairly flexed position to begin
pushoff. (*b*), the spring: The forelegs are being flexed and drawn forward. The
spine is beginning to extend. (*c*), in the air: The forelegs are being drawn forward
and they are beginning to extend. The hind limbs are fully flexed and the spine is
in an extended position. (*d*), forelimb landing: The forelimbs are extended and
braced against impact. The spine is beginning to flex and the hind limbs are
extending. The weight of the body is entirely supported by the forelegs. (*e*), hind
limb landing: The hind limbs are reaching under the body, flexing the spine. The
serratus muscles still support the weight of the body.

717

Stage 1: The forehand is raised by extension of the foreleg and the center of gravity is shifted backward. The triceps, biceps branchii, supraspinatus, superficial digital flexor, and deep digital flexor muscles serve to extend the leg.

The center of gravity is shifted backward and the body lifted to the scapula by contraction of the serratus ventralis muscle and deep pectoral muscles. Contraction of the epaxial muscles in the back flexes the back and aids in raising the forehand.

Stage 2: Extension of the hips, stifles, hocks, and fetlocks occurs as previously discussed for movement of the hind leg. This system generates the propulsive power to lift the horse because it consists of long muscles acting on long levers. However, their action is slow compared with the speed of extension needed by the jumper. To gain the necessary speed of action, the middle gluteal and posterior parts of the adductor and gracilis muscles contract to move the femur. These are short muscles acting on short levers. The vastus muscles aid in stifle-joint extension and the rectus femoris muscle aids hip-joint extension.

Stage 3: The forelegs are flexed at the knee and elbow as previously described. Contraction of the brachiocephalic muscle aids in their flexion. The hind legs are fully extended.

Stage 4: The hocks and stifles are flexed to clear the hind feet over the jump. As they flex, the forelegs become extended in preparation for landing.

Stage 5: One foreleg strikes just prior to the other. After the second foreleg strikes, the first one moves forward so that the first hind leg can strike the same area.

Physiology of Foot Structure

The structures of the foot work together to absorb concussion when the foot strikes the ground (Figure 19-25). As the foot strikes the ground, the heels are expanded by frog action. The frog is pushed upward and, because of its wedge shape, the frog stay forces the digital cushion upward and outward on both sides of the foot. Movement of the digital cushion exerts pressure on the lateral cartilages, which in turn move outward. The lateral cartilages compress the blood veins draining the foot. Compression of the blood veins forces blood toward the heart and also causes blood to pool in the foot. The pooled blood then acts as a hydraulic cushion that absorbs concussion. To aid further in concussion absorption, the third phalanx descends slightly and the sole yields slightly. The laminae also help absorb shock. In addition, the navicular bone helps absorb concussion as a result of its placement in the joint. As the weight is transferred from the second phalanx to the navicular bone, the navicular bone yields slightly before the weight is transferred to

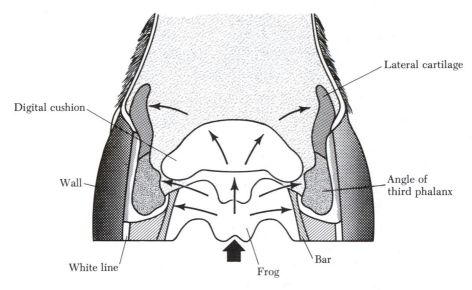

Figure 19-25
Cross section of the posterior aspect of the foot. When the foot takes weight the frog is compressed and expands. This results in pressure on the bars and digital cushion. These in turn press on the lateral cartilages which yield and is followed by expansion of the foot at the heels.

the third phalanx. The deep flexor tendon supports the navicular bone. The remainder of the concussion is absorbed at the pasterns, knee, and shoulder.

The shape of the hind foot is slightly different from that of the forefoot. The toe of the hind foot is more pointed and the sole is more concave.

19.3   Hoof Care

Proper foot care is one of the most important aspects of horsemanship. Lameness of the feet and legs is the cause of most incapacitations of horses. Neglect of a horse's feet can lead to lameness and improper foot action, and good foot care can hasten recovery from lameness or prevent it from occurring.

Cleaning

Horses that are kept in confinement—in stalls and small paddocks—should have their feet cleaned daily. A hoof pick is used to clean out the dirt and debris. The pick should be run from the heels toward the toe. If the pick is being run the opposite way and the horse jerks its foot away, it may bruise or puncture the grog or sole when the foot strikes the ground. The cleft and commissures of the frog should be carefully cleaned. Failure to clean these areas thoroughly is one of the main predisposing causes of thrush.

Daily Inspection

During cleaning, the foot should be inspected for rocks, nails, bruises, loose shoes (if shod), abnormal growth, uneven wear, and hoof condition (that is, cracks, splitting, and inadequate moisture). Close visual inspection may reveal a nail in the foot or a puncture wound. The nail should be removed and the depth of the puncture wound determined. If a veterinarian is not available, the wound should be cleaned out and packed with iodine. If the horse has not had a tetanus toxoid series and yearly boosters, it should be given a tetanus antitoxin shot. Adequate hoof moisture is important because dry hooves tend to contract and lose their physiological function. Moisture may enter a dry hoof from two routes. It may travel up the tubules forming the hoof wall and sole, or it may enter at the coronary band. To increase the entry at the coronary band, blood flow is increased by massaging the coronary-band area. To get moisture to enter from the ground surface of the hoof, horsepeople sometimes allow the water trough to overflow so there is moist ground around it. Other means include packing the feet with mud or wet clay and wrapping the feet in wet burlap sacks. To retain the moisture once it is regained or to prevent its evaporation, a hoof dressing or oil can be applied to the wall, heels, frog, and sole.

Trimming

Mature Horses   The hoof wall grows at the rate of ¼ to ½ inch per month. Therefore, the hooves should be trimmed or shod every 6 to 8 weeks. They may need to be trimmed more frequently if they are growing abnormally fast, such as after founder, or if they show uneven wear. Horses that wear the hoof wall unevenly usually have crooked feet or legs. The uneven wear of the hoof wall accentuates the faulty conformation and way of going. Keeping the hoof wall level will not correct the fault in conformation of mature horses, but it may prevent leg interference. Quite often corrective trimming only requires one side of the hoof wall to be rasped down until it is level with the opposite wall so that the hoof wall is balanced (Figure 19-26).

The outside wall is worn down at a faster rate than the inside wall of horses that are base-narrow and toe in or toe out. Consequently, the inside

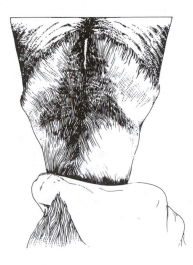

Figure 19-26
Foot balance. The opposite heels and opposite quarters should be equal in length. The hoof wall should be flat all the way around.

wall needs to be rasped level with the outside wall. Additional hoof wall should be removed if the wall is too long after it is leveled. Horses that are base-wide and toe out or toe in wear the inside wall at a faster rate. If the horse only toes out, the inside wall is worn at a faster rate, and if it toes in, the outside wall is worn at a faster rate.

The toe of the hoof wall grows at a faster rate than the heels. A long toe decreases the expansion of the wall when the hoof strikes the ground. Since the contraction forces in a hoof wall are greater than the expansion forces, the walls start to contract inward. Contracted heels cause excess pressure on the third phalanx and thus pain. Failure to trim the hooves when the toes become too long leads to contracted heels, faulty gaits, and limb interference. Long toes and low heels inhibit the rolling over of the foot during forward motion and may cause forging, overreaching, and scalping.

Foals   Many faulty positions of stance or incorrect ways of going in young foals can be helped or corrected by proper trimming of the feet. Corrective trimming of young foals is based on three principles.

First, adequate exercise is important if the foal is to be "straight-legged" when it matures. Support muscles increase in size and strength as the body size increases. However, when foals are kept in close confinement, the muscles required for shoulder propulsion and rotation are not used as much as they should be to stimulate normal development. Consequently, these muscles shrink and weaken as the foal grows, allowing the legs to become base-wide. Then, the feet become toed-out and the inside hoof wall is worn off because the foot rolls over it. To prevent this sequence of events, foals should be kept where they have room to run and play, or they should be vigorously exercised for a few minutes every day.

Second, pressure across the *growth plate* or *epiphysis* of the leg bones influences the rate of bone growth. Uneven distribution of pressure across the epiphyseal plate will cause improper bone growth. The objective of corrective trimming of a young foal's feet is to maintain an even distribution of pressure across the epiphysis and, when necessary, to change the distribution of pressure to stimulate and depress bone growth on the appropriate sides of the bone to straighten the leg. In the toe-wide stance, excess pressure is being applied to the lateral (toward the outside) side of the epiphysis and less pressure is applied to the medial (toward the center of the horse) side. Excess pressure decreases and less pressure increases bone growth, so the medial side of the bone grows at a faster rate than the lateral side. Consequently, the condition worsens if the feet are not correctively trimmed.

Third, corrective trimming must begin when the foal is a few days old and the hoof wall is hard enough to trim. All corrections must be accomplished before the epiphyseal plate closes for a particular bone. Radiographic data indicate that the epiphyses of the first and second phalanges are closed at approximately 9 months (Figure 19-27). The epiphysis at the distal end of the cannon bone closes at 9 to 12 months. Most of the remaining bones' epiphyses are closed at 1 to 2½ years.

Trimming Procedure   Trimming the feet requires a minimum set of tools. In addition to the hoof pick, a farrier's knife, rasp, and nippers (Figure 19-28) are required. (These tools are discussed in the next section on horseshoeing equipment.) After the foot is cleaned, the old shoes are removed (Figure 19-29) if the horse is shod. The excess dead sole is removed with the hoof knife (Figure 19-30). It is necessary to remove more sole from the toe area than from the quarter or heel areas. Removal of too much sole, to the thickness where it can be flexed by pressing on it with your thumbs, does not allow sufficient protection for the ground surface. If trimmed too close, the sole is quite sensitive, and stepping on small rocks or dirt clods causes bruises and pain and thus lameness. After sufficient dead sole is removed the hoof wall is cut with the nippers (Figure 19-30). The wall should be cut to within ¼ inch of the sole and should never be cut below the level of the sole. The wall is then rasped level with the coarse side of the rasp (Figure 19-30). After the wall is level, the outside edge of the wall is rounded to prevent splitting and chipping (Figure 19-30). The frog is trimmed level with the wall. The proper angle of the hoof wall (50 to 55 degrees) should be maintained and is best determined by the angle of the pastern. The angle of the forefeet should be the same as the angle of the shoulder; the angle of the hind feet is usually 2 to 3 degrees greater.

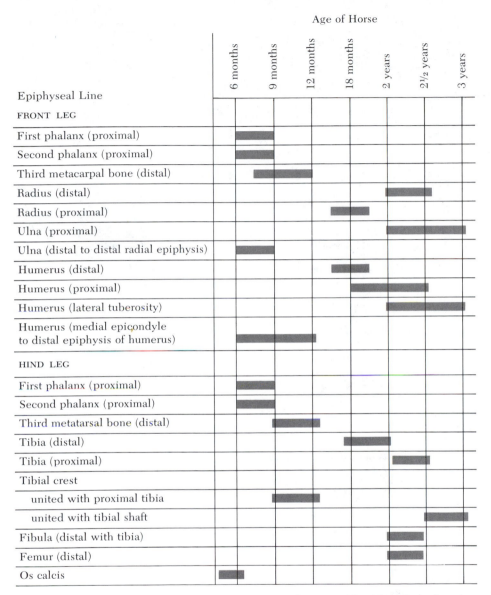

Age of epiphyseal closure based on Quarter Horse and Thoroughbred breeds, both male and female. The age of epiphyseal closure given here is based on radiographic determination, which will be earlier than actual closure determined histologically.

Age during which closure usually takes place.

Figure 19-27
Closure of epiphyseal plates. (From Adams, 1974.).

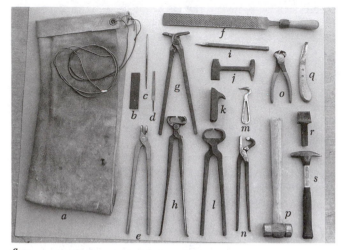

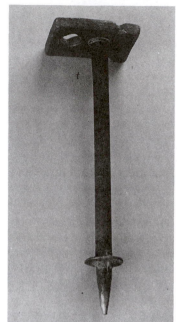

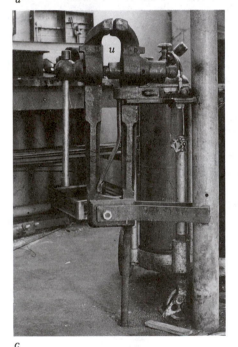

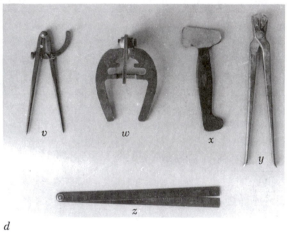

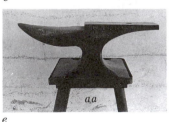

a

b

c

d

e

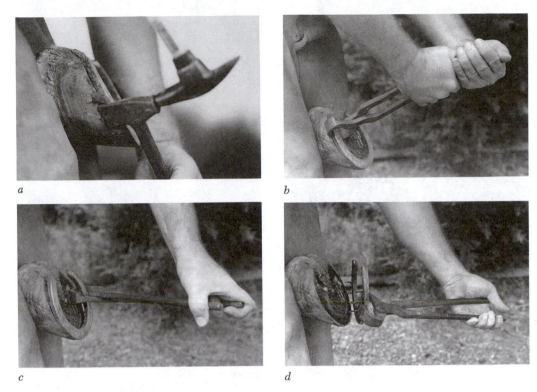

a

b

c

d

Figure 19-29
Removing old shoes: (*a*), clinches are cut; (*b*), pincers are inserted under one heel and rocked forward toward center of toe; (*c*), pincers are inserted under other heel and rocked forward toward center of toe; and (*d*), pincers are moved from side to side of the shoe and rocked forward until the shoe is freed. (Photographs by W. Evans.)

Figure 19-28 (opposite page)
Equipment for trimming feet and horseshoeing: *a*, leather apron; *b*, hone for hoof knife; *c*, round file for hoof knife; *d*, small flat file; *e*, farriers' tongs; *f*, tang rasp; *g*, hoof spreader; *h*, hoof nipper; *i*, pritchel; *j*, clinch cutter; *k*, clinch block; *l*, pinchers; *m*, hoof pick, *n*, nail or alligator clincher; *o*, clinch cutters; *p*, rounding hammer; *q*, farriers' knife; *r*, hardy; *s*, driving hammer; *t*, stall jack; *u*, boxleg vise, 5-inch jaw; *v*, divider; *w*, hoof leveler; *x*, sole knife; *y*, crease nail puller; *z*, metal ruler; *aa*, anvil on stand. (Photographs by W. Evans.)

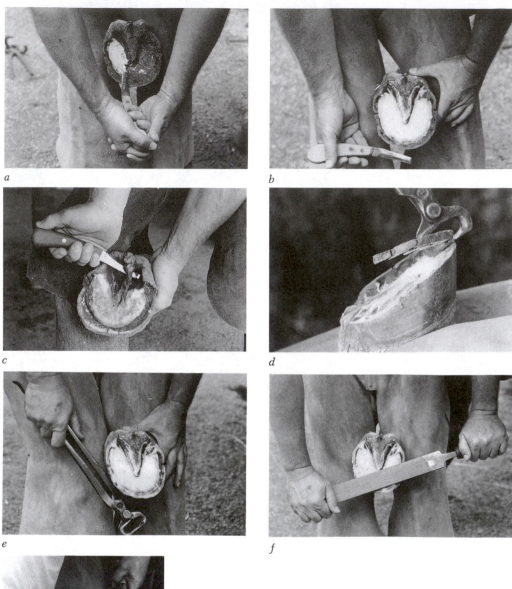

Figure 19-30
Preparing foot for shoeing or to let horse remain without shoes: (*a*), dead sole is removed with hoof knife; (*b*), all dead sole has been removed; (*c*), excess dead frog is removed; (*d*), excess hoof wall is removed with hoof nippers; (*e*), excess hoof wall has been removed; (*f*), hoof wall is rasped level and foot balanced with tang rasp; (*g*), if foot is not to be shod, the edge of the hoof wall is rounded off to prevent chipping. (Photographs by W. Evans.)

726

Shoeing

Horses are shod for various reasons: to increase traction of the feet; to protect the foot from breaking and wearing away at a rate greater than that of horn growth when the horse is kept or worked under circumstances that cause excessive wear of the hoof wall; to modify the action of the feet and legs in order to improve the execution of the gaits; and to improve or correct faults in gaits, such as scalping and overreaching. Shoeing has been termed a necessary evil because it does interfere with proper physiological functioning of the foot structure. Also, each nail destroys a number of horn fibers, thus weakening the hoof wall. In order to keep the interference of physiological function and damage to the hoof wall to a minimum, the farrier must have a working knowledge of the anatomy and physiology of the equine leg and foot and must know how to use his tools properly.

Horseshoeing is almost a lost art, but since about 1960 it has been revived as a result of the increase in the light-horse population. Proper horseshoeing requires considerable knowledge and manual dexterity. Each horse requires its shoes to be shaped or manufactured and fitted to each foot. The farrier must understand each movement of the feet and legs for the various gaits and how to influence foot and leg movement, and must have a knowledge of the therapeutic shoeing required by the various pathological conditions of the feet and legs. Shoeing is not a simple task that can be acquired during a short training period but requires continuing education by experience.

Equipment and Shoeing Aids   The basic horseshoeing tools are shown in Figure 19-28. The anvil is the farrier's workbench, and is used in many ways to prepare the shoe properly. Normally, the anvil weighs 80 to 125 pounds. The farrier's anvil has a thinner heel and a longer, more tapered horn than the blacksmith's anvil. The *horn* may have a chipping block to draw out clips. The *face* or *flat area* has a hole for the hardy and one or two pritchel holes. The *hardy* is used for cutting hot metals, such as the heels of the shoe. It is also used to cut blank bars to the proper length to be made into shoes. A *pritchel* is used to expand the nail holes in a shoe or to remove nails that are broken or cut off even with the web of the shoe. The *stall jack* is a handy tool used by farriers to shape aluminum plates when shoeing racehorses so that the foot does not have to be put down during the shoeing process.

The leather apron (Figure 19-28) protects the farrier against nail cuts and against heat when working at the forge. The tie strings of the apron should be long enough to be crossed behind and tucked under at the sides. When secured in this manner, the farrier may quickly get free by pulling on the strings if a horse hooks a nail in the apron and starts kicking. Many farriers sew a pocket on the apron to hold the hoof knife where it can be reached and to keep the blade from being damaged. The *hoof knife*, or farrier's knife, is used for removing dead sole scales, "tagging-up" the frog, and cutting out corns. During hot, dry weather, the sole may be too hard to cut with the farrier's knife, and a sole knife must be used. The *sole knife* is used in the same manner as a chisel. To protect the blade of the farrier's knife or chisel, the sole of the foot is cleaned of all rocks and dirt with a hoof pick.

The blade end of a *clinch cutter* or *buffer* is used to cut or straighten the clinches before removing a shoe from the foot. After the shoe is removed, the pritchel-shaped end is used to remove seated nails from the shoe after they are cut off.

Many types of hammers are used by farriers, but the rounding and driving hammers are essential. The *rounding hammer* is used for making and shaping shoes. The *driving hammer* is used for driving nails and forming and finishing the clinches. The claws of the driving hammer are used for wringing off the nails.

Cutting nippers have two sharp edges and are used for removing excess hoof wall. Some farriers prefer a *hoof parer*, which is similar to the nippers but has a blunt edge and a sharp edge. The farrier's pincers, commonly called the "pullers," are similar in shape to the cutting nippers but are used to pull shoes or nails from the foot and may be used to clinch the nails. Usually, a clinch block or an alligator clincher is used to set the clinches. If the pincers are used as a clinch block, they are closed, and the backside of the jaw is held against the nail.

The *tang rasp*, fitted with a wooden handle, is used to level the bearing surface of the foot and finish the clinches. The rough side is used for rasping the hoof and the smoother side is used for final leveling and for finishing the clinches. In the final leveling process, a hoof leveler is used to determine the angle of the wall and to determine if the foot is level. The length of the wall is usually checked with hoof calipers or a divider to make sure both feet are the same and are of the proper length.

A 5-inch vise is quite useful, particularly when filing the heels or turning calks.

A shoeing box (Figure 19-31) is used by most farriers to keep their tools organized and handy while working on the horse's feet.

Most competent farriers are skilled in the use of a forge to make and shape shoes. The coal-burning forge (Figure 19-32) is the most common and can be carried in the back of a pickup truck. Gas forges are being used in some farrier's shops. The gas forge has an advantage in that the temperature is constant and will not burn up a shoe that is left in the forge too long. Coal-burning forges require the use of a fire shovel to add coal, and a fire rake to remove clinkers (melting metal particles that bind burned coal into larger masses). Farrier's tongs (Figure 19-28) are used to handle the shoes in the forge and during shaping. A fullering iron (Figure 19-28) is necessary to make a crease in a hand-forged shoe or to repair damaged creases during

Figure 19-31
Shoeing box to carry the horseshoeing tools. (Photographs by W. Evans.)

shaping. If a forge is not available to shape shoes, a shoe spreader is used to spread the heels of the shoe (Figure 19-28).

In addition to the shoeing tools, some other equipment facilitates the shoeing of certain horses. It is essential that the farrier carry a halter and lead rope to use on horses when the owner is not present. In some areas where horses are kept together in large numbers and are not handled regularly, a catch rope is necessary. A 2-foot-long knee strap is handy to hold up a front foot of a problem horse. Often, a horse will stand if a twitch is applied to the muzzle. A couple of long soft ropes may be used to tie up a hind foot or tie down a vicious horse. If it is necessary to tie up a hind foot, a 12-inch strap with D-rings in each end will prevent a rope burn around the pastern (Figure 19-33). Some horses will kick, thus requiring the use of a hoof hook to pick up a hind foot.

If a person shoes horses all day, a 14-inch-high foot stand will prevent excessive back strain. However, such stands are not safe to use when shoeing a nervous horse and will teach a young horse to lean on the farrier. A foot stand is seldom used by a farrier who can properly handle a horse's foot.

Figure 19-32
Coal burning forge that is portable. (Photographs by W. Evans.)

Figure 19-33
A small strap with 2 D-rings
is placed around the pastern
to prevent rope burns.

Shoes

Many types, sizes, and weights of manufactured horseshoes are available, or the farrier may elect to make the shoe from an iron bar. A typical shoe and its parts are shown in Figure 19-34. The kind and weight of the shoe selected for a horse depends on the type of work to be performed. The size of the shoe to be used is determined by the size of the hoof, the position of the nail holes, and length of the heels of the shoe. The last nail hole on a front shoe should not be behind the widest part of the foot. If the nail is placed too far back, expansion of the heels will be inhibited. In the hind foot, the last nail may be a little further back. The branches (Figure 19-34) must be long enough to support the entire hoof wall. If they are too long on the front feet, a hind foot may overreach and pull the front shoe. If too short, the heels of the horse will grow and the heels of the shoe will rest on the sole. This will produce corns.

The most commonly used shoes are keg shoes. *Keg shoes* are pre-sized and often are fitted "cold." There are several types of keg shoes. The *saddle horse, western,* or *cowboy shoe* (Figure 19-35) is used on most pleasure horses and on working cow horses. The cowboy shoe is available in several sizes (Table 19-2) to cover the range of light horse hoof shapes. Standard Diamond brand shoe sizes range from the No. 000, which is 4⁷⁄₁₆ inches long and 3¹¹⁄₁₆ inches wide, to the No. 3, which is 6³⁄₁₆ inches long and 5⁹⁄₁₆ inches wide. The No. 1 shoe fits the average horse. Another size system is used by the Multi-Products Company. In this system the No. 3 is equivalent to the standard No. 00, the No. 6 is equivalent to the standard No. 1, and the No. 9 is equivalent to the standard No. 3. The No. 4 is slightly smaller than the standard No. 0 and the No. 5 is slightly larger than the standard No. 0. The No. 7 is slightly smaller and the No. 8 is slightly larger than the standard

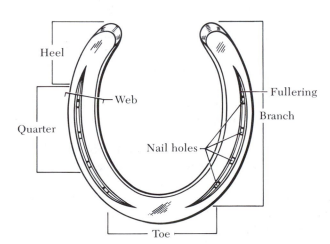

Figure 19-34
Ground surface view of a horseshoe.

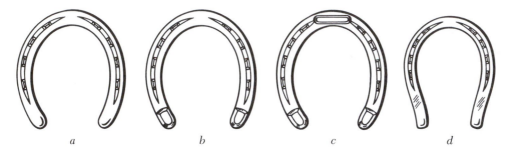

Figure 19-35
Cowboy shoes: (*a*), plain; (*b*), heeled (with heel calks), (*c*), heeled and toed (with heel and toe calks); (*d*), extra extra light (hot shoe).

No. 2. These so-called "half-sizes" almost completely eliminate the need to shape standard shoes to obtain a good fit. Pony shoes come in two sizes (Table 19-3). A No. 2 pony size is equivalent to the standard No. 000.

Cowboy shoes are made in several weights for each size (Table 19-2). They are classified as light, extra light, and extra extra light. A heavy shoe decreases speed and agility and increases leg fatigue. Therefore, most pleasure horses wear the extra light or extra extra light. Keg shoes that are fitted "cold" are extra light. They have the nail holes punched and the heels finished and are usually shaped cold and then nailed on the foot. The extra light shoes (Figure 19-35) are termed "hot shoes" and are not finished. Some need the nail holes punched and/or the heels cut off or used to make calks or any other special structure from the excess heel. They are shaped to the foot after being heated in a forge. Almost all shoeing requiring heavy-weight steel, such as on draft or gaited horses, requires hot shoeing. Cold shoes can also be heated for shaping if excessive reshaping of the shoe is required.

Light and extra light cowboy shoes are also classified as plain, heeled, or toed and heeled (Figure 19-35). The heeled shoes have heel calks, and the heel and toed shoes have heel and toe calks. Calks increase the traction on slick or soft surfaces and in rough terrain. The extra extra light shoes' heels are long enough so that heel calks can be formed without cutting off the heels. To form the calks, it is necessary to heat the shoe with a forge. If only toe calks, or "toe-grabs," are required, they must be welded to the toe.

Mule shoes are a variation of keg shoes and are similar to cowboy shoes. However, their sizes (Table 19-3) are slightly different, and they are shaped to fit a mule's feet (Figure 19-36).

Plates are a type of keg shoe that is used on racehorses. Steel plates are seldom used on running horses because the aluminum racing plate is lighter. The sizes and weights of steel plates are given in Table 19-4. They are

Table 19-2
Specifications for Diamond brand cowboy shoes

Size no.	Length (inches)	Width (inches)	Approx. weight (ounces)	Approx. no. in 50-lb carton
Light, plain				
000	4$\frac{7}{16}$	3$\frac{11}{16}$	7$\frac{1}{2}$	107
00	4$\frac{3}{4}$	4$\frac{9}{32}$	9	90
0	4$\frac{31}{32}$	4$\frac{5}{8}$	10$\frac{3}{4}$	78
1	5$\frac{5}{16}$	4$\frac{13}{16}$	12	66
2	5$\frac{21}{32}$	5$\frac{1}{8}$	14$\frac{1}{4}$	53
3	6$\frac{3}{16}$	5$\frac{9}{16}$	18$\frac{1}{4}$	44
Light, heeled only				
000	4$\frac{7}{16}$	3$\frac{11}{16}$	7$\frac{3}{4}$	103
00	4$\frac{3}{4}$	4$\frac{9}{32}$	9$\frac{3}{4}$	82
0	4$\frac{31}{32}$	4$\frac{5}{8}$	11	73
1	5$\frac{5}{16}$	4$\frac{13}{16}$	12$\frac{1}{2}$	62
2	5$\frac{21}{32}$	5$\frac{1}{8}$	15	53
3	6$\frac{3}{16}$	5$\frac{9}{16}$	19	42
4	6$\frac{21}{32}$	6	22$\frac{1}{2}$	35
Light, toed and heeled				
00	4$\frac{3}{4}$	4$\frac{9}{32}$	9$\frac{3}{4}$	79
0	4$\frac{31}{32}$	4$\frac{5}{8}$	11$\frac{1}{2}$	71
1	5$\frac{5}{16}$	4$\frac{13}{16}$	13$\frac{3}{4}$	61
2	5$\frac{21}{32}$	5$\frac{1}{8}$	16	49
3	6$\frac{3}{16}$	5$\frac{9}{16}$	19$\frac{1}{2}$	42
4	6$\frac{21}{32}$	6	22$\frac{3}{4}$	34
Extra light, plain or heeled				
S00	4$\frac{1}{2}$	4$\frac{1}{8}$	6$\frac{1}{4}$	128
S0	4$\frac{3}{4}$	4$\frac{9}{32}$	7	114
S1	5	4$\frac{21}{32}$	8$\frac{1}{4}$	97
S2	5$\frac{3}{8}$	4$\frac{13}{16}$	9	89
Extra extra light, (hot shoe)				
0	5$\frac{9}{16}$	4$\frac{9}{32}$	8	100
1	5$\frac{13}{16}$	4$\frac{5}{8}$	9	82
2	6$\frac{1}{16}$	4$\frac{25}{32}$	11	79

Table 19-3
Specifications for Diamond brand pony, draft-horse, and mule horse

Size no.	Length (inches)	Width (inches)	Approx. weight (ounces)	Approx. no. in 50-lb carton
Pony shoe				
0	3¾	3⁵⁄₁₆	5¾	139
1	4¹⁄₁₆	3⁷⁄₁₆	6¼	128
Plain mule shoe				
2	5⅛	4	10½	75
3	5½	4¼	12¾	62
Heeled mule shoe				
2	5⅛	4	11¼	70
3	5½	4½	13½	58
4	6	4½	16½	46
Draft-horse shoe				
6	9½	6½	41	(10/carton)
8	9⅞	7⅛	48	(10/carton)

available in two sizes, light or heavy, and can be obtained with a variety of combinations for toe and heel calks (Figure 19-37). Aluminum racing plates (Figures 19-38, 19-39) weigh 2 to 3 ounces. The front plates can be obtained plain, which means that there are no calks on the plate. A regular toe has a toe calk. If the toe calk is lower than the regular height, it is a low toe plate. Heel calks on the front plates are referred to as jar calks. The hind aluminum

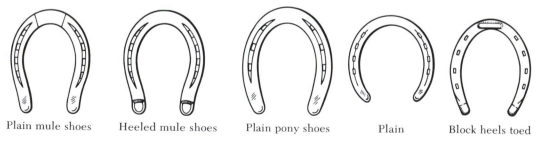

Plain mule shoes   Heeled mule shoes   Plain pony shoes   Plain   Block heels toed

Figure 19-36
Shapes of pony, draft-horse, and mule shoes (Diamond brand).

Table 19-4
Sizes and weights of Phoenix brand steel racing plates

Type	Size	Weight (oz)
Plain, light	00	3¾
	0	4
	1	4¼
	2	4½
	3	4¾
	4	5
Plain, heavy	00	4
	0	4¾
	1	5
	2	5¼
	3	5½
	4	5¾
Hind, toed, and block heels	00	5¼
	0	5½
	1	5¾
	2	6
	3	6¼
	4	6½

racing plates are also available with various toe and heel calk combinations. The heel calk on a hind plate is referred to as a block, or sticker, depending on its shape. The sticker, or "mud-calk," is set on the heel in such a way that it goes across the heel, whereas the block is set lengthwise on the heel. Blocks are used to cause a horse's hind feet to break over faster, thus preventing the track surface from burning the ergot area. Stickers are used to increase the traction on a muddy track surface.

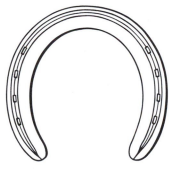

Plain

Block heels toed

Figure 19-37
Shape of steel racing (running) plates.

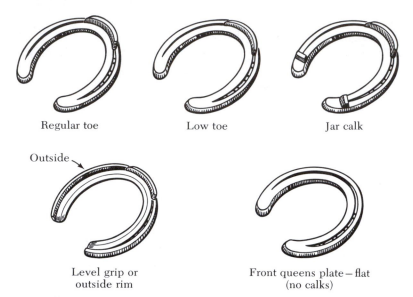

Regular toe   Low toe   Jar calk

Outside

Level grip or
outside rim

Front queens plate—flat
(no calks)

Figure 19-38
Aluminum racing plates for front feet.

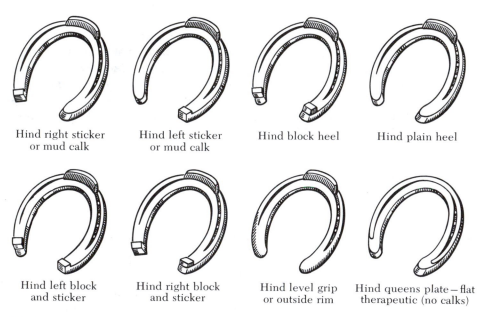

Hind right sticker
or mud calk

Hind left sticker
or mud calk

Hind block heel

Hind plain heel

Hind left block
and sticker

Hind right block
and sticker

Hind level grip
or outside rim

Hind queens plate—flat
therapeutic (no calks)

Figure 19-39
Aluminum racing plates for hind feet.

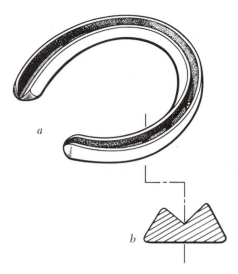

Figure 19-40
Polo shoe (*a*). Cross section (*b*) shows that inside rim is higher than outside rim.

    Polo shoes (Figure 19-40) differ from cowboy shoes in that the inside rim on the web is raised above the outside rim. This shape increases traction, prevents sliding, and enables the foot to roll over faster. Polo horses and western barrel-racing horses make a lot of turns at high speeds. The polo shoe allows the horse to pivot on the shoe and maintain a toe grip regardless of where the foot breaks over. The sizes and weights of polo shoes are given in Table 19-5.

    Toe and/or heel clips (Figure 19-41) can be drawn on shoes to help hold them in position. They are essential for horses that have weak hoof walls and whose hoofs do not hold a nail very well. Gaited and walking horses need toe and heel clips to aid in holding the extra heavy shoes. Horses that are required to make sudden stops and turns are difficult to keep shod unless toe and heel clips are added to prevent the hoof wall from sliding off the shoe

Table 19-5
Specifications of Phoenix brand polo shoes

Size	Weight (oz)	Width (in.)	Height (in.)
0	6½	4¼	4⅝
1	7	4⅝	5
2	7¾	4¾	5½
3	8½	5¼	5¾
4	9¼	5⅜	6¼

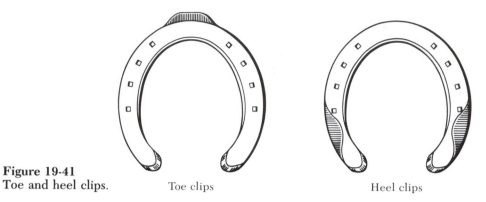

Figure 19-41
Toe and heel clips.   Toe clips   Heel clips

and tearing out the nail holes. They are also used to support the hoof wall in the areas of cracks.

Borium is applied to horseshoes with an oxyacetylene torch to improve the gripping properties and life of the shoe. The roughened surface of the applied borium increases the gripping properties of the shoe on ice, pavement, and dry grass. Borium (a metal alloy) is harder than any substance except diamonds, so it doubles or even triples the life of the shoe. Application of borium is particularly important before long trail rides over rough terrain; borium should also be applied to the shoes of pack horses and mules.

Nails   Horseshoe nails (Figure 19-42) are made so that one side of the shank is flat and the other side is concave, with a bevel near the point. The head of the nail is tapered and roughened on the same side as the bevel. Horseshoe

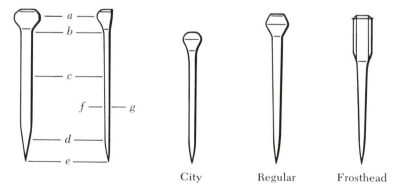

City   Regular   Frosthead

Figure 19-42
Horseshoe nails: parts (*left*) and shapes (*right*). Parts shown are: (*a*), head; (*b*), neck; (*c*), shank; (*d*), bevel; (*e*), point; (*f*), inner face; (*g*) outer face.

nails differ according to size, according to shape of the head, and by the manufacturer. The most commonly used sizes are 4, 5, and 6, but sizes range from 2½ to 12 (small to large). Some companies make a special nail for racing plates. In some areas, sharp beveled heads, called frostheads, are used to increase traction on icy surfaces. A pritchel is used to open the nail holes in a shoe before driving the nails (Figure 19-28). The type of head (regular, city, etc.) on the nail to use depends on the size of the nail hole and/or crease in the shoe. The head and shank of the nail needs to fit snugly in the hole and to have only the very top ¹⁄₁₆ inch of the head be above the surface of the shoe. Nail heads and shanks that do not fit snugly allow the shoe to work loose.

Holding the Foot

Often, the difference between a successful, highly skilled farrier and a mediocre one is the ability to handle a horse's feet correctly. The foot must be held solidly enough to work on without discomfort to either the farrier or the horse.

When the horse is approached, it should not be frightened or excited by any unnecessary noise or quick movements. A confident and deliberate approach usually gives confidence to the horse. Before picking up a foot, the farrier should make sure the horse is standing on all four feet and is aware of the farrier's intention to lift the foot.

When picking up the left front leg (near side) (Figure 19-43), the farrier stands facing the horse's rear quarters and places the left hand on the horse's shoulder. The hand is moved to the fetlock area while remaining in contact with the horse. Pushing the horse with the left shoulder, shifts the horse's weight to its right foot. The horse will lift is foot when the tendon is pinched just above the fetlock. The foot is then straddled and held just above the farrier's knee. If the farrier's toes are turned inward and the knees are flexed, the horse's foot will be held solidly.

The left hind leg is picked up (Figure 19-44) as follows: The left hand is placed on the horse's hip and the right hand is run down the horse's hind-quarters and leg to the pastern. When pressure is applied with the left hand, the horse will shift its weight to the right hind leg and its left hind leg can be picked up by pulling on the pastern. The leg is lifted up and forward and the farrier steps toward, and under, the leg until the cannon is perpendicular to the ground. The farriers hip is placed under the horse's hock so that the hoof can be supported just above the farrier's knees. If the farrier's knees are bent, the toes turned in, and the left arm placed over the hock, the horse's leg is locked into a steady position. The opposite procedure is used to lift the right hind leg.

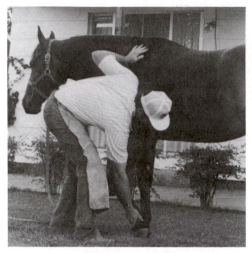

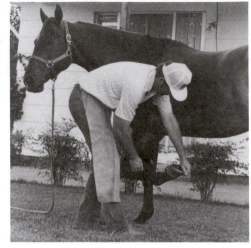

(a)

(b)

(c)

Figure 19-43
Picking up and holding front leg. (Photographs
by W. Evans.)

To hold up a horse's foot, the farrier must crouch (Figures 19-43 and
19-44). If the farrier straightens up, the foot is raised too high and the horse
is uncomfortable and takes its foot away. Injured horses feel increased pain
when their feet are held too high. Recognition of each horse's personality
makes it easier to hold its feet. Some horses tire easily so their feet cannot be
held too long before they start repeatedly to take them away from the
farrier. Others must have a stablemate nearby. Flies are irritating to most
horses. Horses that will not stand while being tied must be held or merely

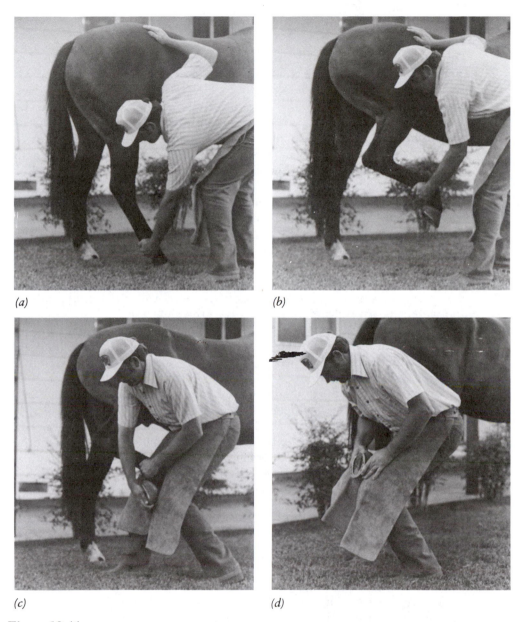

(a)   *(b)*

(c)   *(d)*

Figure 19-44
Picking up and holding hind leg. (Photographs by W. Evans.)

"ground tied." It is much safer for the farrier to have some knowledgeable person hold the horse while it is being shod. Horses are unpredictable and may fall back against the lead shank.

Shoeing the Normal Foot

The first step in the shoeing process (Figure 19-45) is to remove the shoes (Figure 19-29), if the horse is shod. The clinches are cut or straightened with a clinch cutter. Some farriers prefer to file the clinches off. The farrier's pincers, commonly called *pullers*, are inserted under one branch of the shoe toward the heel. The shoe is loosened slightly, and then the pullers are placed under the opposite branch. After the shoe is loose, the pullers are worked from side to side by applying pressure toward the toe and medially until the nails come out and the shoe can be removed. Any nails or stubs remaining in the hoof wall are removed. It is important for the horseman to know how to remove shoes. If a horse pulls a shoe part of the way off, the horseman must remove it to prevent further damage to the hoof wall. Shoes that work loose should be removed or the clinches tightened to prevent the horse from "throwing the shoe" and damaging the hoof wall.

Frequently, a horse's hooves have been permitted to grow too long before they are reshod. Failure to trim the hoof wall or reshoe the horse allows the heels of the shoe to slip inside the hoof wall. Because the sole is a non-weight-bearing structure, the heels of the shoe cause a bruise (corn) in the sole area between the bars and hoof wall. The horse is lame for a few days until it heals. Because the toe grows faster, the angle of the hoof wall decreases, and after 7 to 9 weeks, the change in the angle is sufficient to affect the gaits (see Section 4.4). Long toes retard the breaking over of the hoof, strain the deep flexor tendon (which may result in a "bowed tendon"), and may cause forging and overreaching.

After removal, the shoes are inspected for abnormal wear. Abnormal wear at specific points indicates that the foot is breaking over at a point other than at the toe, or that the foot is landing out of balance. The horse to be reshod or newly shod is then observed to see if it stands straight or crooked, and if the opposite feet are the same size. The angle of the foot should be the same as the angles of the pastern and the shoulder, (Figure 19-46). The horse must also be observed in action to detect lameness (see Section 4.3), faults in gaits, and improper foot action and flight. The overall body conformation should be observed to see if the horse has a short back, long legs, and a tendency to strike the forefeet with the hind feet. Based upon these observations, the farrier prepares the foot and selects the proper shoe.

To prepare the foot (Figure 19-30), the farrier thoroughly cleans it with a hoof pick and examines it for pathological conditions, such as corns or thrush. The frog is "tagged-up" with the knife to remove loose ends. To function properly, the frog must touch the ground after the horse is shod; therefore, it should never be trimmed out. The sole is trimmed with the hoof knife to remove the horn that is flaking away (Figure 19-30). Usually, it is necessary to remove more sole in the toe area than in the quarter or heel areas because the toe grows faster. The excess wall is removed with the

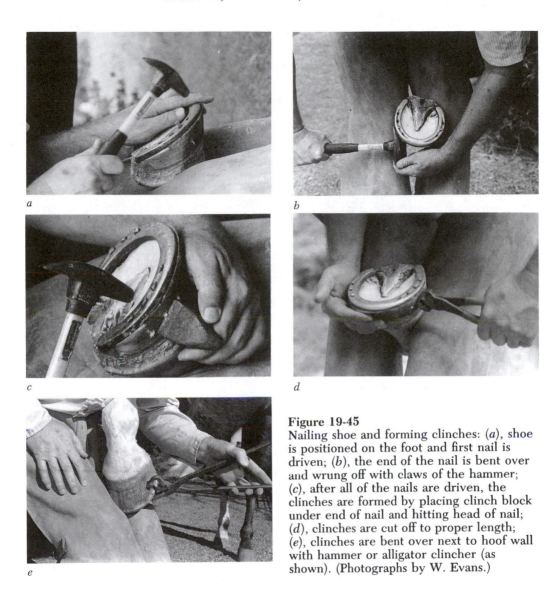

Figure 19-45
Nailing shoe and forming clinches: (a), shoe is positioned on the foot and first nail is driven; (b), the end of the nail is bent over and wrung off with claws of the hammer; (c), after all of the nails are driven, the clinches are formed by placing clinch block under end of nail and hitting head of nail; (d), clinches are cut off to proper length; (e), clinches are bent over next to hoof wall with hammer or alligator clincher (as shown). (Photographs by W. Evans.)

cutting nippers (Figure 19-30). It is customary to begin trimming the excess wall at one heel and to work around toward the other heel. Care must be taken so that the proper amount is removed and the wall remains level. The wall should never be cut lower than the sole. Since the bottom of the foot is concave, more wall will be left projecting above the sole at the quarter of the hoof wall. A common beginner's mistake is to remove too much wall at the quarters, or to trim the wall irregularly. To cut the wall as flat as possible, it

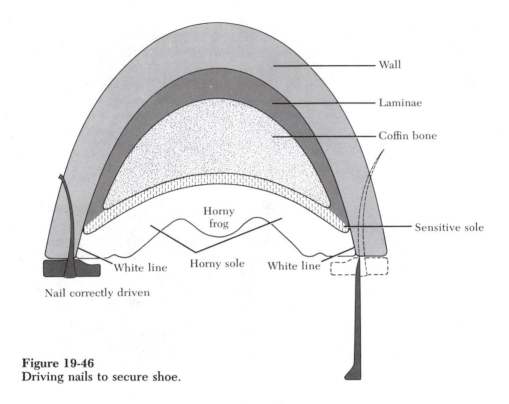

Figure 19-46
Driving nails to secure shoe.

is essential that the angle of the nippers to the wall be correct. The bars should be trimmed to the same level as the wall. If the bars are trimmed out, the heels tend to contract. The tang rasp is then used to rasp the bearing surface of the wall flat (to ensure uniform contact of the wall with the shoe), to smooth the bearing surface of the wall, and to make the final adjustment of the angle and length (Figure 19-30). The foot is then shaped to remove any excess outward flare of the hoof wall. After the foot is balanced and rounded off, the rasp is run around the edge of the wall to remove any burrs. The sole adjacent to the white line must be lowered slightly with the knife so that the shoe does not put pressure on it. If thrush or other diseases are present (see Chapter 16), the foot should be treated before applying the shoe. All cracks should also be treated to prevent further development or to eliminate hoof expansion in that area of the hoof wall before the shoe is applied.

Fitting the shoe is a critical step in shoeing. The shoe must be shaped to fit the hoof and not the reverse. The guideline for the fitting process is actually the white line, since the hoof wall may be distorted from a previous shoeing. The shoe may be fitted hot or cold, but if it is fitted hot, it should not be burned into place. The burned areas that appear on the wall after the hot shoe has been briefly applied indicate the high spots. It is then necessary to rasp the wall to remove them and level the wall. The shoe is centered on a normal foot by using the apex of the frog as a guide. When the shoe is centered, it fits the foot if the outer edge of the hoof wall is flush with the outer edge of the shoe except at the heels. The shoe should be $\frac{1}{16}$ inch wider than the wall on both sides at the heels. The proper length of the shoe is for it to extend $\frac{1}{4}$ inch beyond the heel, and the toe of the shoe is flush with the toe of the foot.

After the shoe has been shaped to the foot and centered, it is nailed. When driving the nails, the point of entry and direction of the nail are important (Figure 19-46). The nail should enter the outer edge of the white line, travel parallel with the horn fibers, and emerge approximately $\frac{3}{4}$ inch above the ground surface. The nail is started with the flat side outside. Light blows with the hammer will allow the nail to travel parallel with the fibers. At the desired depth, a hard blow will force the nail to emerge because of the inside bevel. When the nail is secure, the end is cut off or wrung off with the hammer claws (Figure 19-45). The nail head should extend approximately $\frac{1}{16}$ inch above the crease. If the nail head is too small and fits down into the crease, the shoe will become loose in a few days. If the horse is pricked or the nails are driven too close to the sensitive laminae, as evidenced by flinching, the horse will be lame. The nail should be taken out; if the horse was pricked, iodine should be poured into the opening and a tetanus antitoxin shot should be given.

There is a difference of opinion as to which nail should be driven first. It does not matter as long as the shoe remains centered on the foot. If the shoe moves slightly when the first nail is driven, it can be moved back by tapping it with the hammer. Ordinarily three nails on each side are sufficient to hold the shoe. If the horse makes a lot of sudden stops and turns, four nails on each side are used.

After all the nails are in, the nail heads must be seated in the crease, and the clinches formed. Either the farrier's pincers or a clinch block is held against the end of the nail, and the head of the nail is struck with the driving hammer (Figure 19-45). Finishing and smoothing is accomplished by filing a small groove under each clinch and then seating the clinch (bending nail over adjacent to hoof wall) with the driving hammer. Some farriers prefer to use clinchers (alligator or gooseneck styles) to form and set the clinches. The foot receives a final dressing or is finished off by a light rasping below the clinches if necessary. The hoof wall should not be rasped above the clinches, because if it is, the stratum tectorium (the outer layer of cells, which give the wall a glossy appearance) will be removed. Removal of the stratum tectorium will result in some loss of moisture from the hoof.

Final Inspection

After the horse is shod, the farrier and the owner or handler should inspect the work to ensure proper health and condition of the feet and legs (Figure 19-47). The following points should be noted:

1. The opposite feet should be trimmed to the same size; the toes should be the same length and the heels should be the same height.

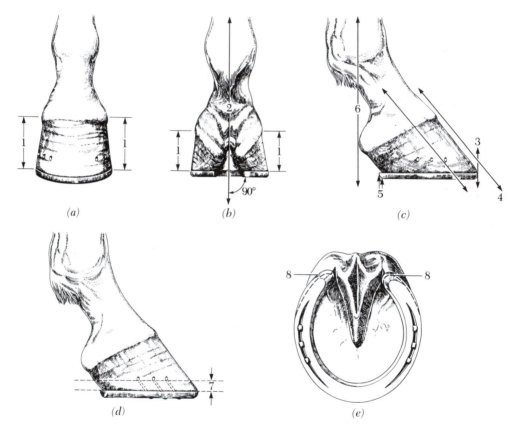

Figure 19-47
Properly shod foot: (*a*), front view: 1's are equal length; (*b*), back view; 1's are equal length, 2 is at right angles to ground surface; (*c*), at 3 shoe is even with edge of hoof wall. Heel of shoe extends up to ¼ inch (6.4 mm) beyond heel of foot at 5. Hoof wall has proper angle at 4 and is continuous with slope of pastern. Line 6 bisects leg, touches bulb of heel, and is perpendicular to the ground surface, (*d*), nails exit (7) ¾ to 1 inch (19.1 to 25.4 mm) above shoe and are in a straight line at 7. Drive nails parallel to tubules in hoof wall. (*e*), shoe extends beyond heels and extends about 1/14 inch (1.6 mm) from hoof wall at 8.

The toes should not be dubbed (rasped perpendicular to the ground). Flares of the hoof wall should be removed.

2. The foot should be balanced in relation to the leg and directly under the leg; that is, an excessive amount of wall should not be removed from one side of the foot, because this will result in an uneven distribution of weight in the foot.

3. The angle of the hoof should be an elongation of the angle of the pastern. This angle is usually the same as the angle of the shoulder. The most common angle is 50 to 55 degrees but depends upon the breed and use of the horse.

4. The nails and clinches should be evenly spaced and placed at the proper height. The clinches should be smooth and not project from the hoof surface. The last nail should not be too far back.

5. The shoe should be fitted to the foot so that excessive rasping on the wall is not necessary. Rasping above the clinches is not necessary. The shoe branches should be the proper length and set for heel expansions.

6. The frog and bars should not be trimmed out. If the ground surface is hard, they are trimmed so they do not touch the ground. This prevents bruising. The hoof should be in contact with the shoe so that there are no air spaces.

7. The horse should not be lame or have defects in gait caused by shoeing.

19.4   Corrective Trimming and Shoeing

To do corrective trimming and shoeing, a thorough understanding of the anatomy and physiology of the horse's foot and leg, as well as a knowledge of the flight of the foot and leg, is necessary. Several horses may have the same defect in their way of going, but each horse may require different corrective measures. Armed with a basic knowledge of the proper functioning of a foot and leg, the farrier is able to observe the horse in action and then determine the best corrective measure. A faulty gait in a mature animal that is caused by conformation cannot be permanently corrected by a farrier; it can at best be modified or improved.

Principles

The basic principle of corrective shoeing is to trim the foot level and then apply corrective shoes to improve the faulty gait. Drastic changes in the flight of the foot and leg can cause excess strain or pressure to be applied to other areas on the foot and leg (bones and structures). The new stresses can lead to pathological changes; thus, the least severe corrective measure should be utilized.

Procedure

Visual Inspection   During the visual inspection of the horse's way of traveling, any "break-over" and "landing" areas of the hoof wall are noted. Watching the horse in motion is important because although the following generalizations on conformation's effect on moving are generally true, exceptions that can only be seen with the horse in motion do exsit. In the base-narrow, toe-in fault in foreleg conformation, the foot breaks over and lands on the outside wall, and if the horse is unshod, the outside wall will be worn down. It is necessary to lower the inside wall to level the foot, and then several corrective measures can be used to force the foot to break over the center of the hoof. One of the simplest and mildest methods of correction is a square-toe shoe (Figure 19-48b). Other measures, such as an outside-half-rim (outside-rim) shoe (Figure 19-48a), calks (Figure 19-48d), and a calk at the first outside nail hole (Figure 19-48c), or a lateral toe extension (Figure 19-48e), encourage the foot to break over the toe. If the horse is base-narrow but toes out, the foot will break over the outside toe, wing inward, and land on the outside wall. Successful correction methods to encourage the foot to break over the toe include: outside-half-rim shoe (Figure 19-48a), square-toe shoe (Figure 19-48b), outside toe extension (Figure 19-48e), and half shoe on outside wall (Figure 19-48g).

Horses with a base-wide, toe-out conformation tend to break over and land on the inside toe. The fault is corrected by modifying the shoe to make it difficult for the foot to break over the inside toe but easier to break over the center of the toe. One of several modifications, such as heel calks (Figure 19-48d), a square toe (Figure 19-48b), or a bar across the break-over point (Figure 19-48h), will force the foot to break over the center of the toe. Base-wide horses that have their toes pointing the opposite way (that is, toe-in) land on the inside wall. To improve their way of going, the inside branch of the shoe must be raised (Figure 19-48a).

Cow-hocked horses tend to break over the inside toe and thus place a strain on the inside of the hock. This fault in conformation is particularly straining when the horse makes a sliding stop. The main objective in correcting the fault is to force the hind foot to break over the center of the toe by braking the inside of the hoof when it lands and rotating the toe inward. A lateral toe extension (Figure 19-48e) or a short inside trailer (Figure 19-48f) will usually accomplish the objective.

The stance of a sickle-hocked or camped-under horse can be improved by shortening the hind toes if they are too long. This correction will cause

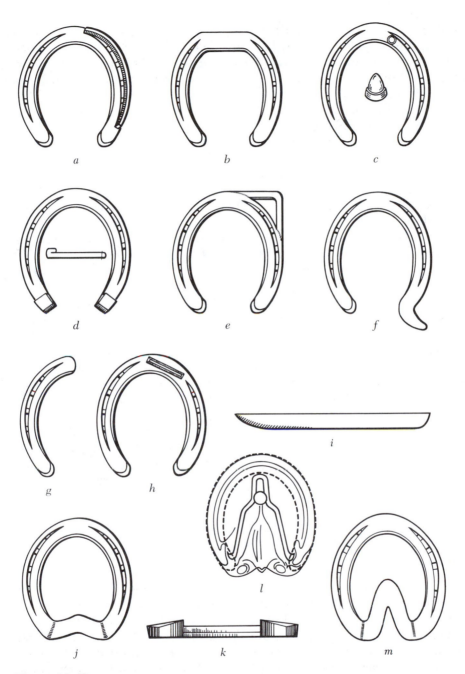

Figure 19-48
Corrective and therapeutic shoes: (*a*), half-rim shoe; (*b*), square toe; (*c*), calk at first outside nail hole; (*d*), heel calks; (*e*), lateral toe extension; (*f*), shoe with trailer; (*g*), half shoe; (*h*), bar across break-over point; (*i*), rolled toe; (*j*), bar shoe; (*k*), slippered heels; (*l*), Chadwick spring; (*m*), heart-bar shoe.

the hind legs to move backward. The opposite effect can be accomplished for the camped-behind horse by lowering its hind heels if they are too long.

Other less obvious faults in conformation may lead to faulty gaits. Forging and overreaching result when the hind foot breaks over faster than the forefoot in such a way that the toe of the hind foot strikes the sole of the heels of the forefoot. Horses that have short backs and long legs or short forelegs and long hind legs, or that stand under in front or behind have a tendency to forge and overreach. The corrective measure is to speed up the breaking over of the forefoot and to retard the breaking over of the hind foot and/or to increase the height of the forefoot flight. A variety of corrective measures can be used. The fault can be corrected in some horses by rolling the toes (and shoes, as in Figure 19-48i) of the forefeet and leaving the toe of the hind foot slightly longer than usual or placing heel calks (Figure 19-48c) on the hind shoes. In an extreme case, a ½-inch bar can be welded across the heel of the front shoes and across the toe of the hind shoes.

When the toe of the forefoot is striking the hairline of the hind foot ("scalping"), rolling the toe will enable the forefoot to break over faster and may prevent contact. The same corrective principle and objectives apply to horses that are scalping and/or shin-hitting as to those that forge and overreach.

Pacers that toe out in front and toe in behind have a tendency to cross fire. To prevent the hind foot from hitting the opposite forefoot, all feet must be shod to encourage breaking over the toe. Corrective measures include those previously discussed for correcting the front toed-out and hind toed-in conditions.

Interference usually results when a horse is base-wide or base-narrow and has a toed-out conformation in the forelimbs or is cow-hocked in the hind limbs. Correction of these conditions has been previously discussed and applies also to interference.

19.5   Therapeutic Shoeing

A farrier often encounters horses with pathological conditions that require therapeutic shoeing. The most common conditions are contracted heels, ringbone, sidebone, navicular disease, founder, and toe or quarter cracks.

There are several ways, or combinations of ways, to re-establish proper foot function in a horse with contracted heels. Frog pressure can be increased with a bar shoe so that the bar maintains constant pressure on the frog (Figure 19-48j). Other types of shoes include shoes with slippered (beveled) heels (Figure 19-48k) or a Chadwick spring (Figure 19-48l) to force the wall outward. The expansion process should be gradual, but additional expansion aids can be used; for example, cutting a horizontal groove below the coronary band, or cutting several vertical grooves in the quarter area, or thinning the wall at the quarters with a rasp.

Horses with ringbone have impaired or no action of the pastern and/or coffin joints. The objective of a corrective measure is to enable the foot to break over more easily so that the required action of the joint or joints is reduced. The simplest corrective measure is to roll the toe (Figure 19-48i). When sidebone is the problem, the action of the coffin joint needs to be decreased, so rolling the toe is helpful. To restore the ability of the foot to expand, and relieve pain if the horse is lame, the quarters of the wall are thinned.

The objective of corrective shoeing of a horse with navicular disease is to make it easier for the foot to break over and to reduce the anticoncussive activity of the deep flexor tendon against the navicular bone. By raising the heels (Figure 19-48c) and rolling the toe (Figure 19-48i), the foot will break over faster and concussion to the navicular bone will be reduced. To further reduce trauma to the area, a bar shoe (Figure 19-48j) or pads are used. This type of correction tends to contract the heels because frog pressure is reduced. The tendency for the heels to contract can be reduced by slippering the heels of the shoe (Figure 19-48k) so that they slope to the outside. This causes the wall to slide outward as the foot strikes the ground. The expansion of the wall can be further aided by thinning the quarters with a rasp.

When shoeing a horse with flat feet, the sole is trimmed slightly but the frog is not trimmed. The sole is prevented from dropping further by making sure that the shoe covers the entire wall and white line and covers but does not touch a small part of the outside edge of the sole. Pads are usually necessary to prevent sole bruising.

Toe, quarter, and heel cracks are encountered quite often. Each crack requires individual analysis with regard to the corrective treatment and shoeing. The general principle is to use a toe clip on each side of the crack to prevent wall expansion and to lower the wall under the crack so the wall will not bear weight. Plastics may be used to seal the crack and prevent expansion and contraction of the crack during foot action.

Founder or laminitis, an inflammation of the laminae of the foot, is one of the most common causes of lameness (see Chapters 4 and 8). There are two basic forms, acute and chronic. The acute case if often referred to as laminitis and the chronic case is often referred to as founder. Shoeing horses with laminitis or founder is based upon the principles of applying and relieving pressure. Pressure is relieved from the sole and laminae and applied slightly behind the point of the frog. The heart-bar shoe (Figure 19-48m) developed by Burney Chapman is designed so that pressure is applied to the frog area and the inner rim of the shoe is beveled to relieve pressure on the sole. The heart-bar shoe aids in preventing or stopping further rotation of the pedal bone, which is one of the primary objectives of therapeutic shoeing of the horse with laminitis or founder. Necrotic laminar tissue is removed by resectioning the hoof wall which allows the re-establishment of normal laminar relationships.

REFERENCES

Adams, O. R. 1974. *Lameness in Horses.* 3rd ed. Philadelphia: Lea and Febiger.

Anonymous. 1979. The horse movers. *Practical Horseman* (July):68–78.

Canfield, D. M. 1968. *Elements of Farrier Science.* Albert Lea, Minnesota: Enderes Tool Co., Inc.

Evans, L. H., J. Jenny, and C. W. Raker. 1966. The repair of hoof cracks with acrylic. *J. A. V. M. A.* 148:355–359.

Goody, P. C. 1976. *Horse Anatomy.* London: J. A. Allen.

Greeley, R. G. 1970. *The Art and Science of Horseshoeing.* Philadelphia: Lippincott.

Hickman, J. 1977. *Farriery: A Complete Illustrated Guide.* London: J. A. Allen.

Horowitz, A. 1965. Guide for the Laboratory Examination of the Anatomy of the Horse. Columbus: Ohio State University.

Julian, L. M. 1970. Functional Comparative Anatomy of the Domestic Animals. Davis: University of California School of Veterinary Medicine, Department of Anatomy.

Kays, D. J. 1969. *How to Shoe a Horse.* New York: A. S. Barnes.

O'Connor, J. T., and J. C. Briggs. 1971. Feet, conformation and motion. *The Morgan Horse* 364:27–32.

Pasquini, C., V. K. Reddy, and M. H. Ratzloff. 1978. *Atlas of Equine Anatomy.* Eureka, California: Sudz Publishing.

Sack, W. O., and R. E. Habel. 1982. *Rooney's Guide to Direction of the Horse.* Ithaca, N.Y.: Veterinary Textbooks.

Sisson, S., and J. D. Grossman. 1953. *Anatomy of the Domestic Animals.* 4th ed. Philadelphia: W. B. Saunders.

Smythe, R. H., and P. C. Goody. 1975. The Horse: Structure and Movement. London: J. A. Allen.

Springhall, J. A. 1964. *Elements of Horseshoeing.* Brisbane: University of Queensland Press.

Technical Manual TMA-220. 1941. *The Horseshoer.* Washington, D.C.: War Department. March 11.

Way, R. F., and D. G. Lee. 1965. The Anatomy of the Horse. Philadelphia: Lippincott.

Wiseman, R. F. 1968. *The Complete Horshoeing Guide.* Norman: University of Oklahoma Press.

CHAPTER 20

FENCES, BUILDINGS, AND EQUIPMENT

The usefulness of the horse lies in its athletic ability, and therefore facilities must be provided that enhance the performance of the animal. Any design of fences, buildings, and facilities must keep in mind the horse's basic nature. For example, fences or stalls that are adequate for cattle may be hazardous or otherwise unsatisfactory for horses. Because the purchase of a horse represents a considerable economic and emotional investment, the primary consideration for any horse facility should be *safety*. There are, of course, other important considerations, such as cost, durability, usefulness, flexibility, expandability, and accessibility. The intent of this chapter is not to provide specific recommendations concerning construction, for no such recommendations can be absolute, but rather to describe various ideas and plans that have been successful.

While the welfare of the horse is the primary concern in developing a horse farm, personal, societal, and geographical considerations also must be addressed. The amount of money available or the extent of the debt to be incurred must be carefully analyzed. Many horse farms, large and small, fail because of unrealistic cost and income projections or insufficient initial capital. The nation is becoming much more conscious of land-use planning. While some communities may welcome the rural and pastoral nature of horse farms others may find them to be a nuisance. Zoning laws are regional in nature, so it is important to explore local zoning before planning a horse facility. Finally, the geographical nature of a particular piece of land can determine its optimum use and development. Some areas are just suited to raising horses; others are ideal. All these factors can have a profound influence on the type of facilities that should be constructed.

20.1   General Considerations for the Design of Horse Facilities

Safety

Facilities must be designed to protect the horse, which, as an active, alert athlete is often injured accidentally because of its inherent reaction to flee from danger whether real or imagined. Many horses are injured as a result of neglect, that is, by protruding nails, sharp edges, unscreened glass, exposed electric wires, sagging fences, broken gates, and barbed wire. Other injuries result from poor planning. Horses have gone through misplaced windows, through fences that could not be seen, and over gates that were too low or poorly designed. O'Dea's survey (1966) of injuries treated by a New York State equine clinic revealed that 29.2 percent were caused by buildings and appointments, 16.2 percent were caused by fences and gates, and 6.3 percent were caused by soil, ground conditions, or pasture hazards. Thus, more than one-half (51.9 percent) of the injuries treated by these veterinarians were the result of the facilities in which the horses were kept. Loading, shipping, teasing, and breeding accounted for only 3.5 percent of the injuries, whereas the use of the horse in such activities as training, racing, hunting, showing, and trail riding accounted for the remaining 44.9 percent. Undoubtedly, there were additional facility-caused injuries that were treated without veterinary assistance or that went undetected by the owners.

Fire is major hazard on horse farms that annually kills and injures hundreds of horses. The main cause of horse-barn fires is human carelessness in the use of smoking material. Stables contain so much combustible material that smoking cannot be tolerated. For example, straw used for bedding is so highly flammable that the horse in a stall where a fire starts has only 30 seconds to be rescued. Other causes of fire are defective wiring and electric equipment, lightning, exposure fires (grass or brush burning, trash barrels, wind-blown sparks), set fires, defective or improperly installed heating equipment, ignited fires (blacksmith's forge, engine exhaust, etc.), and, of course, spontaneous combustion of stored hay. The entire farm must become fire-conscious. An emergency evacuation plan for all buildings and fields should be developed. The local fire department can provide a site review and work on developing a safety plan. Smoke and fire detectors are inexpensive and should be utilized. Fire extinguishers, an adequate water supply, good housekeeping, well-placed telephones, fire alarms, practical security, NO SMOKING signs, and alertness by all farm personnel will contribute to fire safety.

Hazards can be eliminated if safety to the horse and handler is the major consideration in planning horse facilities. There is no excuse for injuries that are caused by human negligence. Many facility-induced injuries can be prevented by proper design, careful construction, and diligent maintenance.

Many Thoroughbred and Standardbred breeding farms are models of safety. Excellent sources of information on horse facilities whose design has stressed safety are the book *A Barn Well Filled* published by *Blood-Horse Magazine*, and Harry M. Harvey's chapter "Stock Farm Management" in the U.S.T.A.'s *Care and Training of the Trotter and Pacer*.

Cost

The most limiting factor in the construction of new horse facilities is cost. When developing a farm it is important that costs not exceed the potential resale value should the business dissolve. Elaborate, well-planned, and well-maintained horse farms are a pleasure to visit, but such facilities are beyond the economic means of the average horseperson. On the other hand, rugged, durable, well-constructed, labor-saving facilities may be less expensive in the long run because they are less expensive to maintain and operate.

Durability Horses are particularly hard on facilities that are not designed correctly. They can chew stalls and gates, break up stalls, and tear down fences at an alarming rate. Warren Evans (1981) describes the horse as "a cross between a large beaver and gopher. It loves to eat everything available including wood and other types of materials, and it likes to dig by pawing. Never forget the horse's fondness for digging, pawing and eating wood when you think of constructing a facility for a horse."

A farm should be so designed and constructed that repairs will be minimal. The common denominator of all poorly managed farms is lack of maintenance. Sagging fences, broken stall doors, and poor gate latches result in wasted labor and, unfortunately, frequently cause serious injuries.

Building materials should also be selected for their low-maintenance characteristics. White board fences, wood clapboard siding, and pine stall walls may be aesthetically appealing but require more maintenance than wire fences, metal siding and concrete stall walls.

Efficient Use of Labor Labor is the most expensive and variable item on a horse farm. The facilities should be so designed that efficient use is made of labor and an attractive and desirable atmosphere is provided in which to work. Too many horse facilities are designed with little thought for the personnel who will work there. Gates that fail to latch, muddy, inaccessible pastures, and inconveniently located barns will discourage most workers.

Flexibility

Planning should include consideration of possible future use. Permanent structures should be so designed that expansion or conversion can be readily accomplished with a minimum of cost and disruption of the farm operation.

Clear-span structures (buildings without interior supporting posts) have the advantage of enabling interior modification without alteration of the structural soundness.

Accessibility

Driveways and lanes should be so designed that pastures, paddocks, and buildings are readily accessible and useable in all types of weather. Plan the facilities so they will not be isolated or destroyed by blizzards, thawing ground, or floods. Barns should have all-weather roads with wide entrance gates for fire protection as well as for ease in the day-to-day management of the stable.

Those who raise horses in the north must make provisions for plowing snow around the barns, lanes, and paddocks. Barn entrances should be planned for easy snow removal and graded to prevent ice buildup. Whenever possible, building entrances should face south so that melting can occur and the slope of the roofs should be designed so that sliding snow will not block building entrances and roadways.

20.2   Farm Planning

Planning is the key to establishing a successful horse farm. The first step is to determine the goals or purpose of the farm. The next decision is whether the farm is to be strictly functional, an elaborate showplace, or somewhere in between. Once these factors are determined, the planning of physical facilities can proceed.

Land

The development of physical facilities starts with the land. Fertile land that will grow an abundance of nutritious grass for grazing and hay is a key to a successful, economical horse farm. Traditional horsebreeding areas of the United States, such as the bluegrass region of Kentucky and Virginia and the great Standardbred regions of Pennsylvania, Maryland, and Ohio, are blessed with well-drained loam soil, fertile, gently rolling land, numerous large shade trees, and an adequate supply of fresh water. The soil of the rapidly growing horse-breeding regions of Florida, California, the Southeast, and the south central states may not be as fertile, but the climate of these areas is ideally suited to raising horses.

Nevertheless, topography and soil are important considerations in selecting a site for raising horses.

The nutrient character of the land can be determined by a soil test. This test will determine the requirement for lime and fertilizer to optimize forage growth. The soil should also be tested for heavy metals, pesticide residues, and toxic wastes. Horses are particularly hard on the land, and the types of soil that produce abundant pastures and withstand wear and tear from horses are the most desirable. Well-drained loam or sandy loam soils are best. Such soils do not pack or become excessively muddy and they support plant growth well. Clay soils are slippery when wet and become rough and hard when dry. However, clay soils withstand drought well. Gravel and sandy soils do not hold nutrients and are unsatisfactory in dry weather. Sand colic may be a problem with sandy soils (see Section 16.5).

The topography can be hilly or flat, but low land with boggy areas is not desirable for horses. Plenty of clean water must be available on a horse farm—either from natural ponds or streams, from drilled wells, or from the municipal water supply. The water supply should be tested for purity to be certain that it is free from agricultural, industrial, and residential pollutants that might interfere with the normal reproduction, growth, or health of the horse herd.

Climate

Climate is also important in choosing the location of a horse farm. Extremes in weather increase management problems. The major increase in the number of horse-breeding farms in the United States has occurred in the South and mid-South rather than in the North, where there is less pasture and breeding seasons are greatly reduced.

Location

Finally, horse farms should be located in rural areas where land is plentiful. Urban sprawl creates additional management problems, and in fact, zoning laws in some regions limit or prohibit horses. On the other hand, to be successful, riding stables generally must be located near population centers. When a piece of land is being considered for acquisition the following questions should be addressed:

1. Size. Is there sufficient area for the present purpose and future expansion?

2. New or renovation? Is this to be a new farm or must you work with existing barns, paddocks, and roads? It is usual to be limited by what is already there.

3. Legal restrictions. Are there any legal restrictions on the site? Deed restrictions? Easements? What are the zoning and sanitary regulations?

4. Social restrictions. Who are the neighbors? How will they react to your intended development?

5. Air and water. Is there adequate clean, fresh water? Adequate drainage? Flooding problems? Clean air? Any pollution problems?

6. Nuisances. Any problems with noise, light, traffic, odors, insects, etc?

7. Farm image. What type of image does the farm want to present? Can it be accomplished at this location?

Once the land proves to be satisfactory the natural environment of the site should be assessed for:

1. Topography (slope and shape of land)
2. Wind intensity and prevailing directions throughout the year
3. Sun angle and intensity at different times of year
4. Precipitation — rain and snow
5. Vegetation — location and type
6. Soil characteristics

This information then provides the basis for locating the proposed buildings, pastures, paddocks, utilities, and roads.

20.3   Pastures

Horses should be pastured whenever possible. Pastures can represent an important part of the farm's feeding program. They supply nutritious feed at low cost while saving labor and improving the health and disposition of horses. Pasture forage provides an excellent source of protein, vitamins, and minerals. (See Sections 7.4 and 21.4 for a discussion of pasture management.) Horses on pasture receive exercise and sunshine and thus are healthier and usually have fewer breeding problems. In addition to the nutritional and physical advantages of pastures, there are psychological benefits. Almost all vices are associated with idle, stabled horses and do not develop on pasture.

Pastures should have shade for horses. Trees are ideal but they must be fenced off, covered with protective wire, or treated to protect them from horses' chewing. In areas where electric storms are common it may be necessary to put lightning rods on trees in horse pastures as horses will seek shelter under them during storms. Pastures should be free of hazards such as abandoned farm machinery, building foundations, or walls. Old fences must be completely removed. Any corner of a pasture is hazardous and should be rounded and fenced off. A lane between pastures is another safety precaution that should be taken when two groups of horses are running side by side. Some type of corral or catch pen is helpful in handling horses. High-traffic areas such as gates and around waterers should be graveled. Waterers should be safely placed and free from sharp edges or corners. Guy wires on electric poles should be guarded or fenced off, as should fence and gate brace wires. Corral, paddock, stallion, and perimeter fences should be higher and of sturdier construction than fences around large pastures or dividing fences.

20.4   Fences

Safe, secure fences that are specifically designed for horses should be the first priority in developing horse facilities. Horses spend most of their time in pastures and many spend their entire life there, so the fences must be built correctly. Even if pasture land is not available on a farm, a series of fenced turn-out areas (also called dry lots, paddocks, corrals, or "turn outs") must be provided. Fences must be built specifically for horses and it is often difficult for contractors to realize that what works for other species won't work for horses. So when planning a fence, keep in mind the nature of the horse, the types of fences available, cost, durability, and the appearance of the finished product.

Nature of the Horse

The horse evolved on the move and by nature is a wanderer and curiosity seeker who survives by fleeing, often irrationally, from any real or perceived danger. Thus, the mere presence of a fence creates a potential hazard to the horse. Philosophically, the owner has to decide if the fence will be designed to keep the horse in or let it go at a time of panic. Because, in most parts of the country, the dangers to a loose horse are greater than the danger of a horse slamming into a fence, it is generally accepted that the best horse fences must be strong, high, tight, and highly visible. It is hoped that such a fence will turn a frightened horse. Horse fences also must be capable of withstanding the passive but powerful attempts of horses to reach through, over, under, or against a fence to get at something on the other side.

Types of Fences

Specific characteristics of horse fences are that they must be high, (at least 60 inches for perimeter and 54 inches for dividing fences) and clearly visible. Strength and tightness is accomplished by correctly setting the posts —particularly the corner anchor, gate, and brace posts. (Figure 20-1). Any corner of a pasture is dangerous, particularly acute angles, so it is best to round corners or fence off sharp angles, (Figure 20-2). Guy wires on electric poles should be guarded or fenced off as should all wire braces for fences and gates. Spacing between planks or strands of wire is also important in all horse fences. The spacings must be either small enough that a horse can't get its

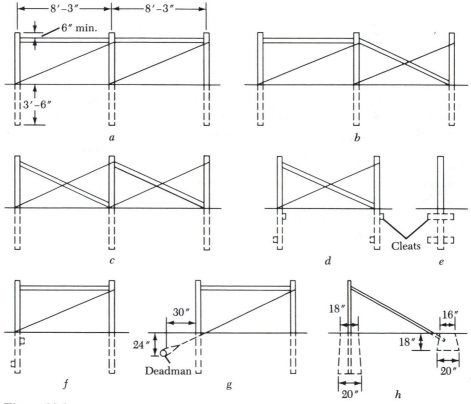

Figure 20-1
Bracing assemblies for end and corner posts: (*a*), (*b*), and (*c*), double-span assemblies with stunts and ties; (*d*) and (*f*), single spans with strut, tie, and cleats—(*f*) requires cleats on corner post only—(*e*), cleats used to resist both horizontal and vertical movement of individual post; (*g*), post braced with stut, and a tie which is anchored to deadman; (*h*), steel post with steel brace set in square concrete pier. Corner or end posts are shaded.

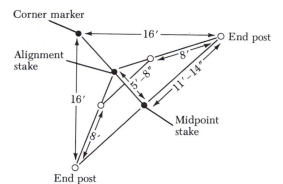

Corner marker

Alignment stake

16′

16′

8′

5′-8″

8″

11′-14″

End post

Midpoint stake

End post

Figure 20-2
Corner layout for fence.

head or foot through the opening or large enough that it can easily get them out. Any type of sharp edge such as protruding nails, gate latches, jagged wire splices, or barbed wire should be avoided on a horse fence.

Traditionally wood fences, either board or post and rail, have been among the safest and are the standard against which other fences are measured. However, horses can be safely contained in a variety of satisfactory fences made of wire, pipe, concrete, cable, and even rubber, so long as they are properly designed and constructed for horses. Wire mesh fences with a wooden top rail and polyvinyl pastic (PVC) fences are rapidly becoming very safe and popular horse fences.

Some fences are particularly unsatisfactory for horses. The development of barbed wire was a boon to the cattlerancher but it is the bane of the horse. There are undoubtedly more miles of barbed wire in the United States than all other types of fencing put together. But *barbed wire is dangerous for horses.* Don't use it and if you have it, start planning now to *eliminate all barbed wire* from your horse pastures. Other fencings that are less than ideal for horses are stone walls, snow fence, wood zigzag or snake fences, and many wire fences, particularly where the wire is lightweight (above 9 gage) and/or corroded.

Wood Fences

Board Fences   Board fences are traditional with horses because they are very attractive and functional. They are rugged, safe, and have the advantage of being easily seen by the horse. They are particularly good where the land is rolling and/or the fence line twists and turns. On the other hand, board fences are expensive to build and maintain, are chewed by horses, will splinter or break if a horse hits or kicks them, and will eventually rot. A white board fence will require painting about every two years (one of the reasons for the increased use of black asphalt paint is that it lasts about four times as long). Spray painting greatly reduces time and labor but wastes paint and leaves residue on the grass. Do not use lead-based paints and be sure all residue is gone and the paint is dry before you use the pasture. Pressure-treated boards are ideal but may be cost-prohibitive for large pastures. Some woods such as oak and cedar will last as long unpainted, and develop an attractive rustic appearance.

The best wood for fence boards is white oak, which is strong, but other hard woods, cypress, hemlock, southern yellow pine, cedar, and western larch can be used. Northern white pine, western pine, and Douglas fir are less desirable because, unless they are treated with a preservative (and even that is only a short-term solution) the horses will devour them quickly.

Boards are usually 6 or 8 inches wide and if the lumber is hardwood and "rough cut" 1×6 or 1×8 are adequate because the planks are full 1-inch thick. In paddocks and heavy traffic areas use 2×6 or 2×8. Finished lumber has been planed and 1-inch planks are actually ¾-inch thick. When using finished lumber 2×6 (actual 1½-inch $\times$ 5½-inch), 2×8 (actual 1½ $\times$ 7¼-inch) or 2×10 (actual 1½-inch $\times$ 9¼-inch) planks should always be used. Oak has the advantage of being too hard for horses to chew, whereas soft wood (pine) must be treated to prevent chewing. Creosote, long popular for this purpose, is no longer available, but motor oil, Ebanol, and liquid asphalt paint (sometimes diluted with motor oil) seem to retard chewing. (Do not employ used motor oil, which may contain high concentrations of lead). The posts can be square, half-round, or round, and at least 4 inches in diameter. They should be no more than 8 feet apart as longer distances will allow the planks to twist and warp, particularly if the boards are freshly cut (green) and not seasoned. Planks 16 feet long will span three posts and the joints between boards can be staggered for added fence strength.

A four-board fence 54 inches to 60 inches high is adequate for most fences but some paddocks are five boards and as high as 72 inches. If the bottom board is 5 to 7 inches above the ground with 8 inches between the remaining boards, the fence will be 53 inches to 55 inches. Some people prefer to put the bottom board high enough (12 inches) that it is easy to trim under the fence, while others prefer to put the bottom board right on the ground and use five boards. A stronger, if less attractive, fence results if the boards are attached on the inside of the post with three ring-shank or screw-shank nails rather than common nails. A vertical cap or facia board covering the joints on every post will increase the strength of the fence and reduce maintenance.

Post and Rail Post and rail fences add beauty to a farm and are easier to install than a board fence, but they are more expensive and not as strong. Neither do they last as long, because the posts, which have holes in them to receive the rails, tend to split and the ends of the rails rot and break.

However, post and rail fences, which are usually made of cedar or split chestnut rails, do not require painting. Other maintenance is also low because individual broken posts and/or rails are easily replaced. Post and rail fences for horses must be heavy-duty, with large posts set at least 30 inches in the ground and 10-foot rails. Because the paddle ends of the rails pass through the posts and overlap 8 to 10 inches the posts are set on 9-foot 4-inch centers. Posts are usually cedar or black locust. The split chestnut rails are excellent because they are strong and long-lived and resist chewing, but they are expensive and difficult to obtain. Cedar rails are the second choice but horses will chew them even if they are treated with penta. Because post and rail fences do not hold up as well as board fences, they are not recommended for stallion paddocks and areas with high concentrations of horses. An electric fence on the inside of the fence will keep horses off the fence and prolong its life. Beware of the many inexpensive lightweight ornamental post and rail fences that will not stand up to horses. Dowelled rails are attractive but weak and short-lived as horse fencing.

Slipboard   Slipboard fences combine some of the advantages of both board and post and rail fences and are very attractive and practical fences. They have the ease of installation and low maintenance of post and rail but the durability, long life, attractive appearance, and high visability of a board fence. Slipboard fence is usually available in pressure-treated lumber, so its initial cost is high but its life expectancy of 30 years and low maintenance make it economical in the long run.

Wire Fences

There are a number of types of wire fences and generally they are less expensive and easier and faster to erect than wood fences, but they are not as safe. Several types of wire mesh fences have been designed specifically for horses and when built with a board on top are considered the premier fence for horses as they are safe, strong, long-lasting, and virtually maintenance-free.

Wire Mesh   Wire mesh fences come in two designs; one is diamond mesh and the other is rectangular with 2-by-4-inch openings. The mesh is too small for a horse to get its foot through and will not injure a horse that runs into it. With posts on 8-foot centers a 1×6 or 2×6 board is attached to the top of the wire to add strength and visibility and to prevent the horses from leaning over and stretching the wire. As with other fences, putting the wire on the horse side of the fence is safer and strengthens the fence. Wire mesh fence is expensive but its safety and durability make it the most popular horse fence today.

Woven Wire Woven-wire or stock fence is the most common wire fence for horses. It is cheaper than mesh fence but not as safe. Woven wire is available in a number of heights, weights, and styles, but all have the disadvantage that a horse can get its leg through the fence. The larger the diameter of the wire (the lower the gage, the larger the wire) the safer the fence. A horse fence should be at least 9 gage. Most woven-wire fences are supposed to hold a variety of livestock, so they are designed with a heavy-gage top and bottom wire, with lighter gage wires "woven" between them. Those fences are usually erected with the larger openings at the top and the smaller openings at the ground. When using woven wire for horses there is some advantage in installing it "upside down" as the smaller openings at the top will strengthen the fence and prevent sagging while the larger openings at the bottom make it easier for a horse to get its leg out if it should happen to get caught in the fence. A woven wire fence with wood posts on 8-foot centers would allow a board on top but such fences are usually built with metal posts set at 12-foot or greater intervals. This reduces the cost but also reduces the fence's visibility and makes it less resilient when hit by a horse. The King Ranch has used woven wire very effectively for horse fencing by using a low-gage woven wire with large openings and threading an aluminum strip in the top of the fence to increase visibility.

High-Tensile Wire High-tensile fence has been called the most significant development in livestock fencing since barbed wire (1874). It is relatively new in the United States but has been used to successfully control sheep, cattle, and horses on the vast ranches of Australia and New Zealand for nearly 30 years. High-tensile wire is smooth wire with, as the name implies, high-tensile strength, which allows it to be erected under tension. High-tensile fence requires the proper anchor posts and braces but when correctly installed it is pulled tighter than other fence wires making it stronger, resilient and more effective than other smooth or twisted-wire fences. A 10- or 12-strand fence with a board on top is recommended for horses, but because one of the advantages of tensile fence is wider post spacing (up to 60 feet between conventional posts with "droppers" in between, see (Figure 20-3), a top board is not generally used. High-tensile fence can be used alone or it can be electrified. It is inexpensive, as it requires fewer posts than any other type of fence. We have used high-tensile wire for horses at the University of Massachusetts for 5 years with good results and no injuries. The poor visibility of the fence remains a concern, but since the wire is under tension the horses seem to "bounce off" it without being injured.

Barbed Wire Barbed wire is inexpensive and *very dangerous* for horses. It is not an appropriate fencing material for horses and should not be used.

Twisted Wire Twisted "barbless" wire has been used successfully with horses. This type of fence should have five strands of wire if it is to be used

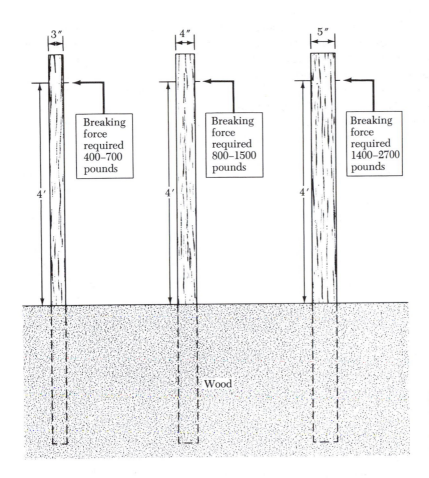

3"

Breaking
force
required
400-700
pounds

4'

4"

Breaking
force
required
800-1500
pounds

4'

5"

Breaking
force
required
1400-2700
pounds

4'

Wood

Figure 20-3
Influence of post
size on breaking
force required to
break the post.

for horses. It is probably the least expensive permanent fence available for horses and many have used this fence with excellent results. However, it is a difficult fence for a horse to see and works best in large pastures where the fence follows a stream, road, tree line or some other type of natural barrier.

Electric Fences

Electrified fences are usually a type of wire fence that can be used either alone or in conjunction with another type of fence. Electric fences have been successfully used with horses but many horsepeople do not like them. They are extremely economical because the posts and wire are inexpensive and installation is fast and easy, but they do require maintenance and will frequently short out. Horses must be trained to an electric fence and it is difficult for them to see the high-gage electric wire, which, moreover, can cut them, particularly if it is corroded. On the other hand, some new electric "wires" on the market are safer, such as a polywire (stainless steel strand-braided plastic). This electroplastic twine is lightweight and easy to work with, and breaks before it breaks the skin. There are also several electric fence "tapes" that are much more visible and less injurious to the horse. Electra tape is a 1½-inch plastic tape with 15 current-carrying stainless steel strands. Glo-guard is a plastic-aluminum electric tape that is highly reflective.

Electric fence is particularly effective for horses when located on top or inside another type of fence. It keeps the horses "off the fences" and thereby reduces maintenance and the chance of injury. Two strands of electric fence supported on crossarms in a dividing fence will reduce "horseplay."

Some managers consider an electric fence a very dangerous fence for horses when it is used alone. It is not an effective permanent horse fence because horses can easily run through it and when it does short out the electric wire or tape is not strong enough to keep a horse from pushing through.

Flexible Fences

Polymer Plastic or PVC PVC (polyvinyl chloride) is a recent addition to horse fencing. This white plastic or polymer fence is extremely durable, safe, will not rot, chip, erode, rust, or peel, is extremely low-maintenance and provides the beauty of traditional white board fence. The PVC fence comes with either round rails or 2 × 6 planks that are lighter but still stronger than wood and much more flexible. The posts are also PVC and can be round or square. The major disadvantage is the initial cost (twice that of board fence) but there is virtually *no* maintenance and PVC will unquestionably outlast all fences with wood posts. The flexibility of the fence prevents injury and the rails will not splinter or cut on impact. The polymer material does expand and contract with temperature changes and some farms have experienced problems with the rails popping out when hit, particularly where the fence was installed around a curve. The posts are set at 8-foot intervals in concrete to make them rigid. This process increases installation costs and makes moving the fence or replacing a post a difficult task. The PVC fence is becoming very popular — except with horses, who find they can't chew the rails.

Polymer fences are undoubtedly the wave of the future. Several other types are about to appear on the market including a four "rail" (ribbon?) solid polymer fence that has the flexibility of rubber-nylon fencing but the beauty of conventional white boards. The polymer fences have guarantees for up to 20 years and a life expectancy exceeding 50 years.

Rubber-Nylon Fences

Rubber-nylon is manufactured from belting material cut into strips 2 inches to 4 inches wide. Because the strips are maintained under tension it is essential that the anchor posts be extremely well braced. When properly erected, the fence is taut, safe, and attractive. It is extremely durable and requires little maintenance. There have been several reports of foreign-body impactions caused by young horses chewing and ingesting ragged rubber and nylon edges but the suppliers are now more careful in the product they are supplying. The rubber-nylon material does not wear or weather so maintenance is minimal. In the north, where there is severe freezing and thawing, the fence has a tendency to loosen, but many in more moderate climates have not experienced this problem. Rubber-nylon fencing is very safe and relatively inexpensive.

Other Types of Fences

Pipe Fences   Pipe fences are strong and safe for horses but prohibitively expensive in many parts of the country. Pipe fences are excellent for corrals, paddocks, and heavily used areas but require welding and painting and the posts must be set in concrete. Pipe fences are most popular in the oil-producing regions where surplus pipe is available. Pipe fences are criticized for being unyielding when a horse hits them but they will not break.

Chain Link Fences   This type of fence is particularly good for stallion pens or other small paddocks because it is extremely strong and safe. The top of the fence should be smooth or protected in some manner because sharp twisted ends are sometimes exposed at the top. Chain link fencing is usually attached to a pipe frame by wire fasteners that have a tendency to break or "pop loose" under pressure. Also, the pipes bend easily from the weight of a horse leaning against the fence. A chain link fence will stretch from kicking, cannot be tightened, and is expensive to install and maintain.

Post and Cable Fences   This combination makes another satisfactory horse fence. The cable is usually threaded through the middle of wooden posts and attached by a turnbuckle or spring to the corner posts. Cable fences are initially expensive but are strong, safe, and long lasting.

Precast Concrete Fences

These fences, made to resemble post and rail, are quite attractive. They are white, never need painting, withstand weather, and are virtually indestructible. However, a concrete fence is difficult to obtain and expensive to ship and install because of its weight. Concrete fences are also unyielding and occasionally rails will pop out when horses hit the fence.

Cost of Fencing

Horse fencing is a major expense in developing a farm. While the cost per foot may be small the many rods of fence necessary represents a major investment. There are tremendous differences in the prices of different types of fencing and the cost of installation can vary depending on type of fence and soil conditions. (See Table 20-1.)

Other factors that can affect the final cost of fencing are the size and shape of the pasture, the life expectancy of the fence, and maintenance cost.

Size and Shape   A square 50-acre field will require 320 rods of fencing for the perimeter. If the same 50 acres is cross-fenced into four square 10-acre

Table 20-1
Relative costs of various types of horse fence (installed)

Type of fence	Relative cost	Life expectancy (years)	Maintenance
Board fence: 4 hardened boards (untreated 8-ft sections)	°	10–15	High
Post and rail: heavy duty, cedar 3-rail, hardwood (10-ft sections)	66% more	15–20	Low
	Same	10–15	Low
Diamond wire mesh with top board (8-ft pressure-treated posts)	33% more	20–25	Low
Slipboard, pressure-treated	same		
Plastic polymer (PVC) 4 square rails (8-ft posts set in concrete)	250% more	30+	Low
High tensile wire horse fence: 10-wire	$\frac{1}{3}$	20	Low
4-wire electric	$\frac{1}{10}$	20	Medium
Twisted wire (no barbs) 5 strands, metal posts	$\frac{1}{4}$	15	Medium
Rubber, 3 strands	$\frac{1}{2}$	25	Low

All fences are compared with the cost of a 4-board installed fence in the northeast (approx. $3.00/ft).
°Actual fence costs will differ considerably depending on available materials in different parts of the country. Installation costs are also highly variable.

pastures it will require 480 rods of fencing. On the other hand four rectangular (20 × 80 rods) 10 acre pastures would require 560 rods of fence (Figure 20-4).

Life Expectancy   Estimates of fence longevity are listed in Table 20-1. Posts cost more to buy and require more labor to install, so their life expectancy is the key to fence's longevity. Osage Orange, black locust, and cedar make the best untreated wood posts. Pressure treating with pentachlorophenol (*penta*) or Copper-Chrome-Arsenate (*CCA*) increases a post's life by from 10 to 30 years but softer pressure-treated wood may not hold staples or nails very well. See Table 20-2. Posts with larger diameters have a significantly longer life and are stronger than small posts.

The minimum fence-post sizes recommended for fences are:

Single-span anchor post assembly:
Corner post 6-inch diameter
Brace post 5-inch diameter
Brace 4-inch diameter
Tie 2 double strands of 9-gage wire
Line posts 4-inch diameter

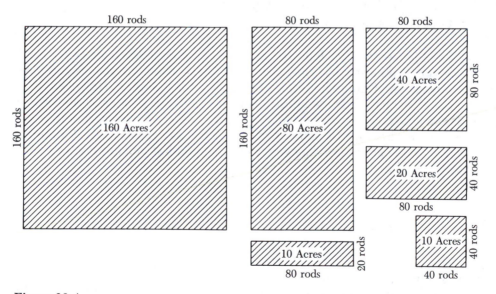

Figure 20-4
The length of fencing required per acre when the size and shape of the fence is varied.

Table 20-2
Life expectancy of untreated and treated fence posts
(years)

Kind of wood	Untreated	Pressure treated
Osage orange	25 to 30 yr	—
Red cedar	15 to 25 yr	20 to 35 yr
Black locus	15 to 25 yr	—
White oak	5 to 10 yr	10 to 15 yr
Cypress	5 to 10 yr	—
Hemlock	4 to 8 yr	20 to 25 yr
Douglas fir	3 to 6 yr	20 to 25 yr
Red oak	2 to 7 yr	10 to 15 yr
Yellow poplar	2 to 7 yr	15 to 20 yr
Cottonwood	2 to 7 yr	15 to 20 yr

Double-span anchor post assembly:
Corner post 5-inch diameter
Each brace post 4-inch diameter
Each brace 4-inch diameter
Each tie 2 double strands of 9-gage wire
Line posts 4-inch diameter

Steel fence posts are lightweight, fireproof, extremely durable, and easily driven into most soils. They are usually used with wire fences and have the advantage of grounding these fences against lightning. However, they are not as attractive as wood posts. They rust and will bend, and their sharp tops can injure a horse. Concrete posts have excellent durability but they are heavy to work with and difficult to attach to fencing materials. Fiberglass and PVC posts are recent innovations and should last indefinitely, provided they do not break.

20.5   Gates

Gates should be strong and as safe as possible, well placed, and easy to use. Generally gates should be placed conveniently in a fence line but not in the corner of a pasture or paddock. The gate should be placed on high ground, and because it is a gathering place the gate area should be filled with gravel to prevent the development of a moat at the pasture entrance. As with a horse fence, the gate must be clearly visible and must not appear to be an opening in the fence. Also, the spacing in the gate panels and braces should be such that a horse can't get a foot hung up.

The width of a gate is determined by its purpose. All pastures should have an equipment gate that is at least 12 feet wide. However, if you are taking horses in and out of a pasture with other horses in it a large gate is hard to handle and a 5-foot-wide gate is better. If haying equipment will be using the gate it should be 16 feet wide.

Wood gates are satisfactory but should be made of oak for strength and to prevent chewing. Such gates are heavy but long lasting. Welded iron or pipe gates are strong and long lasting and a horse is not apt to get caught up in them. Galvanized iron and aluminum gates are available commercially but they are not particularly good as horse gates because they bend and break easily, creating dangerous sharp edges. Woven-wire gates are difficult for a horse to see and not very satisfactory unless the wire is attached to a stout frame.

Gate latches should be durable and easy to use with one hand but not easily opened by the horse or the wind. A chain and snap combination that goes through the gate and around the post is simple and works well.

20.6   Buildings for Horses

Horse barns should be located on sites that are well drained and easily accessible in all types of weather. They should take maximum advantage of prevailing winds in warm weather and be sheltered from winter storms. They should be placed so that the surrounding community is not annoyed by drainage, odor, noise, or flies. The barns should be surrounded by pastures or paddocks so that a loose horse cannot get out onto a main road. The barn area should be readily serviced by a hard-surfaced road and by public utilities (water, electricity, and telephone).

Stables should be pleasant places in which to work and should be designed for ease in feeding and cleaning. Construction considerations include attractiveness, ventilation, and fire resistance, as well as safety, cost, durability, and flexibility of design. Many barns are constructed of wood, but concrete, steel, aluminum, masonry, and other durable, fireproof construction materials can be used. The *Horse Handbook: Housing and Equipment* provides details on the construction of proper ventilation and barn layout. This excellent study is inexpensive and is available from the extension agricultural engineer at your state university.

The type of barn depends upon the climatic conditions and the intended use. The requirements of a breeding farm differ from those of a small pleasure stable or a professional training stable. The type of shelter can range from a simple, open-front building to an elaborate show stable with large box stalls and fancy tackroom and lounge.

Open-Front Shelters

Horses do well in free-access barns. These open-front sheds have become popular in all parts of the country in all phases of the horse industry because of the advantages they offer. They are inexpensive to build and maintain and alleviate the major problems associated with stabling horses: labor, ventilation, and fire. There is no daily stall cleaning and time can be spent on more productive endeavors such as handling and checking every horse every day. The sheds can be cleaned with a tractor and loader, so less labor is necessary. Picking up the sheds regularly and adding clean bedding will reduce flies, odor, and parasite infestation. Run-ins require less bedding and less space per horse than conventional stalls. Open-front sheds are well ventilated and horses housed in them have fewer respiratory and digestive problems. They also seem to have a better mental attitude and improved muscle tone. Some managers believe that injuries may be reduced because the horses get more relaxed exercise and seem less apt to run wildly or fight. Fires, a dread of all horsemen, are not likely to occur in sheds, and if they do start, the horses are not trapped.

The biggest disadvantage of shed housing is that it is impossible to individually control exercise and diet. Every horse should be examined carefully every day, but with shed-housed horses it is easier to overlook an injury or illness. Horses kept in loose housing also grow shaggier hair coats and look rougher and dirtier than stabled horses, particularly during the mud season. Open sheds are not a good place to show off sale horses or to house show, race, or other performing horses.

Horses can withstand the elements so long as they are sheltered from the extremes. Flies, hot weather, cold rains, and strong, cold winds seem to bother horses, and run-in sheds provide adequate protection from these conditions. Charlie Kenney, manager of Bluegrass Farm, thinks that the horses prefer free-access housing. He notes that horses "rarely go into sheds except for food. They will go under a shed in sleeting weather or to get out of a cold rain, but do not seem to be bothered by ordinary snow or rain or wind unless the weather is really bitter. It makes a fellow wonder whether horses are kept up too much" (Hollingsworth, 1971).

Open-front buildings should face away from the prevailing winds and be deep enough (a minimum of 20 feet) to provide sufficient shelter from weather extremes. If it is located in the northern half of this country, the shed should face to the south to take advantage of the low winter sun. The opening should be at least 10 feet high and wide enough for all the horses to run out safely. Unless a clear-span structure is used, the supporting posts are set on 16-foot centers and padded in some manner. Sheds provide flexibility as to the number of horses that can be housed, but a minimum of 75 square feet per horse should be maintained. There must be no sharp edges, and the corners of sheds should be designed so that a horse cannot get trapped by a "kicker."

The roof should slope away from the opening or should be designed in such a way that rainwater or snow will not collect at the entrance to the shed. The sheds should be located on high ground or graded so that water drains away from the building. Because steel and aluminum roofs are noisy in sleet storms and driving rainstorms, they should be insulated to reduce the noise so that it does not drive the horses out in bad weather.

Hay, mineral, salt, and grain feeders can be built into the back wall of the shed, but many managers prefer to feed hay outside — either on the ground or in separate field mangers. Waterers are best placed in the field away from the shed. In paddocks, waterers are located most successfully in the fence lines.

Barns with Stalls

For many horse enterprises, conventional stables with stalls are more satisfactory than open sheds. Horses kept in stalls can receive more individual attention and can be kept cleaner and more presentable, and their feed and exercise program can be controlled. Foaling mares, mares during the breeding season, and young horses can be closely observed. Horses kept in stalls are generally cared for daily, and this consideration is important for young horses and stallions.

Inadequate ventilation is probably the biggest mistake made in the construction of horse facilities. The heat and moisture produced by horses must be removed from the barn to prevent condensation and odor buildup. Horses did not evolve in stuffy, heated barns, but many well-intentioned owners provide low ceilings, small windows, and insufficient ventilation, thereby creating a perfect atmosphere for pneumonia and other respiratory problems. Old, porous buildings with no ceilings or with haylofts overhead provide a better environment than some of the newer, tight, draft-free barns. H. H. Sutton (1963), giving a veterinarian's view of farm management, states that "more respiratory infection is associated with confinement in small, tight, poorly ventilated stalls than in horses confined by gates in drafty barns."

Recent studies at the University of Illinois Veterinary College confirm his observations. They have determined that ammonia fumes (from urine) have an irritating effect on the horse's respiratory tract. Long-term exposure to ammonia fumes can predispose a foal to pneumonia, as it weakens the defense mechanisms that protects the foal from respiratory disease. Foal pneumonia is a major cause of death in young horses, with up to 30 percent of the foal crop being affected by respiratory problems each year.

Closing up a barn in winter will create stuffy, uncomfortable conditions for the horses. Stalls must be well ventilated. Particularly dangerous to foals is the practice of closing up the foaling stall. Some farms even cover the foaling stall with plastic and further exacerbate the condition by adding a heat lamp. This practice creates an environment that is entirely inappropriate. Ammonia fumes and a variety of other contaminants including dust from feed and bedding, mold spores, airborne bacteria, humidity, and methane gas (which replaces oxygen) tend to concentrate at the bottom of the stall in the air the foal breathes. It is now recognized that greater attention must be paid to the air quality in the stall itself. Ground-level louvers in stalls are now recommended, as a window and stall door will not adequately ventilate the floor. Anyone planning a barn should seek professional assistance in developing a ventilation system. With proper design and location of windows, doors, eave openings, ridge vents, adjustable wall louvers, and exhaust fans, ventilation can be developed so that the recommended four air changes per hour without draft can be accomplished.

Because ventilation is such a critical problem in raising horses, many outstanding breeding farms in the bluegrass region continue to use the old, coverted, airy tobacco barns that have served them well for many years. Edward Fallon (1967) notes that "the healthiest barn I can imagine is one that we frequently see, and that is a tobacco barn which contains stalls for horses. They furnish adequate shelter and probably the best ventilation."

Dust is another major irritant in many barns. Indoor arenas are notorious for their dust problems and no completely effective solution has been found. Stables that are connected to arenas are always too dusty. Another major source of dust is the hay and bedding. The best solution is to keep only minimal amounts of hay in the barn and have a separate storage building. This eliminates a source of dust and removes a major fire hazard.

The electric, water, and telephone requirements for buildings should be developed before other construction. Remember to keep a detailed diagram of type, location, and depth of utilities. Shutoffs for the utilities should be prominently marked in case of an emergency. Any barn in which people are working on a regular basis should have a telephone and restroom facilities so there needs to be provision for sewer lines, as well as water. The water supply must be adequate for drinking, washing, and fire protection.

The method of watering horses in stalls is a matter of considerable debate. Automatic stall waterers save labor but require daily cleaning and checking. A plugged "automatic" waterer can leave the horse without any water. Freezing and flooding are also potential problems with automatic waterers. There are some excellent stall waterers available but they must be placed and installed so a horse cannot be injured by them. Some horsemen prefer to water with a bucket because it is easily cleaned, water consumption can be observed, and water can be withheld when necessary. However, bucket watering is time-consuming and horses are often left without water, particularly at night. Rubber buckets are safer than galvanized pails and can be hung in the corner or flat against the wall of the stall with a snap. Water should be available in the aisles of the barn and water should be provided for a wash area, a first aid area, and any laboratory facilities. An unheated barn will have to use frost-free hydrants.

Stalls and aisles should be well lighted. Fluorescent light is not satisfactory in cold regions because it flickers and loses power below 50°F. All wiring in barns should be installed in conduits and all exposed light bulbs protected with dust-free glass covers to reduce fire hazard. Be sure to provide plenty of electrical outlets because they are always in use for a variety of purposes. All wiring, switches, outlets, and electrical devices must be placed out of the reach of horses.

Fire control should be built into the barn with fire and smoke alarms, lightning protection, an inside water supply with adequate hose, and fire extinguishers, and by using fire retardant materials whenever possible. Halters and lead ropes by every stall door can save time and lives in an emergency. Some farms have an exterior door to every stall, as well as the aisle door. If bedding and hay are stored in the barn, they should be separated from the rest of the barn in a fire-controlled area (fire-resistant). Dust and hay chaff have a very low kindling temperature, so the barn should be kept clean and free from flammable materials. When planning facilities, allow plenty of space between buildings.

The aisle in the barn must be wide enough that two horses can pass safely and to accommodate any vehicles that might be used in the barn. You will never regret wide aisles. A minimum width is 12 feet and 14 feet is even better. Narrow aisles are false economy. Concrete aisles are easy to keep clean but they are too slippery for horses. Asphalt is harder to clean but a much safer surface for the aisle. Dirt or clay aisles give excellent traction but tend to stay messy. Sawdust in the aisles provides good footing and a pleasant appearance if raked regularly but it increases the dust in the barn. If you have ceilings in the barn they should be at least 10 feet high as should any overhead pipes, conduits, or light fixtures.

20.7   Stalls

The structure of the stall walls should be considered carefully. Wood is the most popular building material but may be chewed by idle, bored horses. Oak and other hardwoods will withstand chewing longer than the softwoods.

A beautiful new barn of pine was literally devoured by horses during one winter. Wood can be treated with motor oil, liquid asphalt, or other commercial anti-chew products, but covering exposed edges with steel angle iron or sheet metal is a more satisfactory solution. No sharp corners or nail heads should be left exposed. Some stables use an electric fence wire to prevent chewing (and tail rubbing) in the stall, but this is not recommended because it creates a fire hazard and the charged wire can be a nuisance to the handlers.

Stall Walls

Tongue and groove 2-inch pressure-treated yellow pine has become a common material for stall walls, but it is not very durable and will splinter when kicked. The tongue and groove, while providing strength, makes the replacement of boards difficult if the walls are nailed in place. Consequently some contractors install a steel channel on the stall's corner posts and the boards are cut to fit and simply dropped into the channels without nailing. This type of wall can easily be removed and repaired.

The advantage of wood is that it breathes and doesn't get cold and clammy like masonry walls. But wood chewing is a tremendous problem, wood will rot, and it is a fire hazard. Fire-retardant lumber is available but the process increases costs by 50 to 70 percent.

The use of masonry walls is increasing. The walls can be poured concrete, concrete block, or cinder block. Masonry walls are durable, easy to maintain, and fireproof. They are smooth and attractive when covered with a block filler and then painted with epoxy paint. In the winter these walls tend to be cold and can be damp from condensation.

Masonry walls are hard if you have a "stall kicker" and cinderblock will break or crack. To alleviate this disadvantage, the stall walls are lined with oak or covered with a rubber mat. Masonry walls are expensive to begin with and covering them may be an unnecessary additional expense. Colonel Sager, for many years the respected veterinarian at Claiborne Farm says, "Painting the [concrete] stall walls doesn't sound nearly as safe as putting up wood paneling, but we've had no problems. In all the years that we've had these concrete block barns, I can't remember having a horse hurt seriously by kicking in the stall." (Toby, 1980). Enough said!

Stalls are also made of steel link. The advantage of steel walls is that they can be completely prefabricated and just dropped into place. But steel is noisy, is not as attractive as wood or masonry, and tends to be hottest in the summer and coldest in the winter. Chainlink walls are generally used in hot climates where maximum air circulation is desired. Chainlink is durable but will develop bulges with wear and tear.

Stall partitions should be solid to a height of 5 feet with some sort of divider to the ceiling. If there is no ceiling, the stall walls should be 8 feet high. Air circulation may be more important than solid partitions in hot climates. Horses, especially stallions, that can see each other are usually calmer and happier. Thus, a light, airy barn with chain link, welded wire, or pipe dividers at the top of the stall partitions is preferred. However, there are instances when a completely solid stall wall is desirable (that is, for some stallions, for foaling stalls, and for many performing horses). In either case the walls should be smooth — the fewer projections (mangers, feeders, waterers) into the stall, the better (Figure 20-5).

The fronts of the stalls can be open above 5 feet but many owners prefer some sort of screening to keep the horse's head out of the aisle. Doors should slide or open outward. For convenience and safety, never have a stall door open into the stall. The stall doors should be at least 4 feet wide and 8 feet high. Dutch doors are popular: the bottom panel is usually 5 feet high and 4 feet wide and the top panel is 4 feet wide and 3 feet high. Some horsemen insist on having 2 doors for each stall, one into the aisle and the other outside into a paddock to provide fire safety.

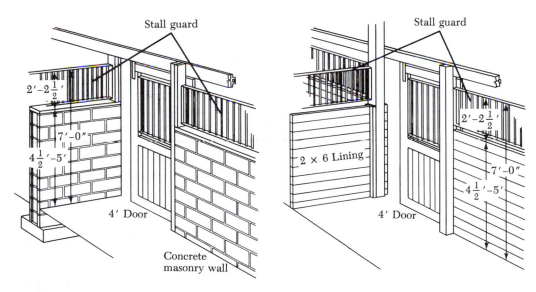

Figure 20-5
Typical box stall construction and arrangements. Open stall guards are made of material such as ½-inch diameter steel rods, ¾-inch pipe spaced not more than 4 inches on center; sections of #4 gage welded steel fence; flattened 1½-inch, 9 gage expanded steel mesh; #6 gage welded wire 4-by-4-inch mesh; chain link fencing; vertical wood slats.

The size of the box stall depends on its use. The most popular size is 12 feet by 12 feet. The more time the horse spends in the stall, the larger it should be. A box stall 10 feet by 10 feet is adequate for young horses and for some of the smaller breeds if the horses are turned out regularly. We may have to rethink stall size with the increased interest in the large warm-blood horses for dressage, Combined Training, and competitive driving. A 17.2- or 18-hand horse may need a stall as large as 16 by 16. Foaling mares and stallions also need larger stalls — 12 by 16, 16 by 16, or even 12 by 20. Sometimes two adjacent stalls will have a removable common wall so that a large stall would be available for foaling or nursing care.

It is desirable to have a window in each stall for light and ventilation. The windows should never be lower than 5 feet from the floor and should slide or open outward. They should *always* be screened in some manner to protect the horse from the glass. There are some high-impact plastics that can be used for windows instead of glass.

The ceilings in stalls should be a minimum of 8 feet high, but 10-foot ceilings or no ceilings are safer and provide better ventilation. Electrical wires and fixtures should be covered and should be safely out of the reach of the horse.

Feeders, Waterers, and Mangers

The use and location of feeders, waterers, and hay mangers is a controversial topic among farm managers. A number of commercial models are available, but care should be used in choosing equipment to make sure that it is safe and designed specifically for horses. Feeders and waterers are usually placed "out of the way" in the corners. Some stable managers prefer to feed hay directly on the floor, a more natural location. Since horses often pull hay out of the mangers and eat it off the ground anyway, these managers consider the manger an unnecessary hazard in the stall. Rubber and plastic feed tubs are rapidly replacing cast iron tubs because they are less expensive, easier to clean, safe, light, and quite durable.

What type of watering facilities the stall should have is also a matter of controversy. Automatic waterers save labor but must be cleaned and checked *regularly*, because a plugged float can leave a horse without water for several days before detection. Freezing and flooding are also potential problems. Some excellent commercial stall waterers are available, but they must be mounted safely in the stall. Some horsepeople prefer to water with a bucket because it is easily cleaned, water consumption can be observed, and the water can be easily withheld when necessary. On the other hand, bucket watering is time-consuming, and horses are often left without water, particularly overnight. Rubber buckets are safer than galvanized pails and can be hung in the corner of the stall with a snap.

Stall Floors

A variety of flooring surfaces have been used successfully in stables. The most popular stall flooring material is hard-packed clay. This is relatively warm and resilient but tends to be slippery when wet, and holes or pockets may develop. A concrete or asphalt apron at the stall door discourages "digging." Clay should be placed over a well-drained subfloor of crushed rock or gravel. Stone dust and sand also make excellent stall floors but are not as durable as clay. Sand requires no additional bedding but must be cleaned regularly and changed occasionally. Dirt may be satisfactory if well tramped but doesn't drain well and can become muddy.

Wooden planks are often used to cover concrete or in elevated floors, but they tend to be slippery when wet, they are difficult to clean completely, and they harbor odors. Wood-block floors were popular in many older barns but are now prohibitively expensive. Concrete, even when roughened, is potentially slippery, cold, and hard, even when plenty of bedding is used, but it is easy to clean and long lasting. Asphalt (macadam or blacktop) is very satisfactory. It provides better traction than concrete and is durable, resilient, and easy to clean. It has been used successfully to cover existing concrete, dirt, clay, and even wood floors but is cold and hard. New composition materials, such as Tartan surfacing, carpeting, and even astroturf, provide fine surfaces but are expensive to install and maintain. Rubber stall mats are used widely and are a good way to cover an existing floor.

Ease of cleaning and of manure disposal are important considerations in planning stalls. Mechanical cleaners built into subfloor gutters with trapdoors are used in some stables but the majority of barns are still cleaned by manual labor. The alley in front should be large enough to accommodate cleaning equipment.

20.8 Specialized Structures

Tie Stalls

Straight stalls or tie stalls are not as widely used for light horses as they were for draft or carriage horses. However, straight stalls provide stabling for more horses in a given area than box stalls and require less bedding and labor. The construction details and flooring of tie stalls are similar to those of box stalls except that a manger is usually built into the front of the stall, provision is made for tying the horse, and a tail chain is often put across the back of the stall. Straight stalls should be 9 feet in length, including the manger, and 5 feet in width. Tie stalls are popular in riding stables.

Foaling Stalls

Foaling facilities should be light, dry, and secluded. The stall is often isolated from the rest of the stalls. On large farms, a separate barn is used as the foaling barn. Foaling stalls should be roomy (12 feet by 16 feet or 16 feet by 20 feet), well ventilated, and designed so that the mare can be observed without disruption. An attendant should always be present during foaling because of potential problems that may arise. Consequently, many foaling stalls have windows facing into adjoining heated "waiting rooms" where the attendant can quietly observe the activities of the expectant mare. Closed-circuit television provides another effective method for surveillance of foaling mares. Adequate lighting is essential in the foaling area because most mares foal after dark. In favorable climates some farms prefer to foal in lighted and observed paddocks.

Broodmare Barns

Broodmares often foal early in the year, so it may be necessary to have facilities for the mare before foaling as well as for the mare and her foal after foaling. Broodmare barns should have large (16 feet by 16 feet) stalls and be located near paddocks or pastures.

Foal Feeders

Many farms provide a foal creep where the nursing foal can have free access to grain while on pasture with its dam. The creep is so designed that foals can enter but mares cannot. A creep may be set up in a part of a shed, a stall, a separate portable building, or a fence line. A pasture creep should provide shade and should be located near an area where mares tend to congregate. The creep should have at least two entrances; the mares are restricted by the height (approximately 46 inches) or width (approximately 16 inches) of the opening of the entrances. There are commercial "foal feeders"—feed boxes with a series of bars across the top close enough together that the foal can get its nose into the feed but the mare cannot. These foal feeders are better than nothing, but a lot of mares find they have awfully long tongues.

Stallion Facilities

Stallions can be kept in the main barn, but they should have their own large, secure stall. Stud farms have separate stallion barns, often with adjacent stud paddocks. There is an increasing interest among horsemen in keeping stallions in individual sheds in stud paddocks. The paddocks should have strong fences. Most managers prefer that the stud fences be 5 feet high and that the paddocks be separated by at least a 10-foot lane to prevent fighting among the stallions.

Breeding Shed

Farms that stand a stallion must have a designated enclosed area to breed mares. A breeding shed should be large enough (at least 20 feet by 20 feet) and high enough (12- to 15-foot ceiling) to be safe for horses and handlers (see Section 11.2). The safety of horses can be further ensured by padding the walls and installing a dust-free, slip-proof floor. Crushed stone or washed gravel is the floor surface most commonly used in sheds, but other materials, including rubber, asphalt, carpeting, and synthetic turf, are also used. The stone floor can be dug up to provide a hole or mound for the mare to stand on depending on her size. The breeding sheds usually include a teasing board that can be folded out of the way. Sometimes provision is made for an observation area for the mare owner or his representative. The breeding shed should have facilities for examining mares and a laboratory equipped with the necessary sterilizing and sanitizing equipment.

Hay, Bedding, and Other Storage

Hay and bedding has traditionally been stored in lofts over the stalls where it is convenient to use. However, it takes labor to put hay into the loft and take it out again and it is expensive to support hay up in the air. But it is the dusty and combustible nature of hay and bedding that makes a compelling reason for a separate hay barn. Architect Stan Gralla points out that the "most universally overlooked and poorly planned structure on any horse farm is the hay barn." He concludes that the hay barn should be convenient to the other barns with plenty of space between buildings for turning equipment. In fact, for fire safety the hay barn should be 100 feet from any other structure. The barn should be designed with doors 12 feet wide and 14 feet high to easily accommodate the vehicles used for delivering and hauling hay. If it is large enough to hold a year's supply of hay, there may be some savings on purchased hay. The hay bays should be sized to conveniently accept the size bales you are using. For example, a 12 by 14 bay would accommodate 36 bales per layer stacked in either direction and that amount represents about a ton of hay. The height of the bays should be 12 to 15 feet as there is no advantage to having to stack more than 8 to 10 layers.

Table 20-3 gives the storage space requirements for feed and bedding. Finally, the hay barn should be built to preserve hay quality. In some areas of

Table 20-3
Storage space requirements for feed and bedding

Grain	lb/bu	lb/cu ft
Corn		
Shelled	56	44.8
Ear	70	28.0
Oats	32	25.6
Barley	48	38.4
Sweet feed		30

Hay and bedding	(cu ft/ton)
Loose:	
Alfalfa	450–500
Grass hay	450–600
Straw	650–1000
Sawdust	
Baled:	
Alfalfa	200–300
Grass	250–330
Straw	450–500
Shavings	100

the country freshly baled hay undergoes a "barn cure" that produces heat. If the temperature of the hay exceeds 100°F dustiness increases and hay quality is reduced. Therefore, it is important that the building be properly ventilated with air intakes at bottom and vents in the eves and ridge (see Figure 20.6).

Feed Room

A feed room in which to formulate and store feed can be located within the stable area. It must be tight, dry, and conveniently located, and must keep out insects, birds, and rodents. The door should be so designed that it cannot be accidently opened by a horse. The feed room may be equipped with metal-lined storage bins, grain crimper, feed cart, and scales. Steel, rodent-proof bulk grain tanks provide good storage and should be so located that the feed grains are augered into the feedroom. On small farms, galvanized cans provide adequate storage.

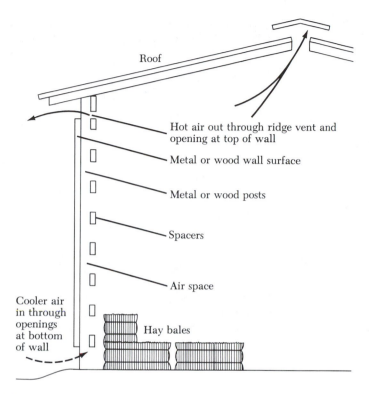

Figure 20-6
Ventilation of enclosed pole-type hay barn using natural air flow. (From Gralla, 1981.)

Barn Office

A stable that serves the public should have a central office where visitors can be greeted, records kept, and business transacted. The office should be comfortable and should be convenient to clean restrooms for the public and employees.

Tack Room

A barn must have a dry, clean room where tack such as bridles, halters, and saddles can be stored. Many tack rooms are show areas where winnings are displayed, but they are also working areas that provide storage for grooming materials, blankets, clippers, farrier's tools, and medications, as well as tack. There should be a refrigerator for storage of pharmaceuticals (see Section 19.6).

Wash Rack

Show, training, and sale stables often have specific areas where horses can be washed. Rubber mats on the floor are recommended, as the area must ensure good footing when wet and must have adequate drains. The wash area must have hot and cold running water turnoffs which should be recessed or located well out of the way of horses.

Swimming Pool

Swimming pools have become popular for conditioning horses, particularly those with leg problems. Swimming increases stamina, wind, and muscle tone while preventing trauma to the feet and legs caused by concussion.

Hot Walker

The hot walker, a mechanical device that leads horses in a circle at a slow walk in order to cool them after training, is another labor-saving device now seen on many farms. The walker has recently gained popularity in show stables for use with performing horses and as a means of developing and conditioning halter horses. Some hot walkers are located under a shelter so they can be used in all weather.

Arena

An indoor arena enables a horse farm to operate regardless of the weather and is indispensable for many operations. The minimum size for a training arena is 60 feet by 100 feet; an arena twice that size is needed for shows and public events.

In planning an arena, special attention must be given to ventilation, the riding surface, insulation, heat, lights, entrances, and location. If the arena is to serve the public, consideration must be given to such items as parking, seating, rest rooms, and food services.

REFERENCES

Blickle, J. D., R. L. Maddex, L. T. Windling, and J. H. Pedersen. 1971. *Horse Handbook; Housing and Equipment.* Midwest Plan Service. Ames: Iowa State University.

Dunn, Norman. 1972. Horse handling facilities. *Stud Managers Handbook.* Vol. 8. Clovis, California: Agriservices Foundation.

Ensminger, M. E. 1969. *Horses and Horsemanship.* Danville, Indiana: Interstate Printers & Publishers.

Fallon, Edward. 1967. Breeding farm hygiene. *Lectures Stud Managers Course.* Lexington, Kentucky: Grayson Foundation.

Gralla, Stan. 1981. *Horseman's Architect.* Oklahoma City: S. Gralla.

Harvey, Harry M. 1968. *"Stock farm management."* Care and Training of the Trotter and Pacer, Chap. 18. Columbus, Ohio: United States Trotting Association.

Hays, F. P. 1984. Fence Sense. *Equus,* 77:56.

Herbert, Kimberly S. 1987. Air for health. *The Blood-Horse* (February): 1292–93.

Herbert, Kimberly S. 1987. Fire safety. *The Blood-Horse* (February) 1430–31.

Hollingworth, Kent. 1971. New-design sheds serve well. In *A Barn Well Filled.* Lexington, Kentucky: Blood-Horse Magazine..

McNamee, Michael A., and Edwin A. Kinne. 1967. *Pasture and Range Fences.* Mountain States Regional Publication No. 2R. Tucson: University of Arizona.

O'Dean, Joseph C. 1966. Veterinarian's view of farm layout. Lexington, Kentucky: The Blood Horse (December):3790–91.

Scarborough, J. N., and C. M. Reitnour. 1985. *Fencing for Horses.* Extension Bulletin 136. Newark: University of Delaware.

Self, Charles R. 1979. Fencing horses in, neighbors out. *Equus* 36:19.

Smith, Peter C. 1967. *The Design and Construction of Stables.* London: J. A. Allen.

Stone, Charles H. 1971. Good Fences Make Good Neighbors and Good Horse Farms. In *A Barn Well Filled.* Lexington, Kentucky: Blood-Horse Magazine.

Sutton, H. H. 1963. The Veterinarian and Horse Farm Management. In *Equine Medicine and Surgery.* 1st ed. Wheaton, Illinois: American Veterinary Publications.

Timmins, Merrill S. Jr. 1971. *Fences for the Farm and Rural Home.* Farmers' Bulletin No. 2247. Washington D.C: U.S.D.A.

Toby, Milton C. 1980. *Col. Sager, Practitioner.* Lexington, Kentucky: Owners and Breeders Association.

Valliere, Donald, W. 1971. A Barn for All Seasons. In *A Barn Well Filled.* Lexington, Kentucky: Blood-Horse Magazine.

Willis, Larryann C. 1973. *The Horse Breeding Farm.* Cranbury, New Jersey: A. S. Barnes.

CHAPTER 21

MANAGING A HORSE FARM

21.1 Care of the Herd

Successful management of a horse operation requires attention to many details and the application of considerable knowledge and skill. Without proper planning of the year's activities and a sufficient labor force, the job can be overwhelming. The major tasks should be distributed throughout the year to ensure that they are completed at the proper time. Although it is impossible to consider all the necessary details of management, a suggested schedule is discussed here.

Midwinter

December and January (midwinter) are good months for inside maintenance and repair work. When the barns and stables are inspected for necessary repairs, they should also be inspected for fire safety.

Stallions are prepared for the breeding season, which usually starts on February 15. In localities with severe winters, the breeding season may start later. Regardless when the season begins, the stallion's feed must be regulated so that the stallion is in good breeding condition. Overweight stallions should be placed on a reducing diet so that the weight is lost before the breeding season. Thin stallions should be gaining weight at the onset of the breeding season. Stallions that will be worked and also bred to mares may need to be a little overweight at the start of the breeding season because they may lose weight as the season progresses. Stallions seem to have a better mental attitude toward breeding mares if they are exercised daily and their muscles have good tone. They should be given a physical examination for breeding soundness and semen quality (see Chapter 10). Unsound horses can be treated, and the book (the mares that are to be bred to a particular stallion) of stallions with poor semen quality can be reduced. Poor semen quality also requires that special attention be given to the number of services per week and that frequent semen evaluations be performed. Stallions that are standing at stud for the first time should be trained to brood mares. If they are returning from performance training, they may need to "let down," as discussed for replacement fillies (below).

The stallion's feet should be trimmed every 6 to 8 weeks. Unsoundness resulting from negligent foot care, particularly of the hind feet, can prevent the use of the stallion if he is unable to mount mares. The stallion's teeth should be checked for "hooks" (Chapter 3) and floated if necessary. A regular worming schedule should be established according to local conditions. As previously discussed (Chapter 17) , stallions should be dewormed at least every 6 months, and it is suggested that they be dewormed every 4 months.

Pregnant mares should be fed so that they are gaining weight. During the last 90 days before foaling, they should gain approximately 5 percent of their normal body weight. Their feet should be trimmed or shod every 6 to 8 weeks. Their teeth should be checked for hooks and other disorders, such as a long tooth growing into the area left by a broken off or lost tooth.

Pregnant mares should be visually checked at least once and preferably twice a day if they are in the last 90 days of pregnancy. Abortions during this period may be complicated and may lead to the mare's death unless she receives veterinary assistance. When deworming, one must be careful and use anthelmintics that are safe for pregnant mares.

Replacement fillies that have been selected to enter the broodmare band are often returning from performance training. After a horse has been in keen competition such as racing, competitive trail riding, or roping for a year or more, it should be given time to relax or let down for 2 to 6 months. This vacation allows the horse sufficient time to relax and readjust its mental attitude, and allows its body structures to recover. Most of the horses are keyed up and quite nervous when they are returned to the farm to let down; therefore, the early stages of the let-down period require time and patience. The ideal situation is to have a box stall with an outside paddock and a small pasture. During the first week or the first few weeks (the time depending on the extent of the nervous condition and the temperament of the horse), it will have to be led out to the pasture and grazed at the end of a lead shank. A couple of weeks may be required before the horse will hold its head down to the ground long enough to graze. The horse will require extra grooming time and frequent washings during the early part of the let-down period. The feet should be shod for normal activities rather than for competition. After the horse stops pacing the fence in its paddock and will graze quietly for a few minutes, it can be released in the pasture. For the first few times, the horse should be released for only an hour or so until it no longer wants to run and is relaxed. Because of its taut muscle tone, the horse should be exercised each day in a relaxing manner until it is sufficiently let down to stay in the pasture all day. The horse's diet is a critical part of the letting-down process. Concentrated feeds must be replaced with hay to prevent digestive disturbances and to prevent the horse from becoming overweight. Because a longer time is required to eat hay, the horse will have something to do and will not get bored. As forage is consumed during the grazing period, the amount of hay can be reduced.

During November, a breeding preparation program should be started for the open and barren mares and replacement fillies. They should be exposed to an artificial light regime (see Chapter 11) and teased at least 3 times a week. When each mare has her first estrous period, she should be given a female genital examination. If the need is indicated from visual inspection, she should be cultured for genital tract infections. Treatment for reproductive disorders should be started during the breeding preparatory period of November, December, and January. The mares' feet should be trimmed or shod every 6 to 8 weeks, and their teeth should be examined and any imperfections corrected.

Two-year-olds, yearlings, and weanlings should be fed so that they continue to grow and develop at an optimal rate, and their feet must be trimmed every 4 to 6 weeks. Young horses that require corrective trimming may have their feet trimmed at more frequent intervals. The yearlings and 2-year-olds are dewormed every 3 months, and the weanlings are dewormed every 2 months. If necessary, some weanlings should be worked with to continue their gentling process. Yearlings that will be raced as 2-year-olds may be placed in light training; if they have finished their preliminary training, they are given a rest period for further development. Two-year-olds returning from the racetrack may be given a rest period before training for their 3-year-old year. Two-year-olds that are to be ridden under the saddle may be placed in light training.

Late Winter

The months of February and March are the beginning of the breeding season for horses. The stallion must be fed properly to maintain good condition. He should be exercised daily. Semen evaluations should be performed weekly if the stallion is breeding several mares each week. During the early part of the season, it may not be necessary to check semen quality as frequently if breeding activity is minimal. One must be careful not to overuse a stallion if several mares are being bred or they come out of winter anestrus.

Pregnant mares are cared for in February and March as they were during November, December, and January. At least 2 weeks before foaling, they should be housed in their foaling area. Mares that are to be foaled and bred on other farms should arrive about 30 days prior to foaling. This allows ample time for them to develop immunity to diseases common to the new farm. This is important for passive immunity of the young foal. Approximately 30 days before their foaling due date, the mares should be checked to see if their vulva was sutured (Caslick's operation, see Chapter 11). If so, the vulva should be opened up. After the mare foals, her feed should be gradually increased to meet her needs for milk production, and she should gain 0.25 to 0.5 pounds of body weight per day. If the mare is to be rebred, she should be teased on a regular schedule; teasing should begin approximately 7 days after foaling.

Barren, open, and maiden mares should be teased daily or at least on alternate days. As the mares come into estrus, their ovaries should be palpated per rectum or ultrasound used on a daily or alternate-day basis to determine the optimal time to breed. All mares should be given a series of pregnancy checks 30 to 40 days after their last breeding if they have not returned to estrus (see Chapter 11).

The 2-year-olds and yearlings are managed as they were in November, December, and January. The young foals should be handled so that they become gentle. Some horsemen prefer to halterbreak them at an early age, but others prefer to wait until just before or after weaning.

Spring Stallions, mares about to foal, and mares still being bred are managed as previously indicated for February and March. Open, barren, and maiden mares that have conceived should be turned out to pasture after they have been dewormed. They should be vaccinated for sleeping sickness, strangles, influenza, and rhinopneumonitis and given their booster shots for tetanus.

Foals should be wormed when 6 to 8 weeks old and then every 2 months until they are yearlings. Foals will start eating grain or a grain-supplement mixture. Foals can be creep-fed to prevent the mare from eating the foal's food. Their feet require trimming at 4-week intervals. Those requiring corrective trimming should be slightly trimmed more frequently. The foals should be immunized against tetanus, sleeping sickness, rhinopneumonitis, strangles, and influenza when they are 2 to 4 months of age.

Stallions are immunized against the common diseases and dewormed. They are fed to maintain proper body condition and exercised on a regular schedule. At the end of the breeding season, it is wise to evaluate the stallion's semen and give him a soundness check. Any disorders can then be treated.

Lactating mares should be fed to maintain body condition and milk production. They should be dewormed on a regular schedule. The schedule will depend upon the degree of internal parasite load as determined by fecal counts of parasite eggs. Dry mares that are open (not bred the previous year) or pregnant should be kept on a maintenance diet and dewormed on a regular schedule with the lactating mares. All mares should have their feet trimmed at 6- to 8-week intervals.

Yearlings are managed as for the late winter period. Those that are to be trained are given their basic handling preparatory to their specific training. They should be immunized with the rest of the horses.

Foals should be weaned at 4 to 5 months of age. Many considerations determine the exact time to wean the foals (feed conditions, condition of mare, mare's milk production, and availability of physical facilities). Foals should be on a full ration by the time they are weaned. Registration applications must be completed and sent to the breed registry.

Decisions should be made concerning to which stallion each mare will be bred the following year. All necessary breeding records for the year must be completed and sent to the breed organization.

21.2   Record Keeping, Identification, and Contracts

Records

A set of records should be kept for each horse regardless of whether it is owned by a one-horse owner, a boarding stable, or a large breeding farm. Records for a horse may be very simple or very sophisticated, depending upon the use of the horse. The records on large establishments use the same type of information as a one-horse owner but integrate the records and business activities into a comprehensive record and bookkeeping system. In the past, most horse records were hand-recorded. Today, progressive horse owners and managers are using a computerized system. The records required for an equine enterprise are the same whether they are hand-entered or computerized. A list of equine computer software systems is given in Appendix 21-1.

Computerized Records   Functions of a computerized record system can be divided into two basic functions: service and management (Truax, 1983). The service function includes all those mundane tasks associated with record keeping, (i.e., accounts receivable, inventory, payroll, etc.). The management function is associated with decision making. The information entered into the computer for the service function is rearranged, analyzed, and evaluated to allow the making of management decisions based upon facts. A farm manager who is able to couple this type of information with a knowledge of the basic biology of the horse has the basic skills to excell.

Computerized record systems offer several advantages. A well-programmed system will allow a savings of clerical time. This occurs when a single entry is integrated into several files and eliminates repetitive hand entries to all the necessary files. Routine reports, such as those to breed registry associations, can be generated with minimal effort. Information necessary for effective and efficient management is readily available to management with minimal preparation time. A computer may provide the incentative to become better organized and to keep better records. This results from the necessity to develop well-defined procedures to insure more complete and improved control of records and flow of documents. The end result is that all activities associated with each horse, as well as all business activities, are documented. Client bills are accurate and complete and the farm does not absorb expenses for drugs, veterinary treatments, tack, and other items that employees fail to account for with handwritten record systems. However, remember that if owners, managers, or employees do not have good records and/or do not take time to put the records into the computer system consistently, poor management decisions will continue to be made and cost accounting will not be effective. Organization is enhanced because most systems allow automated scheduling of routine or special procedures and the generation of a daily, weekly, or monthly scheduled-activities alert report. This alerts managers when immunizations, deworming, weaning, and other procedures should be performed.

One of the most important advantages of a computerized system is the availability of a financial analysis at any time one is required. Cash-flow status can be determined and budgeting or management decisions can be made to meet cash-flow demands. Financial forecasting can be made at any time. Better vendor and client relationships result because bills are prepared or paid on time. Farms are able to provide their clients with more information about the status of their horses at the time of billing. Clients are then more inclined to pay their bill on time and cash flow is improved.

When selecting a computerized system, several steps should be followed. The first step is education. One should learn the general terms used by people working with computers and the basic concepts about computers, and should acquire a general knowledge of what horse-management software systems are capable of doing. The second step is to establish what the computer is to accomplish and prioritize the required tasks. The third step is to select the software system that will perform these tasks. Finally, the hardware necessary to run the selected software system is selected.

Certain characteristics of software systems should be evaluated prior to purchase or lease. Ease of operation and data entry are important and are referred to as being "user-friendly." The program should be no more complicated than is necessary to perform the required tasks. Nonessential features add to the cost and are never used. Also, the program should have an error-checking capability so that it will only accept the type of information the program is asking for. User-friendly programs check all entries and refuse inappropriate data (i.e., horse name for a date or vice versa). A well-written and easy-to-understand description of how to use the program is essential. User-friendly programs contain part of the documentation in the program so that one may receive instruction in a "menu" format (list of options) or by the use of a "help" function key (Figure 21-1). Support services by the retailer and the software company are just as important, because they may be needed to solve problems and answer questions. The above-described characteristics decrease the prior knowledge about computers required to operate a system. Speed of operation or execution can be an important consideration for larger systems. Slow execution times can dramatically increase the time required to generate reports and to make data entries. On time-shared systems, this increases the cost of using the system.

Data security is also important. On large equine enterprises, various personnel such as the veterinarian, farrier, foaling-barn manager, and stallion manager may need to enter and to access data. Only certain personnel or the owner need access to financial records or other specific files. Security codes (passwords) limit the files each person has access to. Other security considerations are the ability to easily make duplicate (back-up) files so that valuable information won't be lost. Some files are duplicated every few minutes while others are copied on a daily or weekly basis. Certain back-up files should be stored in a separate and safe location to prevent loss by fire or theft. Transient power surges are a problem in many areas and the computer system must be protected against them. Finally, the program must be flexible enough to take into account growth and changes in the enterprise.

Most computerized-farm-record systems are divided into at least three systems that are integrated: horse and owner information, horse management, and financial (Figure 21-1). The initial entry of a horse into the system requires many facts about the horse—such as name, registration number, breed, birth date, sire and dam, registration numbers of sire and dam, date

```
                    EQUINE MANAGEMENT SYSTEMS
------------------------------------------------------------------------


          BREEDING INFORMATION ENTRY AND REPORTS

            1. Booking Information and Reports

            2. Teasing Information and Reports

            3. Breeding Information and Reports

            4. Foaling Information and Reports

            5. Broodmare Inventory

            6. Breeding Accounts Receivable

            7. Breeding Year-End Reports

          Enter selection number or RETURN to exit  _____
```

Figure 21-1
Menu for a horse management file regarding breeding. (From Equine Management Systems, Inc.)

acquired, value, depreciation rate, owner, owner's addresses, telephone numbers, and agent, billing information, coat color, height, weight, conformation facts, special feeding and handling instructions, identification number, and location. The facts that are entered will depend on the software utilized and the requirements of the horse enterprise. The facts are used by the program to initialize a number of files in the system for the horse. A typical example of the general file for a horse in the Equicomp Inc. software system is shown in Figure 21-2. After the initial information is entered, all activities related to the horse are entered. Generally, the activities are broken down into entry files related to breeding and reproduction, health (Figure 21-3), farrier, training, sales, and miscellaneous. Barn sheets may be carried by personnel to record their activities (Figure 21-4). They may be for general recording of activities or they may result from daily activity reminder reports. For stallions, information related to semen analysis (Figure 21-5) and the breeding of mares is entered. The response to teasing, ovulation, date bred, palpation data, and foaling are typical data entered for broodmares. The costs for all services and activities are predetermined and entered in a rate file (Figure 21-6). The rates may be changed at any time and special rates may be made for any activity. When the activity code is entered into an activity file, the financial updates are made throughout all the files.

The above files can be sorted in many different ways, depending on the software system, to automatically generate reports that are useful in scheduling routine management activities, making management decisions, and analyzing the effects of changes in management procedures. Automatic scheduling of all routine management activities related to various classes of horses is based on management decisions programmed into the system. Examples of routine management decisions include: days post-foaling for pregnancy checks; intervals between deworming, trimming feet, and shoeing; age at weaning; immunization schedules; and filing stallion breeding reports. A schedule of activities can be printed when needed for the current day or for any time period in the future (Figure 21-7). These activity reminders help make sure that routine activities are performed (Figure 21-8) and prevent the last-minute rush to meet a required report deadline. The reminder reports can be sorted according to personnel responsibilities so that they serve as a daily schedule for specific personnel. Once the procedure has been completed and entered into the system by the person performing the task or by office personnel, it no longer appears on the activity reminder report. Examples of reports that can be automatically generated for one, a group, or all horses include: inventories sorted according to many different parameters (Figure 21-9); complete health (Figure 21-10), immunization, worming, and broodmare veterinarian histories; farrier history; breeding history; horse arrivals, departures or movements; a list of clients and specific information about them (Figure 21-11); owner-horse cross-ref-

```
equicomp  06/29/84  15:00:38                                                    PAGE:  1

                              EQUICOMP NORTH
                    H O R S E   M A S T E R   R E C O R D
                                  Lhuriffa

REF NO: 000203          SEX: M    DATE OF BIRTH: 1/23/62   AGE: 22    ACTIVE STATUSES: MB S

BREED: AR    REG NO: 020709        SECOND BREED:             SECOND REG NO:        DO YOU OWN/LEASE: Y
COLOR: ch    MARKINGS: star, strip, snip, 2 left socks.  freezed  HEIGHT: 14    BLOOD TYPE: unknown    IMPORTED:
SIRE: Lutaf           BREED: AR  REG NO: 002590    REF NO: 000000
DAM: Zeriffa         BREED: AR  REG NO: 010770    REF NO: 000000

OWNER:                                303-443-3726         LESSEE:
# 000001   Eileen S. Hunnes                                # 000000
           Equicomp North
           1722 14th Street, Suite 230
           Boulder           , CO 80302

BILL TO 1:                            303-443-3726         BILL TO 2:
# 000001   Eileen S. Hunnes                                # 000000
%: 100     Equicomp North
           1722 14th Street, Suite 230
           Boulder           , CO 80302

VET:                                  303-435-2434
# 000011   K. M. Brown , D.V.M.       303-454-1478
           Ambulatory Equine Service
           3501 Cottonwood Lane
           Evergreen         , CO 80513

INSURANCE CO: none               POLICY NO:              EXPIRATION: 0/00/00   AMOUNT:      $0

LAST IMMUNIZATIONS (date and schedule):
  EAST/WEST ENCPH   2/13/83 P        2/13/83   P    RHINO - KILLED  LAST NEGATIVE COGGINS: 4/12/83   CID CARRIER:
  INFLUENZA         0/00/00          0/00/00        RHINO - LIVE        9/16/83 P   RHINO  9/16/83 P
                    TETANUS
                    STRANGLES             WORMING METHOD: P        WORMING MEDICATION: Telmin
LAST WORMING (date and schedule):  9/16/83  D   TYPE: TRIM     SPEC. SHOEING INSTR:none
LAST FARRIER (date and schedule): 10/01/83  D   FEEDING INSTR.:1 1/2 lbs. oats + 1 flake grass hay + vitamins + soy - (2x)
HANDLING:
COMMENTS: Difficult to get pregnant and carry foal to term - biopsy 3+

LOCATION: 03           BILLING CODE: NONE         FARM STABLING LOCATION: S. Barn   COLLAR TAG NO: R100
EXP. ARRIVAL DATE: 2/19/83   ACT. ARRIVAL DATE: 2/11/83   EXP. DEPARTURE DATE: 0/00/00  ACT. DEPARTURE DATE: 5/02/83
LOCATION OF HORSE IF NOT ON THIS FARM:                                    #: 000000

LAST FOALING DATE: 3/15/82    REF. NO. OF LAST FOAL: 000000
SIRES STUD FEE WHEN BRED: $400   BIRTH WEIGHT: 0   HEIGHT AT BIRTH:  0.0 hands
CONDITION AT BIRTH: good                                              DAYS INUTERO: 330

DATE OF PURCHASE: 1/22/76    PURCHASE PRICE:    $1500    PREVIOUS OWNER:                         #:000000

DATE DIED: 0/00/00   CAUSE OF DEATH:

DATE ADDED TO FILE: 6/26/84   DATE LAST CHANGED: 6/29/84
```

Figure 21-2
Master record for horse. (From Equicomp, Inc.)

794

```
screen HL.2B                  EQUICOMP NORTH                    06/29/84
                         H E A L T H   R E C O R D S
                            ADD HEALTH RECORDS

   * Enter the date to be updated
       (ie. 06/07/82):               [07/11/83]

   * Enter horse reference number    [000000]
        or horse name and            [Sassys Lark          ]
        identifier (if needed)       [ ]

   * Enter no. for health care type [03]
        1.  FOALING     6.  WEANING
        2.  FARRIER     7.  WORMING
        3.  IMMUNIZE
        4.  MISC
        5.  VET CARE

     Display billing screen (Y/N)    [Y]

     Enter farm no. where work done
         if different than where
         horse currently stabled     [00]

   F1-NEXT SCREEN                                           F10-RETURN
```

Figure 21-3
Prompt for an entry into the health records. (From Equipcomp, Inc.)

erence (Figure 21-12) or horse-owner cross-reference; sales lists sorted by type of horse, price, or other parameters; training reports; and mare breeding summaries including their ages (Figure 21-13) estrous cycles, breeding dates, pregnancy-check dates, and foaling dates that are sorted according to month (Figure 21-14), season (Figure 21-15) or lifetime (Figure 21-16). Analytical reports such as a stallion summary report are important (Figure 21-17). The system can be programmed to alert the manager that mares are failing to conceive when bred to a certain stallion by maintaining a continuous analysis of the average number of estrous cycles per conception. By analyzing the mare records for specific breeding periods, further analysis can aid in determining when the problem started.

Most systems have a completely integrated accounting system, so that all financial records are updated when each activity is entered. The Horse Manager software system developed by Owens and Company, Lexington, Kentucky, has a number of features in the accounting files. Its payroll file prepares a payroll input form, payroll checks, an individual earnings history, a check register, government reports, W-2 forms, an analysis of labor distribution, and automatically posts to general ledger. The accounts payable file edits incoming invoice coding, produces a cash requirements report by period, and prepares vendor checks and remittance advices, account status inquiries, vendor histories, mailing lists and lables, transaction journals, and does a bank reconciliation, automatic posting to general ledger. The general ledger and financial reporting file produces journals, ledgers, subledgers,

DATE 09/07/84
Location: Blk Cyn

WORMING WORKSHEET (282)

NAME	ID#	LAST DATE	TYPE	#DA	LAST DRUG	DATE DONE	DRUG	CHARGE
ZOSIA	287	05/17/84	M	113	EQUAVLEN	.../.../...	....	$.....
INES	305	05/17/84	P	113	Anthelcide BRED 04/09/84....	.../.../...	....	$.....
KANADA	388	05/17/84	P	113	Anthelcide BRED 08/01/84....	.../.../...	....	$.....
KF PAVANE	377	05/31/84	P	99	Strongid-T BRED 04/09/84....	.../.../...	....	$.....
NARLITE	244	05/17/84	P	113	Anthelcide BRED 05/21/84....	.../.../...	....	$.....

DATE 09/08/84
Location: Blk Cyn

FARRIER WORKSHEET (288)

NAME	ID#	LAST DATE	T/S	#DA	COMMENTS:	WORK	CHARGE	DATE DONE
AMBER MYST	(331)	04/13/84	T	148	....	T/S	$..........	.../.../...
AMBROSIA	(253)	11/09/83	T	305	....	T/S	$..........	.../.../...
AN MALAGUENA	(284)	06/08/84	T	92	....	T/S	$..........	.../.../...
ANALLIN	(102)	11/02/83	T	312	....	T/S	$..........	.../.../...
KUZYNA	(343)	08/24/83	S	382	....	T/S	$..........	.../.../...

Figure 21-4
Activities record carried by farm personnel to record their activities that will be entered into computer. (From Howard Kale, Karho Farms.)

equicomp 06/29/84 15:03:31 — EQUICOMP NORTH — PAGE: 1

STALLION SEMEN REPORT

1983

STALLION: NDL Vail — REFNO: 000001 — REG NO:

EJC. #	LOC	DATE	TIME	COLLAR TAG # R COMMENTS	# MTS	GEL FREE SEMEN-ml	CONC. x 10-6	PERCENT MOTILITY	INSEM DOSE	TSO x 10-9	MARES BRED REF NO	MARES BRED NAME
001	02	01/15/83	10:20	Doe collected	02	125.0	60.7	78	0.0	50.0		
002	02	01/17/83	01:50	Doe collected	01	100.0	65.8	90	0.0	80.0		
003	02	01/23/83	11:00	Jones collected	03	130.0	80.0	85	0.0	70.0		
		JANUARY				118.3	68.8	84	0.0	66.7		
004	02	02/01/83	09:00	Doe collected	02	115.0	72.1	85	10.0	80.0	000014 000035	Fanatazee Balarina
005	02	02/03/83	09:30		01	116.0	60.0	75	20.0	50.0	000014 000035	Fanatazee Balarina
006	02	02/05/83	15:20	Doe collected	01	100.0	65.0	60	30.0	15.0	000014	Fanatazee
007	02	02/07/83	12:00	Jones collected	02	125.0	80.0	90	15.0	70.0	000035	Balarina
008	02	02/09/83	08:45	Jones collected	01	140.0	80.0	76	20.0	67.0	000023 000036 000042	Heavenly Flame Brave Lady Rangers Beutya
009	02	02/11/83	11:09		01	100.0	65.0	90	25.0	50.0	000023 000042	Heavenly Flame Rangers Beutya
		FEBRUARY				116.0	70.4	79	20.0	55.3		
		YEAR TO DATE				116.8	69.8	81	13.3	59.1		

Figure 21-5
Computer record of semen evaluations for a stallion. (From Equicomp, Inc.)

```
                    EQUINE MANAGEMENT SYSTEMS
                          RATE LISTING
                          16-Aug-84

       BOARD/TRAINING
       1   BOARD PASTURE                      8.00 DAILY
       2   BOARD STALL                       14.00 DAILY
       3   BOARD SUCKLINGS                    1.50 DAILY
       4   BOARD INDIVIDUAL PADDOCK          11.00 DAILY
       5   BOARD SALES PREPARATION           17.00 DAILY
       6   BOARD STALLION                   400.00 MONTHLY
       7   BOARD LAY UP PASTURE               8.00 DAILY
       8   BOARD LAY UP BARN                 14.00 DAILY
       9   STALLION BOARD - QUARTERLY     3,000.00 QUARTERLY
       77  LAY UP                            14.00 DAILY

       WORMING
       100 WORMING                           10.00 ONE-TIME

       IMMUNIZATIONS
       110 FLU VAC                            6.00 ONE-TIME
       111 TETANUS/SLEEP. SICK. INJEC.        7.00 ONE-TIME
       112 TETANUS INJECTION                  3.50 ONE-TIME
       113 SLEEP. SICKNESS INJ.               3.50 ONE-TIME
       114 STRANGLES INJECTION               10.00 ONE-TIME
       115 RHINOPNEUMONITIS INJ.              7.00 ONE-TIME
       116 PNEUMOBORT K INJECTION             7.00 ONE-TIME

       DENTAL
       120 FLOATING TEETH                 VARIABLE ONE-TIME

       GENERAL VET
       130 WOUND SUTURE                   VARIABLE ONE-TIME
       131 REMOVE SUTURE                  VARIABLE ONE-TIME
       132 COGGINS TEST                      15.00 ONE-TIME
       134 DRAW BLOOD JOCKEY CLUB            20.00 ONE-TIME
       135 FIRED SHINS                    VARIABLE ONE-TIME
       136 BLISTER                        VARIABLE ONE-TIME
       139 X RAY                          VARIABLE ONE-TIME
       140 CASTRATION                     VARIABLE ONE-TIME
       141 UMBILICAL HERNIA REPAIR        VARIABLE ONE-TIME
       146 POST FOALING EXAM              VARIABLE ONE-TIME
       150 INSURANCE EXAM                    15.00 ONE-TIME
       152 SEMEN EVALUATION                  25.00 ONE-TIME

       FOALING
       230 FOALING FEE                      150.00 ONE-TIME

       PALPATION
       231 PALPATION                          7.00 ONE-TIME

       BROODMARE VET
       232 PREGNANCY EXAM                     8.00 ONE-TIME
       233 SPECULUM OVARIAN EXAM             10.00 ONE-TIME
       234 CHORIANIC (HCG)                    8.00 ONE-TIME
       235 CASLICK SUTURE                    15.00 ONE-TIME
       236 CASLICK OPEN                   VARIABLE ONE-TIME
       237 CULTURE                           15.00 ONE-TIME
       238 UTERINE TREATMENT                 20.00 ONE-TIME
       239 PROSTAGLANDIN                     17.50 ONE-TIME
       240 PROGESTERONE                       5.00 ONE-TIME
       241 NAQUAZONE TABLETS              VARIABLE ONE-TIME
       242 GENERAL BROODMARE VET MISC.    VARIABLE ONE-TIME
       299 BRED TO:                       VARIABLE ONE-TIME

       BLACKSMITH
       300 TRIM                               8.00 ONE-TIME
       302 SHOE FRONT TRIM HIND              23.00 ONE-TIME
       303 SHOE ALL FOUR                  VARIABLE ONE-TIME
       304 REPLACE SHOE                       8.00 ONE-TIME
       305 SPECIAL SHOEING                VARIABLE ONE-TIME

       REGISTRATION
       350 REGISTRATION FEE - JOCKEY CLUB VARIABLE ONE-TIME
       351 REGISTRATION FEE - CAL BRED    VARIABLE ONE-TIME
       352 PREPARE REGISTRATION           VARIABLE ONE-TIME

       TACK AND EQUIPMENT
       401 HALTER: - NYLON                   15.00 ONE-TIME
       402 HALTER: - LEATHER                 38.00 ONE-TIME
       403 TACK AND SUPPLIES MISC         VARIABLE ONE-TIME

       ADVERTISING
       500 OTHER CHARGES                  VARIABLE ONE-TIME

       OTHER
       999 MISC.COMMENT FIELD ON BILL     VARIABLE ONE-TIME
```

Figure 21-6
Rate file that contains the codes and billing amounts for all farm activities. (From Equine Management Systems, Inc.)

```
COMPANY NO. 075                THE QUALITY HORSE FARM          RUN DATE-04/15/84
REPORT NO.  HFM310-D          SCHEDULED PROCEDURES REPORT      RUN TIME-10:27:17
                              SCHEDULED 6/25/84-6/31/84        PAGE    001

                   PROCEDURES SCHEDULED FROM 06/25/84 TO 06/31/84
------------------------------------------------------------------------------
  HORSE        HORSE               PROC.                              SCHEDULED
   NO.     ---------NAME---------   NO. --------DESCRIPTION---------     DATE
------------------------------------------------------------------------------

  01009 MR. ROMERO               280 WORM                             06/25/84
  10004 BETTY'S FORTUNE          530 TRIM                             06/26/84
  10004 BETTY'S FORTUNE          830 RECHECK FOR PREGNANCY            06/29/84
  10005 BANQUET BELL             830 RECHECK FOR PREGNANCY            06/29/84
```

Figure 21-7
Scheduled activities reminder for a week's period. (From Equine Management Systems, Inc.)

multilevel financial statements, budget and comparative statements, and cost per horse. The accounting system also automatically generates board bills and common charges, late charges, and special charges by horse; prorates billing to multiple owners; produces owner statements, income analysis, an aged receivable listing, and an owner mailing list and labels; and automatically posts to general ledger. In any accounting system, it is important for the system to be able to allow budget forecasting and analysis and to maintain an audit trail.

Boarding Stables

Because of the booming horse population, an increasing percentage of horses are being kept in boarding stables. The manager of a boarding stable should require the horse owner to sign a contract that clarifies for the owner

```
COMPANY NO. 075                THE QUALITY HORSE FARM          RUN DATE-04/15/84
REPORT NO.  HFM310-F          PROCEDURE EXCEPTION REPORT       RUN TIME-12:14:49
                              HORSES NOT TRIMMED IN 30 DAYS    PAGE    1

        Report of horses who have not had Procedure No. 530   TRIM
        performed in  030 days.

                                          TAG/   LAST DAYS
HORSE
 NO.      HORSE NAME            FRM BRN   STRAP   DATE SINCE   COMMENTS
------------------------------------------------------------------------------

10005 BANQUET BELL              018 003  X       22984  46
10006 NORTHERN LADY             018 006          22984  46 ABCESS RIGHT FORE, LAT
10010 LADY HANOVER              018 008          31084  36
10012 HAVANA                    018 008          31084  36
10002 CATCH ME                  018 010          31584  31
10008 CAREFUL RUNNER            018 010  X77332  31584  31
01009 MR. ROMERO                018 050          31084  36
20002 BUTTON UP                 002 060          22984  46
20001 EASTERN GOLD              002 066          31584  31
```

Figure 21-8
Activities file sorted to alert manager that a certain procedure has not been performed for 30 days. (From Equine Management Systems, Inc.)

COMPANY NO. 075
REPORT NO. HFM600-A

THE QUALITY HORSE FARM
MARE STATUS REPORT
LAST DATE BRED SEQUENCE
84 BREEDING SEASON

RUN DATE-04/15/84
RUN TIME-09:00:32
PAGE 001

CYCLE CODE	MARE NO.	MARE NAME / OWNER NAME	YOB	PRE-BRED STATUS / FOAL DATE	STALLION- CURRENT SEASON / STALLION- PRIOR SEASON	LAST BREED DATE	CF TEST DATE	PREG CHECK DATE	- DAYS SINCE - FOAL TO LDB	LAST BRED	PREG CHECK	PRESENT STATUS
	10012	HAVANA	80	MAIDEN	MR. ROMERO	02/12/84	00/00/00	03/15/84		63	31	IN FOAL
		MRS. L. B. COLEMAN		*01/17/85	NOT AVAILABLE							
BD	10005	BANQUET BELL	74	IN FOAL	MR. ROMERO	02/24/84	00/00/00	04/01/84	32	51	14	IN FOAL
		MR. HAL SAMUELS		01/23/84	LOYAL SANDMAN							
BD	10004	BETTY'S FORTUNE	73	BARREN	NOBEL CAST	03/02/84	04/09/84	04/01/84		44	14	IN FOAL
		DR. EDWIN POLK		*02/07/85	MR. ROMERO							
A	10010	LADY HANOVER	78	IN FOAL	SALUTE	03/19/84	00/00/00	00/00/00	8	27		IN FOAL
		MR. PAUL WILLIAMSON		03/11/84	ONE FOR ONE							
A	10006	NORTHERN LADY	75	IN FOAL	GREEN IVY	04/07/84	00/00/00	00/00/00	9	8		
		MR. HAL SAMUELS		03/29/84	MR. ROMERO							

Figure 21-9
Horse inventory file sorted for a pregnancy status report for all mares. (From Equine Management Systems, Inc.)

```
                          E Q U I C O M P   N O R T H
                       H E A L T H   R E C O R D S
                          Sassys Lark
                          DESC: grey blaze, 2 hind stockings, freeze marked

REF NO: 000201   BREED: AR        REG NO. 121168                     BIRTH DATE: 02/28/75


03  02/19/82  VET CALL   -  VET NAME: Craig
                            SERVICE: Uterine check - ovaries small uterine tone ok   COGGINS TEST TAKE
                                     health certificate filled out-took coggins -
                                     did insurance exam

03  02/19/82  FARRIER    -  FARRIER NAME: Bronzan              TYPE OF WORK: TRIMMED

03  02/21/82  WORMED     -  WORMED BY: Hunnes                  PASTE MEDICATION: febantel - Cutter

03  04/19/82  FARRIER    -  FARRIER NAME: in California        TYPE OF WORK: TRIMMED

03  05/06/82  IMMUNIZE   -  GIVEN BY: Calif. vet               INFLUENZA
                            EAST/WEST ENCPH   TETANUS

03  07/09/82  WORMED     -  WORMED BY: Craig                   TUBE MEDICATION: Strongid-T

03  07/16/82  FARRIER    -  FARRIER NAME: Bronzan              TYPE OF WORK: TRIMMED

03  01/09/83  IMMUNIZE   -  GIVEN BY: Hunnes
                            RHINO - KILLED

03  01/23/83  FARRIER    -  FARRIER NAME: Hunnes               TYPE OF WORK: TRIMMED

03  01/30/83  WORMED     -  WORMED BY: Hunnes                  PASTE MEDICATION: fenbendazole

03  01/31/83  IMMUNIZE   -  GIVEN BY: Hunnes                   INFLUENZA
                            EAST/WEST ENCPH   TETANUS

03  03/17/83  MISC       -  COMMENTS: pin point of wax

03  03/18/83  MISC       -  COMMENTS: goodly amount of colustrum showing on nipples

03  03/19/83  FOALING    -  FOAL REF NO: 000202  SEX: M  DESC: grey star, stripe, snip. Two black nose marks    HT:      WT:
                            FOALS CONDITION:
                            CARE GIVEN FOAL AT BIRTH: Betadined cord, gave enema, vitamin A shot, insurance exam
                            TIME OF DAY FOAL BORN: 001:00            HOURS/MIN UNTIL FOAL NURSES:  000:20
                            COMPATABILITY TEST RESULTS:
                            PARTIAL OR TOTAL FAILURE OF COLOSTRUM TRANSFER: birth CBC 22,000 - low blood counts 6 weeks - not CID
                            MARES CONDITION AFTER FOALING: mare acting like nothing happened - lay down once briefly
                            CONDITION OF UTERUS: vet says everything feels good on month heat
                            PROBLEMS WITH/CONDITION OF PLACENTA: placenta torn - vet found all pieces
                            COMMENTS: foaling took 10 minutes - 1¼" cornet band to middle of knee

03  04/16/83  WORMED     -  WORMED BY: Hunnes                  PASTE MEDICATION: Strongid - T

03  04/23/83  FARRIER    -  FARRIER NAME: Bronzan              TYPE OF WORK: TRIMMED

03  05/07/83  IMMUNIZE   -  GIVEN BY: Varra                    STRANGLES
                            STRANGLES
                            REACTION/COMMENTS: Burrough Welcome Strangles

03  05/27/83  VET CALL   -  VET NAME: Varra                    COGGINS TEST TAKE
                            SERVICE: Health certificate;   teeth floated

03  06/12/83  WEANING    -  COMMENTS: foal Bandoleeta weaned
```

Figure 21-10
Complete health record for a horse. (From Equicomp, Inc.)

equicomp 06/29/84 14:56:36

EQUICOMP NORTH

CLIENT LIST - BY CLIENT

PAGE 1

CLIENT #	ALPHA CODES	NAME/ADDRESS CLIENT NAME / RANCH NAME	ADDRESS	NUMBER PHONES	RECORD TYPES
1	1: HUNNES 2: EQUICOMP N 3:	Eileen S. Hunnes Equicomp North	1722 14th Street, Suite 230 Boulder CO, 80302	303-443-3726	AR - LATE CHARGES =
2	1: BROWN 2: EQUICOMP S 3:	Marian R. Brown Equicomp South	1722 14th Street, Suite #230 Boulder CO, 80302	303-443-3726	AR - LATE CHARGES =
11	1: BROWN 2: AMBULATORY 3:	K. M. Brown , D.V.M. Ambulatory Equine Service	3501 Cottonwood Lane Evergreen CO, 80513	303-435-2434 303-454-1478	
12	1: MOUNTAIN 2: 3:	Mountain High Syndicate	5097 St. Thomas Road Longmont CO, 80501	303-772-8687 303-772-8607	AR - LATE CHARGES = TB - THOROUGHBRED OWNER CC - CURRENT CUSTOMERS CG - HAULING CUSTOMERS
13	1: MOUNTAIN 2: 3:	Mountain Collection Partners	7097 Stroh Road Longmont CO, 80501	303-972-8687 303-972-8607	AR - LATE CHARGES = Y CC - CURRENT CUSTOMERS QH - QUARTER HORSE OWNER CE - ENG. PLS. CUSTOMERS CW - WEST. PLS. CUTSOMERS
14	1: STARRY 2: STARR Y 3:	Richard & Linda Starry Starr Y Arabians	10311 Lemon Ave. Dallas TX, 75229	314-268-1340	AR - LATE CHARGES = CC - CURRENT CUSTOMERS TB - THOROUGHBRED OWNER QH - QUARTER HORSE OWNER CE - ENG. PLS. CUSTOMERS
15	1: ROCKFORD 2: 3:	Ms. Karen Rockford	7931 E. Shoelace Road Parker CO, 80149	313-741-3254	AR - LATE CHARGES = CP - PAST CUSTOMERS TB - THOROUGHBRED OWNER
16	1: CHAMBERLAI 2: 3:	Don and Barb Chamberlain	P.O. Box 550 Aspen CO, 91857	303-949-5007 303-376-7291	AR - LATE CHARGES = Y CC - CURRENT CUSTOMERS QH - QUARTER HORSE OWNER MB - BREEDING MARES
17	1: RIPLEY 2: 3:	Mr. Ron Ripley	228 Bridge Street Aspen CO, 82557	303-476-3109 303-637-9067	AR - LATE CHARGES = Y CC - CURRENT CUSTOMERS TB - THOROUGHBRED OWNER CE - ENG. PLS. CUSTOMERS
18	1: CRYGLE 2: CARAD'N 3:	Dave or Sue Crygle Carad'N Arabians	21158 North 59th Street Broomfield CO, 80401	303-572-8538 303-673-6345	AR - LATE CHARGES = Y CP - PAST CUSTOMERS
19	1: EDEN 2: 3:	Barbara Eden	1245 E. Stanford Lane Lakewood CO, 80110	303-781-8790	AR - LATE CHARGES = CC - CURRENT CUSTOMERS TB - THOROUGHBRED OWNER CW - WEST. PLS. CUTSOMERS

Figure 21-11
Client information file. (From Equicomp, Inc.)

802

```
                  EQUINE MANAGEMENT SYSTEMS        16-Aug-84 10:50 AM
                  OWNER/HORSE CROSS-REFERENCE REPORT

OWNER                               HORSE
NUMBER      NAME OF OWNER           NUMBER   NAME OF HORSE

                                    *************************************
    1   JOHN GOETTER                    21   FLEET NASRULLAH
                                        22   ALABAMA GAL
                                       565   '82 MAYBASKET COLT
                                        13   MARIE ANTOINNETTE
                                        85   '84 MAYBASKET COLT
                                         3   MAYBASKET
                                        86   '84 VICTORIA HILL
                                       200   LINDA
                                       300   CANDY'S BID          50.000%
                                         2   L'ENJOLEUR           12.500%
                                        12   MIA LISA             25.000%
                                        99   BARBARY              50.000%
                                    *************************************
    2   MICHAEL HAYES                    5   OLYMPIAD KING        25.000%
                                         2   L'ENJOLEUR            5.000%
                                        12   MIA LISA             25.000%
                                        99   BARBARY              50.000%
                                    *************************************
    3   SHEILA KAHN                     12   MIA LISA             25.000%
                                    *************************************
    5   DONALD CURTIS                   21   FLEET NASRULLAH      50.000%
                                        22   ALABAMA GAL          50.000%
                                       565   '82 MAYBASKET COLT   50.000%
                                        13   MARIE ANTOINNETTE    50.000%
                                        85   '84 MAYBASKET COLT   50.000%
                                         3   MAYBASKET            50.000%
                                        86   '84 VICTORIA HILL    50.000%
                                       200   LINDA                50.000%
                                         5   OLYMPIAD KING        12.500%
                                         2   L'ENJOLEUR           50.000%
                                    *************************************
    6   HERMAN LYONS                     8   LITTLE TYCOON
                                        16   BANBURY TART
                                        25   BANBERTO
                                        87   '84 BANBURY TART F
                                         5   OLYMPIAD KING         7.500%
                                    *************************************
    7   RONALD OLDFIELD                 15   '83 DUCK FEATHERS
                                        21   FLEET NASRULLAH      50.000%
                                        22   ALABAMA GAL          50.000%
                                       565   '82 MAYBASKET COLT   50.000%
                                        13   MARIE ANTOINNETTE    50.000%
                                        85   '84 MAYBASKET COLT   50.000%
                                         3   MAYBASKET            50.000%
                                        86   '84 VICTORIA HILL    50.000%
                                       200   LINDA                50.000%
                                    *************************************
    8   DAVID RINES                     14   TERRESTORAUNT
                                    *************************************
    9   THOMAS FORD                     17   BEHIND THE LINE
                                        24   DON BEHIND THE LIN
                                         5   OLYMPIAD KING        10.000%
                                    *************************************
   10   ARTHUR FONTAINE                 18   BELLE MERE
```

Figure 21-12
Cross-reference file for owners and horses. (From Equine Management Systems, Inc.)

```
                          EQUINE MANAGEMENT SYSTEMS                              Page 1
                                                                         16-Aug-84 11:00 AM
HORSE TYPE=BROODMARE
SORTED BY: BIRTH DATE
             HORSE NAME            BIRTH DATE   SIRE NAME          DAM NAME
             ALABAMA GAL           01/01/57     DETERMINE          TROJAN LASS
             BREEZIN THRU          01/01/64     ESCADRU            SUMMER BLUES
             BELLE MERE            01/01/66     PROVE IT           LATE AGAIN
             FANFRELUCHE           01/01/67     NORTHERN DANCER    CIBOULETTE
             CONSISTENT PLEA       01/01/69     MR. CONSISTENCY    VERLIS PLEA
             BEHIND THE LINE       01/01/73     ROMAN LINE         SMALL VOICE
             BANBURY TART          01/01/77     RUKEN              STOOP TO CONQUER
             KUTIE PIE             01/01/79                        UNKNOWN
             MARIE ANTOINNETTE     04/04/79     OLYMPIAD KING      DARING LASSIE
             LINDA                 01/01/80     OLYMPIAD KING      MAYBASKET
             WHATSMYNAME           02/15/80     GATO DEL SOL       WHATSUP
             MAYBASKET             03/12/81     L'ENJOLEUR         CANDY'S BID
             CANDY'S BID           05/12/81                        BANBURY TART

Total of 13 items printed
```

Figure 21-13
Broodmares sorted according to their age. (From Equine Management Systems, Inc.)

what is to be done in emergencies and who is to be responsible for certain aspects of the horse's care. It is important that stable personnel have the authority to call a veterinarian in an emergency and are assured that the owner will pay for the services. The responsibility for shoeing, grooming, exercising, and training must be determined between the stable and owner. Both the board rate and the due date should be set. To prevent misunderstandings, the feeding program should be outlined and the type of housing facilities for the horse should be specified. A liability clause is usually included in the contract. The owner usually agrees to abide by the rules and regulations of the stable. At some stables, insurance is provided for tack, equipment, and the horse if the owner desires it. A clear understanding of these points can prevent hard feelings between owner and stable personnel. The owner should obtain the name and address of the stable owner's insurance company for the "care, custody, and control" policy.

Identification

Identifying horses and other livestock has always been a problem for livestock owners. If an identification system is to be successful, the marks used in the system must be visible from a distance, permanent, painless, inexpensive, easy to apply, nondamaging, unalterable, and adaptable to data retrieval.

equicomp 6/29/84 15:17:29

EQUICOMP NORTH
M O N T H L Y B R E E D I N G R E V I E W

PAGE: 1

STALLION: NDL Vail

MARCH - 1983

REF NO: 000001 REG NO: 184345

	1	2	3	4	5	6	7	8	9	10	11	12	13	14	15	16	17	18	19	20	21	22	23	24	25	26	27	28	29	30	31
R02 REF: 000014 Fanatazee Eileen S. Hunnes 303-443-3726						TI PL	TI	TI PN	TI BA	TI PL	TI BA	TI PL	TI BA	TI PL	TI BA	PN									TQ	TI	TI	TI PS	TI CI		
R05 REF: 000023 Heavenly Flame Mountain High Syndicate 303-772-8687											F									TQ	TI	TI	TI PL BA	TI PL	TI PL BA	PN					
R01 REF: 000035 Balarina Bob and Maggie Fortman 303-564-4655	TI	TQ PD	TI	TI PL	BA	TQ PL	TQ	TI PL								TQ	TI	TI PS				PN					I				
R04 REF: 000036 Brave Lady Barbara Eden 303-781-8790	BA	TQ																TQ	TI	TI	BA										
R03 REF: 000042 Rangers Beutya Bob and Maggie Fortman 303-564-4655			TQ	TI	TI	TI PS	TI	TI PL	BA	TQ PL	BA			TI PS	TI	TI	TI	TI PL													

A-arrived D-departed CC-cultured clean CI-cultured infected CP-culture pending CT-treated uterine infection S-sutured
TI-teased in heat TQ-teased questionable TN-teased not in heat BP-pasture bred BN-bred natural cover BA-bred AI F-foaled
OT-ovum transplant PL-palpated: large follicle PS-palpated: small follicle PN-palpated: no follicle PD-palpated:2+ follicles
FF-preg check in foal FP-preg check questionable FN-preg check not in foal AB-aborted/absorbed foal I-infused P-prostin

Figure 21-14
Breeding record sorted in a calender format. (From Equicomp, Inc.)

M A R E   B R E E D I N G   EQUICOMP NORTH   R E V I E W   –   1983

Fanatazee

REF NO: 000014   BREED 1: AR   REG NO: 188231   BRED TO: NDL Vail   REG NO: 184345
                    BREED 2:   REG NO:

JANUARY

1	2	3	4	5	6	7	8	9	10	11	12	13	14	15	16	17	18	19	20	21	22	23	24	25	26	27	28	29	30	31
					TI PL	TI				F						S				TQ	TI	TI	TI	TI	TI	TI PN				

FEBRUARY

1	2	3	4	5	6	7	8	9	10	11	12	13	14	15	16	17	18	19	20	21	22	23	24	25	26	27	28	29	30	31
						TI		TI			TI	TI	TI PL	TI	TI PD	TI BA	PN													

MARCH

1	2	3	4	5	6	7	8	9	10	11	12	13	14	15	16	17	18	19	20	21	22	23	24	25	26	27	28	29	30	31
					TI PL		TI PN	TI BA	TI PL	TI BA	TI PL	TI BA	TI PL	TI PN	PN									TQ	TI		TI PS	TI CI		

APRIL

1	2	3	4	5	6	7	8	9	10	11	12	13	14	15	16	17	18	19	20	21	22	23	24	25	26	27	28	29	30	31
	CT	CT						TI		TI PL	TI PL	TI PL	TI PL BA	TI PL BA	TI PL BA	TI BA	TI PL BA	TI	TI PN BA											

MAY

1	2	3	4	5	6	7	8	9	10	11	12	13	14	15	16	17	18	19	20	21	22	23	24	25	26	27	28	29	30	31
							FF																							

COMMENTS

02/16/83 – Right small, left large
03/28/83 – immature follicle
03/29/83 – E. Coli – Geritoria
04/01/83 – Geritoria
04/02/83 – Geritoria
04/03/83 – Geritoria
04/20/83 – strong heat – bred anyway
05/08/83 – right horn

A-arrived   D-departed   CC-cultured clean   CI-cultured infected   CP-culture pending   CT-treated uterine infection   S-sutured
TI-teased in heat   TQ-teased questionable   TN-teased not in heat   BP-pasture bred   BN-bred natural cover   BA-bred AI   F-foaled
OT-ovum transplant   PL-palpated: large follicle   PS-palpated: small follicle   PN-palpated: no follicle   PD-palpated:2+ follicles
FF-preg check in foal FP-preg check questionable   FN-preg check not in foal   AB-aborted/absorbed foal   I-infused   P-prostin

Figure 21-15
Breeding record sorted for a specific month. (From Equicomp, Inc.)

COMPANY NO. 075
REPORT NO. HFM600-B

THE QUALITY HORSE FARM
MARE BREEDING SUMMARY
ALPHABETICAL SEQUENCE

RUN DATE-04/15/84
RUN TIME-09:00:32
PAGE 001

MARE NO.	NAME / SIRE / DAM/DAM'S SIRE	YOB	SEA	STALLION BRED TO	TIMES BRED	LAST DATE BRED	DATE FOALED	BREEDING RESULTS	RACE HIST (FOAL DATA)	SALES PRICE (FOAL DATA)	LOCATION FARM	LOCATION BARN
10005	BANQUET BELL	74	84	MR. ROMERO	2	02/24/84	*01/29/85	IN FOAL			18	3
	POLYNESIAN		83	LOYAL SANDMAN	1	02/19/83	01/23/84	CHNUT COLT	NR			
	DINNER HORN		82	PICNIC RAIN	3	03/22/82		SLIPPED				
	POT AU FEU		81	MR. ROMERO	3	03/16/81	02/22/82	D BRN COLT	NR	200,000		
10004	BETTY'S FORTUNE	73	84	NOBEL CAST	2	03/02/84	*02/07/85	IN FOAL			18	3
	LUCKY MIKE		83	MR. ROMERO	3	05/01/83		BARREN				
	PIECE OF PIE		82	BOLD NATIVE	1	03/05/82	02/04/83	BAY COLT	NR			
	*THE PIE KING		81	HAMLET RUNNER	1	03/23/81	02/24/82	BAY FILLY	NR	195,000		
			80	NOBEL CAST	2	04/24/80	03/14/81	D BRN FLLY	SW1	180,000		
10012	HAVANA	80	84	MR. ROMERO	1	02/12/84	*01/17/85	IN FOAL			18	8
	RED FOX											
	GENUINE LIGHT											
	CUBAN CIGAR											
10010	LADY HANOVER	78	84	SALUTE	1	03/19/84	*02/23/85	IN FOAL			18	8
	HAND ME DOWN		83	ONE FOR ONE	2	04/06/83	03/11/84	CHNUT FLLY	NR			
	LADY BELL		82	VITZMA	1	03/11/82	02/15/83	GRY FILLY				
	DAN THE MAN											
10006	NORTHERN LADY	75	84	GREEN IVY	1	04/07/84	*03/12/85				18	6
	SOUTHERN DADDY		83	MR. ROMERO	3	05/15/83	03/29/84	BAY FILLY				
	LADY RUNNER		82	HAMLET RUNNER	1	04/03/82		DEAD TWINS				
	PRINCE NORTHERN		81	VITZMA	1	02/11/81	01/27/82	GRY COLT	NR	130,000		

Figure 21-16
Lifetime breeding record for a mare. (From Equine Management Systems, Inc.)

COMPANY NO. 075
REPORT NO. HFM610-B

THE QUALITY HORSE FARM
STALLION SUMMARY REPORT
MR. ROMERO'S SUMMARY FOR 1984

RUN DATE-04/15/84
RUN TIME-09:53:39
PAGE 001

STALLION NUMBER: 01009 NAME: MR. ROMERO

84 BREEDING SEASON

MARE NUMBER	NAME SIRE/DAM/DAM'S SIRE	YOB	PRE-BRED STATUS	MARE OWNER	SHARE NUMBER	SEAS TYPE	DATE LAST BRED	CURRENT STATUS
10005	BANQUET BELL POLYNESIAN DINNER HORN POT AU FEU	74	IN FOAL	MR. HAL SAMUELS	19-00	B'83	02/24/84	IN FOAL
10012	HAVANA RED FOX GENUINE LIGHT CUBAN CIGAR	80	MAIDEN	MRS. L. B. COLEMAN	20-00		02/12/84	IN FOAL

SEASONAL SUMMARY

	1984	- % -	1983	- % -	1982	- % -
NO. OF MARES BRED	2		40		44	
NO. OF COVERS	3		48		61	
AVERAGE COVERS PER MARE	1.50		1.20		1.38	
MARES REPORTED IN FOAL	2	100.00	36	90.00	38	86.00
MARES REPORTED BARREN	0	.00	4	10.00	4	9.00
MARES REPORTED SLIPPED	0	.00	0	.00	2	4.00
MARES UNREPORTED	0	.00	0	.00	0	.00
TOTALS	2	100.00	40	100.00	44	100.00

Figure 21-17
Report analyzing results of stallions' breeding activity for a season. (From Equine Management Systems, Inc.)

Branding

Hot Iron The hot-iron brand is one of the oldest means of identification. It is a permanent brand, but it can be altered. In addition, hair grows over the scar and identification of the brand from a distance becomes difficult. Another problem with hot-iron branding is the stress and pain that must be endured by the horse. Acid brands were introduced but were not successful for similar reasons. Since 1947, Thoroughbred racehorses have been lip-tattooed as a means of identification to prevent the running of "ringers" (look-alikes).

Freeze Branding In 1965 freeze branding, a new method of identifying livestock, was introduced. The procedure is painless. The intense cold of branding irons kept in liquid nitrogen or dry ice and alcohol destroys the pigment cells of the skin. Regrowth of hair on the frozen area is white. Approximately 6 to 8 weeks are required for the white hair to grow sufficiently to make the brand legible, but the brand is permanent and easy to read. Several numberical or code systems have been devised so that several billion horses can be branded, each with a different brand. One system utilizes the 26 letters of the alphabet and the digits 0 through 9. the 36 characters are uniquely designed to prevent alteration of one character to another character (Figure 21-18). Each of the symbols can be arranged in 8 different positions and placed in a quadrant to produce more than 6 billion combinations.

The Y-Tex Corporation has combined freeze branding, a patented angle system, and a computerized data retrieval system to produce an identification system. Using a single iron and a set of alpha characters (Figure 21-19), it is possible to produce more than 20 billion combinations. The angle system makes the brand unalterable. It consists of an alpha character, 2 stacked angles, and 6 angles side by side. The last 6 angles are underscored by a ⅛-line. The line is used as a reference line to read the angles and to

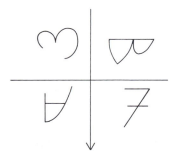

Figure 21-18
The brand for 1A132B473. The number one position of the arrow on the crossbars is for 1; position one for letter A is A1; position two for number 3 is for 32; position four for letter B is for B4; and position three for number 7 is for 73.

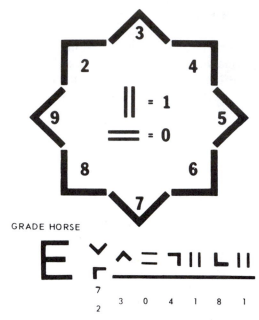

GRADE HORSE

Figure 21-19
The Y-Tex Corporation's system of alpha
characters for freeze branding horses.

TIM HARRIS

FREEZE MARKING DEPT.

provide an unalterable mark. The alpha character is used to identify the
breed of the horse, that is, "A" represents Appaloosa Horse Club registry
and "E" represents unregistered (grade) horses. The stacked angles represent the year of birth. The top angle designates the decade and the bottom
angle designates the year. The last angle designates the identification or
serial number of the horse. Registration numbers are used for registered
horses and a number that is recorded in this program is given to grade
horses.

21.3   Producing and Selling Yearlings

Buying and Selling

When horses are 2 years old, most people make a decision either to keep or
to sell them. the most common way of selling horses is by private treaty. The
parties involved merely reach an agreement regarding the terms and conditions. Horse auctions are also a popular way to buy and sell horses. Buyers

and sellers must understand the conditions of sale as they are printed in the sale catalogue. Horses of higher quality are advertised in other publications, in addition to the sale catalogue. Sellers should have their horses in excellent condition and well mannered for a sale. The stall area assigned to the seller should be kept clean and attractive. Quite often signs or advertising calling attention to the horse are used in the stall area. Public relations at the sale are important, so a place for buyers to relax and have a cup of coffee with the seller is important. Before buying a horse at auction, it is wise to inspect the horse carefully before it enters the sale ring.

Prepurchase Examination   Ever since horses have been bought and sold, the advice of experts has been sought to give an independent opinion on the structural or functional integrity of the individual being sold or purchased. The buyer expects to receive what was bargained for (i.e., the horse is ready to run or ready to be utilized in a breeding program). A veterinary examination is used more and more frequently in the sale of horses. The buyer should contract a veterinarian to perform the examination and clearly spell out the intended use for the horse. The veterinarian's report should be in writing and should include the veterinarian's opinion expressed in layman's terms. The sales contract should contain language to the following effect: The purchase is contingent upon examination by a veterinarian of the buyer's choice and that such veterinarian declare the horse to be sound for its intended use.

Cost Considerations in Producing Yearlings

Regardless whether foals are raised as a hobby or as a commercial enterprise, many costs and factors must be taken into account when agreeing upon a sale price. It is difficult to construct a general cost figure because foals are raised under many different circumstances and in various localities. *At least* the following factors must be considered.

The purchase price of the mare must be allocated among the expected number of saleable foals she will produce. If it is assumed that the average number of foals produced per mare is five and that only approximately 70 percent of live foals can be sold because of deaths, injuries, illness, or poor quality, 28.5 percent of the cost of the mare is allocated to each foal's cost.

Several costs must be borne to keep the mare for the period necessary, on the average, to produce a saleable foal. Approximately 70 percent of the mares bred will conceive, and then approximately 80 percent of the pregnant mares will bear a live foal. Therefore, the chance that a live foal will be born is approximately 56 percent. The average mare will have a foal every 2 years, but only 70 percent of live foals are saleable, so the average mare has a 39 percent chance of producing a saleable yearling. This means that the average mare must be kept 2.6 years per saleable yearling. The average cost of boarding a mare is approximately $150 per month or $4680 for 2.6 years. Veterinary bills will average approximately $150 per year or 390 for 2.6 years. Farrier's charges for 2.6 years will be $312, assuming it costs $20 to trim the mare's feet every 2 months. The total cost for mare care for 2.6 years will be approximately $5382.

Breeding expenses include the stallion fee, the cost of female genital examinations, and the cost of transportation and board when the mare is away from home. Transportation to and from the stallion will cost approximately $400 on the average. The veterinary examination usually costs approximately $70 per breeding season and the extra cost of boarding the mare is approximately $300 for 60 days. These costs total $770, and since 2.6 breeding seasons are required to obtain a saleable yearling, the cost is $2002 per saleable yearling. If the stallion fee is $1000, the breeding fee per saleable yearling will be $1428 because only 70 percent of live foals are saleable as yearlings.

The cost of raising the foal to the age at which it can be sold will include the cost of boarding, veterinary fees, and farrier's fees. If board cost is $100 per month and the average age at sale is 10 months, the board cost will be $1000. Only 70 percent of live foals are saleable, but 85 percent live and must be fed. Therefore, the board cost per saleable yearling is $1176. Farrier's fees will be approximately $60 and veterinary fees will approximate $50. These costs total $1286.

Sale costs vary from nothing for the sale of some yearlings to as much as $1000 or more for those sold at auctions. The sale costs include entry fees, vanning, advertising, extra help at sale, and sale commissions (6 to 20 percent). On the average, it costs approximately $500 to sell a yearling.

Assuming a $5000 purchase price for a mare, the cost per saleable yearling is broken down as follows:

Cost of mares (28.5 percent)	$ 1425
Mare care	5382
Breeding cost	3429
Cost of raising foal	1286
Selling	500
	11,022

The total cost is $11,022 per saleable yearling, and this figure does not include such expenses as insurance, taxes, depreciation of buildings and equipment, interest on investments, and cost of equipment. Table 21-1 provides a comparative estimate of the costs to operate a farm in California.

Table 21-1A
Farm costs by the average per horse per year (PHPY) of 13 California farms participating in a survey and reflecting the farm's standard charges.

Standard charges	PHPY	Percent of PHPY budget
Mortgage/Lease	$ 422	13%
Payroll	831	25½
Feed	506	15½
Advertising	147	4½
Utilities/Water	115	3½
Maintenance/Repair	125	4
Insurance	78	2
Accounting/Computer/Legal	41	1
Auto/Fuel	42	1
Taxes	70	2
Workmen's Compensation	88	3
Entertainment/Travel	108	3
Phone/Mail	68	2
Miscellaneous	618	20
Total	$3259	100%

SOURCE: *The Thoroughbred of California*, 1984, 78(7):4–8.
°Veterinary costs averaged $150–$200 PHPY and farrier costs were $50–$100 PHPY.

Table 21-1B
A comparative look at 13 California farms participating in a survey.

	Range		Occurring most frequently
	High	Low	
Total number of horses	676	120	200–350
Number of stallions	12	2	4
Stallions' stud fees	$50,000	$1000	$2,000–$5000
Number of acres	1,000	60	100–250
Operating costs per year	$3,000,000	$284,000	$450,000–$850,000
Board, mares and yearlings, per day	$15	$6	$10–$12
Board, stallions, per month	$800	$400	$500–$650
Board, lay-ups, per day	$23	$10	$15
Board, horses in training, per day	$36	$18	$25–$30
Board, foals, per day	$2	n/c	$2
Sales prep at farm, per day	$30	$15	$15–$20
Sales prep at sales, per day	$40	$15	$15–$20 plus 5% commission

SOURCE: *The Thoroughbred of California*, 1984, 78(7):4–8.
°Veterinary costs averaged $150–$200 PHPY and farrier costs were $50–$100 PHPY.

Advertising

The quality and scope of the advertising of a horse for sale or a stallion standing at stud can often determine whether there will be a profit or a loss. There are many things to consider, but one of the most important is the photograph of the horse. Peter Winants (1971) has outlined nine hints for a successful photograph:

1. An attractive head is essential. It should be turned to an angle where the off eye is barely visible.

2. A clean, well-fitting halter helps to create a favorable impression of both the horse and the farm.

3. A still day helps to insure orderly appearance of the mane and tail, but hair setting gel may be necessary.

4. A bright day and side lighting help to provide brilliant highlights in the horse's coat and more attractive detail.

5. A relatively simple, uncluttered background helps to focus attention on the horse and not the scenery.

6. A straight line from the point of the hock to the ground shows the favorable points of the hind leg.

7. Freshly mowed grass helps to expose the feet and creates a neat appearance for the farm and horse.

8. Forelegs should be slightly separated with the near in front to allow prospective buyers to view both legs. "Daylight to the knees" is a thumb rule.

9. A straight line from the shoulder to the ground is attractive and shows the way a horse should be built.

Another consideration is which trade magazines should be chosen for advertising. Obviously, magazines that reach appropriate interested parties should be utilized. The value of the horse determines the scope of coverage. Inexpensive horses do not require nationwide coverage, which is quite expensive. With regard to stallions, mailing lists from breeders' associations are available, so that a personal advertisement may be sent.

Other, indirect means of advertising are just as important. Neat, well-appointed, and functional facilities are indicative of good management. Attractive horses that are well formed, healthy, in good condition, well mannered, sound, well groomed, and adapted to the purposes desired create buyer demand. Field days and tours for interested groups, including 4-H and Future Farmers of America groups, create potential buyers. Appearance at and participation in breeders' association meetings, acceptance of speaking engagements at service clubs, and appearance on television or radio shows are good means of subtle advertising. Horsepeople that follow up on horses they have sold create the impression that they are interested in their product.

Loading and Hauling

Before a horse is sold, the animal should be taught to be loaded into a trailer. When problems arise in the loading of a horse, the buyer usually knows he is loading a green horse or a problem horse. This situation is poor advertisement for the farm. The green horse is reluctant to enter a trailer because it is afraid. Confidence must be instilled in the horse through patience. If the green horse is gently urged up the ramp, it will usually load. The gentle urging requires the use of two ropes attached to each side of the trailer. The horse is kept directly behind the trailer with its head forward while one person pulls on each rope and a third person leads the horse into the trailer (Figure 21-20). After several practice sessions, the horse should enter the

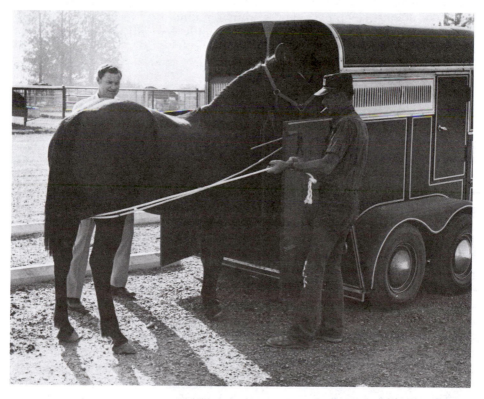

Figure 21-20
Use of butt ropes to load a reluctant horse into a trailer.

trailer without hesitation. Problem horses can be handled in the same way, but a broom or whip can be used to let the horse know that the horsepeople is in command of the situation.

If the horse is to be hauled a long distance, it is wise to give the animal a rest every 3 or 4 hours. Some stallions or geldings will not urinate in a trailer. It is advisable to haul a water bucket with which the horse is familiar because many horses will not drink strange water or from strange water troughs. If a stallion must be hauled next to a mare or if two horses that fight must be hauled together, Vicks Vaporub rubbed into their nostrils prevents them from smelling each other. Before going on a long trip, it is advisable to check the trailer for tire damage, rotted floorboards, safety chains, lights, brakes and wheel bearings. The driver should drive at an even rate of speed, avoid weaving in and out of traffic, and avoid sudden stops, starts, and turns.

Health Care

The 2-year-olds that will be kept or sold and the yearlings should be vaccinated for tetanus, sleeping sickness, rhinopneumonitis, strangles, and influenza, and dewormed regularly. The 2-year-olds that will be kept should be marked for permanent identification.

21.4   Pasture Management

Good pastures can be an important part of the horse care program on many horse farms. However, pastures mean different things to different farms. On some farms, pastures are used as exercise and exhibition paddocks and play no role in the nutrition program. On other farms, high-quality pastures supply most or all of the nutrient requirements during most of the year. High-quality pastures offer several advantages. The forage is high in digestible nutrients, is perennial, is harvested by the horses, requires no storage, provides exercise, and requires limited labor. The forage produced by pastures is usually less expensive and provides more nutrients than hay and other purchased feed.

The pasture should have good shade (either trees or artificial shelter) for the horses. A source of fresh, clean water must be available. Salt should also be available, and it may be necessary to have a source of minerals. The pasture should be clean, that is, free from all sharp obstacles and holes. It should be fenced with one of the types of fence mentioned in Section 20.4.

The amount of pasture required for a horse farm can be calculated from the number of horses kept on the farm and the expected forage production from the pasture. The average mature horse will consume approximately 25 pounds of dry matter per day or 750 pounds per month. Forage production per acre varies according to the type of pasture and its location. In many counties of the United States, county extension atgents have conducted pasture demonstrations. These results are available and can be used as guidelines for calculating forage production. Pastures grow irregularly throughout the year and this must be taken into account. During the period of maximum growth, the excess forage can be made into hay, which can be used to supplement the horses' feed during the period of minimum forage production.

Horse owners are somewhat limited as to the kinds of pasture grasses and legumes that can be grown in each geographical area and that are suitable for horses. Certain species of forage grow profusely while others grow little or not at all. There are warm-season perennials, such as Bermuda grass, Johnson grass, bluestems, and native grasses. There are also warm-season annuals, which include sorghum-Sudan hybrids, Sudans, and millets. The sorghum-Sudan hybrids and Sudans have been known to cause a disorder known as cystitis syndrome, so the county extension agent or local veterinarian should be consulted before they are used as pasture. The cool-season perennials include fescue, orchard grass, Timothy, smooth bromegrass, perennial rye-grass, and bluegrass. Varieties of fescue that are contaminated with the endophytic fungus, *Acremonium coenophialum*, should be avoided because they can cause abortions, stillbirths, thick placentas and/or a failure to lac-tate. Usually, the foals are carried full term and are alive in the uterus until birth; they may fail to breathe after birth. The cool-season annuals include oats, barley, wheat, rye, and annual ryegrass. Legumes that are suitable for horses are alfalfa, clover, and birdsfoot trefoil.

Productive grasses, such as orchard grass, which is a bunch grass that produces regrowth rapidly following grazing, should be chosen for irrigated or subirrigated pastures. Smooth bromegrass forms a sod that has a distinct value in the prevention of compaction due to trampling. On wet areas or areas that are subject to heavy travel, tall fescue is quite productive though less palatable than orchard grass or smooth bromegrass. Where a high water table is present, reed canary grass or creeping meadow foxtail are both well adapted. Kentucky bluegrass will usually invade an irrigated grass pasture, so it may not be necessary to include the seed in a seedling mixture. Ladino clover is a selection of white clover that is productive and quite palatable to horses. It requires frequent irrigation and does not withstand shading by tall grass growth. Alsike clover is adapted to wet, poorly drained sites and is short-lived. Birdsfoot trefoil is difficult to establish when competition from other plants is intense. It does withstand close grazing and is not as exacting in its water requirements as Ladino clover. Trefoil is not winter hardy in areas with extremely cold temperatures and little snow cover. Alfalfa can be maintained as a pasture if grazing is followed by long rest periods. The rambler or rhizoma varieties have low creeping growth and are more suitable for grazing than the hay varieties.

Dryland pastures are usually seeded to crested wheat grass, intermediate wheat grass, or Russian wild rye. Russian wild rye is leafy and palatable throughout the grazing season. Crested wheat grass is best during the spring and after fall regrowth, whereas intermediate wheat grass provides good grazing in the early part of the summer. Often, the dryland pasture will be more productive if Ladak or Orenberg alfalfa are seeded with the grass.

The specific seed mixture should be chosen in consultation with the local county extension agent. He can help the horseman select a legumegrass mixture on the basis of adaptation to soil and climate. The legumes and grasses must be compatible. An aggressive legume cannot be planted with a slow-growing grass and vice versa. Horses are selective grazers, so a highly palatable plant should not be planted with one that is unpalatable.

All necessary land leveling, grading, and shaping should be completed before seedbed preparation. The field should be plowed, deseeded, harrowed, and packed with a roller if necessary to make a firm seedbed. The seedbed should be free from old sod and weeds. If old plants are not completely removed, they will aggressively compete with new seedlings. A firm seedbed is one that allows a mare's foot to sink no deeper than ½ inch. Seeds germinate when the absorb soil moisture. This moisture usually comes from below the seed and moves upward through a well-packed seedbed. Packing the seedbed will prevent early drying of the soil and death of the seedlings. The soil should be tested to determine the plant nutrient needs for optimal production. Such a test is inexpensive and allows the horseperson to make meaningful decisions about fertilizers.

Proper fertilization will improve protein content, increase tolerance to dry weather, increase regrowth after grazing, and increase resistance to disease. Grass and legume seeds should be planted between ¼ and ¾ inches deep. If planted deeper, the seedlings do not have the vigor to emerge and establish themselves. Seeding rates vary with the geographical area, but 6 to 8 pounds of grass seed plus 1 to 3 pounds of legume seed per acre are common rates.

After the pasture is established and is being grazed, several important management practices should be followed. The dung piles should be scattered periodically or removed. This will help reduce the parasite population. The pasture should be mowed periodically to remove weeds and rank vegetation around manure and urine spots. A new pasture should not be grazed until it is at least 6 inches high. At no time should the pasture be overgrazed to the point where not enough leaf material is left to permit rapid regrowth. A new pasture should be irrigated frequently so that the soil is not dry near the seed. After emergence, the seedling should not be allowed to become stressed. The roots of established pastures are usually in the upper 3 feet of soil. After this depth is watered, excessive water is wasteful. The pasture should not be irrigated when the horses are on it. The wet soil will compact, water penetration will be restricted, and plant growth will be reduced. If possible, pastures should be grazed on a rotational basis. They should be grazed for one week and rested for 2 weeks. During the rest period, manure is scattered, and the pasture is mowed to a height of 4 to 5 inches and irrigated. Horses tend to selectively overgraze portions of a pasture and undergraze others. Alternate grazing with cattle will result in better forage utilization. Cattle and horses tend to avoid grazing grass that grows around their own droppings, but will graze around each other's droppings. Grazing with cattle will also tend to reduce the parasite problem. The parasite problem will be minimized if all horses are dewormed before they are turned out to pasture and then dewormed on a regular basis. Fertilization rates should be established in consultation with the local county extension agent.

Horses establish a hierarchy, so the horseperson must be certain that adequate pasture area is available to prevent aggressive horses from injuring the less aggressive ones. If insects become a problem, the horses should be sprayed or dusted.

21.5   Selection of Bedding

Bedding for horses should absorb urine and odors, be free from dust, be readily available, be inexpensive, provide a comfortable bed and secure footing, and be easy to handle. Several materials are commonly used, but the straws from barley and oats are the most common. The straw should be left in long lengths to keep it free from dust, even though the absorptive capacity of chopped straw is much higher. If it is left long, the stable personnel can quickly separate the clean areas from the dirty areas in the stall and thus less straw is used each day. Wood shavings are preferred by many horsemen. It is essential that such shavings be free from sawdust to prevent respiratory problems. The shavings from the softer woods have a greater absorption capacity. Shavings offer a definite advantage for people traveling to horse shows or sales in that they can be purchased in bales. Rice hulls and peanut hulls provide excellent bedding when they are available. However, they do not absorb water and the stall floor soon becomes saturated with urine. Recycled newspaper that has been shredded and repulped to a cottonlike texture has a high absorbancy and is being used in areas where other types of bedding are difficult to obtain. When other products are not available, peat moss, can be used. It can be purchaed from garden centers.

Sand is a poor bedding. It absorbs very little urine, and horses that pick around in it for hay may develop sand colic. It is also quite heavy to handle.

Rubber mats make good stall floor surfaces. The corners of heavy-duty mats do not curl upward and are designed to prevent slipping. The cost of bedding is usually dramatically reduced because less is required.

21.6   Fire Safety

The problem of stable fires is acute, but the most people's attitude toward them is similar to their attitude toward fatal wrecks: They always seem to happen to someone else—always, that is, until they happen to you. Each year there are several reported stable fires in the United States in which horses are burned to death, and property damage amounts to millions of dollars. The prevention of stable fires is of concern to all horsepeople.

Protecting a horse in a stall from the danger of fire is a totally different problem than fire prevention in a person's home. The horse is usually standing in some type of bedding that is kept dry. Oat straw will develop a temperature of approximately 300 degrees at the head of the fire approximately one minute after it starts. Barley straw takes approximately 5 minutes to develop to 300 degrees. The point is that horses are standing in material that develops as much heat at the same rate as gasoline. All that is required to start a stable fire is a match or a spark.

A common problem is that there is usually only one door to the stall and the horse cannot escape into a paddock. The size of the stall is approximately 10 feet by 10 feet or 12 feet by 12 feet. Approximately 2 or 3 minutes are required for a straw fire in a stall to burn an area 10 feet in diameter. By the time the fire covers an area 4 feet in diameter, most horses have been injured. Their lungs are seared when the fires has covered an area 6 feet in diameter. They start to suffocate when the area is 8 feet in diameter and are dead by the time an area 10 feet in diameter has burned.

If a horse is to survive, the fire must be extinguished in the stall within approximately 30 seconds. How does the horseperson cope with this special problem?

The most common answer is proper stable construction. Cement-block barns or similar types of construction are meant to save the building and not the horse in an individual stall. If the internal spread of fire through a barn is to be prevented, some construction features are mandatory. Fire-retardant paints are effective in delaying the combusion of wood framework inside the barn. Every third or fourth stall should have partition walls that extend to the ceiling to help delay the spread of fire through the barn.

Sprinkler systems can also be used but most sprinklers were not designed to put out a fire under the circumstances that exist in a stall.

Automatic fire-extinguishing systems must meet at least two requirements to be used in a stable. They must not suffocate the horses when they are activated, and they must react within seconds after a fire starts. Most automatic sprinklers were designed to throw a circular pattern instead of a square pattern. The spray must be strong enough to reach the corners of the stall. If a fog-type system is used, the fog must suffocate the fire to extinguish it, otherwise the horse will suffocate along with the fire. Water can cause extensive smoke formation, which in turn will suffocate the horse or at least cuase lung damage. This can be prevented if the sprinkler is activated before the fire covers an area 1 foot in diameter. A thermal lag system with sensors located above the stalls takes 4 to 5 minutes to activate, but this is too long. Recent developments have resulted in sprinkler systems that begin to operate approximately 5 seconds after a fire starts.

Of course, not all fires start inside a stall. A layer of chaff and dust covers the floor of the loft of many barns. The chaff and dust is like gasoline when it starts to burn. The loft and other storage areas of a barn should be swept on a regular schedule. Periodically, the inside of a barn should be hosed down with water to remove all the dust and cobwebs. This is just as important as keeping the alleyways clean.

An adequate number of exits from the barn are necessary so that fire will not block the only exit. All exits should open into enclosed areas. In case of fire, the horses can be turned loose but they must not flee the stable area. The fleeing horses may be killed by automobiles when they cross or travel down roads. They may also ruin other property, such as lawns and crops.

The doors on the stalls should open outward into the alley. When horses flee through an emergency exit, their hips and other parts can be caught on doors that open into a stall. When the doors open into the alleyway, it is possible for someone to run through the stable, opening all the doors and letting the horses out. If a door accidentally closes and remains unlocked, the hose will open it if it runs into it trying to flee.

Because time is important in emergencies, the stall door latches should be strong and easy for people (but not for horses) to open.

If a barn is constructed in this manner, all the stall doors can be opened by the rescuer on the first trip through the alley. The horses that leave by themselves will be caught in the outside enclosed area. However, someone should prevent them from returning inside the barn because a scared horse likes to return to its familiar stall. Because of this trait, many horses will not leave their stalls unless they are led out. It may be necessary to cover their head with a sack or something similar to prevent them from seeing the fire and smoke. They are easier to lead this way, and if led any other way, they try to climb the back wall of their stalls.

All electricity in a barn should be turned off at night, but the master switch should be located close to an entrance. Then the lights can be turned on and one can see what is taking place inside during an emergency. Flash-lights and the bouncing, moving shadows and lighted areas they cause are frightening to a horse. Do not use flashlights unless absolutely essential. The electrical wire in a barn must be protected from mice and rats. They will chew the insulation off, leaving two bare wires that can easily start an electrical fire. At the same location as the electrical switch, which should be at an exit, a fire extinguisher should be available in case of a small fire. At this location, there should also be a water hose that will reach the length of the barn. This hose should remain attached to a water hydrant so that one can quickly get water to all parts of a barn.

As another protective measure, each stall should have a hook beside it with a halter and lead rope attached. Then it is not necessary to waste time taking a halter off one horse and putting it on another while rescuing horses. Time is of the essence and a halter is thus waiting at the next stall door.

Personnel working at the stables should obey a strict set of fire preven-tion rules. Smoking should not be allowed in or adjacent to buildings. For convenience, large receptacles should be placed at building entrances. All electrical equipment should be in safe working condition and inspected regularly. Such appliances as hot plates, coffee pots, and radios should not be left unattended when in use. If they are used they should be used only in a special area. Inflammable materials such as lighter fluid, solvents, or clean-ing fluids should not be used in the stable area. The owner or manager should not tolerate rule infractions by his employees or visitors. The rules must be rigidly enforced at all times.

21.7   Obtaining Horse Information

It is essential that the horse owner or manager involved in the horse business have rapid access to information about specific horses and keep informed of the latest information being developed about the care and management of horses.

To keep informed about the care and management of horses, you can attend educational meetings sponsored by universities, breed associations, farms, and the Cooperative Extension Service. There are numerous popular horse magazines from a variety of sources that regularly contain education articles as well as numerous publications from university sources.

People actively involved in the technical aspects of horse production may stay abreast of the technical literature by subscribing to computerized information services such as Dialog and the Bibliographic Retrieval Service (Appendix 21-1). Through these services, you can search three of the major databases and obtain the scientific information being published about all aspects of horse husbandry. Agricola (Agricultural Online Access) is compiled by the U.S. Department of Agriculture and provides comprehensive coverage of newly acquired worldwide publications in agriculture. CAB Abstracts (Commonwealth Agricultural Bureau) is a comprehensive file of agricultural information containing all the records in the 31 journals published by CAB. The Biosis Previews database contains citations from the major publications of BioScience Information Services.

Bibliographies in libraries are another good source of information. *Horsemanship: A Guide to Information Sources* (1979) is an annotated list of horse books organized by subject. *Equestrian Studies: The Salem College Guide to Sources in English, 1950–1980*, is arranged by subject with 12 categories. It does include some magazine article citations from general interest magazines. A good British bibliography is *The Horse, A Bibliography of British Books 1851–1976*. The annual *Books in Print* provides information on all books that are currently available. To find horse related magazine articles, two sources are very useful: *A Reader's Guide to Periodical Literature* under "Horses" and related subject headings and *The Magazine Index* (back to 1976). A comprehensive listing of magazines currently in print and their addresses can be found in a two-volume set called *Ulrich's International Periodicals Directory*.

Finding information about an individual horse can be rather easy for some breeds and may be impossible for others. The pedigrees of individuals can be traced through the stud books maintained by each registry.

The Bloodstock Research Statistical Bureau provides a variety of information services related to Thoroughbreds (Appendix 21-1). Some of the references they publish include: *Auctions* of each year, which gives the buyer and the amount paid for every Thoroughbred sold at public auction for a specific year; *Stakes Winners* of each year, which includes the pedigrees and racing statistics for all stakes winners in the United States; *Principal Winners Abraod* for each year, which covers European races; *Dams of Stakes Winners* for each year; *Broodmare Produce Records*, updated yearly; and *Dams of Stakes Winners*, also updated yearly. The records are also offered on microfiche. A computerized record service is available via personal computers on a 24-hour basis, and is updated weekly. The bureau's databank can be accessed to obtain pedigrees, production, and race records dating back to the early 1900s in a variety of formats, such as broodmare produce, stallion progeny, broodmare sire reports, previous week's stakes races, and previous week's foreign stakes races. The annual *American Racing Manual* published by the *Daily Racing Form* is a source of information and statistical data encompassing many phases of the Thoroughbred racing industry. The statistics are official for conditions of races and records of horses, jockeys, owners, trainers, and breeders.

Records for American Quarter horses are kept by the American Quarter Horse Association. A variety of records—such as pedigree, performance record, get-of-sire, and mare produce — can be obtained from their computerized databank. The results of major sales held for running Quarter Horses can be obtained from Q/T Analysis (see Appendix 21-1). The sales results can be analyzed many ways by their system to provide information to aid in accessing the value of a horse. *The Quarter Horse Running Chart Book* contains information about all the Quarter Horse races that have been held.

REFERENCES

Breuer, L. H., T. L. Bullard, and B. F. Yeates. 1970. A suggested schedule for management for a horse breeding farm. *Proc. Tenth Annual Horse Short Course.* Texas A & M University.

Davies, J. 1971. Fire protection for horses is different. *Thoroughbred Record* 193:556–557, 560.

Dolan, D., and V. W. Henry. 1984. Expanding horizons on computer software for the horseman. *Quarter Horse of the Pacific Coast* 21(2):143–145.

Farrell, R. K., G. A. Laisner, and T. S. Russell. 1969. An international freezemark animal identification system. *J. A. V. M. A.* 154:1561–1572.

Guenthner, H. R. 1974. *Horse Pastures.* Horse Science Series. Cooperative Extension Service C151. Reno: University of Nevada.

Henry, V. W. 1984. The theory is great, but how do they work? *Quarter Horse of the Pacific Coast* 21(2):146–147.

Jeffries, N. W. 1970. *Better Pastures for Horses and Ponies.* Montana Cooperative Extension Circular 294 [May]. Bozeman: Montana State University.

Platt, J. N. 1974. Let's grow better pastures. *The Quarter Horse Journal* 26(9):162, 192.

Stover, J. 1984. Costs of horses down on the farm. *Thoroughbred of California* 78(7):4–8.

Trimmer, L. 1974. Successful horse photography. *American Horseman* 4:20–22.

Truax, Coleen. 1983. High tech in the barn. The *Quarter Horse Journal* 36(1):274–279.

Vance, J. D. 1969. Cost factors in commercial thoroughbred breeding. *The Washington Horse* 23(8):940–942.

Vasiloff, M. J. 1970. Letting down. The Chronicle of the Horse 33:36 [January 9].

Walding, M. 1974. Horse photography. *The Quarter Horse Journal* 26:178–180.

Winants, P. 1971. Photographing your horse. *The Canadian Horse* 11(6):18–20.

APPENDIX 21-1

Farm Management Software Companies

Agnet
105 Miller Hall
University of Nebraska
Lincoln, NE 68583
402-472-1892

Computerized Management Network
272 University City Office Building
Blacksburg, VA 24061
703-961-5184

DeOr Software, Inc.
831 Oxbow Lake Rd.
Union Lake, MI 48085
313-363-8840

Equicomp Inc.
4650 Pleasant Ridge Road
Boulder, CO 80301
303-443-3726

Equine Computer Software
P.O. Box 96
Otwell, IN 47564
812-354-2223

Equine Management Systems
21363 Lassen St., Suite 110
Chatsworth, CA 91311
818-700-1124

Farm Management Systems
499 South Capital St., Suite 407
Washington, DC 20003
202-484-1880

Joe McGee & Associates
P.O. Box 501
Pilot Point, TX 76258
817-686-2283

Oklahoma Management Systems
8404 Blackberry Ridge
Edmond, OK 73034
405-341-1876

Owens & Company
1750 Alexandria Drive
Lexington, KY 40504
606-278-2342

Howard Kale
Karho Farms
5900 East Bell Road
Scottsdale, AZ 85254
602-971-1115

The Software Collection
4583 Camino del Mirasol
Los Olivos, CA 93110
805-688-0205

Universal Microsystems, Inc.
P.O. Box 1343
Duluth, GA 30136
404-449-0690

Performance, Pedigree, and Other Horse Records

American Quarter Horse Association
P.O. Box 200
Amarillo, TX 79168
806-376-4811

Arabian Horse Research Services
108 Factory Ave. N., Suite 3
Renton, WA 98055
206-271-3058

Bloodstock Research
801 Corporate Dr.
Lexington, KY 40544
606-223-4444

Compu-Equine Services
2822 West Lancaster
Fort Worth, TX 76107
817-877-5075

Equine Line
821 Corporate Drive
Lexington, KY 40503
606-224-2800

Equine Services Inc.
4037 W. 50th
Amarillo, TX 79109
806-352-4949

Thoroughbred Data Services
P.O. Box 11613
Marina Del Rey, CA 90291
213-264-1785

Scientific Information and Data Base Services

Bibliographic Retrieval Service
1200 Route 7
Latham, N.Y. 12110
800-553-5566

Dialog Information Services, Inc.
3460 Hillview Avenue
Palo Alto, CA 94304
800-334-2564

INDEX